U0275586

寰宇文献 Universal Library | SINOLOGY 系列

SELECTED WORKS OF BERTHOLD LAUFER

劳费尔著作集

第十卷

[美] 劳费尔 著

黄曙辉 编

中西书局
ZHONGXI BOOK COMPANY

图书在版编目(CIP)数据

劳费尔著作集 / (美) 劳费尔著；黄曙辉编.—上
海：中西书局，2022
　　(寰宇文献)
　　ISBN 978-7-5475-2015-4

　　Ⅰ.①劳… Ⅱ.①劳…②黄… Ⅲ.①劳费尔－人类
学－文集 Ⅳ.①Q98-53

中国版本图书馆CIP数据核字（2022）第207067号

第 10 卷

150

中国伊朗编——中国对古代伊朗文明的贡献，
着重于栽培植物及产品的历史
附：书评三则

FIELD MUSEUM OF NATURAL HISTORY

PUBLICATION 201

ANTHROPOLOGICAL SERIES   VOL. XV, No. 3

SINO-IRANICA

Chinese Contributions to the History of Civilization in Ancient Iran

With Special Reference to the History of Cultivated Plants and Products

BY

BERTHOLD LAUFER
Curator of Anthropology

The Blackstone Expedition

CHICAGO
1919

CONTENTS

iii

Sino-Iranica

By Berthold Laufer

INTRODUCTION

If we knew as much about the culture of ancient Iran as about ancient Egypt or Babylonia, or even as much as about India or China, our notions of cultural developments in Asia would probably be widely different from what they are at present. The few literary remains left to us in the Old-Persian inscriptions and in the Avesta are insufficient to retrace an adequate picture of Iranian life and civilization; and, although the records of the classical authors add a few touches here and there to this fragment, any attempts at reconstruction, even combined with these sources, will remain unsatisfactory. During the last decade or so, thanks to a benign dispensation of fate, the Iranian horizon has considerably widened: important discoveries made in Chinese Turkistan have revealed an abundant literature in two hitherto unknown Iranian languages,— the Sogdian and the so-called Eastern Iranian.[1] We now know that Iranian peoples once covered an immense territory, extending all over Chinese Turkistan, migrating into China, coming in contact with Chinese, and exerting a profound influence on nations of other stock, notably Turks and Chinese. The Iranians were the great mediators between the West and the East, conveying the heritage of Hellenistic ideas to central and eastern Asia and transmitting valuable plants and goods of China to the Mediterranean area. Their activity is of world-historical significance, but without the records of the Chinese we should be unable to grasp the situation thoroughly. The Chinese were positive utilitarians and always interested in matters of reality: they have bequeathed to us a great amount of useful information on Iranian plants, products, animals, minerals, customs, and institutions, which is bound to be of great service to science.

The following pages represent Chinese contributions to the history of civilization in Iran, which aptly fill a lacune in our knowledge of Iranian tradition. Chinese records dealing with the history of Iranian peoples also contain numerous transcriptions of ancient Iranian words,

---

[1] Cf., for instance, P. Pelliot, Influences iraniennes en Asie centrale et en Extrême-Orient (Paris, 1911).

185

part of which have tested the ingenuity of several sinologues and historians; but few of these Sino-Iranian terms have been dealt with accurately and adequately. While a system for the study of Sino-Sanskrit has been successfully established, Sino-Iranian has been woefully neglected. The honor of having been the first to apply the laws of the phonology of Old Chinese to the study of Sino-Iranica is due to ROBERT GAUTHIOT.[1] It is to the memory of this great Iranian scholar that I wish to dedicate this volume, as a tribute of homage not only to the scholar, but no less to the man and hero who gave his life for France.[2] Gauthiot was a superior man, a *kiün-tse* 君子 in the sense of Confucius, and every line he has written breathes the mind of a thinker and a genius. I had long cherished the thought and the hope that I might have the privilege of discussing with him the problems treated on these pages, which would have considerably gained from his sagacity and wide experience — 非夫人之爲慟而誰爲.

Iranian geographical and tribal names have hitherto been identified on historical grounds, some correctly, others inexactly, but an attempt to restore the Chinese transcriptions to their correct Iranian prototypes has hardly been made. A great amount of hard work remains to be done in this field.[3] In my opinion, it must be our foremost object first to record the Chinese transcriptions as exactly as possible in their ancient phonetic garb, according to the method so successfully inaugurated and applied by P. Pelliot and H. Maspero, and then to proceed from this secure basis to the reconstruction of the Iranian model. The accurate restoration of the Chinese form in accordance with

---

[1] Cf. his Quelques termes techniques bouddhiques et manichéens, *Journal asiatique*, 1911, II, pp. 49–67 (particularly pp. 59 *et seq.*), and his contributions to Chavannes and Pelliot, Traité manichéen, pp. 27, 42, 58, 132.

[2] Gauthiot died on September 11, 1916, at the age of forty, from the effects of a wound received as captain of infantry while gallantly leading his company to a grand attack, during the first offensive of Artois in the spring of 1915. Cf. the obituary notice by A. MEILLET in *Bull. de la Société de Linguistique*, No. 65, pp. 127–132.

[3] I hope to take up this subject in another place, and so give only a few examples here. Ta-ho šwi 達曷水 is the Ta-ho River on which Su-li, the capital of Persia, was situated (*Sui šu*, Ch. 83, p. 7 b). HIRTH (China and the Roman Orient, pp. 198, 313; also *Journal Am. Or. Soc.*, Vol. XXXIII, 1913, p. 197), by means of a Cantonese Tat-hot, has arrived at the identification with the Tigris, adding an Armenian Deklath and Pliny's Diglito. Chinese *ta*, however, corresponds neither to ancient *ti* nor *de*, but only to *tat, dat, dad, dar, d'ar, while *ho* 曷 represents *hat, kat, kad, kar, kal. We accordingly have *Dar-kat, or, on the probable assumption that a metathesis has taken place, *Dak-rat. Hence, as to the identification with the Tigris, the vocalism of the first syllable brings difficulties: it is *i* both in Old Persian and in Babylonian. Old Persian Tigram (with an alteration due to popular etymology, cf. Avestan *tiγriš*, Persian *ßr*, "arrow") is borrowed from Babylonian Di-ik-lat (that

rigid phonetic principles is the essential point, and means much more than any haphazardly made guesses at identification. Thus Mu-lu 木鹿, name of a city on the eastern frontier of An-si (Parthia),[1] has been identified with Mouru (Muru, Merw) of the Avesta.[2] Whether this is historically correct, I do not wish to discuss here; from an historical viewpoint the identification may be correct, but from a phonetic viewpoint it is not acceptable, for Mu-lu corresponds to ancient *Muk-luk, Mug-ruk, Bug-luk, Bug-rug, to be restored perhaps to *Bux-rux.[3] The scarcity of linguistic material on the Iranian side has imposed certain restrictions: names for Iranian plants, one of the chief subjects of this study, have been handed down to us to a very moderate extent, so that in many cases no identification can be attempted. I hope, however, that Iranian scholars will appreciate the philological contributions of the Chinese to Iranian and particularly Middle-Persian lexicography, for in almost every instance it is possible to restore with a very high degree of certainty the primeval Iranian forms from which the Chinese transcriptions were accurately made. The Chinese scholars had developed a rational method and a fixed system in reproducing words of foreign languages, in the study of which, as is well known, they took a profound interest; and from day to day, as our experience widens, we have occasion to admire the soundness, solidity, and consistency of this system. The same laws of transcription worked out for Sanskrit, Malayan, Turkish, Mongol, and Tibetan, hold good also for Iranian. I have only to ask Iranian scholars to have confidence in our method, which has successfully stood many tests. I am convinced that this plea is unnecessary for the savants of France, who are the

---

is, Dik-lat, Dik-rat), which has passed into Greek Τίγρης and Τίγρις and Elamite Ti-ig-ra (A. MEILLET, Grammaire du vieux perse, p. 72). It will thus be seen that the Chinese transcription *Dak-rat corresponds to Babylonian Dik-rat, save the vowel of the first element, which cannot yet be explained, but which will surely be traced some day to an Iranian dialect.—The T'ai p'iñ hwan yü ki (Ch. 185, p. 19) gives four geographical names of Persia, which have not yet been indicated. The first of these is the name of a city in the form 褐婆竭 Ho-p'o-kie, *Hat(r, l)-bwa-g'iat. The first two elements *Har-bwa correspond to Old Persian Haraiva (Babylonian Hariva), Avestan Haraēva, Pahlavi *Harēw, Armenian Hrew,—the modern Herat. The third element appears to contain a word with the meaning "city." The same character is used in 竭離別 Kie-li-pie, *G'iat-li-b'iet, name of a pass in the north-eastern part of Persia; here *g'iat, *g'iar, seems to represent Sogdian γr, *γara ("mountain"). Fan-tou 番 or 蕃兜 (Ts'ien Han šu, Ch. 96 A), anciently *Pan-tav, *Par-tav, corresponds exactly to Old Persian Parθava, Middle Persian Parθu.

[1] Hou Han šu, Ch. 116, p. 8 b.
[2] HIRTH, China and the Roman Orient, p. 143.
[3] Cf. also the observation of E. H. PARKER (Imp. and As. Quarterly Review, 1903, p. 154), who noticed the phonetic difficulty in the proposed identification.

most advanced and most competent representatives of the sinological field in all its varied and extensive branches, as well as in other domains of Oriental research. It would have been very tempting to summarize in a special chapter the Chinese method of transcribing Iranian and to discuss the phonology of Iranian in the light of Chinese contributions. Such an effort, however, appears to me premature at this moment: our knowledge of Sino-Iranian is in its infancy, and plenty of fresh evidence will come forward sooner or later from Turkistan manuscripts. There is no doubt that many hundreds of new Iranian terms of various dialects will be revived, and will considerably enrich our now scanty knowledge of the Iranian onomasticon and phonology. In view of the character of this publication, it was necessary to resort to a phonetic transcription of both ancient and modern Chinese on the same basis, as is now customary in all Oriental languages. The backwardness of Chinese research is illustrated by the fact that we slavishly adhere to a clumsy and antiquated system of romanization in which two and even three letters are wasted for the expression of a single sound. My system of transliteration will be easily grasped from the following comparative table.

OLD STYLE	PHONETIC STYLE
ng	*ñ*
ch	*č*
ch'	*č'*
j	*ž* (while *j* serves to indicate the palatal
sh	*š* sonant, written also *dž*).

Other slight deviations from the old style, for instance, in the vowels, are self-explanatory. For the sake of the numerous comparative series including a large number of diverse Oriental languages it has been my aim to standardize the transcription as far as possible, with the exception of Sanskrit, for which the commonly adopted method remains. The letter *x* in Oriental words is never intended for the combination *ks*, but for the spirant surd, sometimes written *kh*. In proper names where we are generally accustomed to *kh*, I have allowed the latter to pass, perhaps also in other cases. I do not believe in super-consistency in purely technical matters.

The linguistic phenomena, important as they may be, form merely a side-issue of this investigation. My main task is to trace the history of all objects of material culture, pre-eminently cultivated plants, drugs, products, minerals, metals, precious stones, and textiles, in their migration from Persia to China (Sino-Iranica), and others transmitted from China to Persia (Irano-Sinica). There are other groups of Sino-Iranica not included in this publication, particularly the animal world, games, and musical instruments.[1] The manuscript dealing with the fauna of Iran is ready, but will appear in another article the object of which is to treat all foreign animals known to the Chinese according to geographical areas and from the viewpoint of zoögeography in ancient and modern times. My notes on the games (particularly polo) and musical instruments of Persia adopted by the Chinese, as well as a study of Sino-Iranian geographical and tribal names, must likewise be reserved for another occasion. I hope that the chapter on the titles of the Sasanian government will be welcome, as those preserved in the Chinese Annals have been identified here for the first time. New results are also offered in the notice of Persian textiles.

As to Iranian plants of which the Chinese have preserved notices, we must distinguish the following groups: (1) cultivated plants actually disseminated from Iranian to Chinese soil, (2) cultivated and wild plants of Iran merely noticed and described by Chinese authors, (3) drugs and aromatics of vegetable origin imported from Iran to China. The material, as far as possible, is arranged from this point of view and in chronological order. The single items are numbered. Apart from the five appendices, a hundred and thirty-five subjects are treated. At the outset it should be clearly understood that it is by no means the intention of these studies to convey the impression that the Chinese owe a portion of their material culture to Persia. Stress is laid on the point that the Chinese furnish us with immensely useful material for elaborating a history of cultivated plants. The foundation of Chinese civilization with its immense resources is no more affected by these introductions than that of Europe, which received numerous plants from the Orient and more recently from America. The Chinese merit our admiration for their far-sighted economic policy in making so many useful foreign plants tributary to themselves and amalgamating them with their sound system of agriculture. The Chinese were thinking, sensible, and broad-minded people, and never declined to accept gratefully whatever good things foreigners had to offer. In plant-economy they are the foremost masters of the world, and China presents a unique spectacle in that all useful plants of the universe are cultivated there. Naturally, these cultivations were adopted and absorbed by a gradual process: it took the Chinese many centuries to become familiar with the flora of their own country, and the long series of their herbals (*Pen ts'ao*) shows us well how their knowledge of species increased from the T'ang to the present time, each of these works stating the

---

[1] Iranian influences on China in the matter of warfare, armor, and tactics have been discussed in Chinese Clay Figures, Part I.

number of additional species as compared with its predecessor. The introduction of foreign plants begins from the latter part of the second century B.C., and it was two plants of Iranian origin, the alfalfa and the grape-vine, which were the first exotic guests in the land of Han. These were followed by a long line of other Iranian and Central-Asiatic plants, and this great movement continued down to the fourteenth century in the Yüan period. The introduction of American species in the sixteenth and seventeenth centuries denotes the last phase in this economic development, which I hope to set forth in a special monograph. Aside from Iran, it was Indo-China, the Malayan region, and India which contributed a large quota to Chinese cultivations. It is essential to realize that the great Iranian plant-movement extends over a period of a millennium and a half; for a learned legend has been spread broadcast that most of these plants were acclimatized during the Han period, and even simultaneously by a single man, the well-known general, Čaṅ K'ien. It is one of my objects to destroy this myth. Čaṅ K'ien, as a matter of fact, brought to China solely two plants,—alfalfa and the grape-vine. No other plant is attributed to him in the contemporaneous annals. Only late and untrustworthy (chiefly Taoist) authors credit him also with the introduction of other Iranian plants. As time advanced, he was made the centre of legendary fabrication, and almost any plant hailing from Central Asia and of doubtful or obscure history was passed off under his name: thus he was ultimately canonized as the great plant-introducer. Such types will spring up everywhere under similar conditions. A detailed discussion of this point will be found under the heading of each plant which by dint of mere fantasy or misunderstanding has been connected with Čaṅ K'ien by Chinese or European writers. In the case of the spinach I have furnished proof that this vegetable cannot have been culti-vated in Persia before the sixth century A.D., so that Čaṅ K'ien could not have had any knowledge of it. All the alleged Čaṅ-K'ien plants were introduced into China from the third or fourth century A.D. down to the T'ang period inclusively (618–906). The erroneous reconstruction alluded to above was chiefly championed by Bretschneider and Hirth; and A. de Candolle, the father of the science of historical botany, who, as far as China is concerned, depended exclusively on Bretschneider, fell victim to the same error.

F. v. RICHTHOFEN,[1] reproducing the long list of Bretschneider's Čaṅ-K'ien plants, observes, "It cannot be assumed that Čaṅ K'ien himself brought along all these plants and seeds, for he had to travel

---

[1] China, Vol. I, p. 459.

with caution, and for a year was kept prisoner by the Hiun-nu." When he adds, however, "but the relations which he had started brought the cultivated plants to China in the course of the next years," he goes on guessing or speculating.

In his recent study of Čan K'ien, HIRTH[1] admits that of cultivated plants only the vine and alfalfa are mentioned in the *Ši ki*.[2] He is unfortunate, however, in the attempt to safeguard his former position on this question when he continues to argue that "nevertheless, the one hero who must be looked upon as the pioneer of all that came from the West was Chang K'ien." This is at best a personal view, but an unhistorical and uncritical attitude. Nothing allows us to read more from our sources than they contain. The *Ts'i min yao šu*, to which Hirth takes refuge, can prove nothing whatever in favor of his theory that the pomegranate, sesame, garlic,[3] and coriander were introduced by Čan K'ien. The work in question was written at least half a millennium after his death, most probably in the sixth century A.D., and does not fall back on traditions coeval with the Han and now lost, but merely resorts to popular traditions evolved long after the Han period. In no authentic document of the Han is any allusion made to any of these plants. Moreover, there is no dependence on the *Ts'i min yao šu* in the form in which we have this book at present. BRETSCHNEIDER[4] said wisely and advisedly, "The original work was in ninety-two sections. A part of it was lost a long time ago, and much additional matter by later authors is found in the edition now current, which is in ten chapters. . . . According to an author of the twelfth century, quoted in the *Wen hien t'un k'ao*, the edition then extant was already provided with the interpolated notes; and according to Li Tao, also an author of the Sung, these notes had been added by Sun Kun of the Sung dynasty."[5] What such a work would be able to teach us on actual conditions of the Han era, I for my part am unable to see.

---

[1] *Journal Am. Or. Soc.*, Vol. XXXVII, 1917, p. 92. The new translation of this chapter of the *Ši ki* denotes a great advance, and is an admirable piece of work. It should be read by every one as an introduction to this volume. It is only on points of interpretation that in some cases I am compelled to dissent from Hirth's opinions.

[2] This seems to be the direct outcome of a conversation I had with the author during the Christmas week of 1916, when I pointed out this fact to him and remarked that the alleged attributions to Čan K'ien of other plants are merely the outcome of later traditions.

[3] This is a double error (see below, p. 302).

[4] Bot. Sin., pt. I, p. 77.

[5] Cf. also PELLIOT (*Bull. de l'Ecole française*, Vol. IX, p. 434), who remarks, "Ce vieil et précieux ouvrage nous est parvenu en assez mauvais état."

It has been my endeavor to correlate the Chinese data first of all with what we know from Iranian sources, and further with classical, Semitic, and Indian traditions. Unfortunately we have only fragments of Iranian literature. Chapter XXVII of the Bûndahišn[1] contains a disquisition on plants, which is characteristic of the treatment of this subject in ancient Persia. As it is not only interesting from this point of view, but also contains a great deal of material to which reference will be made in the investigations to follow, an extract taken from E. W. West's translation[2] may be welcome.

"These are as many genera of plants as exist: trees and shrubs, fruit-trees, corn, flowers, aromatic herbs, salads, spices, grass, wild plants, medicinal plants, gum plants, and all producing oil, dyes, and clothing. I will mention them also a second time: all whose fruit is not welcome as food of men, and are perennial, as the cypress, the plane, the white poplar, the box, and others of this genus, they call trees and shrubs (dār va diraxt). The produce of everything welcome as food of men, that is perennial, as the date, the myrtle, the lote-plum (kūnār, a thorny tree, allied to the jujube, which bears a small plum-like fruit), the grape, the quince, the apple, the citron, the pomegranate, the peach, the fig, the walnut, the almond, and others in this genus, they call fruit (mīvak). Whatever requires labor with the spade, and is perennial, they call a shrub (diraxt). Whatever requires that they take its crop through labor, and its root withers away, such as wheat, barley, grain, various kinds of pulse, vetches, and others of this genus, they call corn (jūrdāk). Every plant with fragrant leaves, which is cultivated by the hand-labor of men, and is perennial, they call an aromatic herb (siparam). Whatever sweet-scented blossom arises at various seasons through the hand-labor of men, or has a perennial root and blossoms in its season with new shoots and sweet-scented blossoms, as the rose, the narcissus, the jasmine, the dog-rose (nēstarūn), the tulip, the colocynth (kavastīk), the pandanus (kēdi), the camba, the ox-eye (hēri), the crocus, the swallow-wort (zarda), the violet, the kārda, and others of this genus, they call a flower (gūl). Everything whose sweet-scented fruit, or sweet-scented blossom, arises in its season, without the hand-labor of men, they call a wild plant (vahār or nihāl). Whatever is welcome as food of cattle and beasts of burden they call grass (giyāh). Whatever enters into cakes (pēs-pārakihā) they call spices (āvzārihā). Whatever is welcome in eating of bread, as torn shoots of the coriander, water-cress (kakīj), the leek, and

---

[1] Cf. E. W. West, Pahlavi Literature, p. 98 (in Grundriss iran. Phil., Vol. II).
[2] Pahlavi Texts, pt. I, p. 100 (Sacred Books of the East, Vol V).

others of this genus, they call salad (*tērak* or *tārak*, Persian *tarah*). Whatever is like spinning cotton, and others of this genus, they call clothing plants (*jāmak*). Whatever lentil (*mačag*) is greasy, as sesame, *dūšdāṅ*, hemp, *vandak* (perhaps for *zētō*, 'olive,' as Anquetil supposes, and Justi assumes), and others of this genus, they call an oil-seed (*rōkanō*). Whatever one can dye clothing with, as saffron, sapan-wood, *začava*, *vaha*, and others of this genus, they call a dye-plant (*rag*). Whatever root, or gum (*tūf*), or wood is scented, as frankincense (Pazand *kendri* for Pahlavi *kundur*), *varāst* (Persian *barghast*), *kust*, sandalwood, cardamom (Pāzand *kākura*, Persian *qaqulah*, 'cardamoms, or *kākul*, *kākūl*, 'marjoram'), camphor, orange-scented mint, and others of this genus, they call a scent (*bod*). Whatever stickiness comes out from plants they call gummy (*vadak*). The timber which proceeds from the trees, when it is either dry or wet, they call wood (*čibā*). Every one of all these plants which is so, they call medicinal (*dārūk*).

"The principal fruits are of thirty kinds, and there are ten species the inside and outside of which are fit to eat, as the fig, the apple, the quince, the citron, the grape, the mulberry, the pear, and others of this kind. There are ten the outside of which is fit to eat, but not the inside, as the date, the peach, the white apricot, and others of this kind; those the inside of which is fit to eat, but not the outside, are the walnut, the almond, the pomegranate, the coco-nut,[1] the filbert (*funduk*), the chestnut (*šahbalūt*), the pistachio nut, the *vargān*, and whatever else of this description are very remarkable.

"This, too, it says, that every single flower is appropriate to an angel (*amešōspend*),[2] as the white jasmine (*saman*) is for Vohūman, the myrtle and jasmine (*yāsmin*) are Auharmazd's own, the mouse-ear (or sweet marjoram) is Ašavahist's own, the basil-royal is Šatvīrō's own, the musk flower is Spendarmad's, the lily is Horvadad's, the *čamba* is Amerōdad's, Dīn-pavan-Atarō has the orange-scented mint (*vādrang-bōd*), Atarō has the marigold (*ādargun*), the water-lily is Avān's, the white *marv* is Xūršed's, the *ranges* (probably *rand*, 'laurel') is Māh's, the violet is Tīr's, the *mēren* is Gōs's, the *kārda* is Dīn-pavan-Mitrō's, all violets are Mitrō's, the red chrysanthemum (*xēr*) is Srōš's, the dog-rose (*nestran*) is Rašnū's, the cockscomb is Fravardīn's, the *sisebar* is Vāhrām's, the yellow chrysanthemum is Rām's, the orange-

---

[1] Pazand *anārsar* is a misreading of Pahlavi *anārgīl* (Persian *nārgīl*), from Sanskrit *nārikela*.

[2] These are the thirty archangels and angels whose names are applied to the thirty days of the Parsi month, in the order in which they are mentioned here, except that Auharmazd is the first day, and Vohūṅan is the second.

scented mint is Vād's, the trigonella is Dīn-pavan-Dīn's, the hundred-petalled rose is Dīn's, all kinds of wild flowers (*vahār*) are Ard's, Āçtād has all the white Hōm, the bread-baker's basil is Āsmān's, Zamyād has the crocus, Māraspend has the flower of Ardašīr, Anīrān has this Hōm of the angel Hōm, of three kinds."

From this extract it becomes evident that the ancient Persians paid attention to their flora, and, being fond of systematizing, possessed a classification of their plants; but any of their botanical literature, if it ever existed, is lost.

The most important of the Persian works on pharmacology is the *Kitāb-ulabniyat 'an haqā'iq-uladviyat* or "Book of the Foundations of the True Properties of the Remedies," written about A.D. 970 by the physician Abū Mansūr Muvaffaq bin 'Alī alharavī, who during one of his journeys visited also India. He wrote for Mansūr Ibn Nūh II of the house of the Samanides, who reigned from 961 to 976 or 977. This is not only the earliest Persian work on the subject, but the oldest extant production in prose of New-Persian literature. The text has been edited by R. SELIGMANN from a unique manuscript of Vienna dated A.D. 1055, the oldest extant Persian manuscript.[1] There is a translation by a Persian physician, ABDUL-CHALIG ACHUNDOW from Baku.[2] The translation in general seems good, and is provided with an elaborate commentary, but in view of the importance of the work a new critical edition would be desirable. The sources from which Abū Mansūr derived his materials should be carefully sifted: we should like to know in detail what he owes to the Arabs, the Syrians, and the Indians, and what is due to his own observations. Altogether Arabic influence is pre-eminent. Cf. Appendix III.

A good many Chinese plant-names introduced from Iran have the word Hu 胡 prefixed to them. Hu is one of those general Chinese designations without specific ethnic value for certain groups of foreign tribes. Under the Han it appears mainly to refer to Turkish tribes; thus the Hiun-nu are termed Hu in the *Ši ki*. From the fourth century onward it relates to Central Asia and more particularly to peoples of

---

[1] Codex Vindobonensis sive Medici Abu Mansur Muwaffak Bin Alī Heratensis liber Fundamentorum Pharamacologiae Pars I Prolegomena et textum continens (Vienna, 1859).

[2] Die pharmakologischen Grundsätze des A. M. Muwaffak, in R. Kobert's Historische Studien aus dem Pharmakologischen Institute der Universität Dorpat, 1873. Quoted as "Achundow, Abu Mansur." The author's name is properly 'Abdu'l-Khaliq, son of the Akhund or schoolmaster. Cf. E. G. BROWNE, Literary History of Persia, pp. 11, 478.

Iranian extraction.[1] BRETSCHNEIDER[2] annotated, "If the character *hu* occurs in the name of a plant, it can be assumed that the plant is of foreign origin and especially from western Asia, for by *Hu žen* the ancient Chinese denoted the peoples of western Asia." This is but partially correct. The attribute *hu* is by no means a safe criterion in stamping a plant as foreign, neither does *hu* in the names of plants which really are of foreign origin apply to West-Asiatic or Iranian plants exclusively.

1. The word *hu* appears in a number of names of indigenous and partially wild plants without any apparent connection with the tribal designation Hu or without allusion to their provenience from the Hu. In the *Li Sao*, the famous elegies by K'ü Yüan of the fourth century B.C., a plant is mentioned under the name *hu šen* 胡繩, said to be a fragrant grass from which long cords were made. This plant is not identified.[3]

2. The acid variety of *yu* 柚 (*Citrus grandis*) is styled *hu kan* 胡甘,[4] apparently an ironical nickname, which may mean "sweet like the Hu." The tree itself is a native of China.

3. The term *hu hien* 胡莧 occurs only in the *T'u kiñ pen ts'ao* of Su Suñ of the eleventh century as a variety of *hien* (*Amarantus*), which is indigenous to China. It is not stated that this variety came from abroad, nor is it known what it really was.

4. *Hu mien mañ* 胡面莽 is a variety of *Rehmannia*,[5] a native of China and Japan. The name possibly means "the *mañ* with the face of a Hu."[6] Č'en Ts'añ-k'i of the T'ang says in regard to this plant that it grows in Liñ-nan (Kwañ-tuñ), and is like *ti hwañ* 地黄 (*Rehmannia glutinosa*).

5. The plant known as *ku-sui-pu* 骨碎補 (*Polypodium fortunei*) is indigenous to China, and, according to Č'en Ts'añ-k'i, was called

---

[1] "Le terme est bien en principe, vers l'an 800, une désignation des Iraniens et en particulier des Sogdiens" (CHAVANNES and PELLIOT, Traité manichéen, p. 231). This in general is certainly true, but we have well authenticated instances, traceable to the fourth century at least, of specifically Iranian plants the names of which are combined with the element *Hu*, that can but apply to Iranians.

[2] *Chinese Recorder*, 1871, p. 221.

[3] BRETSCHNEIDER, Bot. Sin., pt. II, No. 420; and *Li sao ts'ao mu su* (Ch. 2, p. 16 b, ed. of *Či pu tsu čai ts'uñ šu*) by Wu Žen-kie 吳仁傑 of the Sung period. See also *T'ai p'iñ yü lan*, Ch. 994, p. 6 b.

[4] BRETSCHNEIDER, *op. cit.*, No. 236; W. T. SWINGLE in Plantæ Wilsonianæ, Vol. II, p. 130.

[5] STUART, Chinese Materia Medica, p. 372.

[6] Cf. analogous plant-names like our Jews-mallow, Jews-thorn, Jews-ear, Jews-apple.

by the people of Kiaṅ-si 胡孫薑 *hu-sun-kiaṅ*, a purely local name which does not hint at any relation to the Hu.

6. Another botanical name in which the word *hu* appears without reference to the Hu is *č'ui-hu-ken* 鎚胡根, unidentified, a wild plant diffused all over China, and first mentioned by Č'en Ts'aṅ-k'i as growing in the river-valleys of Kiaṅ-nan.[1]

7–8. The same remark holds good for *ts'e-hu* 茈 (柴) 胡[2] (*Bupleurum falcatum*), a wild plant of all northern provinces and already described in the *Pie lu*, and for *ts'ien-hu* 前胡[3] (*Angelica decursiva*), growing in damp soil in central and northern China.

9. *Šu-hu-lan* 蜀胡爛 is an unidentified plant, first and solely mentioned by Č'en Ts'aṅ-k'i,[4] the seeds of which, resembling those of *Pimpinella anisum*, are eatable and medicinally employed. It grows in Annam. One might be tempted to take the term as *hu-lan* of Šu (Se-č'wan), but *šu-hu-lan* may be the transcription of a foreign word.

10. The *ma-k'in* 馬蘄 or *niu* 牛 *k'in* (*Viola pinnata*), a wild violet, is termed *hu k'in* 胡芹 in the *T'uṅ či* 通志 by Čeṅ Tsiao 鄭樵 (1108–62) and in the *T'u kiṅ pen ts'ao* of Su Suṅ.[5] No explanation as to the meaning of this *hu* is on record.

11. The *hu-man (wan)* 胡蔓 is a poisonous plant, identified with *Gelsemium elegans*.[6] It is mentioned in the *Pei hu lu*[7] with the synonyme *ye-ko* 冶葛,[8] the vegetable *yuṅ* 蕹 (*Ipomoea aquatica*) being regarded as an antidote for poisoning by *hu-man*. Č'en Ts'aṅ-k'i is cited as authority for this statement. The *Liṅ piao lu i*[9] writes the name 野葛, and defines it as a poisonous grass; *hu-man* grass is the common colloquial name. The same work further says, "When one has eaten of this plant by mistake, one should use a broth made from sheep's blood which will neutralize the poison. According to some, this plant grows as a creeper. Its leaves are like those of the *lan hiaṅ* 蘭香, bright and thick. Its poison largely penetrates into the leaves, and is not employed

---

[1] *Pen ts'ao kaṅ mu*, Ch. 16, p. 7 b.

[2] *Op. cit.*, Ch. 13, p. 6 b.

[3] *Op. cit.*, Ch. 13, p. 7 b.

[4] *Op. cit.*, Ch. 26, p. 22 b.

[5] *Op. cit.*, Ch. 26, p. 21; *Či wu miṅ ši l'u k'ao*, Ch. 14, p. 76.

[6] Cf. C. FORD, *China Review*, Vol. XV, 1887, pp. 215–220. STUART (Chinese Materia Medica, p. 220) says that the plant is unidentified, nevertheless he describes it on p. 185.

[7] Ch. 2, p. 18 b (ed. of Lu Sin-yúan).

[8] According to MATSUMURA (Shokubutsu mei-i, No. 2689), *Rhus toxicodendron* (Japanese *tsula-uruši*).

[9] Ch. B, p. 2 (ed. of *Wu yiṅ tien*).

as a drug. Even if an antidote is taken, this poison will cause death within a half day. The goats feeding on the sprouts of this plant will fatten and grow." Fan Č'eṅ-ta 范成大 (1126–93), in his *Kwei hai yü heṅ či*,[1] mentions this plant under the name *hu-man t'eṅ* 藤 ("*hu-man* creeper"), saying that it is a poisonous herb, which, rubbed and soaked in water, will result in instantaneous death as soon as this liquid enters the mouth. The plant is indigenous to southern China, and no reason is given for the word *hu* being prefixed to it.

12. *Hu t'ui-tse* 胡頹子 (literally, "chin of the Hu") is the name of an evergreen tree or shrub indigenous throughout China, even to Annam. The name is not explained, and there are no data in Chinese records to indicate that it was introduced from abroad.[2] It is mentioned by Č'en Ts'aṅ-k'i as a tree growing in P'iṅ-lin 平林, and it is said to be alluded to in the chapter *Wu hiṅ či* 五行志 of the *Suṅ šu*. The synonyme *k'io'r-su* 雀兒酥 ("sparrow-curd," because the birds are fond of the fruit) first appears in the *Pao či lun* of Lei Hiao of the fifth century. The people of Yüe call the plant *p'u-t'ui-tse* 蒲頹子; the southerners, *lu-tu-tse* 盧都子, which according to Liu Tsi 劉績 of the Ming, in his *Fei süe lu* 霏雪錄, is a word from the speech of the Man. The people of Wu term the tree *pan-han-č'un* 半含春, because its fruit ripens at an early date. The people of Siaṅ 襄 style it *hwaṅ-p'o-nai* 黃婆嬭 ("yellow woman's breast"), because the fruit resembles a nipple.

13. In *hu-lu* 胡 or 葫蘆 (*Lagenaria vulgaris*) the first character is a substitute for 瓠 *hu*. The gourd is a native of China.

14. *Hui-hui tou* 回回豆 (literally, "Mohammedan bean") is a plant everywhere growing wild in the fields.[3] The same remark holds good for *hu tou* 胡豆, a kind of bean which is roasted or made into flour, according to the *Pen ts'ao ši i*, a weed growing in rice-fields. Wu K'i-ts'ün, author of the *Či wu miṅ ši t'u k'ao*, says, "What is now *hu tou*, grows wild, and is not the *hu tou* of ancient times."[4]

15. *Yen hu su* 延胡素 denotes tubers of *Corydalis ambigua*: they are little, hard, brown tubers, of somewhat flattened spherical form, averaging half an inch in diameter. The plant is a native of Siberia,

---

[1] Ed. of *Či pu tsu čai ts'uṅ šu*, p. 30.

[2] STUART (Chinese Materia Medica, p. 161) is mistaken in saying that several names of this plant are "possibly transliterations of Turkic or Mongol names." There are no such names on record. The tree is identified with *Elæagnus longipes* or *pungens*.

[3] *Či wu miṅ ši t'u k'ao*, Ch. 2, p. 11 b. It is first mentioned in the *Kiu hwaṅ pen ts'ao*, being also called *na-ho-tou* 那合豆

[4] See, further, below, p. 305.

Kamchatka, and the Amur region, and flowers upon the melting of the snow in early spring.[1] According to the *Pen ts'ao kaṅ mu*,[2] the plant is first mentioned by Č'en Ts'aṅ-k'i of the T'ang period as growing in the country Hi 奚, and came from Ṅan-tuṅ 安東 (in Korea). Li Ši-čen annotates that by Hi the north-eastern barbarians should be understood. Waṅ Hao-ku 王好古, a physician of the thirteenth century, remarks that the name of the plant was originally *hüan* 玄 *hu-su*, but that on account of a taboo (to avoid the name of the Emperor Čen-tsuṅ of the Sung) it was altered into *yen-hu-su;* but this explanation cannot be correct, as the latter designation is already ascribed to Č'en Ts'aṅ-k'i of the T'ang. It is not known whether *hu* in this case would allude to the provenience of the plant from Korea. In the following example, however, the allusion to Korea is clear.

The mint, 薄荷 *po-ho*, *bak-xa (*Mentha arvensis* or *aquatica*), occurs in China both spontaneously and in the cultivated state. The plant is regarded as indigenous by the Chinese, but also a foreign variety is known as *hu pa-ho* (*bwat-xa) 胡菝蔄.[3] Č'en Ši-liaṅ 陳士良, in his *Ši siṅ pen ts'ao* 食性本草, published in the tenth century, introduced the term *wu* 吳 *pa-ho*, "mint of Wu" (that is, Su-čou, where the best mint was cultivated), in distinction from *hu pa-ho*, "mint of the Hu." Su Suṅ, in his *T'u kiṅ pen ts'ao*, written at the end of the eleventh century, affirms that this foreign mint is similar to the native species, the only difference being that it is somewhat sweeter in taste; it grows on the border of Kiaṅ-su and Če-kiaṅ, where the people make it into tea; commonly it is styled *Sin-lo* 新羅 *po-ho*, "mint of Sinra" (in Korea). Thus this variety may have been introduced under the Sung from Korea, and it is to this country that the term *hu* may refer.

Li Ši-čen relates that Sun Se-miao 孫思邈, in his *Ts'ien kin faṅ* 千金方,[4] writes the word 蕃荷 *fan-ho*, but that this is erroneously due to a dialectic pronunciation. This means, in other words, that the first character *fan* is merely a variant of 菝,[5] and, like the latter, had the phonetic equivalent *bwat, bat.[6]

---

[1] HANBURY, Science Papers, p. 256.

[2] Ch. 13, p. 13.

[3] The word *po-ho* is Chinese, not foreign. The Persian word for "peppermint" is *pūdene, pudina, budenk* (Kurd *punk*); in Hindī it is *pūdīnā* or *pūdinēkā*, derived from the Persian. In Tibetan (Ladākh) it is *p'o-lo-liṅ;* in the Tibetan written language, *byi-rug-pa*, hence Mongol *jirukba;* in Manchu it is *farsa*.

[4] See below, p. 306.

[5] As Sun Se-miao lived in the seventh century, when the Korean mint was not yet introduced, his term *fan-ho* could, of course, not be construed to mean "foreign mint."

[6] In *T'oung Pao* (1915, p. 18) PELLIOT has endeavored to show that the char-

In the following example there is no positive evidence as to the significance of *hu*. *Hu waṅ ši če* 胡王使者 ("envoy of the king of the Hu") is a synonyme of *tu hwo* 獨活 (*Peucedanum decursivum*).[1] As the same plant is also styled *k'iaṅ ts'iṅ* 羌青, *k'iaṅ hwo*, and *hu k'iaṅ ši če* 護羌使者, the term K'iaṅ (*Giaṅ) alluding to Tibetan tribes, it may be inferred that the king of the Hu likewise hints at Tibetans. In general, however, the term Hu does not include Tibetans, and the present case is not conclusive in showing that it does. In the chapter on the walnut it will be seen that there are two introduced varieties,— an Iranian (*hu t'ao*) and a Tibetan one (*k'iaṅ t'ao*).

In *hu ts'ai* (*Brassica rapa*) the element *hu*, according to Chinese tradition, relates to Mongolia, while it is very likely that the vegetable itself was merely introduced there from Iran.[2]

In other instances, plants have some relation to the Hu; but what this relation is, or what group of tribes should be understood by Hu, is not revealed.

There is a plant, termed *hu hwaṅ lien* 胡黃連, the *hwaṅ-lien* (*Coptis teeta*) of the Hu, because, as Li Ši-čen says, its physical characteristics, taste, virtue, and employment are similar to those of *hwaṅ-lien*. It has been identified with *Barkhausia repens*. As evidenced by the

---

acter *fan*, on the authority of K'aṅ-hi, could never have had the pronunciation *po* nor a final consonant, and that, accordingly, in the tribal name T'u-fan (Tibet) the character *fan*, as had previously been assumed, could not transcribe the Tibetan word *bod*. True it is that under the character in question K'aṅ-hi has nothing to say about *po*, but 蕃 is merely a graphic variant of 番, with which it is phonetically identical. Now under this character, K'aṅ-hi indicates plainly that, according to the *Tsi yün* and *Čeṅ yün*, *fan* in geographical names is to be read *p'o* (anciently *bwa) 婆 (*fan-ts'ie* 蒲 波), and that, according to the dictionary *Ši wen*, the same character was pronounced *p'o* (*bwa) 婆, *p'u* 蒲, and *p'an* 盤 (cf. also SCHLEGEL, Secret of the Chinese Method, pp. 21–22). In the ancient transcription 番 or 蕃 兜 *fan-tou*, *par-tav*, reproduction of Old Persian Parθava (see above, p. 187), *fan* corresponds very well to *par* or *bar;* and if it could interchange with the phonetic 拔 *pa*, *bwat, bwar, it is perfectly clear that, contrary to Pelliot's theory, there were at least dialectic cases, where 番 was possessed of a final consonant, being sounded *bwat* or *bwar.* Consequently it could have very well served for the reproduction of Tibetan *bod*. From another phonetic viewpoint the above case is of interest: we have *bak-xa and *bwat-xa as ancient names for the mint, which goes to show that the final consonants of the first element were vacillating or varied in different dialects (cf. *T'oung Pao*, 1916, pp. 110–114).

[1] *T'uṅ či* (above, p. 196), Ch. 75, p. 12 b.

[2] See below, p. 381. In the term *hu yen* ("swallow of the Hu"), *hu* appears to refer to Mongolia, as shown by the Manchu translation *monggo čibin* and the Turkī equivalent *qalmaq qarlogač* (Mongol *xatun xariyatsai*, Tibetan *gyi-gyi k'ug-rta;* cf. Ross, Polyglot List of Birds, No. 267). The bird occurs not only in Mongolia, but also in Če-kiaṅ Province, China (see *Kwei ki san fu ču* 會稽三賦註, Ch. 2, p. 8; ed. of *Si yin hüan ts'uṅ šu*).

attribute Hu, it may be of foreign origin, its foreign name being 割狐
露澤 ko-hu-lu-tse (*kat-wu-lou-dzak). Unfortunately it is not indicated
at what time this transcription was adopted, nor does Li Ši-čen state
the source from which he derived it. The only T'ang author who
mentions the plant, Su Kuṅ, does not give this foreign name. At all
events, it does not convey the impression of representing a T'ang
transcription; on the contrary, it bears the ear-marks of a transcription
made under the Yüan. Su Kuṅ observes, "Hu hwaṅ-lien is produced
in the country Po-se and grows on dry land near the sea-shore. Its
sprouts are like those of the hia-ku ts'ao 夏枯草 (Brunella vulgaris).
The root resembles a bird's bill; and the cross-section, the eyes of the
mainah. The best is gathered in the first decade of the eighth month."
Su Suṅ of the Sung period remarks that the plant now occurs in Nan-hai
(Kwaṅ-tuṅ), as well as in Ts'in-luṅ 秦隴 (Šen-si and Kan-su). This
seems to be all the information on record.[1] It is not known to me that
Barkhausia grows in Persia; at least, Schlimmer, in his extensive dictionary of Persian plants, does not note it.

Šou-ti 數低 is mentioned by Č'en Ts'aṅ-k'i as a plant (not yet
identified) with seeds of sweet and warm flavor and not poisonous, and
growing in Si-fan (Western Barbarians or Tibet) and in northern China
北土, resembling hwai hiaṅ 懷香 (Pimpinella anisum). The Hu make
the seeds into a soup and eat them.[2] In this case the term Hu may be
equated with Si-fan, but among the Chinese naturalists the latter term
is somewhat loosely used, and does not necessarily designate Tibet.[3]

Hiuṅ-k'iuṅ 芎藭 (Conioselinum univittatum) is an umbelliferous
plant, which is a native of China. As early as the third century A.D.
it is stated in the Wu ši pen ts'ao[4] that some varieties of this plant grow
among the Hu; and Li Ši-čen annotates that the varieties from the Hu
and Žuṅ are excellent, and are hence styled hu k'iuṅ 胡藭.[5] It is stated
that this genus is found in mountain districts in Central Europe,
Siberia, and north-western America.[6]

---

[1] What STUART (Chinese Materia Medica, p. 65) says regarding this plant is
very inexact. He arbitrarily identifies the term Hu with the Kukunor, and wrongly
ascribes Su Kuṅ's statement to T'ao Huṅ-kiṅ. Such an assertion as, "the drug is
now said to be produced in Nan-hai, and also in Šen-si and Kan-su," is misleading,
as this "now" comes from an author of the Sung period, and does not necessarily
hold good for the present time.

[2] Pen ts'ao kaṅ mu, Ch. 26, p. 22 b.

[3] Cf. below, p. 344.

[4] Cf. Beginnings of Porcelain, p. 115.

[5] He also imparts a Sanskrit name from the Suvarnaprabhāsa-sūtra in the form
闍莫迦 še-mo-k'ie, *ja-mak-gia. The genus is not contained in WATT's Dictionary.

[6] Treasury of Botany, Vol. I, p. 322.

In *hu tsiao* ("pepper") the attribute *hu* distinctly refers to India.[1] Another example in which *hu* alludes to India is presented by the term *hu kan kian* 胡乾薑 ("dried ginger of the Hu"), which is a synonyme of *T'ien-ču* 天竺 *kan kian* ("dried ginger of India"), "produced in the country of the Brahmans."[2]

In the term *hu fen* 胡粉 (a cosmetic or facial powder of white lead), the element *hu* bears no relation to the Hu, although it is mentioned as a product of Kuča[3] and subsequently as one of the city of Ili (Yi-li-pa-li).[4] In fact, there is no Chinese tradition to the effect that this substance ever came from the Hu.[5] F. P. SMITH[6] observed with reference to this subject, "The word *hu* does not denote that the substance was formerly obtained from some foreign source, but is the result of a mistaken character." This evidently refers to the definition of the dictionary *Ši min* 釋名 by Liu Hi of the Han, who explains this *hu* by 糊 *hu* ("gruel, congee"), which is mixed with grease to be rubbed into the face. The process of making this powder from lead is a thoroughly Chinese affair.

In the term *hu yen* 胡鹽 ("salt of the Hu") the word Hu refers to barbarous, chiefly Tibetan, tribes bordering on China in the west; for there are also the synonymes *žun* 戎 *yen* and *k'ian* 羌 *yen*, the former already occurring in the *Pie lu*. Su Kun of the seventh century equalizes the terms *žun yen* and *hu yen*, and gives *t'u-ten* 禿登 *yen* as the word used in Ša-čou 沙州. Ta Min 大明, who wrote in A.D. 970, says that this is the salt consumed by the Tibetans (Si-fan), and hence receives the designation *žun* or *k'ian yen*. Other texts, however, seem to make a distinction between *hu yen* and *žun yen:* thus it is said in the biography of Li Hiao-po 李孝伯 in the *Wei šu*, "The salt of the Hu cures pain of the eye, the salt of the Žun heals ulcers."

The preceding examples are sufficient to illustrate the fact that the element *hu* in botanical terms demands caution, and that each case must be judged on its own merits. No hard and fast rule, as deduced by Bretschneider, can be laid down: the mere addition of *hu* proves neither that a plant is foreign, nor that it is West-Asiatic or Iranian. There are native plants equipped with this attribute, and there are foreign plants thus characterized, which hail from Korea, India, or

---

[1] See below, p. 374.

[2] *Čen lei pen ts'ao*, Ch. 6, p. 67 b.

[3] *Čou šu*, Ch. 50, p. 5; *Sui šu*, Ch. 83, p. 5 b.

[4] *Ta Min i t'un či*, Ch. 89, p. 22; *Kwan yü ki*, Ch. 24, p. 6 b.

[5] *Pen ts'ao kan mu*, Ch. 8, p. 6; GEERTS (Produits, pp. 596–601), whose translation "poudre des pays barbares" is out of place.

[6] Contributions towards the Materia Medica of China, p. 231.

some vaguely defined region of Central Asia. The fact, however, remains that there are a number of introduced, cultivated Hu plants
coming from Iranian lands, but in each and every case it has been my
endeavor to furnish proof for the fact that these actually represent
Iranian cultivations. With the sole exception of the walnut, the history of which may tolerably well be traced, the records of these Hu
plants are rather vague, and for none of them is there any specific
account of the introduction. It is for botanical rather than historical
reasons that the fact of the introduction becomes evident. It is this
hazy character of the traditions which renders it impossible to connect
these plants in any way with Čaṅ K'ien. Moreover, it cannot be
proved with certainty that any names of plants or products formed
with the element *hu* existed under the Han. The sole exception would
be *hu ts'ai*,[1] but its occurrence in the *T'uṅ su wen* of the Han is not
certain either; and this *hu*, according to Chinese tradition, refers to
Mongolia, not to Iran. Another merely seeming exception is presented
by *hu t'uṅ-lei*,[2] but this is a wild, not a cultivated tree; and *hu*, in this
case, has a geographical rather than an ethnographical significance.   In
the wooden documents discovered in Turkistan we have one good,
datable instance of a Hu product; and this is *hu t'ie* ("iron of the Hu"
and implements made of such iron). These tablets belong to the Tsin
period (A.D. 265–419),[3] while in no wooden document of the Han has
any compound with Hu as yet been traced. Again, all available evidence goes to show that these Hu plants were not introduced earlier
than the Tsin dynasty, or, generally speaking, during what is known
as the Leu č'ao or six minor dynasties, covering the time from the
downfall of the Han to the rise of the T'ang dynasty. It is noteworthy
that of none of these plants is an Iranian name on record.

The element *hu*, in a few cases, serves also the purpose of a transcription: thus probably in the name of the coriander, *hu-swi*,[4] and
quite evidently in the name of the fenugreek, *hu-lu-pa*.[5]

Imported fruits and products have been named by many nations
for the countries from which they hailed or from the people by whom
they were first brought. The Greeks had their "Persian apple" (μῆλον
Περσικόν, "peach"), their "Medic apple" (μῆλον Μηδικόν, "citron"),
their "Medic grass" (Μηδικὴ πόα, "alfalfa"), and their "Armenian

---

[1] Below, p. 381.

[2] Below, p. 339.

[3] CHAVANNES, Documents chinois découverts par Aurel Stein, pp. 168, 169.

[4] Below, p. 298.

[5] Below, p. 446. It thus occurs also in geographical names, as in Hu-č'a-la
(Guzerat); see HIRTH and ROCKHILL, Chao Ju-kua, p. 92.

apple" (μῆλον Ἀρμενιακόν, "apricot"). RABELAIS (1483–1553)[1] has already made the following just observation on this point, "Les autres [plantes] ont retenu le nom des regions des quelles furent ailleurs transportées, comme pommes medices, ce sont pommes de Medie, en laquelle furent premierement trouvées; pommes puniques, ce sont grenades, apportées de Punicie, c'est Carthage. *Ligusticum*, c'est livesche, apportée de Ligurie, c'est la couste de Genes: rhabarbe, du fleuve Barbare nommé Rha, comme atteste Ammianus: santonique, fenu grec; castanes, persiques, sabine; stoechas, de mes isles Hieres, antiquement dites Stoechades; spica celtica et autres." The Tibetans, as I have shown,[2] form many names of plants and products with Bal (Nepal), Mon (Himalayan Region), rGya (China), and Li (Khotan).

In the same manner we have numerous botanical terms preceded by "American, Indian, Turkish, Turkey, Guinea," etc.

Aside from the general term Hu, the Chinese characterize Iranian plants also by the attribute Po-se (Parsa, Persia): thus *Po-se tsao* ("Persian jujube") serves for the designation of the date. The term Po-se requires great caution, as it denotes two different countries, Persia and a certain Malayan region. This duplicity of the name caused grave confusion among both Chinese and European scholars, so that I was compelled to devote to this problem a special chapter in which all available sources relative to the Malayan Po-se and its products are discussed. Another tribal name that quite frequently occurs in connection with Iranian plant-names is Si-žun 西戎 ("the Western Žun"). These tribes appear as early as the epoch of the *Ši kin* and *Šu kin*, and seem to be people of Hiun-nu descent. In post-Christian times Si-žun developed into a generic term without ethnic significance, and vaguely hints at Central-Asiatic regions. Combined with botanical names, it appears to be synonymous with Hu.[3] It is a matter of course that all these geographical and tribal allusions in plant-names have merely a relative, not an absolute value; that is, if the Chinese, for instance, designate a plant as Persian (Po-se) or Hu, this signifies that from their viewpoint the plant under notice hailed from Iran, or in some way was associated with the activity of Iranian nations, but it does not mean that the plant itself or its cultivation is peculiar or due to Iranians. This may be the case or not, yet this point remains to be determined by a special investigation in each particular instance. While the Chinese, as will be seen, are better informed on the history

---

[1] Le Gargantua et le Pantagruel, Livre III, chap. L.

[2] *T'oung Pao,* 1916, pp. 409, 448, 456.

[3] For examples of its occurrence consult Index.

of important plants than any other people of Asia (and I should even
venture to add, of Europe), the exact and critical history of a plant-
cultivation can be written only by heeding all data and consulting all
sources that can be gathered from every quarter. The evidence accruing
from the Semites, from Egypt, Greece, and Rome, from the Arabs,
India, Camboja, Annam, Malayans, Japan, etc., must be equally
requisitioned. Only by such co-ordination may an authentic result be
hoped for.

The reader desirous of information on the scientific literature
of the Chinese utilized in this publication may be referred to Bretschneider's "Botanicon Sinicum" (part I).[1] It is regrettable that no
Pen ts'ao (Herbal) of the T'ang period has as yet come to light, and
that for these works we have to depend on the extracts given in later
books. The loss of the *Hu pen ts'ao* ("Materia Medica of the Hu")
and the *Č'u hu kwo fan* ("Prescriptions from the Hu Countries") is
especially deplorable. I have directly consulted the *Čen lei pen ts'ao*,
written by T'an Šen-wei in 1108 (editions printed in 1521 and 1587),
the *Pen ts'ao yen i* by K'ou Tsun-ši of 1116 in the edition of Lu Sin-
yüan, and the well-known and inexhaustible *Pen ts'ao kan mu* by Li
Ši-čen, completed in 1578. With all its errors and inexact quotations,
this remains a monumental work of great erudition and much solid
information. Of Japanese *Pen ts'ao* (*Honzō*) I have used the *Yamato
honzō*, written by Kaibara Ekken in 1709, and the *Honzō kōmoku keimō*
by Ono Ranzan. Wherever possible, I have resorted to the original
source-books. Of botanical works, the *Kwan k'ün fan p'u*, the *Hwa p'u*,
the *Či wu min ši t'u k'ao*, and several Japanese works, have been utilized.
The *Yu yan tsa tsu* has yielded a good many contributions to the plants
of Po-se and Fu-lin; several Fu-lin botanical names hitherto unexplained
I have been able to identify with their Aramaic equivalents. Although
these do not fall within the subject of Sino-Iranica, but Sino-Semitica,
it is justifiable to treat them in this connection, as the Fu-lin names
are given side by side with the Po-se names. Needless to say, I have
carefully read all accounts of Persia and the Iranian nations of Central
Asia contained in the Chinese Annals, and the material to be found
there constitutes the basis and backbone of this investigation.[2]

There is a class of literature which has not yet been enlisted for the

---

[1] We are in need, however, of a far more complete and critical history of the
scientific literature of the Chinese.

[2] The non-sinological reader may consult to advantage E. H. PARKER, Chinese
Knowledge of Early Persia (*Imp. and Asiatic Quarterly Review*, Vol. XV, 1903,
pp. 144–169) for the general contents of the documents relating to Persia. Most
names of plants and other products have been omitted in Parker's article.

study of cultivated plants, and this is the early literature on medicine. Prominent are the books of the physician Čań Čuń-kiń 張仲景 or Čań Ki 張機, who is supposed to have lived under the Later Han at the end of the second century A.D. A goodly number of cultivated plants is mentioned in his book *Kin kwei yü han yao lio fań lun* 金匱玉函要畧方論 or abbreviated *Kin kwei yao lio*.[1] This is a very interesting hand-book of dietetics giving detailed rules as to the avoidance of certain foods at certain times or in certain combinations, poisonous effects of articles of diet, and prescriptions to counteract this poison. Neither this nor any other medical writer gives descriptions of plants or notes regarding their introduction; they are simply enumerated in the text of the prescriptions. But it is readily seen that, if such a work can be exactly dated, it has a chronological value in determining whether a given plant was known at that period. Thus Čań Ki mentions, of plants that interest us in this investigation, the walnut, the pomegranate, the coriander, and *Allium scorodoprasum* (*hu swan*). Unfortunately, however, we do not know that we possess his work in its original shape, and Chinese scholars admit that it has suffered from interpolations which it is no longer possible to unravel. The data of such a work must be utilized with care whenever points of chronology are emphasized. It was rather tempting to add to the original prescriptions of Čań Ki, and there is no doubt that the subsequent editions have blended primeval text with later comments. The earliest commentary is by Wań Šu-ho 王叔和 of the Tsin. Now, if we note that the plants in question are otherwise not mentioned under the Han, but in other books are recorded only several centuries later, we can hardly refrain from entertaining serious doubts as to Čań Ki's acquaintance with them. A critical bibliographical study of early Chinese medical literature is an earnest desideratum.

A. DE CANDOLLE's monumental work on the "Origin of Cultivated Plants" is still the only comprehensive book on this subject that we have. It was a masterpiece for his time, and still merits being made the basis and starting-point for any investigation of this kind. De Candolle possessed a really critical and historical spirit, which cannot be said of other botanists who tried to follow him on the path of historical research; and the history of many cultivated plants has been outlined by him perfectly well and exactly. Of many others, our conceptions are now somewhat different. Above all, it must be said that

---

[1] Reprinted in the *Yü tswan i tsuń kin kien* of 1739 (WYLIE, Notes on Chinese Literature, p. 101). A good edition of this and the other works of the same author on the basis of a Sung edition is contained in the medical Ts'uń-šu, the *I t'uń čeń mo ts'üan šu*, published by the Če-kiań šu kū.

since his days Oriental studies have made such rapid strides, that his notes with regard to India, China, and Japan, are thoroughly out of date. As to China, he possessed no other information than the superficial remarks of BRETSCHNEIDER in his "Study and Value of Chinese Botanical Works,"[1] which teem with misunderstandings and errors.[2] De Candolle's conclusions as to things Chinese are no longer acceptable. The same holds good for India and probably also for Egypt and western Asia. In point of method, de Candolle has set a dangerous precedent to botanists in whose writings this effect is still visible, and this is his over-valuation of purely linguistic data. The existence of a native name for a plant is apt to prove little or nothing for the history of the plant, which must be based on documentary and botanical evidence. Names, as is well known, in many cases are misleading or deceptive; they constitute a welcome accessory in the chain of evidence, but they cannot be relied upon exclusively. It is a different case, of course, if the Chinese offer us plant-names which can be proved to be of Iranian origin. If on several occasions I feel obliged to uphold V. Hehn against his botanical critic A. Engler, such pleas must not be construed to mean that I am an unconditional admirer of Hehn; on the contrary, I am wide awake to his weak points and the shortcomings of his method, but wherever in my estimation he is right, it is my duty to say that he is right. A book to which I owe much information is CHARLES JORET's "Les Plantes dans l'antiquité et au moyen âge" (2 vols., Paris, 1897, 1904), which contains a sober and clear account of the plants of ancient Iran.[3]

A work to which I am greatly indebted is "Terminologie médico-pharmaceutique et anthropologique française-persane," by J. L. SCHLIMMER, lithographed at Teheran, 1874.[4] This comprehensive work of over 600 pages folio embodies the lifelong labors of an instructor at the Polytechnic College of Persia, and treats in alphabetical order of animal and vegetable products, drugs, minerals, mineral waters, native

---

[1] Published in the *Chinese Recorder* for 1870 and 1871.

[2] They represent the fruit of a first hasty and superficial reading of the *Pen ts'ao kan mu* without the application of any criticism. In Chinese literature we can reach a conclusion only by consulting and sifting all documents bearing on a problem. Bretschneider's Botanicon Sinicum, much quoted by sinologues and looked upon as a sort of gospel by those who are unable to control his data, has now a merely relative value, and is uncritical and unsatisfactory both from a botanical and a sinological viewpoint; it is simply a translation of the botanical section of the *Pen ts'ao kan mu* without criticism and with many errors, the most interesting plants being omitted.

[3] Joret died in Paris on December 26, 1914, at the age of eighty-five years (cf. obituary notice by H. CORDIER, *La Géographie*, 1914, p. 239).

[4] Quoted "SCHLIMMER, Terminologie." I wish to express my obligation to the Surgeon General's Library in Washington for the loan of this now very rare book.

therapeutics and diseases, with a wealth of solid information that has hardly ever been utilized by our science.

It is hoped that these researches will chiefly appeal to botanists and to students of human civilization; but, as it can hardly be expected that the individual botanist will be equally interested in the history of every plant here presented, each subject is treated as a unit and as an independent essay, so that any one, according to his inclination and choice, may approach any chapter he desires. Repetitions have therefore not been shunned, and cross-references are liberally interspersed; it should be borne in mind, however, that my object is not to outline merely the history of this or that plant, but what I wish to present is a synthetic and comprehensive picture of a great and unique plant-migration in the sense of a cultural movement, and simultaneously an attempt to determine the Iranian stratum in the structure of Chinese civilization. It is not easy to combine botanical, oriental, philological, and historical knowledge, but no pains have been spared to render justice to both the botanical and the historical side of each problem. All data have been sifted critically, whether they come from Chinese, Japanese, Indian, Persian, Arabic, or classical sources, and in no instance have I depended on a second-hand or dogmatic statement. The various criticisms of A. de Candolle, A. Engler, E. Bretschneider, and other eminent authorities, arise from the critical attitude toward the subject, and merely aim at the furtherance of the cause.

I wish to express my thanks to Dr. Tanaka Tyōzaburō in the Bureau of Plant Industry of the Department of Agriculture, Washington, for having kindly prepared a translation of the notices on the grape-vine and the walnut from Japanese sources, which are appended to the chapters on the history of these plants. The manuscript of this publication was completed in April, 1918.

The generosity of Mrs. T. B. Blackstone and Mr. Charles R. Crane in contributing a fund toward the printing of this volume is gratefully acknowledged.

ALFALFA

1. The earliest extant literary allusion to alfalfa[1] (*Medicago sativa*) is made in 424 B.C. in the Equites ("The Knights") of Aristophanes, who says (V, 606):

"Ἦσθιον δὲ τοὺς παγούρους ἀντὶ ποίας μηδικῆς.
"The horses ate the crabs of Corinth as a substitute for the Medic."

The term "Mēdikē" is derived from the name of the country Media. In his description of Media, Strabo[2] states that the plant constituting the chief food of the horses is called by the Greeks "Mēdikē" from its growing in Media in great abundance. He also mentions as a product of Media *silphion*, from which is obtained the Medic juice.[3] Pliny[4] intimates that "Medica" is by nature foreign to Greece, and that it was first introduced there from Media in consequence of the Persian wars under King Darius. Dioscorides[5] describes the plant without referring to a locality, and adds that it is used as forage by the cattle-breeders. In Italy, the plant was disseminated from the middle of the second century B.C. to the middle of the first century A.D.,[6]—almost coeval with its propagation to China. The Assyriologists claim that *aspasti* or *aspastu*, the Iranian designation of alfalfa, is mentioned in a Babylonian text of ca. 700 B.C.;[7] and it would not be impossible that its favorite fodder followed the horse at the time of its introduction from Iran into Mesopotamia. A. DE CANDOLLE[8] states that *Medicago*

---

[1] I use this term (not lucerne) in accordance with the practice of the U. S. Department of Agriculture; it is also the term generally used and understood by the people of the United States. The word is of Arabic origin, and was adopted by the Spaniards, who introduced it with the plant into Mexico and South America in the sixteenth century. In 1854 it was taken to San Francisco from Chile (J. M. WEST-GATE, Alfalfa, p. 5, Washington, 1908).

[2] XI. XIII, 7.

[3] Theophrastus (Hist. plant., VIII. VII, 7) mentions alfalfa but casually by saying that it is destroyed by the dung and urine of sheep. Regarding *silphion* see p. 355.

[4] XIII, 43.

[5] II, 176.

[6] HEHN, Kulturpflanzen, 8th ed., p. 412.

[7] SCHRADER in Hehn, p. 416; C. JORET (Plantes dans l'antiquité, Vol. II, p. 68) states after J. Halévy that *aspasti* figures in the list drawn up by the gardener of the Babylonian king Mardukbalidin (Merodach-Baladan), a contemporary of Ezechias King of Juda.

[8] Origin of Cultivated Plants, p. 103.

208

sativa has been found wild, with every appearance of an indigenous plant, in several provinces of Anatolia, to the south of the Caucasus, in several parts of Persia, in Afghanistan, Baluchistan, and in Kashmir.[1] Hence the Greeks, he concludes, may have introduced the plant from Asia Minor as well as from India, which extended from the north of Persia. This theory seems to me inadmissible and superfluous, for the Greeks allude solely to Media in this connection, not to India. Moreover, the cultivation of the plant is not ancient in India, but is of recent date, and hardly plays any rôle in Indian agriculture and economy.

In ancient Iran, alfalfa was a highly important crop closely associated with the breeding of superior races of horses. Pahlavi *aspast* or *aspist* New Persian *aspust, uspust, aspist, ispist*, or *isfist* (Puštu or Afghan *spastu, špēšta*), is traceable to an Avestan or Old-Iranian *aspō-asti (from the root *ad*, "to eat"), and literally means "horse-fodder."[2] This word has penetrated into Syriac in the form *aspestā* or *pespestā* (the latter in the Geoponica). Khosrau I (A.D. 531–578) of the Sasanian dynasty included alfalfa in his new organization of the land-tax:[3] the tax laid on alfalfa was seven times as high as that on wheat and barley, which gives an idea of the high valuation of that forage-plant. It was also employed in the pharmacopœia, being dealt with by Abu Mansur in his book on pharmacology.[4] The seeds are still used medicinally.[5] The Arabs derived from the Persians the word *isfist*, Arabicized into *fisfisa;* Arabic designations being *ratba* and *qatt*, the former for the plant in its natural state, the latter for the dried plant.[6]

The mere fact that the Greeks received *Medicago* from the Persians, and christened it "Medic grass," by no means signifies or proves at the outset that *Medicago* represents a genuinely Iranian cultivation. It is well known how fallacious such names are: the Greeks also had the peach under the name "Persian apple," and the apricot as "Armenian apple;" yet peach and apricot are not originally Persian or Armenian, but Chinese cultivations: Iranians and Armenians in this case merely

---

[1] As to Kashmir, it will be seen, we receive a confirmation from an ancient Chinese document. See also G. WATT, Dictionary of the Economic Products of India, Vol. V, pp. 199–203.

[2] NELDEKE, ZDMG, Vol. XXXII, 1878, p. 408. Regarding some analogous plant-names, see R. v. STACKELBERG, *ibid.*, Vol. LIV, 1900, pp. 108, 109.

[3] NÖLDEKE, Tabari, p. 244.

[4] ACHUNDOW, Abu Mansur, p. 73 (cf. above, p. 194).

[5] SCHLIMMER, Terminologie, p. 365. He gives *yondže* as the Persian name, which, however, is of Turkish origin (from *yont*, "horse"). In Asia Minor there is a place Yonjali ("rich in alfalfa").

[6] LECLERC, Traité des simples, Vol. III, p. 35.

acted as mediators between the far east and the Mediterranean. However, the case of alfalfa presents a different problem. The Chinese, who cultivate alfalfa to a great extent, do not claim it as an element of their agriculture, but have a circumstantial tradition as to when and how it was received by them from Iranian quarters in the second century B.C. As any antiquity for this plant is lacking in India or any other Asiatic country, the verdict as to the centre of its primeval cultivation is decidedly in favor of Iran. The contribution which the Chinese have to make to the history of *Medicago* is of fundamental importance and sheds new light on the whole subject: in fact, the history of no cultivated plant is so well authenticated and so solidly founded.

In the inscription of Persepolis, King Darius says, "This land Persia which Auramazda has bestowed on me, being beautiful, populous, and abundant in horses — according to the will of Auramazda and my own, King Darius — it does not tremble before any enemy." I have alluded in the introduction to the results of General Čaṅ Kʻien's memorable expedition to Central Asia. The desire to possess the fine Iranian thoroughbreds, more massively built than the small Mongolian horse, and distinguished by their noble proportions and slenderness of feet as well as by the development of chest, neck, and croup, was one of the strongest motives for the Emperor Wu (140–87 B.C.) to maintain regular missions to Iranian countries, which led to a regular caravan trade with Fergana and Parthia. Even more than ten such missions were dispatched in the course of a year, the minimum being five or six. At first, this superior breed of horse was obtained from the Wu-sun, but then it was found by Čaṅ Kʻien that the breed of Fergana was far superior. These horses were called "blood-sweating" (*han-hüe* 汗血),[1] and were believed to be the offspring of a heavenly horse (*tʻien ma* 天馬). The favorite fodder of this noble breed consisted in *Medicago sativa*; and it was a sound conclusion of General Čaṅ Kʻien, who was a practical man and possessed of good judgment in economic matters, that, if these much-coveted horses were to continue to thrive on Chinese soil, their staple food had to go along with them. Thus he obtained the seeds of alfalfa in Fergana,[2] and presented them in 126 B.C. to his imperial master, who had wide tracts of land near his palaces covered

----

[1] This name doubtless represents the echo of some Iranian mythical concept, but I have not yet succeeded in tracing it in Iranian mythology.

[2] In Fergana as well as in the remainder of Russian Turkistan *Medicago sativa* is still propagated on an immense scale, and represents the only forage-plant of that country, without which any economy would be impossible, for pasture-land and hay are lacking. Alfalfa yields four or five harvests there a year, and is used for the feeding of cattle either in the fresh or dry state. In the mountains it is cultivated up to an elevation of five thousand feet; wild or as an escape from cultivation it reaches

with this novel plant, and enjoyed the possession of large numbers of celestial horses.[1] From the palaces this fodder-plant soon spread to the people, and was rapidly diffused throughout northern China. According to Yen Ši-ku (A.D. 579-645), this was already an accomplished fact during the Han period. As an officinal plant, alfalfa appears in the early work *Pie lu*.[2] The *Ts'i min yao šu* of the sixth century A.D. gives rules for its cultivation; and T'ao Huṅ-kiṅ (A.D. 451-536) remarks that "it is grown in gardens at Č'aṅ-ñan (the ancient capital in Šen-si), and is much valued by the northerners, while the people of Kiaṅ-nan do not indulge in it much, as it is devoid of flavor. Abroad there is another *mu-su* plant for healing eye-diseases, but different from this species."[3]

Čaṅ K'ien was sent out by the Emperor Wu to search for the Yūe-či and to close an alliance with them against the Turkish Hiuṅ-nu. The Yūe-či, in my opinion, were an Indo-European people, speaking a North-Iranian language related to Scythian, Sogdian, Yagnōbi, and Ossetic. In the course of his mission, Čaṅ K'ien visited Fergana, Sogdiana, and Bactria, all strongholds of an Iranian population. The "West" for the first time revealed by him to his astounded countrymen was Iranian civilization, and the products which he brought back were thoroughly and typically Iranian. The two cultivated plants (and only these two) introduced by him into his fatherland hailed from Fergana: Ferganian was an Iranian language; and the words for the alfalfa and grape, *mu-su* and *p'u-t'ao*, were noted by Čaṅ K'ien in Fergana and transmitted to China along with the new cultivations. These words were Ferganian; that is, Iranian.[4] Čaṅ K'ien himself was

---

an altitude up to nine thousand feet. Cf. S. KORŽINSKI, Vegetation of Turkistan (in Russian), p. 51. Russian Turkistan produces the largest supply of alfalfa-seed for export (E. BROWN, Bull. Dep. of Agriculture, No. 138, 1914).

[1] *Ši ki*, Ch. 123.

[2] Cf. Chinese Clay Figures, p. 135.

[3] *Čeṅ lei pen ts'ao*, Ch. 27, p. 23. It is not known what this foreign species is.

[4] HIRTH's theory (*Journal Am. Or. Soc.*, Vol. XXXVII, 1917, p. 149), that the element *yüan* of Ta-yüan (Fergana) might represent a "fair linguistic equivalent" of Yavan (Yavana, the Indian name of the Greeks), had already been advanced by J. EDKINS (*Journal China Branch Roy. As. Soc.*, Vol. XVIII, 1884, p. 5). To me it seems eccentric, and I regret being unable to accept it. In the T'ang period we have from Hüan Tsaṅ a reproduction of the name Yavana in the form 閻摩那 Yen-mo-na, *Yam-mwa-na (PELLIOT, *Bull. de l'Ecole française*, Vol. IV, p. 278). For the Han period we should expect, after the analogy of 葉調 Ye-tiao, *Yap (Džap)-div (Yavadvīpa, Java), a transcription 葉那 Ye-na, *Yap-na, for Yavana. The term 於越 Yü-yüe, *Yu-vat(var), does not represent a transcription of Yavana, as supposed by CHAVANNES (Mémoires historiques de Se-ma Ts'ien, Vol. IV, 1901, pp. 558-559), but is intended to transcribe the name Yuan (*Yuvar, Yuar), still employed by the Čam and other peoples of Indo-China as a designation of

very well aware of the fact that the speech of the people of Fergana was
Iranian, for he stated in his report, that, although there were different
dialects in the tract of land stretching from Fergana westward as far
as Parthia (An-si), yet their resemblance was so great that the people
could make themselves intelligible to each other.[1] This is a plain
allusion to the differentiation and at the same time the unity of Iranian
speech;[2] and if the Ferganians were able to understand the Parthians,
I do not see in what other language than Iranian they could have
conversed. Certainly they did not speak Greek or Turkish, as some
prejudiced theorists are inclined to imagine.

The word brought back by Čaṅ Kʻien for the designation of alfalfa,
and still used everywhere in China for this plant, was *mu-su* 目宿,
consisting of two plain phonetic elements,[3] anciently *muk-suk (Japanese *moku-šuku*), subsequently written 苜蓿 with the addition of the
classifier No. 140. I recently had occasion to indicate an ancient Tibetan
transcription of the Chinese word in the form *bug-sug*,[4] and this appears
to come very near to the Iranian prototype to be restored, which was
*buksuk or *buxsux, perhaps *buxsuk. The only sensible explanation
ever given of this word, which unfortunately escaped the sinologues,
was advanced by W. TOMASCHEK,[5] who tentatively compared it with
Gīlakī (a Caspian dialect) *būso* ("alfalfa"). This would be satisfactory
if it could be demonstrated that this *būso* is evolved from *bux-sox* or
the like. Further progress in our knowledge of Iranian dialectology

---

Annam and the Annamese (cf. Čam Yuan or Yụởn, Bahnar, Juởn, Khmer Yuon,
Stieṅ Juởn). This native name, however, was adapted to or assimilated with Sanskrit
Yavana; for in the Sanskrit inscriptions of Campā, particularly in one of the reign
of Jaya-Rudravarman dated A.D. 1092, Annam is styled Yavana (A. BERGAIGNE,
L'Ancien royaume de Campā, p. 61 of the reprint from *Journal asiatique*, 1888).
In the Old-Javanese poem Nāgarakṛtāgama, completed in A.D. 1365, Yavana
occurs twice as a name for Annam (H. KERN, *Bijdragen tot de taal- land- en volkenkunde*,
Vol. LXXII, 1916, p. 399). Kern says that the question as to how the name of the
Greeks was applied to Annam has not been raised or answered by any one; he over-
looked the contribution of Bergaigne, who discussed the problem.

[1] Strabo (XV. II, 8) observes, "The name of Ariana is extended so as to include
some part of Persia, Media, and the north of Bactria and Sogdiana; for these peoples
speak nearly the same language."

[2] Emphasized by R. GAUTHIOT in his posthumous work Trois Mémoires sur
l'unité linguistique des parlers iraniens (reprinted from the *Mémoires de la Société
de Linguistique de Paris*, Vol. XX, 1916).

[3] The two characters are thus indeed written without the classifiers in the Han
Annals. The writings 牧蓿 *muk-suk of Kwo Pʻo and 木粟 *muk-swok of Lo
Yüan, author of the *Er ya i* (simply inspired by attempts at reading certain mean-
ings into the characters), have the same phonetic value. In Annamese it is *muk-tuk*.

[4] *Tʻoung Pao*, 1916, p. 500, No. 206.

[5] Pamir-Dialekte (*Sitzber. Wiener Akad.*, 1880, p. 792).

will no doubt supply the correct form of this word. We have to be mindful of the fact that the speech of those East-Iranian tribes, the advance-guard of Iran proper, with whom the Chinese first came in contact, has never been committed to writing, and is practically lost to us. Only secluded dialects may still harbor remnants of that lost treasure. We have to be the more grateful to the Chinese for having rescued for us a few words of that extinct language, and to place *buksuk or *buxsux on record as the ancient Ferganian appellation of *Medicago sativa*. The first element of this word may survive in Sariqolī (a Pamir dialect) *wux* ("grass"). In Waxī, another Pamir idiom, alfalfa is styled *wujerk;* and grass, *wüš.* "Horse" is *yaš* in Waxī, and *vurj* in Sariqolī.[1]

BRETSCHNEIDER[2] was content to say that *mu-su* is not Chinese, but most probably a foreign name. WATTERS, in his treatment of foreign words in Chinese, has dodged this term. T. W. KINGSMILL[3] is responsible for the hypothesis that *mu-su* "may have some connection with the Mηδικὴ βοτάνη of Strabo." This is adopted by the Chinese Dictionary of GILES.[4] This Greek designation had certainly not penetrated to Fergana, nor did the Iranian Ferganians use a Greek name for a plant indigenous to their country. It is also impossible to see what the phonetic coincidence between *muk-suk or *buk-suk and *mēdikē* is supposed to be.

The least acceptable explanation of *mu-su* is that recently propounded by HIRTH,[5] who identifies it with a Turkish *burčak*, which is Osmanli, and refers to the pea.[6] Now, it is universally known that a language like Osmanli was not in existence in the second century B.C., but is a comparatively modern form of Turkish speech; and how Čan K'ien should have picked up an Osmanli or any other Turkish word for a typically Iranian plant in Fergana, where there were no Turks at that time, is unintelligible. Nor is the alleged identification phonetically correct: Chinese *mu*, *muk, *buk, cannot represent *bur*, nor can *su*,

---

[1] Cf. R. B. SHAW, On the Ghalchah Languages (*Journal As. Soc. Bengal*, 1876, pp. 221, 231). According to TOMASCHEK (*op. cit.*, p. 763), this word is evolved from *bharaka, Ossetic *bairāg* ("good foal").

[2] Bot. Sin., pt. III, p. 404.

[3] *Journal China Branch Roy. As. Soc.*, Vol. XIV, 1879, p. 19.

[4] No. 8081, wrongly printed Μεδικἡ. The word βοτάνη is not connected with the name of the plant, but in the text of Strabo is separated from Μηδικήν by eleven words. Μηδικἡ is to be explained as scil. πόα, "Medic grass or fodder."

[5] *Journal Am. Or. Soc.*, Vol. XXXVII, 1917, p. 145.

[6] *Kara burčak* means the "black pea" and denotes the vetch.

*suk, stand for čak.[1] The entire speculation is deplorable, and we are even expected "to allow for a change the word may have undergone from the original meaning within the last two thousand years"; but there is no trace of evidence that the Osmanli word has existed that length of time, neither can it be reasonably admitted that the significance of a word can change from "pea" to "alfalfa." The universal term in Central Asia for alfalfa is bidā[2] or bēdä,[3] Djagatai bidä. This word means simply "fodder, clover, hay."[4] According to TOMASCHEK,[5] this word is of Iranian origin (Persian beda). It is found also in Sariqolī, a Pamir dialect.[6] This would indicate very well that the Persians (and it could hardly be expected otherwise) disseminated the alfalfa to Turkistan.

According to VÁMBÉRY,[7] alfalfa appears to have been indigenous among the Turks from all times; this opinion, however, is only based on linguistic evidence, which is not convincing: a genuine Turkish name exists in Djagatai jonuška (read yonučka) and Osmanli yondza[8] (add Kasak-Kirgiz yonurčka), which simply means "green fodder, clover." Now, these dialects represent such recent forms of Turkish speech, that so far-reaching a conclusion cannot be based on them. As far as I know, in the older Turkish languages no word for alfalfa has as yet been found.

A Sanskrit 塞鼻力迦 sai-pi-li-k'ie, *sak-bi-lik-kya, for the designation of mu-su, is indicated by Li Ši-čen,[9] who states that this is the word for mu-su used in the Kin kwaṅ miṅ kiṅ 金光明經 (Suvarṇaprabhāsa-sūtra). This is somewhat surprising, in view of the fact that there is no Sanskrit word for this plant known to us;[10] and there can be no doubt that the latter was introduced into India from Iran in comparatively recent times. BRETSCHNEIDER's suggestion,[11] that in

---

[1] Final k in transcriptions never answers to a final r, but only to k, g, or x (cf. also PELLIOT, T'oung Pao, 1912, p. 476).

[2] A. STEIN, Khotan, Vol. I, p. 130.

[3] LE COQ, Sprichwörter und Lieder aus Turfan, p. 85.

[4] I. KUNOS, Sulejman Efendi's Čagataj-Osman. Wörterbuch, p. 26.

[5] Pamir-Dialekte, p. 792.

[6] R. B. SHAW, Journal As. Soc. Bengal, 1876, p. 231.

[7] Primitive Cultur des turko-tatarischen Volkes, p. 220.

[8] The etymology given of this word by Vámbéry is fantastic and unacceptable.

[9] Pen ts'ao kaṅ mu, Ch. 27, p. 3 b. Mu-su is classified by him under ts'ai ("vegetables").

[10] This was already remarked by A. DE CANDOLLE (Origin of Cultivated Plants, p. 104). Also WATT gives only modern Indian vernacular names, three of which, spastu, sebist, and beda, are of Iranian origin.

[11] Bot. Sin., pt. III, p. 404.

Kabul the *Trifolium giganteum* is called *sibarga*, and *Medicago sativa* is styled *riška*, is unsatisfactory. The word *sibarga* means "trefoil" (*si*, "three;" *barga* = Persian *barak*, *varak*, "leaf"), and is Iranian, not Sanskrit; the corresponding Sanskrit word is *tripatra* or *triparṇa*. The word *riška* is Afghan; that is, likewise Iranian.[1] Considering the fact that nothing is known about the plant in question in early Indian sources, it is highly improbable that it should figure in a Buddhist Sūtra of the type of the Suvarṇaprabhāsa; and I think that Li Ši-čen is mistaken as to the meaning of the word, which he says he encountered there.

The above transcription occurs also in the *Fan yi miṅ yi tsi* (section 27) and answers to Sanskrit *çāka-vṛika*, the word *çāka* denoting any eatable herb or vegetable, and *vṛika* (or *baka*) referring to a certain plant not yet identified (cf. the analogous formation *çāka-bilva*, "eggplant"). It is not known what herb is to be understood by *çāka-vṛika*, and the Chinese translation *mu-su* may be merely a makeshift, though it is not impossible that the Sanskrit compound refers to some species of *Medicago*. We must not lose sight of the fact that the equations established in the Chinese-Sanskrit dictionaries are for the greater part merely bookish or lexicographical, and do not relate to plant introductions. The Buddhist translators were merely anxious to find a suitable equivalent for an Indian term. This process is radically different from the plant-names introduced together with the plants from Iranian, Indian, or Southeast-Asiatic regions: here we face living realities, there we have to do with literary productions. Two other examples may suffice. The *Fan yi miṅ yi tsi* (section 24) offers a Sanskrit botanical name in the form 鎮頭迦 *čen-t'ou-kia*, anciently **tsin(tin)-du-k'ie*, answering to Sanskrit *tinduka* (*Diospyros embryopteris*), a dense evergreen small tree common throughout India and Burma. The Chinese gloss explains the Indian word by *ši* 柿, which is the well-known *Diospyros kaki* of China and Japan, not, however, found in ancient India; it was but recently introduced into the Botanical Garden of Calcutta by Col. Kyd, and the Chinese gardeners employed there call it *čin* ("Chinese").[2] In this case it signifies only the *Diospyros embryopteris* of India. Under the heading *kan-suṅ hiaṅ* (see p. 455), which denotes the spikenard (*Nardostachys jatamansi*), Li Ši-čen gives a Sanskrit term 苦彌哆 *k'u-mi-č'e*, **ku-mi-či*, likewise taken from the Suvarṇaprabhāsasūtra; this corresponds to Sanskrit *kuñci* or *kuñcika*, which applies to three different plants,— 1. *Abrus precatorius*, 2. *Nigella indica*,

---

[1] There are, further, in Afghan *sebist* (connected with Persian *supust*) and *durešta*.

[2] W. ROXBURGH, Flora Indica, p. 412.

3. *Trigonella fœnum graecum*. In this case the compromise is a failure, or the identification of *kuñci* with *kan-suñ* even results from an error; the Sanskrit term for the spikenard is *gandhamāmsī*.

We must not draw inferences from mere Sanskrit names, either, as to the origin of Chinese plants, unless there is more substantial evidence. Thus STUART[1] remarks under *li* 李 (*Prunus domestica*) that the Sanskrit equivalent 居陵迦 *kü-liñ-kia* indicates that this plum may have been introduced from India or Persia. *Prunus domestica*, however, is a native of China, mentioned in the *Ši kiñ, Li ki*, and in *Moñ-tse*. The Sino-Indian word is given in the *Fan yi miñ yi tsi* (section 24) with the translation *li*. The only corresponding Sanskrit word is *kuliñgā*, which denotes a kind of gall. The question is merely of explaining a Sanskrit term to the Chinese, but this has no botanical or historical value for the Chinese species.

Thus the records of the Chinese felicitously supplement the meagre notices of alfalfa on the part of the ancients, and lend its history the proper perspective: we recognize the why and how of the world-wide propagation of this useful economic plant.[2] Aside from Fergana, the Chinese of the Han period discovered *mu-su* also in Ki-pin (Kashmir),[3] and this fact is of some importance in regard to the early geographical distribution of the species; for in Kashmir, as well as in Afghanistan and Baluchistan, it is probably spontaneous.[4]

Mu-su gardens are mentioned under the Emperor Wu (A.D. 265-290) of the Tsin dynasty, and the post-horses of the T'ang dynasty were fed with alfalfa.[5]

The fact that alfalfa was used as an article of human food under the T'ang we note from the story of Sie Liñ-či 薛令之, preceptor at the Court of the Emperor Yüan Tsuñ (A.D. 713-755), who wrote a versified complaint of the too meagre food allotted to him, in which alfalfas with long stems were the chief ingredient.[6] The good teacher, of course, was not familiar with the highly nutritive food-values of the plant.

---

[1] Chinese Materia Medica, p. 358.

[2] It is singular that A. DE CANDOLLE, in his Origin of Cultivated Plants, while he has conscientiously reproduced from Bretschneider all his plants wrongly ascribed to Čañ K'ien, does not make any reference to China in speaking of *Medicago* (pp. 102-104). In fact, its history has never before been outlined correctly.

[3] *Ts'ien Han šu*, Ch. 96 A.

[4] A. DE CANDOLLE, *op. cit.*, p. 103; G. T. VIGNE, Travels in Kashmir, Vol. II, p. 455.

[5] S. MATSUDA 松田定久, On Medicago sativa and the Species of Medicago in China (*Botanical Magazine* 植物學雜誌, Tōkyō, Vol. XXI, 1907, p. 243). This is a very interesting and valuable study written in Japanese.

[6] Cf. C. PÉTILLON, Allusions littéraires, p. 350.

According to the *Šu i ki* 述異記, written by Žen Faṅ 任昉 in the beginning of the sixth century, "the *mu-su* (alfalfa) gardens of Čaṅ K'ien are situated in what is now Lo-yaṅ; *mu-su* was originally a vegetable in the land of the Hu, and K'ien was the first to obtain it in the Western Countries." A work, *Kiu č'i ki* 仇池記,[1] says that east of the capital there were *mu-su* gardens, in which there were three pestles driven by water-power.

The *Si kiṅ tsa ki* 西京雜記[2] states, "In the Lo-yu gardens 樂遊苑 (in the capital Č'aṅ-ṅan) there are rose-bushes 玫瑰樹 (*Rosa rugosa*), which grow spontaneously. At the foot of these, there is abundance of *mu-su*, called also *hwai fuṅ* 懷風 ('embracing the wind'), sometimes *kwaṅ fuṅ* 光風 ('brilliant wind').[3] The people of Mou-liṅ 茂陵[4] style the plant *lien-či ts'ao* 連枝草 ('herb with connected branches')."[5]

The *Lo yaṅ k'ie lan ki* 洛陽伽藍記, a record of the Buddhist monasteries in the capital Lo-yaṅ, written by Yaṅ Hüan-či 楊衒之 in A.D. 547 or shortly afterwards, says that "Hüan-wu 宣武 is situated north-east of the Ta-hia Gate 大夏門; now it is called Kwaṅ-fuṅ Garden 光風園, producing *mu-su*." *Kwaṅ-fuṅ*, as shown by the *Si kiṅ tsa ki*, is a synonyme of *mu-su*.

K'ou Tsuṅ-ši, in his *Pen ts'ao yen i*,[6] written in A.D. 1116, notes that alfalfa is abundant in Šen-si, being used for feeding cattle and horses, and is also consumed by the population, but it should not be eaten in large quantity. Under the Mongols, the cultivation of alfalfa was much encouraged, especially in order to avert the danger of famines;[7] and gardens were maintained to raise alfalfa for the feeding of horses.[8] According to Li Ši-čen (latter part of the sixteenth century),[9] it was in his time a common, wild plant in the fields everywhere, but was cultivated in Šen-si and Kan-su. He apparently means, however, *Medicago denticulata*, which is a wild species and a native of China. FORBES

---

[1] *T'ai p'iṅ yü lan*, Ch. 824, p. 9.

[2] That is, Miscellaneous Records of the Western Capital (Č'aṅ-ṅan in Šen-si), written by Wu Kūn 吳均 of the sixth century A.D.

[3] The explanation given for these names is thus: the wind constantly whistles in these gardens, and the sunlight lends brilliancy to the flowers.

[4] Ancient name for the present district of Hiṅ-p'iṅ 興平 in the prefecture of Si-ṅan, Šen-si.

[5] *T'ai p'iṅ yü lan*, Ch. 996, p. 4 b.

[6] Ch. 19, p. 3 (ed. of Lu Sin-yüan).

[7] *Yüan ši*, Ch. 93, p. 5 b.

[8] *Ibid.*, Ch. 91, p. 6 b.

[9] *Pen ts'ao kaṅ mu*, Ch. 28, p. 3 b.

and HEMSLEY[1] give as Chinese species *Medicago denticulata, falcata,*[2] and *lupulina* (the black Medick or nonsuch), *M. lupulina* "apparently common, and from the most distant parts," and say with reference to *Medicago sativa* that it is cultivated in northern China, and also occurs in a wild state, though it is probably not indigenous. This "wild" *Medicago sativa* may be an escape from cultivation. It is an interesting point that those wild species are named *ye mu-su* ("wild alfalfa"), which goes to show that these were observed by the Chinese only after the introduction of the imported cultivated species.[3] Wu K'i-tsün[4] has figured two *ye mu-su,* following his illustration of the *mu-su,*— one being *Medicago lupulina,* the other *M. denticulata.*

The Japanese call the plant *uma-goyaši* ("horse-nourishing").[5] MATSUMURA[6] enumerates four species: *M. sativa: murasaki* ("purple") *umagoyaši;*[7] *M. denticulata: umagoyaši; M. lupulina: kometsubu-umagoyaši;* and *M. minima: ko-umagoyaši.*

In the Tibetan dialect of Ladākh, alfalfa is known as *ol.* This word refers to the *Medicago sativa* indigenous to Kashmir or possibly introduced there from Iran. In Tibet proper the plant is unknown. In Armenia occur *Medicago sativa, M. falcata, M. agrestis,* and *M. lupulina.*[8]

Under the title "Notice sur la plante mou-sou ou luzerne chinoise par C. de Skattschkoff, suivie d'une autre notice sur la même plante traduite du chinois par G. PAUTHIER,"a brief article of 16 pages appeared in Paris, 1864, as a reprint from the *Revue de l'Orient.*[9] Skattschkoff, who had spent seven years in Peking, subsequently became Russian consul in Dsungaria, and he communicates valuable information on the agriculture of *Medicago* in that region. He states that seeds of this

---

[1] *Journal Linnean Soc.,* Vol. XXIII, p. 154.

[2] Attempts are being made to introduce and to cultivate this species in the United States (cf. OAKLEY and GARVER, Medicago Falcata, U. S. Department of Agriculture, Bull. No. 428, 1917).

[3] We shall renew this experience in the case of the grape-vine and the walnut.

[4] *Či wu miṅ ši t'u k'ao,* Ch. 3, pp. 58, 59.

[5] In the same manner, Manchu *morxo* is formed from *morin* ("horse") and *orxo* ("grass").

[6] Shoku butsu-mei-i, Nos. 183–184.

[7] The flower of this species is purple-colored.

[8] A. BÉGUINOT and P. N. DIRATZSUYAN, Contributo alla flora dell' Armenia, p. 57.

[9] The work of Pauthier is limited to a translation of the notice on the plant in the *Či wu miṅ ši t'u k'ao.* The name Yü-lou nuṅ frequently occurring in this work does not refer to a treatise on agriculture, as conceived by Pauthier, but is the literary style of Wu K'i-tsün, author of that work.

plant were for the first time sent from China to Russia in 1840, and that he himself has been active for six years in propagating it in Russia, Livonia, Esthonia, and Finland. This is not to be doubted, but the point I venture to question is that the plant should not have been known in Russia prior to 1840. Not only do we find in the Russian language the words *medunka* (from Greek *mēdikē*) and the European *l'utserna* (lucerne) for the designation of *Medicago sativa*, but also *krasni* ("red") *burkun, lečuxa, lugovoi v'azel* ("*Coronilla* of the meadows"); the word *burkun, burundŭk*, referring to *Medicago falcata* (called also *yŭmorki*), *burunčik* to *M. lupulina*. It is hard to realize that all these terms should have sprung up since 1840, and that the Russians should not have received information about this useful plant from European, Iranian, or Turkish peoples. A. DE CANDOLLE[1] observes, "In the south of Russia, a locality mentioned by some authors, it is perhaps the result of cultivation as well as in the south of Europe." Judging from the report of N. E. HANSEN,[2] it appears that three species of *Medicago* (*M. falcata, M. platycarpa,* and *M. ruthenica*) are indigenous to Siberia.

The efforts of our Department of Agriculture to promote and to improve the cultivation of alfalfa in this country are well known; for this purpose also seeds from China have been introduced. Argentine chiefly owes to alfalfa a great amount of its cattle-breeding.[3]

---

[1] Origin of Cultivated Plants, p. 103.

[2] The Wild Alfalfas and Clovers of Siberia, pp. 11-15 (Bureau of Plant Industry, Bull. No. 150, Washington, 1909).

[3] Cf. I. B. LORENZETTI, La Alfafa en la Argentina (Buenos Aires, 1913, 360 p.)·

THE GRAPE-VINE

2. The grape-vine (*Vitis vinifera*) belongs to the ancient cultivated plants of western Asia and Egypt. It is not one of the most ancient cultivations, for cereals and many kinds of pulse are surely far earlier, but it is old enough to have its beginnings lost in the dawn of history. Viticulture represents such a complexity of ideas, of a uniform and persistent character throughout the ancient world, that it can have been disseminated but from a single centre. Opinions as to the location of this focus are of course divided, and our present knowledge of the subject does not permit us to go beyond more or less probable theories. Certain it is that the primeval home of vine-growing is to be sought in the Orient, and that it was propagated thence to Hellas and Italy, while the Romans (according to others, the Greeks) transplanted the vine to Gaul and the banks of the Rhine.[1] For botanical reasons, A. DE CANDOLLE[2] was inclined to regard the region south of the Caucasus as "the central and perhaps the most ancient home of the species." In view of the Biblical tradition of Noah planting the grape-vine near the Ararat,[3] it is a rather attractive hypothesis to conceive of Armenia as the country from which the knowledge of the grape took its starting-point.[4] However, we must not lose sight of the fact that both vine and wine were known in Egypt for at least three or four millenniums B.C.,[5] and were likewise familiar in Mesopotamia at a very early date. This is not the place for a discussion of O. SCHRADER'S theory[6] that the name and cultivation of the vine are due to Indo-Europeans of anterior Asia; the word for "wine" may well be of Indo-European or, more specifically, Armenian origin, but this does not

---

[1] Cf. the excellent study of G. CURTEL, La Vigne et le vin chez les Romains (Paris, 1903). See also A. STUMMER, Zur Urgeschichte der Rebe und des Weinbaues (*Mitt. Anthr. Ges. Wien*, 1911, pp. 283–296).

[2] Origin of Cultivated Plants, p. 192.

[3] Genesis, IX, 20.

[4] Cf. R. BILLIARD, La Vigne dans l'antiquité, p. 31 (Lyon, 1913). This is a well illustrated and artistic volume of 560 pages and one of the best monographs on the subject. As the French are masters in the art of viticulture, so they have also produced the best literature on the science of vine and wine. Of botanical works, J.-M. GUILLON, Étude générale de la vigne (Paris, 1905), may be recommended.

[5] V. LORET, Flore pharaonique, p. 99.

[6] In HEHN, Kulturpflanzen, pp. 91–95.

220

prove that the origin of viticulture itself is traceable to Indo-Europeans. The Semitic origin seems to me to be more probable. The Chinese received the grape-vine in late historical times from Fergana, an Iranian country, as a cultivation entirely unknown in previous epochs; and it is therefore sufficient for our purpose to emphasize the fact that vine-culture in its entire range was at that time firmly established in Western Asia, inclusive of Iran.

The first knowledge of the cultivated vine (*Vitis vinifera*) and of wine produced from its grapes was likewise obtained by the Chinese through the memorable mission of General Čaṅ K'ien, when in 128 B.C. he travelled through Fergana and Sogdiana on his way to the Yüe-či and spent a year in Bactria. As to the people of Fergana (Ta-yüan), he reported, "They have wine made of grapes." The same fact he learned regarding the Parthians (An-si). It is further stated in the same chapter of the *Si ki* that the wealthy among the people of Fergana stored grape-wine in large quantity up to ten thousand gallons (石, a dry measure) for a long time, keeping it for several decades without risk of deterioration; they were fond of drinking wine in the same manner as their horses relished alfalfa. The Chinese envoys took the seeds of both plants along to their country, and the Son of Heaven was the first to plant alfalfa and the vine in fertile soil; and when envoys from abroad arrived at the Court, they beheld extensive cultivations of these plants not far from the imperial palace. The introduction of the vine is as well authenticated as that of alfalfa. The main point to be noted is that the grape, in like manner as alfalfa, and the art of making wine, were encountered by the Chinese strictly among peoples of Aryan descent, principally of the Iranian family, not, however, among any Turkish tribes.

According to the Han Annals, the kingdom Li-yi 栗弋, which depended on Sogdiana, produced grapes; and, as the water of that country is excellent, its wine had a particular reputation.[2]

K'aṅ (Sogdiana) is credited with grapes in the Annals of the Tsin Dynasty.[3] Also grape-wine was abundant there, and the rich kept up to a thousand gallons of it.[4] The Sogdians relished wine, and were fond of songs and dances.[5] Likewise in Ši (Tashkend) it was a favorite bever-

---

[1] This is also the conclusion of J. Hoops (Waldbäume und Kulturpflanzen, p. 561).

[2] *Hou Han šu*, Ch. 118, p. 6 (cf. CHAVANNES, *T'oung Pao*, 1907, p. 195).

[3] *Tsin šu*, Ch. 97, p. 6 b (*ibid.*, p. 6: grape-wine in Ta-yüan or Fergana).

[4] *Sui šu*, Ch. 83, p. 4 b.

[5] *T'aṅ šu*, Ch. 221 B, p. 1.

age.[1] When the Sogdian K'añ Yen-tien in the first part of the seventh century A.D. established a Sogdian colony south of the Lob Nor, he founded four new cities, one of which was called "Grape City" (P'u-t'ao č'eñ); for the vine was planted in the midst of the town.[2]

The Iranian Ta Yüe-či or Indo-Scythians must also have been in possession of the vine, as we are informed by a curious text in the *Kin lou tse* 金樓子,[3] written by the Emperor Yüan 元 (A.D. 552–555) of the Liang dynasty. "The people in the country of the Great Yüe-či are clever in making wine from grapes, flowers, and leaves. Sometimes they also use roots and vegetable juice, which they cause to ferment.[4] These flowers resemble those of the clove-tree (*tiñ-hiañ* 丁香, *Caryophyllus aromaticus*), but are green or bright-blue. At the time of spring and summer, the stamens of the flowers are carried away and scattered around by the wind like the feathers of the bird *lwan* 鸞. In the eighth month, when the storm blows over the leaves, they are so much damaged and torn that they resemble silk rags: hence people speak of a grape-storm (*p'u-t'ao fuñ*), or also call it 'leaves-tearing storm' (*lie ye fuñ* 裂葉風)."

Finally we know also that the Aryan people of Kuča, renowned for their musical ability, songs, and dances, were admirers of grape-wine, some families even storing in their houses up to a thousand *hu* 斛 of the beverage. This item appears to have been contained in the report of General Lū Kwañ 呂光, who set out for the conquest of Kuča in A.D. 384.[5]

In the same manner as the Chinese discovered alfalfa in Ki-pin (Kashmir), they encountered there also the vine.[6] Further, they found it in the countries Tsiü-mo 且末[7] and Nan-tou 難兜.

---

[1] *T'ai p'iñ hwan yü ki*, Ch. 186, p. 7 b; also in Yen-k'i (Karašar): *Čou šu*, Ch. 50, p. 4 b.

[2] PELLIOT, *Journal asiatique*, 1916, I, p. 122.   [3] Ch. 5, p. 23.

[4] Strabo (XI. XIII, 11) states that the inhabitants of the mountainous region of northern Media made a wine from some kind of roots.

[5] Other sources fix the date in the year 382 (see SYLVAIN LÉVI, Le "Tokharien B," langue de Koutcha, *Journal asiatique*, 1913, II, p. 333). The above fact is derived from the *Hou liañ lu* 後涼錄, quoted in the *T'ai p'iñ yü lan* (Ch. 972, p. 3); see also *T'añ šu*, Ch. 221 A, p. 8. We owe to S. Lévi the proof that the people of Kuča belong to the Indo-European family, and that their language is identical with what was hitherto known from the manuscripts discovered in Turkistan as Tokharian B.

[6] *Ts'ien Han šu*, Ch. 96 A, p. 5. Kashmir was still famed for its grapes in the days of the Emperor Akbar (H. BLOCHMANN, Ain I Akbari, Vol. I, p. 65), but at present viticulture is on the decline there (WATT, Commerical Products of India, pp. 1112, 1114).

[7] Regarding this name, see CHAVANNES, Les Pays d'occident d'après le Wei lio (*T'oung Pao*, 1905, p. 536).

In the T'ang period the Chinese learned also that the people of Fu-lin (Syria) relished grape-wine,[1] and that the country of the Arabs (Ta-ši) produced grapes, the largest of the size of fowl's eggs.[2] In other texts such grapes are also ascribed to Persia.[3] At that epoch, Turkistan had fallen into the hands of Turkish tribes, who absorbed the culture of their Iranian predecessors; and it became known to the Chinese that the Uigur had vine and wine.

Viticulture was in a high state of development in ancient Iran. Strabo[4] attributes to Margiana (in the present province of Khorasan) vines whose stock it would require two men with outstretched arms to clasp, and clusters of grapes two cubits long. Aria, he continues, is described as similarly fertile, the wine being still richer, and keeping perfectly for three generations in unpitched casks. Bactriana, which adjoins Aria, abounds in the same productions, except the olive.

The ancient Persians were great lovers of wine. The best vintage-wines were served at the royal table.[5] The couch of Darius was over-shadowed by a golden vine, presented by Pythius, a Lydian.[6] The inscription of Persepolis informs us that fifty congius[7] of sweet wine and five thousand congius of ordinary wine were daily delivered to the royal house.[8] The office of cup-bearer in the palace was one of importance.[9] The younger Cyrus, when he had wine of a peculiarly fine flavor, was in the habit of sending half-emptied flagons of it to some of his friends, with a message to this effect: "For some time Cyrus has not found a pleasanter wine than this one; and he therefore sends some to you, begging you to drink it to-day with those whom you love best."[10]

Strabo[11] relates that the produce of Carmania is like that of Persia, and that among other productions there is the vine. "The Carmanian

---

[1] HIRTH, China and the Roman Orient, pp. 58, 63.

[2] T'ai p'iṅ hwan yü ki, Ch. 186, p. 15 b.

[3] For instance, Pen ts'ao yen i, Ch. 18, p. 1 (ed. of Lu Sin-yüan).

[4] II. 1, 14, and XI. x, 2.

[5] Esther, 1, 7 ("And they gave them drink in vessels of gold, the vessels being diverse one from another, and royal wine in abundance, according to the state of the king").

[6] Herodotus, VII, 27; Athenaeus, XII, 514 f. According to G. W. ELDERKIN (Am. Journal of Archaeology, Vol. XXI, 1917, p. 407), the ultimate source of this motive would be Assyrian.

[7] A measure of capacity equal to about six pints.

[8] JORET, Plantes dans l'antiquité, Vol. II, p. 95.

[9] Xenophon, Cyropædia, I. III, 8–9.

[10] Xenophon, Anabasis, I. IX, 25.

[11] XV. II, 14.

vine, as we call it, often bears bunches of grapes of two cubits in size, the seeds being very numerous and very large; probably the plant grows in its native soil with great luxuriance." The kings of Persia were not content, however, with wines of native growth; but when Syria was united with their empire, the Chalybonian wine of Syria became their privileged beverage.[1] This wine, according to Posidonius, was made in Damascus, Syria, from vines planted there by the Persians.[2]

Herodotus[3] informs us that the Persians are very fond of wine and consume it in large quantities. It is also their custom to discuss important affairs in a state of intoxication; and on the following morning their decisions are put before them by the master of the house where the deliberations have been held. If they approve of the decision in the state of sobriety, they act accordingly; if not, they set it aside. When sober at their first deliberation, they always reconsider the matter under the influence of wine. In a similar manner, Strabo[4] says that their consultations on the most important affairs are carried on while drinking, and that they consider the resolutions made at that time more to be depended upon than those made when sober. In the Šāhnāmeh, the Persian epic, deliberations are held during drinking-bouts, but decision is postponed till the following day.[5] Cambyses was ill reputed for his propensity for wine.[6] Deploring the degeneracy of the Persians, Xenophon[7] remarks, "They continue eating and drinking till those who sit up latest go to retire. It was a rule among them not to bring large cups to their banquets, evidently thinking that abstinence from drinking to excess would less impair their bodies and minds. The custom of not bringing such vessels still continues; but they drink so excessively that instead of bringing in, they are themselves carried out, as they are no longer able to walk upright." Procopius, the great Byzantine historian of the sixth century,[8] says that of all men the Massagetae (an Iranian tribe) are the most intemperate drinkers. So

---

[1] Strabo, XV. III, 22.

[2] Athenaeus, I.

[3] I, 133.

[4] XV. III, 20.

[5] F. SPIEGEL, Eranische Altertumskunde, Vol. III, p. 672. Cf. what JOHN FRYER (New Account of East India and Persia being Nine Years' Travels 1672–81, Vol. II, p. 210, ed. of Hakluyt Society) says of the modern Persians: "It is incredible to see what quantities they drink at a merry-meeting, and how unconcerned the next day they appear, and brisk about their business, and will quaff you thus a whole week together."

[6] Herodotus, III, 34.

[7] Cyropædia, VIII. VIII, 9–10.

[8] Historikon, III. XII, 8.

were also the Sacae, who, maddened with wine, were defeated by Cyrus.[1] In the same passage, Strabo speaks of a Bacchanalian festival of the Persians, in which men and women, dressed in Scythian style, passed day and night in drinking and wanton play. On the other hand, it must not be forgotten that such judgments passed by one nation on another are usually colored or exaggerated, and must be accepted only at a liberal discount; also temperance was preached in ancient Persia, and intemperance was severely punished.[2] With all the evils of over-indulgence in wine and the social dangers of alcohol, the historian, whose duty it is to represent and to interpret phenomena as they are, must not lose sight of the fact that wine constitutes a factor of economic, social, and cultural value. It has largely contributed to refine and to intensify social customs and to heighten sociability, as well as to promote poetry, music, and dancing. It has developed into an element of human civilization, which must not be underrated. Temperance literature is a fine thing, but who would miss the odes of Anakreon, Horace, or Hāfiz?

The word for the grape, brought back by Čaṅ K'ien and still current in China and Japan (*budō*), is 蒲桃 (ancient phonetic spelling of the Han Annals, subsequently 葡萄)[3] *p'u-t'ao*, *bu-daw, "grape, vine". Since Čaṅ K'ien made the acquaintance of the grape in Ta-yüan (Fergana) and took its seeds along from there to China, it is certain that he also learned the word in Fergana; hence we are compelled to assume that *bu-daw is Ferganian, and corresponds to an Iranian *budāwa or *buδawa, formed with a suffix *wa* or *awa*, from a stem *buda*, which in my opinion may be connected with New Persian *bāda* ("wine") and Old Persian βατιάκη ("wine-vessel") = Middle Persian *bātak*, New Persian *bādye*.[4] The Sino-Iranian word might also be conceived as a dialectic form of Avestan *maδav* ("wine from berries").

It is well known that attempts have been made to derive the Chinese word from Greek βότρυς ("a bunch of grapes"). TOMASCHEK[5] was the first to offer this suggestion; T. KINGSMILL[6] followed in 1879, and

---

[1] Strabo, XI. VIII, 5.

[2] Cf. JACKSON, in Grundriss der iranischen Philologie, Vol. II, p. 679.

[3] The graphic development is the same as in the case of *mu-su* (see above, p. 212).

[4] Cf. HORN, Neupersische Etymologie, No. 155. The Chinese are fond of etymologizing, and Li Ši-čen explains the word *p'u-t'ao* thus: "When people drink (*p'u* 醏) it, they become intoxicated (*t'ao* 酶)." The joke is not so bad, but it is no more than a joke.

[5] Sogdiana, *Sitzungsber. Wiener Akad.*, 1877, p. 133.

[6] *Journal China Branch Roy. As. Soc.*, Vol. XIV, pp. 5, 19.

HIRTH[1] endorsed Kingsmill. No one gave a real demonstration of the case. Tomaschek argued that the dissemination of the vine in Central Asia is connected with Macedonian-Greek rule and Hellenic influence. This is decidedly wrong, for the vine grows spontaneously in all northern Iranian regions; and its cultivation in Iran is traceable to a great antiquity, and is certainly older there than in Greece. The Greeks received vine and wine from western Asia.[2] Greek βότρυς, in all likelihood, is a Semitic loan-word.[3] It is highly improbable that the people of Fergana would have employed a Greek word for the designation of a plant which had been cultivated in their dominion for ages, nor is there any evidence for the silent admission that Greek was ever known or spoken in Fergana at the time of Čaṅ K'ien's travels. The influence of Greek in the Iranian domain is extremely slight: nothing Greek has as yet been found in any ancient manuscripts from Turkistan. In my opinion, there is no connection between p'u-t'ao and βότρυς, nor between the latter and Iranian *budawa.

It is well known that several species of wild vine occur in China, in the Amur region, and Japan.[4] The ancient work Pie lu is credited with the observation that the vine (p'u-t'ao) grows in Luṅ-si (Kan-su), Wu-yüan 五原 (north of the Ordos), and in Tun-hwaṅ (in Kan-su).[5] Li Ši-čen therefore argues that in view of this fact the vine must of old have existed in Luṅ-si in pre-Han times, but had not yet advanced into Šen-si. It is inconceivable how BRETSCHNEIDER[6] can say that the introduction of the grape by Čaṅ K'ien is inconsistent with the notice of the grape in the earliest Chinese materia medica. There is, in fact, nothing alarming about it: the two are different plants; wild vines are natives of northern

---

[1] Fremde Einflüsse in der chin. Kunst, p. 28; and Journal Am. Or. Soc., Vol. XXXVII, 1917, p. 146. Hirth's arguments are based on unproved premises. The grape-design on the so-called grape mirrors has nothing to do with Greek or Bactrian art, but comes from Iranian-Sasanian art. No grape mirrors were turned out under the Han, they originated in the so-called Leu-č'ao period from the fourth to the seventh century. The attribution "Han" simply rests on the puerile assumption made in the Po ku t'u lu that, because Čaṅ K'ien introduced the grape, the artistic designs of grapes must also have come along with the same movement.

[2] Only a "sinologue" could assert that the grape was "originally introduced from Greece, viâ Bactria, about 130 B.C." (GILES, Chinese Dictionary, No. 9497).

[3] MUSS-ARNOLT, Transactions Am. Phil. Assoc., Vol. XXIII, 1892, p. 142. The variants in spelling βόστρυχος, βότρυχος, plainly indicate the status of a loan-word. In Dioscorides (III, 120) it denotes an altogether different plant,—Chenopodium botrys.

[4] The Lo-lo of Yün-nan know a wild grape by the name ko-p'i-ma, with large, black, oblong berries (P. VIAL, Dictionnaire français-lolo, p. 276). The grape is ze-mu-se-ma in Nyi Lo-lo, sa-lu-zo or sa-šo-zo in Ahi Lo-lo.

[5] Pen ts'ao kaṅ mu, Ch. 33, p. 3.

[6] Bot. Sin., pt. III, p. 438.

China, but have never resulted in a cultivation; the cultivated species (*Vitis vinifera*) was introduced from Iran, and never had any relation to the Chinese wild species (*Vitis bryoniaefolia*). In a modern work, *Muṅ ts'üan tsa yen* 叢泉雜言,[1] which gives an intelligent discussion of this question, the conclusion is reached that the species from Fergana is certainly different from that indigenous to China. The only singular point is that the *Pie lu* employs the Ferganian word *p'u-t'ao* with reference to the native species; but this is not an anachronism, for the *Pie lu* was written in post-Christian times, centuries after Čaṅ K'ien; and it is most probable that it was only the introduced species which gave the impetus to the discovery of the wild species, so that the latter received the same name.[2]

Another wild vine is styled *yin-yü* 蘡薁 (*Vitis bryoniaefolia* or *V. labrusca*), which appears in the writings of T'ao Huṅ-kiṅ (A.D. 451–536) and in the *T'aṅ pen ts'ao* of Su Kuṅ, but this designation has reference only to a wild vine of middle and northern China. Yen Ši-ku (A.D. 579–645), in his *K'an miu čeṅ su* 刊謬正俗,[3] ironically remarks that regarding the *yin-yü* as a grape is like comparing the *či* 枳 (*Poncirus trifoliata*) of northern China with an orange (*kü* 橘); that the *yin-yü*, although a kind of *p'u-t'ao*, is widely different from the latter; and that the *yin-yü* of Kiaṅ-nan differs again from the *yin-yü* of northern China. Hirth's theory,[4] that this word might represent a transcription of New Persian *angur*, is inadmissible. We have no right to regard Chinese words as of foreign origin, unless these are expressly so indicated by the Chinese philologists who never fail to call attention to such borrowing. If this is not the case, specific and convincing reasons must be adduced for the assumption that the word in question cannot be Chinese. There is no tradition whatever that would make *yin-yü* an Iranian or a foreign word. The opposite demonstration lacks any sound basis: New Persian, which starts its career from the end of the tenth century, could not come into question here, but at the best Middle Persian, and *angur* is a strictly New-Persian type. A word like *angur* would have been dissected by the Chinese into *an+gut* (*gur*), but not into *aṅ+uk*; moreover, it is erroneous to suppose that final *k* can transcribe final *r*;[5] in Iranian transcriptions, Chinese final *k* corresponds to Iranian *k*, *g*, or the spirant *x*. It is further inconceivable that the Chinese might

---

[1] *T'u šu tsi č'eṅ*, xx, Ch. 113.

[2] Compare the analogous case of the walnut.

[3] Ch. 8, p. 8 b (ed. of *Hu pei ts'uṅ šu*).

[4] Fremde Einflüsse in der chinesischen Kunst, p. 17.

[5] Compare above, p. 214.

have applied a Persian word designating the cultivated grape to a wild vine which is a native of their country, and which particularly grows in the two Kiaṅ provinces of eastern China. The Gazetteer of Su-čou[1] says expressly that the name for the wild grape, *šan p'u-t'ao*, in the Kiaṅ provinces, is *yiṅ-yü*. Accordingly it may be an ancient term of the language of Wu. The *Pen ts'ao kaṅ mu*[2] has treated *yiṅ-yü* as a separate item, and Li Ši-čen annotates that the meaning of the term is unexplained. It seems to me that for the time being we have to acquiesce in this verdict. *Yen-yü* 燕薁 and *yiṅ-še* 嬰舌 are added by him as synonymes, after the *Mao ši* 毛詩 and the *Kwaṅ ya*, while *ye p'u-t'ao* ("wild grape") is the common colloquial term (also *t'eṅ miṅ* or *mu luṅ* 藤名木龍). It is interesting to note that the earliest notices of this plant come only from Su Kuṅ and Č'en Ts'aṅ-k'i of the T'ang dynasty. In other words, it was noted by the Chinese naturalists more than seven centuries later than the introduction of the cultivated grape,— sufficient evidence for the fact that the two are not in any way interrelated.

It must not be imagined that with Čaṅ K'ien's deed the introduction of the vine into China was an accomplished fact; but introductions of seeds were subsequently repeated, and new varieties were still imported from Turkistan by K'aṅ-hi. There are so many varieties of the grape in China, that it is hardly credible that all these should have at once been brought over by a single man. It is related in the Han Annals that Li Kwaṅ-li 李廣利, being General of Er-ši 二師 (*Ni-š'i), after the subjugation of Ta-yüan, obtained grapes which he took along to China.

Three varieties of grape are indicated in the *Kwaṅ či*,[3] written before A.D. 527,— yellow, black, and white. The same varieties are enumerated in the *Yu yaṅ tsa tsu*, while Li Ši-čen speaks of four varieties,— a round one, called *ts'ao luṅ ču* 草龍珠 ("vegetable dragon-pearls"); a long one, *ma žu p'u-t'ao* (see below); a white one, called "crystal grapes" (*šwi tsiṅ p'u-t'ao*); and a black one, called "purple grapes" (*tse* 紫 *p'u-t'ao*),—and assigns to Se-č'wan a green (綠) grape, to Yün-nan grapes of the size of a jujube.[4] Su Suṅ of the Sung mentions a variety of seedless grapes.

---

[1] *Su čou fu či*, Ch. 20, p. 7 b.

[2] Ch. 33, p. 4.

[3] *T'ai p'iṅ yü lan*, Ch. 972, p. 3.

[4] T'an Ts'ui 檀萃, in his valuable description of Yün-nan (*Tien haɪ yü heṅ či*, published in 1799, Ch. 10, p. 2, ed. of *Wen yiṅ lou yü ti ts'uṅ šu*), states that the grapes of southern Yün-nan are excellent, but that they cannot be dried or sent to distant places.

In Haṅ-čou yellow and bright white grapes were styled *ču-tse* 珠子 ("beads, pearls"); another kind, styled "rock-crystal" (*šwi-tsiṅ*), excelled in sweetness; those of purple and agate color ripened at a little later date.[1]

To Turkistan a special variety is attributed under the name *so-so* 瑣瑣 grape, as large as *wu-wei-tse* 五味子 ("five flavors," *Schizandra chinensis*) and without kernels 無核. A lengthy dissertation on this fruit is inserted in the *Pen ts'ao kaṅ mu ši i*.[2] The essential points are the following. It is produced in Turfan and traded to Peking; in appearance it is like a pepper-corn, and represents a distinct variety of grape. Its color is purple. According to the *Wu tsa tsu* 五雜俎, written in 1610, when eaten by infants, it is capable of neutralizing the poison of small-pox. The name *so-so* is not the reproduction of a foreign word, but simply means "small." This is expressly stated in the *Pen kiṅ fuṅ yüan* 本經逢原, which says that the *so-so* grapes resemble ordinary grapes, but are smaller and finer, and hence are so called (而瑣細 故名). The *Pi č'en* 筆塵 of Yü-wen Tiṅ 于文定 annotates, however, that *so-so* is an error for *sa-so* 吸娑, without giving reasons for this opinion. *Sa-so* was the name of a palace of the Han emperors, and this substitution is surely fantastic. Whether *so-so* really is a vine-grape seems doubtful. It is said that *so-so* are planted everywhere in China to be dried and marketed, being called in Kiaṅ-nan *fan p'u-t'ao* ("foreign grape").[3]

The Emperor K'aṅ-hi (1662–1722), who knew very well that grapes had come to China from the west, tells that he caused three new varieties to be introduced into his country from Hami and adjoining territories,— one red or greenish, and long like mare-nipples; one not very large, but of agreeable taste and aroma; and another not larger than a pea, the most delicate, aromatic, and sweetest kind. These three varieties of grape degenerate in the southern provinces, where they lose their aroma. They persist fairly well in the north, provided they are planted in a dry and stony soil. "I would procure for my subjects," the Emperor concludes, "a novel kind of fruit or grain, rather than build a hundred porcelain kilns."[4]

Turkistan is well known to the Chinese as producing many varieties

---

[1] *Moṅ liaṅ lu* 夢粱錄, by Wu Tse-mu 吳自牧 of the Sung (Ch. 18, p. 5 b; ed. of *Či pu tsu čai ts'uṅ šu*).

[2] Ch. 7, p. 69. This valuable supplement to the *Pen ts'ao kaṅ mu* was first published in 1650 (reprinted 1765 and appended to several modern editions of the *Pen ts'ao*) by Čao Hio-min 趙學敏 (*hao* Šu-hien 恕軒) of Haṅ-čou.

[3] *Muṅ ts'üan tsa yen* 蒙泉雜言, cited in *T'u šu tsi č'eṅ*, XX, Ch. 130.

[4] *Mémoires concernant les Chinois*, Vol. IV, 1779, pp. 471–472.

of grape. According to the *Hui k'ian či* 回疆志 ("Records of Turkistan"), written in 1772 by the two Manchu officers Fusambô and Surde, "there are purple, white, blue, and black varieties; further, round and long, large and small, sour and sweet ones. There is a green and seedless variety, comparable to a soy-bean, but somewhat larger, and of very sweet and agreeable flavor [then the *so-so* is mentioned]. Another kind is black and more than an inch long; another is white and large. All varieties ripen in the seventh or eighth month, when they are dried and can be transported to distant places." According to the *Wu tsa tsu*, previously quoted, Turkistan has a seedless variety of grape, called *tu yen* 兎眼 *p'u-t'ao* ("hare-eye grape").

A. v. LE COQ[1] mentions under the name *sōzuq saivī* a cylindrical, whitish-yellow grape, the best from Toyoq and Bulayiq, red ones of the same shape from Manas and Shichō. Sir AUREL STEIN[2] says that throughout Chinese Turkistan the vines are trained along low fences, ranged in parallel rows, and that the dried grapes and currants of Ujat find their way as far as the markets of Aksu, Kashgar, and Turfan.

Every one who has resided in Peking knows that it is possible to obtain there during the summer seemingly fresh grapes, preserved from the crop of the previous autumn, and that the Chinese have a method of preserving them. The late F. H. KING,[3] whose studies of the agriculture of China belong to the very best we have, observed regarding this point, "These old people have acquired the skill and practice of storing and preserving such perishable fruits as pears and grapes so as to enable them to keep them on the market almost continuously. Pears were very common in the latter part of June, and Consul-General Williams informed me that grapes are regularly carried into July. In talking with my interpreter as to the methods employed, I could only learn that the growers depend simply upon dry earth cellars which can be maintained at a very uniform temperature, the separate fruits being wrapped in paper. No foreigner with whom we talked knew their methods." This method is described in the *Ts'i min yao šu*, an ancient work on husbandry, probably from the beginning of the sixth century,[4] although teeming with interpolations. A large pit is dug in a room of the farmhouse for storing the grapes, and holes are bored in the walls near the surface of the ground and stuffed with branches. Some of these holes are filled with mud to secure proper support for the room.

---

[1] Sprichwörter und Lieder aus Turfan, p. 92.

[2] Sand-Buried Ruins of Khotan, p. 228.

[3] Farmers of Forty Centuries, p. 343 (Madison, Wis., 1911).

[4] See BRETSCHNEIDER, Bot. Sin., pt. I, p. 77; HIRTH, *T'oung Pao*, 1895, p. 436; PELLIOT, *Bulletin de l'Ecole française*, Vol. IX, p. 434.

The pit in which the grapes are stored is covered with loam, and thus an even temperature is secured throughout the winter.[1]

The Jesuit missionaries of the eighteenth century praise the raisins of Hoai-lai-hien[2] on account of their size: "Nous parlons d'après le témoignage de nos yeux: les grains de ces grappes de raisins sont gros comme des prunes damas-violet, et la grappe longue et grande à proportion. Le climat peut y faire; mais si les livres disent vrai, cela vient originairement de ce qu'on a enté des vignes sur des jujubiers; et l'épaisseur de la peau de ces raisins nous le ferait croire."[3]

Raisins are first mentioned as being abundant in Yün-nan in the *Yün-nan ki*[4] ("Memoirs regarding Yün-nan"), a work written in the beginning of the ninth century. Li Ši-čen remarks that raisins are made by the people of the West as well as in T'ai-yüan and P'in-yan in Šan-si Province, whence they are traded to all parts of China. Hami in Turkistan sends large quantities of raisins to Peking.[5] In certain parts of northern China the Turkish word *kišmiš* for a small kind of raisin is known. It is obtained from a green, seedless variety, said to originate from Bokhara, whence it was long ago transplanted to Yarkand. After the subjugation of Turkistan under K'ien-lun, it was brought to Jehol, and is still cultivated there.[6]

Although the Chinese eagerly seized the grape at the first opportunity offered to them, they were slow in accepting the Iranian custom of making and drinking wine.[7] The Arabic merchant Soleiman (or whoever may be responsible for this account), writing in A.D. 851, reports that "the wine taken by the Chinese is made from rice; they do not make wine from grapes, nor is it brought to them from abroad;

---

[1] A similar contrivance for the storage of oranges is described in the Mémoires concernant les Chinois, Vol. IV, p. 489.

[2] I presume that Hwai (or Hwo)-lu hien in the prefecture of Čen-tin, Či-li Province, is meant.

[3] Mémoires concernant les Chinois, Vol. III, 1778, p. 498.

[4] *T'ai p'in yü lan*, Ch. 972, p. 3.

[5] An article on Hami raisins is inserted in the Mémoires concernant les Chinois (Vol. V, 1780, pp. 481–486). The introduction to this article is rather strange, an effort being made to prove that grapes have been known in China since times of earliest antiquity; this is due to a confusion of the wild and the cultivated vine. In Vol. II, p. 423, of the same collection, it is correctly stated that vine and wine became known under the reign of the Emperor Wu.

[6] Cf. O. FRANKE, Beschreibung des Jehol-Gebietes, p. 76.

[7] The statement that Čan K'ien taught his countrymen the art of making wine, as asserted by GILES (Biographical Dictionary, p. 12) and L. WIEGER (Textes historiques, p. 499), is erroneous. There is nothing to this effect in the *Ši ki* or in the Han Annals.

they do not know it, accordingly, and make no use of it."[1] This doubt-less was correct for southern China, where the information of the Arabic navigators was gathered. The grape, however, is chiefly to be found in northern China,[2] and at the time of Soleiman the manu-facture of grape-wine was known in the north. The principal document bearing on this subject is extant in the history of the T'ang dynasty.

In A.D. 647 a peculiar variety of grapes, styled *ma žu p'u t'ao* 馬乳葡萄 ("mare-nipple grapes") were sent to the Emperor T'ai Tsuṅ 太宗 by the (Turkish) country of the Yabgu 葉護. It was a bunch of grapes two feet long, of purple color.[3] On the same occasion it is stated, "Wine is used in the Western Countries, and under the former dynasties it was sometimes sent as tribute, but only after the destruction of Kao-č'aṅ 高昌 (Turfan), when 'mare-nipple grapes' cultivated in orchards were received, also the method of making wine was simultaneously introduced into China (A.D. 640). T'ai Tsuṅ experienced both its injurious and beneficial effects. Grape-wine, when ready, shines in all colors, is fragrant, very fiery, and tastes like the finest oil. The Emperor bestowed it on his officials, and then for the first time they had a taste of it in the capital."[4]

These former tributes of wine are alluded to in a verse of the poet Li Po of the eighth century, "The Hu people annually offered grape-wine."[5] Si Waṅ Mu, according to the *Han Wu ti nei čwan* of the third century or later, is said to have presented grape-wine to the Han Emperor Wu, which certainly is an unhistorical and retrospective tradition.

A certain Čaṅ Huṅ-mao 張洪茂, a native of Tun-hwaṅ in Kan-su, is said to have devoted to grape-wine a poem of distinct quality.[6] The locality Tun-hwaṅ is of significance, for it was situated on the

---

[1] M. REINAUD, Relation des voyages faits par les Arabes et les Persans dans l'Inde et à la Chine, Vol. I, p. 23.

[2] In the south, I am under the impression it is rather isolated. It occurs, for instance, in Šaṅ-se čou 上思州 in the prefecture of T'ai-p'iṅ, Kwaṅ-si Province, in three varieties,—green, purple, and crystal,—together with an uneatable wild grape (*Šaṅ se čou či*, Ch. 14, p. 8, ed. published in 1835). "Grapes in the neighbor-hood of Canton are often unsuccessful, the alternations of dry heat and rain being too much in excess, while occasional typhoons tear the vines to pieces" (J. F. DAVIS, China, Vol. II, p. 305). They occur in places of Fu-kien and in the Chusan Archi-pelago (cf. *T'u šu tsi č'eṅ*, VI, Ch. 1041).

[3] *T'aṅ hui yao*, Ch. 200, p. 14; also *Fuṅ ši wen kien ki* 封氏聞見記, Ch. 7, p. 1 b (ed. of *Ki fu ts'uṅ šu*), by Fuṅ Yen 封演 of the T'ang.

[4] *Ibid.*, p. 15.

[5] *Pen ts'ao yen i*, Ch. 18, p. 1.

[6] This is quoted from the *Ts'ien liaṅ lu* 前凉錄, a work of the Tsin dynasty, in the *Ši leu kwo č'un ts'iu* (*T'ai p'iṅ yü lan*, Ch. 972, p. 1 b).

road to Turkistan, and was the centre from which Iranian ideas radiated into China.

The curious point is that the Chinese, while they received the grape in the era of the Han from an Iranian nation, and observed the habit of wine-drinking among Iranians at large, acquired the art of wine-making as late as the T'ang from a Turkish tribe of Turkistan. The Turks of the Han period knew nothing of grapes or wine, quite naturally, as they were then restricted to what is now Mongolia, where soil and climatic conditions exclude this plant. Vine-growing, as a matter of course, is compatible solely with a sedentary mode of life; and only after settling in Turkistan, where they usurped the heritage of their Iranian predecessors,[1] did the Turks become acquainted with grape and wine as a gift of Iranians. The Turkish word for the grape, Uigur *özüm* (other dialects *üzüm*), proves nothing along the line of historical facts, as speculated by VÁMBÉRY.[2] It is even doubtful whether the word in question originally had the meaning "grape"; on the contrary, it merely seems to have signified any berry, as it still refers to the berries and seeds of various plants. The Turks were simply epigones and usurpers, and added nothing new to the business of vine-culture.

In accordance with the introduction of the manufacture of grape-wine into China, we find this product duly noted in the *Pen ts'ao* of the T'ang,[3] published about the middle of the seventh century; further, in the *Ši liao pen ts'ao* by Moṅ Šen 孟詵 (second half of the seventh century), and in the *Pen ts'ao ši i* by Č'en Ts'aṅ-k'i 陳藏器, who wrote in the K'ai-yüan period (713–741). The *T'aṅ pen ts'ao* also refers to the manufacture of vinegar from grapes.[4] The *Pen ts'ao yen i*, published in 1116, likewise enumerates grape-wine among the numerous brands of alcoholic beverages.

The *Lian se kuṅ tse ki* by Čaṅ Yüe (667–730)[5] contains an anecdote to the effect that Kao-č'aṅ offered to the Court frozen wine made from dried raisins, on which Mr. Kie made this comment: "The taste of grapes with thin shells is excellent, while grapes with thick shells are bitter of taste. They are congealed in the Valley of Eight Winds (Pa fuṅ ku 八風谷). This wine does not spoil in the course of years."[6]

---

[1] This was an accomplished fact by the end of the fourth century A.D.

[2] Primitive Cultur des turko-tatarischen Volkes, p. 218.

[3] *Čeṅ lei pen ts'ao*, Ch. 23, p. 7.

[4] *Ibid.*, Ch. 26, p. 1 b.

[5] See The Diamond, this volume, p. 6.

[6] *Pen ts'ao kaṅ mu*, Ch. 25, p. 14 b. A different version of this story is quoted in the *T'ai p'iṅ yü lan* (Ch. 845, p. 6 b).

A recipe for making grape-wine is contained in the *Pei šan tsiu kiṅ* 北山酒經,[1] a work on the different kinds of wine, written early in the twelfth century by Ču Yi-čuṅ 朱翼中, known as Ta-yin Weṅ 大隱翁. Sour rice is placed in an earthen vessel and steamed. Five ounces of apricot-kernels (after removing the shells) and two catties of grapes (after being washed and dried, and seeds and shells removed) are put together in a bowl of thin clay (*ša p'en* 砂盆),[2] pounded, and strained. Three pecks of a cooked broth are poured over the rice, which is placed on a table, leaven being added to it. This mass, I suppose, is used to cause the grape-juice to ferment, but the description is too abrupt and by no means clear. So much seems certain that the question is of a rather crude process of fermentation, but not of distillation (see below).

Sü T'iṅ 徐霆, who lived under the Emperor Li Tsuṅ (1224-63) of the Southern Sung, went as ambassador to the Court of the Mongol Emperor Ogotai (1229-45). His memoranda, which represent the earliest account we possess of Mongol customs and manners, were edited by P'eṅ Ta-ya 彭大雅 of the Sung under the title *Hei Ta ši lio* 黑韃事略 ("Outline of the Affairs of the Black Tatars"), and published in 1908 by Li Wen-t'ien and Hu Se in the *Wen yiṅ lou yü ti ts'uṅ šu*.[3] Sü T'iṅ informs us that grape-wine put in glass bottles and sent as tribute from Mohammedan countries figured at the headquarters of the Mongol Khan; one bottle contained about ten small cups, and the color of the beverage resembled the juice of the *Diospyros kaki* [known in this country as Japanese persimmons] of southern China. It was accordingly a kind of claret. The Chinese envoy was told that excessive indulgence in it might result in intoxication.

---

[1] Ch. c, p. 19 b (ed. of *Či pu tsu čai ts'uṅ šu*). The work is noted by WYLIE (Notes on Chinese Literature, p. 150).

[2] Literally, "sand-pot." This is a kind of thin pottery (colloquially called *ša kwo* 砂鍋) peculiar to China, and turned out at Hwai-lu (Či-li), P'iṅ-tiṅ čou and Lu-ṅan (San-si), and Yao-čou (Šen-si). Made of clay and sand with an admixture of coal-dust, so that its appearance presents a glossy black, it is extremely light and fragile; but, on account of their thin walls, water may be heated in these pots with a very small quantity of fuel. They are a money and time saving device, and hence in great demand among the poor, who depend upon straw and dried grass for their kitchen fire. With careful handling, such pots and pans may endure a long time. The proverb runs, "The sand-pot will last a generation if you do not hit it"; and there is another popular saying, "You may pound garlic in a sand-pan, but you can do so but once" (A. H. SMITH, Proverbs and Common Sayings from the Chinese, p. 204). Specimens of this ware from Yao-čou may be seen in the Field Museum, others from Hwai-lu are in the American Museum of New York (likewise collected by the writer). The above text of the Sung period is the first thus far found by me which contains an allusion to this pottery.

[3] This important work has not yet attracted the attention of our science. I hope to be able to publish a complete translation of it in the future.

In his interesting notice "Le Nom turc du vin dans Odoric de Pordenone,"[1] P. PELLIOT has called attention to the word *bor* as a Turkish designation of grape-wine, adding also that this word occurs in a Mongol letter found in Turfan and dated 1398.[2] I can furnish additional proof for the fact that *bor* is an old Mongol word in the sense of wine, although, of course, it may have been borrowed from Turkish. In the Mongol version of the epic romance of Geser or Gesar Khan we find an enumeration of eight names of liquor, all supposed to be magically distilled from *araki* ("arrack, brandy"). These are: *aradsa* (*araja*), *xoradsa* or *xuradsa*, *širadsa*, *boradsa*, *takpa*, *tikpa*, *marba*, *mirba*.[3] These terms have never been studied, and, with the exception of the first and third, are not even listed in Kovalevski's and Golstuntki's Mongol Dictionaries. The four last words are characterized as Tibetan by the Tibetan suffix *pa* or *ba*. *Marwa* (corresponding in meaning to Tibetan *č'aṅ*) is well known as a word generally used throughout Sikkim and other Himalayan regions for an alcoholic beverage.[4] As to *tikpa*, it seems to be formed after the model of Tibetan *tig-č'aṅ*, the liquor for settling (*tig*) the marriage-affair, presented by the future bridegroom to the parents of his intended.[5]

The terms *aradsa*, *xoradsa* or *xuradsa*, *širadsa*, and *boradsa*, are all provided with the same ending. The first is given by KOVALEVSKI[6] with the meaning "very strong koumiss, spirit of wine." A parallel is offered by Manchu in *arčan* ("a liquor prepared from milk"), while Manchu *arjan* denotes any alcoholic drink. The term *xoradsa* or *xuradsa* may be derived from Mongol *xuru-t* (*-t* being suffix of the plural), corresponding to Manchu *kuru*, which designates "a kind of cheese made from fermented mare's milk, or cheese prepared from cow's or mare's milk with the addition of sugar and sometimes pressed into forms." The word *širadsa* has been adopted by Schmidt and Kovalevski in their respective dictionaries as "wine distilled for the fourth time" or "esprit de vin quadruple;" but these explanations are simply based on the above passage of Geser, in which one drink is supposed to be

---

[1] *T'oung Pao*, 1914, pp. 448-453.

[2] Ramstedt's tentative rendering of this word by "beaver" is a double error: first, the beaver does not occur in Mongolia and is unknown to the Mongols, its easternmost boundary is formed by the Yenisei; second, *bor* as an animal-name means "an otter cub," and otter and beaver are entirely distinct creatures.

[3] Text, ed. I. J. SCHMIDT, p. 65; translation, p. 99. Schmidt transcribes *arasa*, *chorasa*, etc., but the palatal sibilant is preferable.

[4] Cf. H. H. RISLEY, Gazetteer of Sikkim, p. 75, where also the preparation is described.

[5] JÄSCHKE, Tibetan Dictionary, p. 364.

[6] Dictionnaire mongol, p. 143.

distilled from the other. This process, of course, is purely fantastic, and described as a magical feat; there is no reality underlying it.

The word *boradsa*, in my opinion, is derived from the Turkish word *bor* discussed by Pelliot; there is no Mongol word from which it could be explained. In this connection, the early Chinese account given above of foreign grape-wine among the Mongols gains a renewed significance. Naturally it was a rare article in Mongolia, and for this reason we hear but little about it. Likewise in Tibet grape-wine is scarcely used, being restricted to religious offerings in the temples.[1]

The text of the Geser Romance referred to is also important from another point of view. It contains the loan-word *ariki*, from Arabic *'araq*, which appears in eastern Asia as late as the Mongol epoch (below, p. 237). Consequently our work has experienced the influence of this period, which is visible also in other instances.[2] The foundation of the present recension, first printed at Peking in 1716, is indeed traceable to the thirteenth and fourteenth centuries; many legends and motives, of course, are of a much older date.

MARCO POLO relates in regard to T'ai-yüan fu, called by him Taianfu, the capital of Šan-si Province, "There grow here many excellent vines, supplying a great plenty of wine; and in all Cathay this is the only place where wine is produced. It is carried hence all over the country."[3] Marco Polo is upheld by contemporary Chinese writers. Grape-wine is mentioned in the Statutes of the Yüan Dynasty.[4] The *Yin šan čen yao* 飲膳正要, written in 1331 (in 3 chapters) by Ho Se-hwi 和斯輝, contains this account:[5] "There are numerous brands of wine: that coming from Qarā-Khoja (Ha-la-hwo 哈喇火)[6] is very strong, that coming from Tibet ranks next. Also the wines from P'in-yan and T'ai-

---

[1] Cf. *T'oung Pao*, 1914, p. 412.

[2] Cf. *ibid.*, 1908, p. 436.

[3] YULE and CORDIER, The Book of Ser Marco Polo, Vol. II, p. 13. KLAPROTH (cf. Yule's notes, *ibid.*, p. 16) was quite right in saying that the wine of that locality was celebrated in the days of the T'ang dynasty, and used to be sent in tribute to the emperors. Under the Mongols the use of this wine spread greatly. The founder of the Ming accepted the offering of wine from T'ai-yüan in 1373, but prohibited its being presented again. This fact is contained in the Ming Annals (cf. L. WIEGER, Textes historiques, p. 2011).

[4] *Yüan tien čan* 元典章, Ch. 22, p. 65 (ed. 1908).

[5] *Pen ts'ao kan mu*, Ch. 25, p. 14 b. Regarding that work, cf. the Imperial Catalogue, Ch. 116, p. 27 b.

[6] Regarding this name and its history see PELLIOT, *Journal asiatique*, 1912, I, p. 582. Qarā-Khoja was celebrated for its abundance of grapes (BRETSCHNEIDER, Mediæval Researches, Vol. I, p. 65). J. DUDGEON (The Beverages of the Chinese, p. 27), misreading the name Ha-so-hwo, took it for the designation of a sort of wine. Stuart (Chinese Materia Medica, p. 459) mistakes it for a transliteration of "hol-

yüan (in Šan-si) take the second rank. According to some statements, grapes, when stored for a long time, will develop into wine through a natural process. This wine is fragrant, sweet, and exceedingly strong: this is the genuine grape-wine."[1] The *Ts'ao mu tse* 草木子, written in 1378 by Ye Tse-k'i 葉子奇, contains the following information: "Under the Yüan dynasty grape-wine was manufactured in Ki-niṅ 冀寧 and other circuits 路 of Šan-si Province. In the eighth month they went to the T'ai-haṅ Mountain 太行山[2] in order to test the genuine and adulterated brands: the genuine kind when water is poured on it, will float; the adulterated sort, when thus treated, will freeze.[3] In wine which has long been stored, there is a certain portion which even in extreme cold will never freeze, while all the remainder is frozen: this is the spirit and fluid secretion of wine.[4] If this is drunk, the essence will penetrate into a man's arm-pits 腋 , and he will die. Wine kept for two or three years develops great poison."

The first author who offers a coherent notice and intelligent discussion of the subject of grape-wine is Li Ši-čen at the end of the sixteenth century.[5] He is well acquainted with the fact that this kind of wine was anciently made only in the Western Countries, and that the method of manufacturing it was but introduced under the T'ang after the subjugation of Kao-č'aṅ. He discriminates between two types of grape-wine,— the fermented 釀成者, of excellent taste, made from grape-juice with the addition of leaven in the same fashion as the ordinary native rice-wine (or, if no juice is available, dried raisins may be used), and the distilled 燒酒. In the latter method "ten catties of grapes are taken with an equal quantity of great leaven (distillers' grains) and subjected to a process of fermentation. The whole is then placed in an earthen kettle and steamed. The drops are received in a vessel, and this liquid is of red color, and very pleasing." There is one question, however, left open by Li Ši-čen. In a preceding notice on distillation 燒酒 he states that this is not an ancient method, but was practised only from the Yüan period; he then describes it in its application to rice-

---

lands," or maybe "alcohol." The latter word has never penetrated into China in any form. Chinese *a-la-ki* does not represent the word "alcohol," as conceived by some authors, for instance, J. MACGOWAN (*Journal China Branch Roy. As. Soc.*, Vol. VII, 1873, p. 237); see the following note.

[1] This work is also the first that contains the word *a-la-ki* 阿剌吉, from Arabic *'araq* (see *T'oung Pao*, 1916, p. 483).

[2] A range of mountains separating Šan-si from Či-li and Ho-nan.

[3] This is probably a fantasy. We can make nothing of it, as it is not stated how the adulterated wine was made.

[4] This possibly is the earliest Chinese allusion to alcohol.

[5] *Pen ts'ao kaṅ mu*, Ch. 25, p. 14 b.

wine in the same manner as for grape-wine. Certain it is that distillation is a Western invention, and was unknown to the ancient Chinese.[1] Li Ši-čen fails to inform us as to the time when the distillation of grape-wine came into existence. If this process had become known in China under the T'ang in connection with grape-wine, it would be strange if the Chinese did not then apply it to their native spirits, but should have waited for another foreign impulse until the Mongol period. On the other hand, if the method due to the Uigur under the T'ang merely applied to fermented grape-wine, we may justly wonder that the Chinese had to learn such a simple affair from the Uigur, while centuries earlier they must have had occasion to observe this process among many Iranian peoples. It would therefore be of great interest to seize upon a document that would tell us more in detail what this method of manufacture was, to which the T'ang history obviously attaches so great importance. It is not very likely that distillation was involved; for it is now generally conceded that the Arabs possessed no knowledge of alcohol, and that distillation is not mentioned in any relevant literature of the Arabs and Persians from the tenth to the thirteenth century.[2] The statement of Li Ši-čen, that distillation was first practised under the Mongols, is historically logical and in keeping with our present knowledge of the subject. It is hence reasonable to hold (at least for the present) also that distilled grape-wine was not made earlier in China than in the epoch of the Yüan. Moṅ Šen of the T'ang says advisedly that grapes can be fermented into wine, and the recipe of the Sung does not allude to distillation.

In the eighteenth century European wine also reached China. A chest of grape-wine figures among the presents made to the Emperor K'aṅ-hi on the occasion of his sixtieth birthday in 1715 by the Jesuits Bernard Kilian Stumpf, Joseph Suarez, Joachim Bouvet, and Dominicus Parrenin.[3]

P. OSBECK,[4] the pupil of Linné, has the following notice on the importation of European wine into China: "The Chinese wine, which our East India traders call Mandarin wine, is squeezed out of a fruit which is here called *Pausio*,[5] and reckoned the same with our grapes.

---

[1] Cf. BRETSCHNEIDER, Bot. Sin., pt. II, p. 155; J. DUDGEON, The Beverages of the Chinese, pp. 19–20; EDKINS, *China Review*, Vol. VI, p. 211. The process of distillation is described by H. B. GRUPPY, Samshu-Brewing in North China (*Journal China Branch Roy. As. Soc.*, Vol. XVIII, 1884, pp. 163–164).

[2] E. O. v. LIPPMANN, Abhandlungen, Vol. II, pp. 206–209; cf. also my remarks in *American Anthropologist*, 1917, p. 75.

[3] Cf. *Wan šou šeṅ tien* 萬 壽 盛 典, Ch. 56, p. 12.

[4] A Voyage to China and the East Indies, Vol. I, p. 315 (London, 1771).

[5] Apparently a bad or misprinted reproduction of *p'u-t'ao*.

This wine was so disagreeable to us, that none of us would drink it. The East India ships never fail taking wine to China, where they often sell it to considerable advantage. The Xeres (sherry) wine, for which at Cadiz we paid thirteen piastres an anchor, we sold here at thirty-three piastres an anchor. But in this case you stand a chance of having your tons split by the heat during the voyage. I have since been told, that in 1754, the price of wine was so much lowered at Canton, that our people could with difficulty reimburse themselves. The Spaniards send wines to Manilla and Macao, whence the Chinese fetch a considerable quantity, especially for the court of Peking. The wine of Xeres is more agreeable here than any other sort, on account of its strength, and because it is not liable to change by heat. The Chinese are very temperate in regard to wine, and many dare not empty a single glass, at least not at once. Some, however, have learned from foreigners to exceed the limits of temperance, especially when they drink with them at free cost."

Grape-wine is attributed by the Chinese to the Arabs.[1] The Arabs cultivated the vine and made wine in the pre-Islamic epoch. Good information on this subject is given by G. JACOB.[2]

Theophrastus[3] states that in India only the mountain-country has the vine and the olive. Apparently he hints at a wild vine, as does also Strabo,[4] who says after Aristobulus that in the country of Musicanus (Sindh) there grows spontaneously grain resembling wheat, and a vine producing wine, whereas other authors affirm that there is no wine in India. Again, he states[5] that on the mountain Meron near the city Nysa, founded by Bacchus, there grows a vine which does not ripen its fruit; for, in consequence of excessive rains, the grapes drop before arriving at maturity. They say also that the Sydracae or Oxydracae are descendants of Bacchus, because the vine grows in their country. The element -dracae (drakai) is probably connected with Sanskrit drākṣā ("grape"). These data of the ancients are vague, and do not prove at all that the grape-vine has been cultivated in India from time immemorial, as inferred by JORET.[6] Geographically they only refer to the regions bordering on Iran. The ancient Chinese knew only of grapes in Kashmir (above, p. 222). The *Wei šu*[7] states that grapes were ex-

---

[1] HIRTH, Chao Ju-kua, pp. 115, 121.

[2] Altarabisches Beduinenleben, 2d ed., pp. 96–109.

[3] Hist. plant., IV. IV, 11.

[4] xv, 22.

[5] XV. 1, 8.

[6] Plantes dans l'antiquité, Vol. II, p. 280.

[7] Ch. 102, p. 8.

ported from Pa-lai 拔頼 (*Bwat-lai) in southern India. Hüan Tsaṅ[1]
enumerates grapes together with pears, crab-apples, peaches, and
apricots,[2] as the fruits which, from Kashmir on, are planted here and
there in India. The grape, accordingly, was by no means common in
India in his time (seventh century).

The grape is not mentioned in Vedic literature, and Sanskrit *drākṣā*
I regard with SPIEGEL[3] as a loan-word. Viticulture never was extensive
or of any importance in Indian agriculture. Prior to the Mohammedan conquest, we have little precise knowledge of the cultivation of
the vine, which was much fostered by Akbar. In modern times it is
only in Kashmir that it has been received with some measure of
success.

Hüan Tsaṅ[4] states that there are several brands of alcoholic and
non-alcoholic beverages in India, differing according to the castes.
The Kṣatriya indulge in grape and sugar-cane wine. The Vaiçya take
rich wines fermented with yeast. The Buddhists and Brahmans partake
of a syrup of grapes or sugar-cane, which does not share the nature
of any wine.[5] In Jātaka No. 183, grape-juice (*muddikāpānam*) of intoxicating properties is mentioned.

Hüan Yiṅ[6] gives three Sanskrit words for various kinds of wine:—
(1) 窣羅 *su-lo*, *suδ-la*, Sanskrit *surā*, explained as rice-wine
米酒.[7]

---

[1] *Ta T'aṅ si yü ki*, Ch. 2, p. 8.

[2] Not almond-tree, as erroneously translated by JULIEN (Mémoires, Vol. I,
p. 92). Regarding peach and apricot, see below, p. 539.

[3] Arische Periode, p. 41.

[4] *Ta T'aṅ si yü ki*, Ch. 2, p. 8 b.

[5] S. JULIEN (Mémoires, Vol. I, p. 93) translates wrongly, "qui diffèrent tout à
fait du vin distillé." Distilled wine was then unknown both to the Chinese and in
India, and the term is not in the text. "Distillation of wines" is surely not spoken
of in the Çukranīti, as conceived by B. K. SARKAR (The Sukraniti, p. 157; and Hindu
Sociology, p. 166).

[6] *Yi ts'ie kiṅ yin i*, Ch. 24, p. 8 b.

[7] This definition is of some importance, for in BOEHTLINGK's Sanskrit Dictionary
the word is explained as meaning "a kind of beer in ancient times, subsequently,
however, in most cases brandy," which is certainly wrong. Thus also O. SCHRADER's
speculation (Sprachvergleichung, Vol. II, p. 256), connecting Finno-Ugrian *sara,
sur*, etc. ("beer") with this word, necessarily falls to the ground. MACDONELL and
KEITH (Vedic Index, Vol. II, p. 458) admit that "the exact nature of *surā* is not
certain, it may have been a strong spirit prepared from fermented grains and plants,
as Eggeling holds, or, as Whitney thought, a kind of beer or ale." It follows also
from Jātaka No. 512 that *surā* was prepared from rice. In Cosmas' Christian
Topography (p. 362, ed. of Hakluyt Society) we have ῥογχοσοβρα ("coconut-
wine"); here *sura* means "wine," while the first element may be connected with
Arabic *ranej* or *ranj* ("coco-nut").

(2) 迷麗邪 *mi-li-ye*, *mei-li(ri)-ya, answering to Sanskrit *maireya*, explained as a wine mixed from roots, stems, flowers, and leaves.[1]

(3) 末陀 *mo-t'o*, *mwaδ-do, Sanskrit *madhu*, explained as "grape-wine" (*p'u-t'ao tsiu*). The latter word, as is well known, is connected with Avestan *maδa* (Middle Persian *mai*, New Persian *mei*), Greek μέθυ, Latin *temetum*. Knowledge of grape-wine was conveyed to India from the West, as we see from the Periplus and Tamil poems alluding to the importation of Yavana (Greek) wines.[2] In the Raghuvaṃça (IV, 65), *madhu* doubtless refers to grape-wine; for King Raghu vanquished the Yavana, and his soldiers relieve their fatigue by enjoying *madhu* in the vine regions of the Yavana country.

According to W. Ainslie,[3] the French at Pondicherry, in spite of the great heat of the Carnatic, are particularly successful in cultivating grapes; but no wine is made in India, nor is the fruit dried into raisins as in Europe and Persia. The Arabians and Persians, particularly the latter, though they are forbidden wine by the Koran, bestow much pains on the cultivation of the grape, and suppose that the different kinds possess distinguishing medicinal qualities. Wine is brought to India from Persia, where, according to Tavernier (1605–89), three sorts are made: that of Yezd, being very delicate; the Ispahan produce, being not so good; and the Shiraz, being the best, rich, sweet, and generous, and being obtained from the small grapes called *kišmiš*, which are sent for sale to Hindustan when dried into raisins.[4] There are two brands of Shiraz wine, a red and a white, both of which are excellent, and find a ready market in India. Not less than four thousand tuns of Shiraz wine is said to be annually sent from Persia to different parts of the world.[5] The greatest quantity is produced in the district of Korbal, near the village of Bend Emir.[6] In regard to Assam,

---

[1] Compare above (p. 222) the wine of the Yüe-či. According to Boehtlingk, *maireya* is an intoxicating drink prepared from sugar and other substances.

[2] V. A. Smith, Early History of India, p. 444 (3d ed.).

[3] Materia Indica, Vol. I, p. 157.

[4] Compare above, p. 231.

[5]   "Wines too, of every clime and hue,
       Around their liquid lustre threw;
       Amber Rosolli,—the bright dew
       From vineyards of the Green-Sea gushing;
       And Shiraz wine, that richly ran
         As if that jewel, large and rare,
       The ruby, for which Kublai-Khan
       Offer'd a city's wealth, was blushing
        Melted within the goblets there!"
                Thomas Moore, Lalla Rookh.

[6] Ainslee, *l.c.*, p. 473.

TAVERNIER[1] states that there are quantities of vines and good grapes, but no wine, the grapes being merely dried to distil spirits from. Wild vine grows in upper Siam and on the Malay Peninsula, and is said to furnish a rather good wine.[2]

A wine-yielding plant of Central Asia is described in the *Ku kin ču* 古今注[3] by Ts'ui Pao 崔豹 of the fourth century, as follows: "The *tsiu-pei-t'en* 酒杯藤 ("wine-cup creeper") has its habitat in the Western Regions (Si-yü). The creeper is as large as an arm; its leaves are like those of the *ko* 葛 (*Pachyrhizus thunbergianus*, a wild-growing creeper); flowers and fruits resemble those of the *wu-t'un* (*Sterculia platanifolia*), and are hard; wine can be pressed out of them. The fruits are as large as a finger and in taste somewhat similar to the *tou-k'ou* 荳蔲 (*Alpinia globosum*); their fragrance is fine, and they help to digest wine. In order to secure wine, the natives get beneath the creepers, pluck the flowers, press the wine out, eat the fruit for digestion, and become intoxicated. The people of those countries esteem this wine, but it is not sent to China. Čan K'ien obtained it when he left Ta-yüan (Fergana). This affair is contained in the *Čan K'ien ču kwan či* 張騫 出關志 ('Memoirs of Čan K'ien's Journey')."[4] This account is restricted to the *Ku kin ču*, and is not confirmed by any other book. Li Ši-čen's work is the only *Pen ts'ao* which has adopted this text in an abridged form.[5] Accordingly the plant itself has never been introduced into China; and this fact is sufficient to discard the possibility of an introduction by Čan K'ien. If he had done so, the plant would have been disseminated over China and mentioned in the various early *Pen ts'ao;* it would have been traced and identified by our botanists. Possibly the plant spoken of is a wild vine, possibly another genus. The description, though by no means clear in detail, is too specific to be regarded as a mystification.

The history of the grape-vine in China has a decidedly methodological value. We know exactly the date of the introduction and

---

[1] Travels in India, Vol. II, p. 282.

[2] DILOCK PRINZ VON SIAM, Landwirtschaft in Siam, p. 167.

[3] Ch. c, p. 2 b. The text has been adopted by the *Sü po wu či* (Ch. 5, p. 2 b) and in a much abbreviated form by the *Yu yan tsa tsu* (Ch. 18, p. 6 b). It is not in the *Pen ts'ao kan mu*, but in the *Pen ts'ao kan mu ši i* (Ch. 8, p. 27).

[4] HIRTH (*Journal Am. Or. Soc.*, Vol. XXXVII, 1917, p. 91) states that this work is mentioned in the catalogue of the library of the Sui dynasty, but not in the later dynastic catalogues. We do not know when and by whom this alleged book was written; it may have been an historical romance. Surely it was not produced by Čan K'ien himself.

[5] See also *T'u šu tsi č'en*, XX, Ch. 112, where no other text on the subject is quoted.

the circumstances which accompanied this important event. We have likewise ascertained that the art of making grape-wine was not learned by the Chinese before A.D. 640. There are in China several species of wild vine which bear no relation to the imported cultivated species. Were we left without the records of the Chinese, a botanist of the type of Engler would correlate the cultivated with the wild forms and assure us that the Chinese are original and independent viticulturists. In fact, he has stated[1] that *Vitis thunbergii*, a wild vine occurring in Japan, Korea, and China, seems to have a share in the development of Japanese varieties of vine, and that *Vitis filifolia* of North China seems to have influenced Chinese and Japanese vines. Nothing of the kind can be inferred from Chinese records, or has ever been established by direct observation. The fact of the introduction of the cultivated grape into China is wholly unknown to Engler. The botanical notes appended by him to HEHN's history of the grape[2] have nothing whatever to do with the history of the cultivated species, but refer exclusively to wild forms. It is not botany, but historical research, that is able to solve the problems connected with the history of our cultivated plants.

Dr. T. TANAKA of the Bureau of Plant Industry, U. S. Department of Agriculture, Washington, has been good enough to contribute the following notes on the history of the grape-vine in Japan:—

"The early history of the cultivation of the grape-vine (*Vitis vinifera*) in Japan is very obscure. Most of the early Japanese medical and botanical works refer to *budō* 葡萄 (Chinese *p'u-t'ao*) as *ebi*, the name occurring in the *Kojiki* (compiled in A.D. 712, first printed in 1644) as *yebikadzura*,[3] which is identified by J. MATSUMURA[4] as *Vitis vinifera*. It seems quite incomprehensible that the grape-vine, which is now found only in cultivated form, should have occurred during the mythological period as early as 660 B.C. The *Honzō-wamyō* 本草倭名 (compiled during the period 897–930, first printed 1796) mentions *ō-ebi-kadzura* as vine-grape, distinguishing it from ordinary *ebi-kadzura*, but the former is no longer in common use in distinction from the latter. The *ebi-dzuru* which should correctly be termed *inu-ebi* (false *ebi* plant), as suggested by Ono Ranzan,[5] is widely applied in Japan for 蘡薁 (Chinese *yin-yü*), and is usually identified as *Vitis thunbergii*,

---

[1] Erläuterungen zu den Nutzpflanzen der gemässigten Zonen, p. 30.

[2] Kulturpflanzen, pp. 85–91.

[3] B. H. CHAMBERLAIN, Ko-ji-ki, p. XXXIV.

[4] *Botanical Magazine*, Tōkyō, Vol. VII, 1893, p. 139.

[5] *Honzō kōmoku keimō*, ed. 1847, Ch. 29, p. 3.

but is an entirely different plant, with small, deeply-lobed leaves, copiously villose beneath. *Ebi-kadzura* is mentioned again in the *Wamyō-ruijušō* 和名類聚鈔 (compiled during the period 923-931, first edited in 1617), which gives *budō* as the fruit of *šikwatsu* or *Vitis coignetiae*[1], as growing wild in northern Japan.

"These three plants are apparently mixed up in early Japanese literature, as pointed out by Arai Kimiyoši.[2] Describing *budō* as a food plant, the *Hončō šokukan* 本朝食鑑[3] mentions that the fruit was not greatly appreciated in ancient times; for this reason no mention was made of it in the Imperial chronicles, nor has any appropriate Japanese term been coined to designate the vine-grape proper.

"In the principal vine-grape district of Japan, Yamanaši-ken (previously called Kai Province), were found a few old records, an account of which is given in Viscount Y. Fukuba's excellent discourse on Pomology.[4] An article on the same subject was published by J. DAUTREMER.[5] This relates to a tradition regarding the accidental discovery by a villager, Amenomiya Kageyu (not two persons), of the vine-grape in 1186 (Dautremer erroneously makes it 1195) at the mountain of Kamiiwasaki 上岩崎, not far from Kōfu 甲府. Its cultivation must have followed soon afterward, for in 1197 a few choice fruits were presented to the Šōgun Yoritomo (1147-99). At the time of Takeda Harunobu (1521-73) a sword was presented to the Amenomiya family as a reward for excellent fruits which they presented to the Lord. Viscount Fukuba saw the original document relative to the official presentation of the sword, and bearing the date 1549.[6] The descendants of this historical grape-vine are still thriving in the same locality around the original grove, widely recognized among horticulturists as a true *Vitis vinifera*. According to a later publication of Fukuba,[7] there is but one variety of it. Several introductions of *Vitis vinifera* took place in the early Meiji period (beginning 1868) from Europe and America.

"The following species of *Vitis* are mentioned in Umemura's work *Inošokukwai-no-šokubutsu-ši* 飲食界之植物誌[8] as being edible:

---

[1] MATSUMURA, Shokubutsu Mei-i, p. 380.

[2] *Tōga* 東雅 (completed in 1719), ed. 1906, p. 272.

[3] Ch. 4, p. 50 (ed. of 1698).

[4] *Kwaju engei-ron* 果樹園藝論, privately published in 1892.

[5] Situation de la vigne dans l'empire du Japon, *Transactions Asiatic Society of Japan*, Vol. XIV, 1886, pp. 176-185.

[6] Fukuba, *op. cit.*, pp. 461-462.

[7] *Kwaju saibai jenšo* 果樹栽培全書, Vol. IV, 1896, pp. 119-120.

[8] Vol. 4, 1906.

"Yama-budō (*Vitis coignetiae*): fruit eaten raw and used for wine; leaves substituted for tobacco.

"Ebi-dzuru (*V. thunbergii*): fruit eaten raw, leaves cleaned and cooked; worm inside the cane baked and eaten by children as remedy for convulsions.

"Sankaku-dzuru (*V. flexuosa*): fruit eaten raw.

"Ama-dzuru (*V. saccharifera*): fruit eaten raw; children are very fond of eating the leaves, as they contain sugar."

THE PISTACHIO

3. *Pistacia* is a genus of trees or shrubs of the family *Anacardiaceae*, containing some six species, natives of Iran and western Asia, and also transplanted to the Mediterranean region. At least three species (*Pistacia vera, P. terebinthus*, and *P. acuminata*) are natives of Persia, and from ancient times have occupied a prominent place in the life of the Iranians. Pistachio-nuts are still exported in large quantities from Afghanistan to India, where they form a common article of food among the well-to-do classes. The species found in Afghanistan and Baluchistan do not cross the Indian frontier.[1] The pistachio (*Pistacia vera*) in particular is indigenous to ancient Sogdiana and Khorasan,[2] and still is a tree of great importance in Russian Turkistan.[3]

When Alexander crossed the mountains into Bactriana, the road was bare of vegetation save a few trees of the bushy terminthus or terebinthus.[4] On the basis of the information furnished by Alexander's scientific staff, the tree is mentioned by Theophrastus[5] as growing in the country of the Bactrians; the nuts resembling almonds in size and shape, but surpassing them in taste and sweetness, wherefore the people of the country use them in preference to almonds. Nicandrus of Colophon[6] (third century B.C.), who calls the fruit βιστάκιον or φιττάκιον, a word derived from an Iranian language (see below), says that it grows in the valley of the Xoaspes in Susiana. Posidonius, Dioscorides, Pliny, and Galenus know it also in Syria. Vitellius introduced the tree into Italy; and Flaccus Pompeius, who served with him, introduced it at the same time into Spain.[7]

The youths of the Persians were taught to endure heat, cold, and rain; to cross torrents and to keep their armor and clothes dry; to pasture animals, to watch all night in the open air, and to subsist on wild fruit, as terebinths (*Pistacia terebinthus*), acorns, and wild pears.[8]

---

[1] WATT, Dictionary of the Economic Products of India, Vol. VI, p. 268.

[2] JORET, Plantes dans l'antiquité, Vol. II, pp. 47, 76.

[3] S. KORŽINSKI, Vegetation of Turkistan (in Russian), pp. 20, 21.

[4] Strabo, XV. II, 10.

[5] Hist. plant., IV. IV, 7.

[6] Theriaka, 890.

[7] Pliny, XV, 22, §91. A. DE CANDOLLE (Origin of Cultivated Plants, p. 316) traces *Pistacia vera* only to Syria, without mentioning its occurrence in Persia.

[8] Strabo, XV. III, 18.

246

The Persians appeared to the ancients as terebinth-eaters, and this title seems to have developed into a sort of nickname: when Astyages, King of the Medians, seated on his throne, looked on the defeat of his men through the army of Cyrus, he exclaimed, "Woe, how brave are these terebinth-eating Persians!"[1] According to Polyaenus,[2] terebinth-oil was among the articles to be furnished daily for the table of the Persian kings. In the Būndahišn, the pistachio-nut is mentioned together with other fruits the inside of which is fit to eat, but not the outside.[3] "The fruits of the country are dates, pistachios, and apples of Paradise, with other of the like not found in our cold climate."[4]

Twan Č'eṅ-ši 段成式, in his *Yu yaṅ tsa tsu* 酉陽雜俎, written about A.D. 860 and containing a great amount of useful information on the plants of Persia and Fu-lin, has the following:—

"The hazel-nut (*Corylus heterophylla*) of the Hu (Iranians), styled *a-yüe* 阿月, grows in the countries of the West.[5] According to the statement of the barbarians, *a-yüe* is identical with the hazel-nuts of the Hu. In the first year the tree bears hazel-nuts, in the second year it bears *a-yüe*."[6]

Č'en Ts'aṅ-k'i 陳藏器, who in the K'ai-yüan period (A.D. 713–741) wrote the Materia Medica *Pen ts'ao ši i* 本草拾遺, states that "the fruits of the plant *a-yüe-hun* 阿月渾 are warm and acrid of flavor, non-poisonous, cure catarrh of the bowels, remove cold feeling, and make people stout and robust, that they grow in the western countries, the barbarians saying that they are identical with the hazel-nut of the Hu 胡榛子. During the first year the tree bears hazel-nuts, in the second year it bears *a-yüe-hun*."

Li Sün 李珣, in his *Hai yao pen ts'ao* 海藥本草 (second half of the eighth century), states, "According to the *Nan čou ki* 南洲記 by Sü Piao 徐表,[7] the Nameless Tree (*wu miṅ mu* 無名木) grows in the mountainous valleys of Liṅ-nan (Kwaṅ-tuṅ). Its fruits resemble in appearance the hazel-nut, and are styled Nameless Fruits (*wu miṅ tse* 無名

---

[1] Nicolaus of Damaskus (first century B.C.), cited by HEHN, Kulturpflanzen, p. 424.

[2] Strategica, IV. III, 32.

[3] These fruits are walnut, almond, pomegranate, coconut, filbert, and chestnut. See WEST, Pahlavi Texts, Vol. I, p. 103.

[4] MARCO POLO, Yule's edition, Vol. I, p. 97.

[5] The editions of the *Yu yaṅ tsa tsu* write 西圃, "in the gardens of the West"; but the *T'u šu tsi č'eṅ* (section botany, Ch. 311) and *Či wu miṅ ši t'u k'ao*, in reproducing this text, offer the reading 西圉, which seems to me preferable.

[6] *Yu yaṅ tsa tsu* 續集, Ch. 10, p. 3 b (ed. of *Tsin tai pi šu*).

[7] This work is quoted in the *Ts'i miṅ yao šu*, written by Kia Se-niu under the Hou Wei dynasty (A.D. 386–534).

子). Persians 波斯家 designate them *a-yüe-hun* fruits."[1] For the same period we have the testimony of the Arabic merchant Soleiman, who wrote in A.D. 851, to the effect that pistachios grow in China.[2]

As shown by the two forms, *a-yüe* of the *Yu yan tsa tsu* and *a-yüe-hun* of the *Pen ts'ao ši i* and *Hai yao pen ts'ao*, the fuller form must represent a compound consisting of the elements *a-yüe* and *hun*. In order to understand the transcription *a-yüe*, consideration of the following facts is necessary.

The Old-Iranian word for the walnut has not been handed down to us, but there is good evidence to prompt the conclusion that it must have been of the type **agōza* or **angōza*. On the one hand, we have Armenian *engoiz*, Ossetic *ängozä* or *änguz*, and Hebrew *egōz*;[3] on the other hand, we meet in Yidgha, a Hindu-Kush language, the form *ogūzo*, as compared with New Persian *kōz* and *gōz*.[4] The signification of this word is "nut" in general, and "walnut" in particular. Further, there is in Sanskrit the Iranian loan-word *ākhōṭa*, *akṣōṭa*, or *akṣōḍa*, which must have been borrowed at an early date, as, in the last-named form, the word occurs twice in the Bower Manuscript.[5] It has survived in Hindustānī as *axrōt* or *ākrōt*. The actual existence of an East-Iranian form with the ancient initial *a-* is guaranteed by the Chinese transcription *a-yüe*; for *a-yüe* 阿月 answers to an ancient **a-ṅwieδ* (*ṅw'eδ*) or **a-gwieδ, a-gwüδ*;[6] and this, in my opinion, is intended to represent the Iranian word for "nut" with initial *a-*, mentioned above; that is, **aṅgwīz, aṅgwōz, agōz*.

Chinese *hun* 渾 answers to an ancient **γwun* or *wun*. In regard to this Iranian word, the following information may be helpful. E.

---

[1] If it is correct that the transcription *a-yüe-hun* was already contained in the *Nan čou ki* (which it is impossible to prove, as we do not possess the text of this work), the transcription must have been based on an original prototype of early Sasanian times or on an early Middle-Persian form. This, in fact, is confirmed by the very character of the Sino-Iranian word, which has preserved the initial *a-*, while this one became lost in New Persian. It may hence be inferred that Li Sün's information is correct, and that the transcription *a-yüe-hun* may really have been contained in the *Nan čou ki*, and would accordingly be pre-T'aṅ.

[2] M. REINAUD, Relation des voyages faits par les Arabes et les Persans dans l'Inde et à la Chine, Vol. I, p. 22.

[3] Whether Georgian *nigozi* and the local name Νίγουζα of Ptolemy (W. TOMASCHEK, Pamirdialekte, *Sitzber. Wiener Akad.*, 1880, p. 790) belong here, I do not feel certain. Cf. HÜBSCHMANN, Armenische Grammatik, p. 393.

[4] In regard to the elision of initial *a* in New Persian, see HÜBSCHMANN, Persische Studien, p. 120.

[5] HOERNLE's edition, pp. 32, 90, 121.

[6] Regarding the phonetic value of 月, see the detailed study of PELLIOT (*Bull. de l'Ecole française*, Vol. V, p. 443) and the writer's Language of the Yüe-chi or Indo-Scythians.

KAEMPFER[1] speaks of *Terebinthus* or *Pistacea sylvestris* in Persia thus: "Ea Pistaceae hortensi, quam Theophrastus Therebinthum Indicam vocat, tum magnitudine, tum totius ac partium figurâ persimilis est, nisi quod flosculos ferat fragrantiores, nuces vero praeparvas, insipidas; unde a descriptione botanica abstinemus. Copiosa crescit in recessibus montium brumalis genii, petrosis ac desertis, circa Schamachiam Mediae, Schirasum Persidis, in Luristano et Larensi territoriis. Mihi nullibi conspecta est copiosior quam in petroso monte circa Majin, pagum celebrem, unâ diaeta dissitum Sjirasô: in quo mihi duplicis varietatis indicarunt arborem; unam vulgariorem, quae generis sui retineat appellationem *Diracht* [*diraxt*, 'tree'] *Ben* seu *Wen;* alteram rariorem, in specie *Kasudaan* [kasu-dān], vel, ut rustici pronunciant, *Kasudèn* dictam, quae a priori fructuum rubedine differat." ROEDIGER and POTT[2] have added to this *ben* or *wen* a Middle-Persian form *ven* ("wild pistachio"). In the Persian Dictionary edited by STEINGASS (p. 200) this word is given as *ban* or *wan* (also *banak*), with the translation "Persian turpentine seed."[3] VULLERS[4] writes it *ban*. SCHLIMMER[5] transcribes this word *beneh*. He identifies the tree with *Pistacia acuminata* and observes, "C'est l'arbre qui fournit en Perse un produit assez semblable à la trémentine, mais plutôt mou que liquide, vu qu'on l'obtient par des découpures, dont le produit se rassemble durant les grandes chaleurs dans un creux fait en terre glaise au pied de l'arbre, de façon à ce que la matière sécrétée perd une grande partie de son huile essentielle avant d'être enlevée. Le même produit, obtenu à Kerman dans un outre, fixé à l'arbre et enlevé aussitôt plein, était à peu près aussi liquide que la térébenthine de Venise. . . . La *Pistacia acuminata* est sauvage au Kordesthan persan et, d'après Buhse, aussi à Reshm, Damghan et Dereghum (province de Yezd); Haussknecht la vit aussi à Kuh Kiluye et dans le Luristan."

The same word we meet also in Kurd *dariben*, *dar-i-ben* ("the tree *ben*"), and in all probability in Greek τερέβινθος, older forms τέρμινθος and τρέμιθος.[6] Finally WATT[7] gives a Baluči word *ban, wan, wana, gwa,*

---

[1] Amoenitatum exoticarum fasciculi V, p. 413 (Lemgoviae, 1712).

[2] *Zeitschr. Kunde d. Morgenl.*, Vol. V, 1844, p. 64.

[3] This notion is also expressed by *banāsīb* (cf. *bināst*, "turpentine").

[4] Lexicon persico-latinum, Vol. I, p. 184.

[5] Terminologie, p. 465.

[6] The Greek ending, therefore, is -θος, not -νθος, as stated by SCHRADER (in Hehn, Kulturpflanzen, 8th ed., p. 221); *n* adheres to the stem: *tere-bin-θos*.

[7] Commercial Products of India, p. 902; and Dictionary of the Economic Products of India, Vol. VI, p. 271.

gwan, gwana, for *Pistacia mutica* (or *P. terebinthus,* var. *mutica*); this form comes nearest to the Chinese transcription.

While a compound **agoz-van(vun),* that is, "nut of pistachio," as far as I know, has not yet been traced in Iranian directly, its existence follows from the Chinese record of the term. An analogy to this compound is presented by Kurd *kizvan, kezvān, kazu-van, kasu-van* ("pistachio" or "terebinthus-tree").[1]

The *Honzō kōmoku keimō* (Ch. 25, fol. 24), written by Ono Ranzan 小野蘭山, first published in 1804, revised in 1847 by Iguči Bōši 井口望之, his grandson, mentions the same plant 阿月渾子, which reads in Japanese *agetsu-konši.* He gives also in Kana the names *fusudasiu* or *fusudasu.*[2] He states, "The plant is not known in Japan to grow wild. It used to come from foreign countries, but not so at present. A book called *Zōkyōhi furoku* 象敎皮附錄 mentions this plant, stating that *agetsu-konši* is the fruit of the tree *č'a mu* 柵木 (in Japanese *sakuboku*)."[3]

---

[1] A. JABA, Dictionnaire kurde-français, p. 333. Cf. above the *kasu-dān* of Kaempfer.

[2] These terms are also given by the eminent Japanese botanist MATSUMURA in his Shokubutsu mei-i (No. 2386), accompanied by the identification *Pistacia vera.*

[3] This tradition is indeed traceable to an ancient Chinese record, which will be found in the *Čeň lei pen ts'ao* of 1108 (Ch. 12, p. 55, ed. of 1583). Here the question is of the bark of the *san* or *č'a* tree 柵木皮, mentioned as early as the sixth century in the *Kwaň či* 廣志 of Kwo Yi-kuň as growing in wild country of Kwaň-nan 廣南 (the present province of Kwaň-tuň and part of Kwaň-si), and described in a commentary of the *Er ya* as resembling the mulberry-tree. This, of course, is a wild tree indigenous to a certain region of southern China, but, as far as I know, not yet identified, presumably as the ancient name is now obsolete. The *Nan čou ki* by Sŭ Piao (see above) says that the fruits of this tree are styled *wu miň tse* 無名子 ("nameless fruits"); hence the conclusion is offered by T'aň Šen-wei, author of the *Čeň lei pen ts'ao,* that this is the tree termed *a-yüe-hun* by the Persians (that is, a cultivated *Pistacia*). This inference is obviously erroneous, as the latter was introduced from Persia into China either under the T'ang or a few centuries earlier, while the *san* or *č'a* tree pre-existed spontaneously in the Chinese flora. The only basis for this hazardous identification is given by the attribute "nameless." A solution of this problem is possible if we remember the fact that there is a wild Pistacia, *Pistacia chinensis,* indigenous to China, and if we identify with it the tree *san* or *č'a;* then it is conceivable that the wild and the imported, cultivated species were correlated and combined under the same popular term *wu miň.* MATSUMURA (*op. cit.,* No. 2382) calls *P. chinensis* in Japanese *ōrenju,* adding the characters 黃楝. The word *lien* refers in China to *Melia azedarach.* The modern Chinese equivalent for *P. chinensis* is not known to me. The peculiar beauty of this tree, and the great age to which it lives, have attracted the attention of the indefatigable workers of our Department of Agriculture, who have already distributed thousands of young trees to parks throughout the country (see Yearbook of the U. S. Department of Agriculture 1916, p. 140, Washington, 1917). In the English and Chinese Standard Dictionary, the word "pistachio" is rendered by *fei* 榧, which, however, denotes a quite dif-

G. A. STUART[1] has identified *a-yüe hun-tse*[2] with *Pistacia vera*, and this is confirmed by Matsumura.

The Japanese name *fusudasiu* or *fusudasu* is doubtless connected with Persian *pista*, from Old Iranian *pistaka*, Middle Persian *pistak*,[3] from which is derived Greek βιστάκιον, φιττάκιον, πιστάκιον or ψιστάκιον, Latin *psittacium*, and our *pistacia* or *pistachio*. It is not known to me, however, to what date the Japanese word goes back, or through what channels it was received. In all likelihood it is of modern origin, the introduction into Japan being due to Europeans.

In Chinese literature, the Persian word appears in the Geography of the Ming Dynasty,[4] in the transcription [*ki-*] *pi-se-tan* [劑] 芘思檀, stated to be a product of Samarkand, the leaves of the tree resembling those of the *šan č'a* 山茶 (*Camellia oleifera*), and its fruit that of the *yin hiṅ* 銀杏 (*Salisburia adiantifolia*).

The Persian word, further, occurs in the new edition of the *Kwaṅ yü ki*, entitled *Tseṅ tiṅ kwaṅ yü ki* 增訂廣輿記. The original, the *Kwaṅ yü ki*, was written by Lu Yiṅ-yaṅ 陸應暘,[5] and published during the Wan-li period in 1600. The revised and enlarged edition was prepared by Ts'ai Faṅ-piṅ 蔡方炳 (*hao* Kiu-hia 九霞) in 1686; a reprint of this text was issued in 1744 by the publishing-house Se-mei t'aṅ 四美堂. Both this edition and the original are before me. The latter[6] mentions only three products under the heading "Samarkand"; namely, coral, amber, and ornamented cloth (*hwa žui pu* 花蕊布). The new edition, however, has fifteen additional items, the first of these being [*ki-*] *pi-se-t'an*, written as above,[7] stated to be a tree growing in the region of Samarkand. "The leaves of the tree," it is said, "resemble those of the *šan č'a* (*Camelia oleifera*); the fruits have the appearance of the nut-like seeds of the *yin hiṅ* (*Salisburia adiantifolia*), but are smaller." The word *pi-se-t'an* doubtless represents the transcription of Persian

---

ferent plant,—*Torreya nucifera*. A revival on the part of the Chinese, of the good, old terms of their own language, would be very desirable, not only in this case, but likewise in many others.

[1] Chinese Materia Medica, p. 334.

[2] Wrongly transcribed by him *o-yüeh-chün-tzŭ*.

[3] These reconstructions logically result from the phonetic history of Iranian, and are necessitated by the existence of the Greek loan-word. Cf., further, Byzantine *pustux* and *fustox*, Comanian *pistac*, and the forms given below (p. 252). Persian *pista* is identified with *Pistacia vera* by SCHLIMMER (Terminologie, p. 465).

[4] *Ta Miṅ i t'uṅ či*, Ch. 89, p. 23.

[5] WYLIE, Notes on Chinese Literature, p. 59.

[6] Ch. 24, p. 6 b.

[7] The addition of *ki* surely rests on an error (SCHOTT also reads *pi-se-t'an*, which he presumably found in his text; see the following note).

pistān ("a place abounding with pistachio-nuts").[1] Again, the Persian word in the transcription *pi-se-ta* 必思答 appears in the *Pen ts'ao kan mu ši i*[2] by Čao Hio-min, who states that the habitat of the plant is in the land of the Mohammedans, and refers to the work *Yin šan čen yao*[3] of 1331, ascribed by him to Hu-pi-lie 忽必烈; that is, the Emperor Kubilai of the Yüan dynasty. We know, however, that this book was written in 1331 by Ho Se-hwi.[4] Not having access to this, I am unable to state whether it contains a reference to *pi-se-ta*, nor do I know whether the text of Čao Hio-min, as printed in the second edition of 1765, was thus contained in the first edition of his work, which was published in 1650. It would not be impossible that the transcription *pi-se-ta*, accurately corresponding to Persian *pista*, was made in the Mongol period; for it bears the ear-marks of the Yüan style of transcription.

The Persian word *pista* (also *pasta*) has been widely disseminated: we find it in Kurd *fystiq*, Armenian *fesdux* and *fstoül*, Arabic *fistaq* or *fustaq*, Osmanli *fistiq*,[5] and Russian *fistaška*.

In the Yüan period the Chinese also made the acquaintance of mastic, the resinous product of *Pistacia lentiscus*.[6] It is mentioned in the *Yin šan čen yao*, written in 1331, under its Arabic name *mastaki*, in the transcription 馬思答吉 *ma-se-ta-ki*.[7] Li Ši-čen knew only the medical properties of the product, but confessed his ignorance regarding the nature of the plant; hence he placed his notice of it as an appendix to cummin (*ži-lo*). The *Wu tsa tsu* 五雜組, written in 1610, says that *mastaki* is produced in Turkistan and resembles the *tsiao* 椒 (*Zanthoxylum*, the fruit yielding a pepper-like condiment); its odor is very strong; it takes the place there of a condiment like pepper, and is beneficial to digestion.[8] The Persian word for "mastic" is *kundurak* (from *kundur*, "incense"), besides the Arabic loan-word *mastaki* or

---

[1] As already recognized by W. SCHOTT (Topographie der Producte des chinesischen Reiches, *Abh. Berl. Akad.*, 1842, p. 371), who made use only of the new edition.

[2] Ch. 8, p. 19; ed. of 1765 (see above, p. 229).

[3] Cf. above, p. 236.

[4] BRETSCHNEIDER, Bot. Sin., pt. 1, p. 213.

[5] Hence Pegoletti's *fistuchi* (YULE, Cathay, new ed. by CORDIER, Vol. III p. 167).

[6] Greek σχῖνος (Herodotus, IV, 177).

[7] The Arabic word itself is derived from Greek μαστίχη (from μαστάζειν, "to chew"), because the resin was used as a masticatory. Hence also Armenian *maztak'ē*. Spanish *almáciga* is derived from the Arabic, as indicated by the Arabic article *al*, while the Spanish form *másticis* is based on Latin *mastix*.

[8] Quoted in the *Pen ts'ao kan mu ši i*, Ch. 6, p. 12 b. The digestive property is already emphasized by Dioscorides (I, 90).

mästäkï.[1] The Persianized form is *masdax;* in Kurd it is *mstekki.* "On these mountains the *Mastich* Tree brings forth plenty of that gum, of which the country people make good profit. . . . As for the Mastick Trees, they bore red berries, and if wounded would spew out the liquid resin from the branches; they are not very tall, of the bigness of our Bully Trees: Whether they bring forth a cod or not, this season would not inform me, nor can I say it agrees in all respects with the Lentisk Tree of Clusius."[2] The resin (mastic) occurs in small, irregular, yellowish tears, brittle, and of a vitreous fracture, but soft and ductile when chewed. It is used as a masticatory by people of high rank in India to preserve the teeth and sweeten the breath, and also in the preparation of a perfume.[3] It is still known in India as the "gum mastic of Rûm."[4]

The case of the pistachio (and there are several others) is interesting in showing that the Chinese closely followed the development of Iranian speech, and in course of time replaced the Middle-Persian terms by the corresponding New-Persian words.

---

[1] ACHUNDOW, Abu Mansur, pp. 137, 267.

[2] JOHN FRYER, New Account of East India and Persia, Vol. II, p. 202 (Hakluyt Soc., 1912).

[3] WATT, Commercial Products of India, p. 902.

[4] D. C. PHILLOTT, *Journal As. Soc. Bengal,* Vol. VI, 1910, p. 81.

THE WALNUT

4. The Buddhist dictionary *Fan yi min yi tsi* 翻譯名義集, compiled by Fa Yün 法雲,[1] contains a Chinese-Sanskrit name for the walnut (*hu t'ao* 胡桃, *Juglans regia*) in the transcription *po-lo-ši* 播囉師, which, as far as I know, has not yet been identified with its Sanskrit equivalent.[2] According to the laws established for the Buddhist transcriptions, this formation is to be restored to Sanskrit *pārasī*, which I regard as the feminine form of the adjective *pārasa*, meaning "Persian" (derived from *Parsa*, "Persia"). The walnut, accordingly, as expressed by this term, was regarded in India as a tree or fruit suspected of Persian provenience. The designation *pārasī* for the walnut is not recorded in Boehtlingk's Sanskrit Dictionary, which, by the way, contains many other lacunes. The common Sanskrit word for "walnut" is *ākhōṭa, akṣōṭa, akṣōṣa*,[3] which for a long time has been regarded as a loan-word received from Iranian.[4]

Pliny has invoked the Greek names bestowed on this fruit as testimony for the fact that it was originally introduced from Persia, the

---

[1] Ch. 24, p. 27 (edition of Nanking).—BUNYIU NANJIO (Catalogue of the Buddhist Tripiṭaka, No. 1640) sets the date of the work at 1151. WYLIE (Notes on Chinese Literature, p. 210) and BRETSCHNEIDER (Bot. Sin., pt. 1, p. 94) say that it was completed in 1143. According to S. JULIEN (Méthode, p. 13), it was compiled from 1143 to 1157.

[2] BRETSCHNEIDER (Study and Value of Chinese Botanical Works, *Chinese Recorder*, Vol. III, 1871, p. 222) has given the name after the *Pen ts'ao kan mu*, but has left it without explanation.

[3] The last-named form occurs twice in the Bower Manuscript (HOERNLE's edition, pp. 32, 90, 121). In Hindustānī we have *axrōt* or *ākrōt*.

[4] F. SPIEGEL, Arische Periode, p. 40. The fact that the ancient Iranian name for the walnut is still unknown does not allow us to explain the Sanskrit word satisfactorily. Its relation to Hebrew *egōz*, and Persian *kōz, gōz* (see below), is perspicuous. Among the Hindu-Kush languages, we meet in Yidgha the word *oghūzoh* (J. BIDDULPH, Tribes of the Hindoo Koosh, Appendices, p. CLXVII), which appears as a missing link between Sanskrit on the one hand and the Semitic-Armenian forms on the other hand: hence we may conjecture that the ancient Iranian word was something like *agōza, angōza*; and this supposition is fully confirmed by the Chinese transcription *a-yüe* (above, p. 248). Large walnuts of India are mentioned by the traveller Č'an Te toward the middle of the thirteenth century (BRETSCHNEIDER, Mediæval Researches, Vol. I, p. 146). The walnuts of the province of Kusistan in Persia, which are much esteemed, are sent in great quantities to India (W. AINSLIE, Materia Indica, Vol. I, p. 464).

best kinds being styled in Greek *Persicum* and *basilicon*,[1] and these being the actual names by which they first became known in Italy.[2] Pliny himself employs the name *nuces iuglandes*. Although *Juglans regia* is indigenous to the Mediterranean region, the Greeks seem to have received better varieties from anterior Asia, hence Greek names like κάρυα περσικά or κάρυα σινωπικά.[3]

In fact, *Juglans regia* grows spontaneously in northern Persia and in Baluchistan; it has been found in the valleys of the Pskem and Ablatun at altitudes varying from 1000 to 1500 m. Another species (*Juglans pterocarpa*, "*Juglans* with winged fruits") is met in the provinces of Ghilan and Mazanderan and in the vicinity of Astrabad.[4] A. ENGLER[5] states that the walnut occurs wild also in eastern Afghanistan at altitudes of from 2200 to 2800 m. Ibn Haukal extols the walnuts of Arrajān, Muqaddasī those of Kirmān, and Istaxrī those of the province of Jīruft.[6]

In Fergana, Russian Turkistan, the walnut is cultivated in gardens; but the nuts offered for sale are usually derived from wild-growing trees which form complete forests in the mountains.[7] According to A. STEIN,[8] walnuts abound at Khotan. The same explorer found them at Yūl-arik and neighboring villages.[9]

---

[1] That is, "Persian nut" and "nut of the king," respectively, the king being the Basileus of Persia. These two designations are also given by Dioscorides (1, 178).

[2] Et has e Perside regibus translatas indicio sunt Graeca nomina: optimum quippe genus earum Persicum atque basilicon vocant, et haec fuere prima nomina (*Nat. hist.*, XV, 22, § 87).

[3] J. HOOPS, Waldbäume und Kulturpflanzen, p. 553. The Romans transplanted the walnut into Gallia and Germania during the first centuries of our era. Numerous walnuts have been brought to light from the wells of the Saalburg, testifying to the favor in which they were held by the Romans. The cultivation of the tree is commended in Charles the Great's *Capitulare de villis* and Garden Inventories. Its planting in Gaul is shown by the late Latin term *nux gallica*, Old French *nois gauge*, which survives in our "walnut" (German *walnuss*, Danish *valnöd*, Old Norse *valhnot*, Anglo-Saxon *wealh-hnutu*); *walh*, *wal*, was the Germanic designation of the Celts (derived from the Celtic tribe Volcae), subsequently transferred to the Romanic peoples of France and Italy.

[4] C. JORET, Plantes dans l'antiquité, Vol. II, p. 44. Joret (p. 92) states that the Persians cultivated nut-trees and consumed the nuts, both fresh and dried. The walnut is twice mentioned in the Būndahišn among the fruits serving as food, and among fruits the inside of which is fit to eat, but not the outside (WEST, Pahlavi Texts, Vol. I, pp. 101, 103; cf. also p. 275).

[5] Erläuterungen zu den Nutzpflanzen der gemässigten Zonen, p. 22.

[6] P. SCHWARZ, Iran im Mittelalter, pp. 114, 218, 241.

[7] S. KORŽINSKI, Sketches of the Flora of Turkistan, in Russian (*Memoirs Imp. Russ. Ac.*, 8th ser., Vol. IV, No. 4, pp. 39, 53).

[8] Ancient Khotan, Vol. I, p. 131.

[9] Ruins of Desert Cathay, Vol. I, p. 152.

The New-Persian name for the walnut is *kōz* and *gōz*.[1] According to HÜBSCHMANN, this word comes from Armenian.[2] The Armenian word is *ĕngoiz;* in the same category belongs Hebrew *egōz,*[3] Ossetic *ängozä,* Yidghal *oγuza,* Kurd *egvīz,* Gruzinian *nigozi.*[4] The Persian word we meet as a loan in Turkish *koz* and *xoz.*[5]

The earliest designation in Chinese for the cultivated walnut is *hu t'ao* 胡 桃 ("peach of the Hu": Hu being a general term for peoples of Central Asia, particularly Iranians). As is set forth in the Introduction, the term *hu* is prefixed to a large number of names of cultivated plants introduced from abroad. The later substitution *hu* or *ho t'ao* 核 桃 signifies "peach containing a kernel," or "seed-peach," so called because, while resembling a peach when in the husk, only the kernel is eaten.[6] In view of the wide dissemination of the Persian word, the question might be raised whether it would not be justifiable to recognize it also in the Chinese term *hu t'ao* 胡 桃, although, of course, in the first line it means "peach of the Hu (Iranians)." There are a number of cases on record where Chinese designations of foreign products may simultaneously convey a meaning and represent phonetic transcriptions. When we consider that the word *hu* 胡 was formerly possessed of an initial guttural sonant, being sounded *gu (γu) or *go,[7] the possibility that this word might have been chosen in imitation of, or with especial regard to, an Iranian form of the type *gōz,* cannot be denied: the twofold thought that this was the "peach styled *go*" and the "peach of the Go or Hu peoples" may have been present simultaneously in the minds of those who formed the novel term; but this is merely an hypothesis, which cannot actually be proved, and to which no great importance is to be attached.

---

[1] Arabic *jōz;* Middle Persian *joz, joj.* Kurd *gwīz (guwīz),* from *govz, gōz* (SOCIN, Grundr. iran. Phil., Vol. I, pt. 2, p. 268). Sariqolī *ghauz* (SHAW, *Journal As. Soc. Bengal,* 1876, p. 267). Puštu *ughz, waghz.* Another Persian designation for "walnut" is *girdū* or *girdgān.*

[2] Grundr. iran. Phil., Vol. I, pt. 2, p. 8; Armen. Gram., p. 393.

[3] Canticle VI, 10. Cf. Syriac *gauzā.*

[4] W. MILLER, Sprache der Osseten, p. 10; HÜBSCHMANN, Arm. Gram., p. 393.

[5] RADLOFF, Wörterbuch der Türk-Dialecte, Vol. II, col. 628, 1710. In Osmanli *jeviz.*

[6] The term *ho t'ao* is of recent date. It occurs neither under the T'ang nor under the Sung. It is employed in the *Kwo su* 果 疏, a work on garden-fruits by Waṅ Ši-mou 王 世 懋, who died in 1591, and in the *Pen ts'ao kaṅ mu.* The latter remarks that the word *ho* 核 is sounded in the north like *hu* 胡, and that the substitution thus took place, citing a work *Miṅ wu či* 名 物 志 as the first to apply this term.

[7] Compare Japanese *go-ma* 胡 麻 and *go-fun* 胡 粉.

There is a tradition to the effect that the walnut was introduced into China by General Čaṅ K'ien.[1] This attribution of the walnut to Čaṅ K'ien, however, is a purely retrospective thought, which is not contained in the contemporaneous documents of the Han Annals. There are, in fact, as we have seen, only two cultivated plants which can directly be credited to the mission of Čaṅ K'ien to the west,—the grape and the alfalfa. All others are ascribed to him in subsequent books. BRETSCHNEIDER, in his long enumeration of Čaṅ-K'ien plants,[2] has been somewhat uncritical in adopting the statements of such a recent work as the *Pen ts'ao kaṅ mu* without even taking pains to examine the sources there referred to. This subject requires a renewed critical investigation for each particular plant. As regards the walnut, Bretschneider was exposed to singular errors, which should be rectified, as they have passed into and still prominently figure in classical botanical and historical books of our time. According to Bretschneider, the walnut was brought from K'iang-hu 羌胡, and "K'iang" was at the time of the Han dynasty the name for Tibet. There is, of course, no such geographical name as "K'iaṅ-hu"; but we have here the two ethnical terms, "K'iaṅ" and "Hu," joined into a compound. Moreover, the K'iaṅ (anciently *Giaṅ) of the Han period, while they may be regarded as the forefathers of the subsequent Tibetan tribes, did not inhabit the country which we now designate as Tibet; and the term "Hu" as a rule does not include Tibetans. What is said in this respect in the *Pen ts'ao kaṅ mu*[3] is vague enough: it is a single sentence culled from the *T'u kiṅ pen ts'ao* 圖經本草 of Su Suṅ 蘇頌 (latter part of the eleventh century) of the Sung period, which reads, "The original habitat of this fruit was in the countries of the K'iaṅ and the Hu" (此果本出羌胡). Any conclusion like an introduction of the walnut from "Tibet" cannot be based on this statement.

Bretschneider's first victim was the father of the science of historical and geographical botany, A. DE CANDOLLE,[4] who stated, referring to him as his authority, "Chinese authors say that the walnut was introduced among them from Tibet, under the Han dynasty, by Chang-

---

[1] The first to reveal this tradition from the *Pen ts'ao kaṅ mu* was W. SCHOTT (*Abh. Berl. Akad.*, 1842, p. 270).

[2] *Chinese Recorder*, 1871, pp. 221–223; and Bot. Sin., pt. 1, p. 25. Likewise Hirth, *T'oung Pao*, Vol. VI, 1895, p. 439. Also GILES (Biographical Dictionary, p. 12) connects the walnut with Čaṅ K'ien.

[3] Ch. 30, p. 16.

[4] Origin of Cultivated Plants, p. 427.

kien, about the year 140–150 B.C."[1] In Hehn's "Kulturpflanzen"[2]
we still read in a postscript from the hand of the botanist A. ENGLER,
"Whether the walnut occurs wild in North China may be doubted, as
according to Bretschneider it is said to have been imported there from
Tibet." As will be seen below, a wild-growing species of *Juglans* is
indeed indigenous to North China. As to the alleged feat of Čaṅ K'ien,
the above-mentioned Su Suṅ, who lived during the Sung period in the
latter part of the eleventh century, represents the source of this purely
traditional opinion recorded by Bretschneider. Su Suṅ, after the above
statement, continues, "At the time of the Han, when Čaṅ K'ien was
sent on his mission into the Western Regions, he first obtained the
seeds of this fruit, which was then planted in Ts'in (Kan-su); at a later
date it gradually spread to the eastern parts of our country; hence it
was named *hu t'ao*."[3] Su Suṅ's information is principally based on the
Pen ts'ao of the Kia-yu period (1056–64) 嘉祐補註本草; this work
was preceded by the *Pen ts'ao* of the K'ai-pao period (968–976) 開寶
本草; and in the latter we meet the assertion that Čaṅ K'ien should
have brought the walnut along from the Western Regions, but cautiously
preceded by an *on dit* (云).[4] The oldest text to which I am able to trace
this tradition is the *Po wu či* 博物志 of Čaṅ Hwa 張華 (A.D. 232–300).[5]
The spurious character of this work is well known. The passage, at any
rate, existed, and was accepted in the Sung period, for it is reproduced
in the *T'ai p'iṅ yü lan*.[6] We even find it quoted in the Buddhist dictionary *Yi ts'ie kiṅ yin i* 一切經音義,[7] compiled by Yüan Yiṅ 元應
about A.D. 649, so that this tradition must have been credited in the

---

[1] Besides Bretschneider's article in the *Chinese Recorder*, de Candolle refers to
a letter of his of Aug. 23, 1881, which shows that Bretschneider had not changed
his view during that decade. Needless to add, that Čaṅ K'ien never was in Tibet,
and that Tibet as a political unit did not exist in his time. Two distinct traditions
are welded together in Bretschneider's statement.

[2] Eighth edition (1911), p. 400.

[3] *Čeṅ lei pen ts'ao*, Ch. 23, p. 45 (edition of 1521). G. A. STUART (Chinese
Materia Medica, p. 223) regards the "Tangut country about the Kukunor" as the
locality of the tree pointed out in the *Pen ts'ao*.

[4] The text of the *K'ai-pao pen ts'ao* is not reproduced in the *Pen ts'ao kaṅ mu,*
but will be found in the *Či wu miṅ ši t'u k'ao*, Ch. 17, p. 33. T'aṅ Šen-wei 唐慎微,
in his *Čeṅ lei pen ts'ao* (Ch. 23, p. 44 b), has reproduced the same text in his own
name.

[5] 張騫使西域還乃 (or 返) 得胡桃種 (Ch. 6, p. 4, of the Wu-č'aṅ
print).

[6] Ch. 971, p. 8.

[7] Ch. 6, p. 8 b (ed. of Nanking). In this text the pomegranate and grape are
added to the walnut. In the same form, the text of the *Po wu či* is cited in the modern
editions of the *Ts'i min yao šu* (Ch. 10, p. 4).

beginning of the T'ang dynasty. It is not impossible, however, that this text was actually written by Čan Hwa himself, or at least that the tradition underlying it was formed during the fourth century; for, as will be seen, it is at that time that the walnut is first placed on record. Surely this legend is not older than that period, and this means that it sprang into existence five centuries after Čan K'ien's lifetime. It should be called to mind that the *Po wu či* entertains rather fantastic notions of this hero, and permits him to cross the Western Sea and even to reach Ta Ts'in.[1] It is, moreover, the *Po wu či* which also credits to Čan K'ien the introduction of the pomegranate and of *ta* or *hu swan* 大 (胡) 蒜 or *hu* 葫 (*Allium scorodoprasum*).[2] Neither is this tradition contained in the texts of the Han period. The notion that Čan K'ien really introduced the walnut in the second century B.C. must be positively rejected as being merely based on a retrospective and unauthentic account.[3]

The question now arises, Is there any truth in Su Sun's allegation that the walnut was originally produced in the country of the K'ian? Or, in other words, are we entitled to assume the co-existence of two Chinese traditions,— first, that the walnut was introduced into China from the regions of the Hu (Iranians); and, second, that another introduction took place from the land of the K'ian, the forefathers of the Tibetans?[4] There is indeed an ancient text of the Tsin period from the first part of the fourth century, one of the earliest datable references to the walnut, in which its origin from the K'ian is formally admitted. This text is preserved in the *T'ai p'in yü lan* as follows:—

"The mother of Liu T'ao 劉沼,[5] in her reply to the letter of Yü 虞, princess of the country of Wu 吳國, said, 'In the period Hien-ho 咸和 (A.D. 326-335, of the Tsin dynasty) I escaped from the rebellion

---

[1] Ch. 1, p. 3 b.

[2] See below, p. 302.

[3] The Čan-K'ien legend is also known in Korea (*Korea Review*, Vol. II, 1902, p. 393).

[4] The term *k'ian t'ao* 羌 桃 for the walnut is given, for instance, in the *Hwa kin* 花 鏡, "Mirror of Flowers" (Ch. 3, p. 49), written by Č'en Hao-tse 陳 淏 子 in 1688. He gives as synonyme also *wan swi tse* 萬 歲 子 ("fruits of ten thousand years"). The term *k'ian t'ao* is cited also in the *P'ei wen čai kwan k'ün fan p'u* (Ch. 58, p. 24; regarding this work cf. BRETSCHNEIDER, Bot. Sin., pt. 1, p. 70), and in the *P'an šan či* 盤 山 志 (Ch.15, p. 2 b; published in 1755 by order of K'ien-lun).

[5] The *T'u šu tsi č'en* and *Kwan k'ün fan p'u* (Ch. 58, p. 25) write this name Niu 紐. The *Ko či kin yüan* (Ch. 76, p. 5), which ascribes this text to the *Tsin šu*, gives it as 鈕. The *T'an Sun pai k'un leu t'ie* 唐 宋 白 孔 六 帖 (Ch. 99, p. 12) has, "The mother of Liu T'ao of the Tsin dynasty said, in reply to a state document, 'walnuts were originally grown in the country of the Western K'ian.'"

of Su Tsûn 蘇峻[1] into the Lin-ńan mountains 臨安山. The country
of Wu sent a messenger with provisions, stating in the accompanying
letter: 'These fruits are walnuts 胡桃 and *fei-žaṅ* 飛穄.[2] The latter
come from southern China. The walnuts were originally grown abroad
among the Western K'iaṅ (胡桃本生西羌外圏). Their exterior is hard,
while the interior is soft and sweet. Owing to their durability I wish to
present them to you as a gift.' "[3] It is worthy of note, that, while the
walnut is said in this text to hail from the Western K'iaṅ, the term
hu t'ao (not *k'iaṅ t'ao*) is employed; so that we may infer that the introduction of the fruit from the Hu preceded in time the introduction
from the K'iaṅ. It is manifest also that in this narrative the walnut
appears as a novelty.

The Tibetan name of the walnut in general corresponds to a type
tar-ka, as pronounced in Central Tibetan, written *star-ka*, *star-ga*,
and *dar-sga*.[4] The last-named spelling is given in the Polyglot Dictionary of K'ien-luṅ,[5] also in Jäschke's Tibetan Dictionary. The element
ka or *ga* is not the well-known suffix used in connection with nouns,[6]
but is an independent base with the meaning "walnut," as evidenced
by Kanaurī *kā* ("walnut").[7] The various modes of writing lead to a
restitution *tar*, *dar*, *d'ar* (with aspirate sonant). This word is found
also in an Iranian dialect of the Pāmir: in Waxī the walnut is called

---

[1] He died in A.D. 328. His biography is in the *Tsin šu*, Ch. 100, p. 9. See also
L. WIEGER, Textes historiques, p. 1086.

[2] Literally, "flying stalk of grain." Bretschneider and Stuart do not mention
this plant. Dr. T. Tanaka, assistant in the Bureau of Plant Industry, Department
of Agriculture, Washington, tells me that *fei-žaṅ* is a synonyme of the fingered citrus
(*fu šou kan* 佛手柑, *Citrus chirocarpus*). He found this statement in the *Honzō
kōmoku keimō* (Ch. 26, p. 18, ed. 1847) by Ono Ranzan, who on his part quotes the
T'uṅ ya 通雅 by Faṅ I-či.

[3] The *T'ai p'iṅ yü lan* reads 質以堅欲以奉貢. The *T'aṅ Suṅ pai k'uṅ
leu t'ie* and the *T'u šu tsi čeṅ*, however, have 質似古賢欲以奉貢, "their
substance resembles the ancient sages, and I wish to present them,"—apparently a
corruption of the text.

[4] W. W. ROCKHILL (Diary of a Journey through Mongolia and Tibet, p. 340)
gives *taga* as pronunciation in eastern Tibet. J. D. HOOKER (Himalayan Journals,
p. 237) offers *taga-šiṅ* (*šiṅ*, "tree") as Bhutia name.

[5] Ch. 28, p. 55.

[6] SCHIEFNER, *Mélanges asiatiques*, Vol. I, pp. 380–382.

[7] Given both by T. R. JOSHI (Grammar and Dictionary of the Kanāwari Language, p. 80) and T. G. BAILEY (Kanauri-English Vocabulary, *Journal Royal As.
Soc.*, 1911, p. 332). Bailey adds to the word also the botanical term *Juglans regia*.
The same author, further, gives a word *gē* as meaning "kernel of walnut; edible part
of *Pinus gerardiana*"; while Joshi (p. 67) explains the same word as the "wild
chestnut." Thus it seems that *ge*, *ka*, originally referred to an indigenous wild-growing fruit, and subsequently was transferred to the cultivated walnut.

tar.[1] This apparently is a loan-word received from the Tibetan, for in Sariqolī and other Pāmir dialects we find the Iranian word *ghōz*.[2] *Tarka* is a genuine Tibetan word relating to the indigenous walnut, wild and cultivated, of Tibetan regions. In view of this state of affairs, it is certainly possible that the Chinese, in the beginning of the fourth century or somewhat earlier, received walnuts and their seeds also from Tibetan tribes, which resulted in the name K'iaṅ t'ao. The Lepcha of Sikkim are acquainted with the walnut, for which they have an indigenous term, *kŏl-pŏt*, and one of their villages is even called "Walnut-Tree Foundation" (Kŏl-baṅ).[3]

G. WATT[4] informs us that the walnut-tree occurs wild and cultivated in the temperate Himalaya and Western Tibet, from Kashmir and Nubra eastwards. W. ROXBURGH[5] says about *Juglans regia*, "A native of the mountainous countries immediately to the north and north-east of Hindustan, on the plains of Bengal it grows pretty well, but is not fruitful there." Another species of the same genus, *J. plerococca* Roxb., is indigenous in the vast forests which cover the hills to the north and east of the province of Silhet, the bark being employed for tanning, while *J. regia* is enlisted among the oil-yielding products.[6] J. D. HOOKER[7] is authority for the information that the walnut occurs wild in Sikkim, and is cultivated in Bhūtān, where also Captain TURNER[8] found it growing in abundance. KIRKPATRICK[9] met it in Nepal. In Burma it grows in the Ava Hills. In the Shan states east of Ava grows another species of *Juglans*, with smaller, almost globose, quite smooth nuts, but nothing is known about the tree itself.[10]

The Tibetans certainly cultivate the walnut and appreciate it

---

[1] R. B. SHAW, On the Ghalchah Languages (*Journal As. Soc. Bengal*, 1876, p. 267), writes the word *tor*. A. HUJLER (The Languages Spoken in the Western Pamir, p. 36, Copenhagen, 1912) writes *tar*, explaining the letter *a* as a "dark deep *a*, as in the French *pas*."

[2] W. TOMASCHEK (Pamirdialekte, p. 790) has expressed the opinion that Waxī *tor*, as he writes, is hardly related to Tibetan *star-ga*; this is not correct.

[3] G. MAINWARING, Dictionary of the Lepcha Language, p. 30.

[4] Dictionary of the Economic Products of India, Vol. IV, p. 550.

[5] Flora Indica, p. 670.

[6] N. G. MUKERJI, Handbook of Indian Agriculture, p. 233.

[7] Himalayan Journals, p. 235; also RISLEY, Gazetteer of Sikkim, p. 92 (compare DARWIN, Variation of Animals and Plants under Domestication, Vol. I, p. 445).

[8] Account of an Embassy to the Court of the Teshoo Lama, p. 273. Also EDEN and PEMBERTON (Political Missions to Bootan, p. 198, Calcutta, 1895) mention the walnut in Bhūtān.

[9] Account of Nepaul, p. 81.

[10] S. KURZ, Forest Flora of British Burma, Vol. II, p. 490 (Calcutta, 1877).

much. The tree is found everywhere in eastern Tibet where horticulture is possible, and among the Tibetan tribes settled on the soil
of Se-č'wan Province. W. W. ROCKHILL[1] even mentions that in the
Ba-t'aṅ region barley and walnuts are used in lieu of subsidiary coinage.
Lieut.-Col. WADDELL[2] makes two references to cultivated walnut-trees
in Central Tibet. The Chinese authors mention "Tibetan walnuts"
as products of the Lhasa district.[3]

While the Čaṅ-K'ien tradition is devoid of historical value, and
must be discarded as an historical fact, yet it is interesting from a
psychological point of view; for it shows at least that, at the time when
this fiction sprang into existence, the Chinese were under the impression
that the walnut was not an indigenous tree, but imported from abroad.
An autochthonous plant could not have been made the object of such a
legend. A direct reference to the introduction of the cultivated walnut
with an exact date is not extant in Chinese records, but the fact of such
an introduction cannot reasonably be called into doubt. It is supported
not only by the terms *hu t'ao* and *k'iaṅ t'ao* ("peach of the Hu," "peach
of the K'iaṅ"), but also by the circumstantial evidence that in times
of antiquity, and even under the Han, no mention is made of the
walnut. True it is, it is mentioned in the *Kin kwei yao lio* of the second
century; but, as stated, this may be an interpolation.[4] Of all the data
relating to this fruit, there is only one that may have a faint chance to
be referred to the Han period, but even this possibility is very slight.
In the *Si kiṅ tsa ki* 西京雜記[5] it is said that in the gardens of the
Šaṅ-lin Park 上林苑 of the Han emperors there were walnuts which
had come from the Western Regions or Central Asia. The *Si kiṅ tsa ki*,
however, is the work of Wu Kün 吳均, who lived in the sixth century
A.D.,[6] and cannot be regarded as a pure source for tracing the culture
of the Han. It is not difficult to see how this tradition arose. When the
Šaṅ-lin Park was established, the high dignitaries of the empire were
called upon to contribute famed fruits and extraordinary trees of distant
lands. We know that after the conquest of Nan-yüe in 111 B.C. the
Emperor Wu ordered southern products, like oranges, areca-nuts,

---

[1] Diary of a Journey through Mongolia and Tibet, p. 347.

[2] Lhasa and its Mysteries, pp. 307, 315. See also N. V. KÜNER, Description of
Tibet (in Russian), Vol. I, pt. 2, p. 137.

[3] ROCKHILL, *Journal Royal As. Soc.*, 1891, p. 273.

[4] Above, p. 205. Čaṅ Ki says or is made to say, "Walnuts must not be eaten in
large quantity, for they rouse mucus and cause man to drink" (Ch. c, p. 27).

[5] Ch. I, p. 6 (ed. of *Han Wei ts'uṅ šu*).

[6] WYLIE, Notes on Chinese Literature, p. 189; and CHAVANNES, *T'oung Pao*,
1906, p. 102.

luṅ ṅan, li-či, etc., to be brought to the capital Č'aṅ-ṅan, and to be planted in the Fu-li Palace 扶荔宮, founded in commemoration of the conquest of Nan-yūe, whereupon many gardeners lost their lives when the crops of the *li-či* proved a failure.[1] Several of his palaces were named for the fruits cultivated around them: thus there were a Grape-Palace and a Pear-Palace. Hence the thought that in this exposition of foreign fruits the walnut should not be wanting, easily impressed itself on the mind of a subsequent writer. Wu Kün may also have had knowledge of the Čaṅ-K'ien tradition of the *Po wu či*, and thus believed himself consistent in ascribing walnuts to the Han palaces. Despite his anachronism, it is interesting to note Wu Kün's opinion that the walnut came from Central Asia or Turkistan.

It is not probable that the walnut was generally known in China earlier than the fourth century A.D., under the Eastern Tsin 東晉 dynasty (265–419).[2] In the *Tsin kuṅ ko miṅ* 晉宮閣名, a description of the palaces of the Tsin emperors, written during that dynasty,[3] it is stated that there were eighty-four walnut-trees in the Hwa-lin Park

---

[1] The palace Fu-li was named for the *li-či* 荔枝 (see *San fu hwaṅ t'u* 三輔黃圖, Ch. 3, p. 9 b, ed. of *Han Wei ts'uṅ šu*).

[2] BRETSCHNEIDER (Bot. Sin., pt. 1, p. 39) asserts that *Juglans regia* figures among the plants mentioned passingly in the *Nan faṅ ts'ao mu čwaṅ* by Ki Han 稽含, a minister of state under the Emperor Hui 惠 of the Tsin dynasty (A.D. 290–306). He does not give any particulars. There are only two allusions to the walnut, that I am able to trace in this work: in the description of the coco-nut, the taste of this fruit is likened to that of the walnut; and the flavor of the "stone chestnut" (*ši-li* 石栗, *Aleurites triloba*) is compared with that of the same fruit. We know at present that the book in question contains interpolations of later date (see L. AUROUSSEAU, *Bull. de l'Ecole française*, Vol. XIV, 1914, p. 10); but to these the incidental mention of the walnut does not necessarily belong, as Ki Han lived under the Tsin. It is likewise of interest that the walnut is not dealt with as a special item in the *Ts'i min yao šu*, a work on husbandry and economic botany, written by Kia Se-niu 賈思勰 of the Hou Wei dynasty (A.D. 386–534); see the enumeration of plants described in this book in BRETSCHNEIDER (*op. cit.*, p. 78). In this case, the omission does not mean that the tree was unknown to the author, but it means only that it had then not attained any large economic importance. It had reached the palace-gardens, but not the people. In fact, Kia Se-niu, at least in one passage (Ch. 10, p. 48 b, ed. 1896), incidentally mentions the walnut in a quotation from the *Kiao čou ki* 交州記 by Liu Hin-k'i 劉欣期, where it is said, "The white *yüan* tree 白緣樹 [evidently = 櫞] is ten feet high, its fruits being sweeter and finer than walnuts 胡桃." As the *Kiao čou ki* is a work relating to the products of Annam, it is curious, of course, that it should allude to the cultivated walnut, which is almost absent in southern China and Annam; thus it is possible that this clause may be an interpolation, but possibly it is not. The fact that the same work likewise contains the tradition connecting the walnut with Čaṅ K'ien has been pointed out above. The tree *pai yüan* is mentioned again in the *Pen ts'ao kaṅ mu ši i* (Ch. 8, p. 23), where elaborate rules for the medicinal employment of the fruit are given.

[3] BRETSCHNEIDER, Bot. Sin., pt. 1, p. 202, No. 945.

華林園.[1] Another allusion to the walnut relative to the period Hien-ho (A.D. 326–335) has been noted above (p. 259). There is, further, a reference to the fruit in the history of Šu 蜀 , when, after the death of Li Hiuṅ 李雄 in A.D. 334, Han Pao 韓豹 from Fu-fuṅ 扶風 in Šen-si was appointed Grand Tutor (*t'ai fu* 太傅) of his son Li K'i 李期, and asked the latter to grant him seeds for the planting of walnut-trees, which, on account of his advanced age, he was anxious to have in his garden.[2]

During the third or fourth century, the Chinese knew also that walnuts grew in the Hellenistic Orient. "In Ta Ts'in there are jujubes, jasmine, and walnuts," it is stated in the *Wu ši wai kwo či* 吳時外國志 ("Memoirs of Foreign Countries at the time of the Wu").[3]

The *Kwaṅ či* 廣志 by Kwo Yi-kuṅ 郭義恭[4] contains the following account: "The walnuts of Č'en-ts'aṅ 陳倉[5] have a thin shell and a large kernel; those of Yin-p'iṅ 陰平[6] are large, but their shells are brittle, and, when quickly pinched, will break."[7]

Coming to the T'ang period, we encounter a description of the walnut in the *Yu yaṅ tsa tsu* 酉陽雜俎, written about A.D. 860,[8] from which the fact may be gleaned that the fruit was then much cultivated

---

[1] *T'ai p'iṅ yü lan, l.c.*

[2] This story is contained in the *Kwaṅ wu hiṅ ki* 廣五行記 (according to BRETSCHNEIDER, a work of the Sung literature). As the text is embodied in the *T'ai p'iṅ yü lan*, it must have been extant prior to A.D. 983, the date of Li Faṅ's cyclopædia.

[3] Presumably identical with the *Wu ši wai kwo čwan* noted by PELLIOT (*Bull. de l'Ecole française*, Vol. IV, p. 270) as containing information secured by the mission of K'aṅ T'ai in the first part of the third century A.D. Cf. also *Journal asiatique*, 1918, II, p. 24. The *Miṅ ši* ascribes walnuts to Ormuz (BRETSCHNEIDER, Notices of the Mediæval Geography, p. 294).

[4] This work is anterior to the year A.D. 527, as it is cited in the *Šwi kiṅ ču* of Li Tao-yüan, who died in that year. Kwo Yi-kuṅ is supposed to have lived under the Tsin (A.D. 265–419). Cf. PELLIOT, *Bull. de l'Ecole française*, Vol. IV, p. 412.

[5] Now the district of Pao-ki in the prefecture of Fuṅ-siaṅ, Šen-si Province.

[6] At the time of the Han period, Yin-p'iṅ was the name for the present prefecture of Luṅ-ṅan 龍安 in the province of Se-č'wan. There was also a locality of the same name in the prefecture of Kiai in the province of Kan-su, inhabited by the Ti, a Tibetan tribe (CHAVANNES, *T'oung Pao*, 1905, p. 525).

[7] *T'ai p'iṅ yü lan, l. c.; Ko či kiṅ yüan,* Ch. 76, p. 5; *Či wu miṅ ši t'u k'ao, l. c.* This text is cited also by Su Suṅ in his *T'u kiṅ pen ts'ao.* The earliest quotation that I can trace of it occurs in the *Pei hu lu*, written by Twan Kuṅ-lu about A.D. 875 (Ch. 3, p. 4 b, ed. of Lu Sin-yüan), where, however, only the last clause in regard to the walnuts of Yin-p'iṅ is given (see below, p. 268).

[8] PELLIOT, *T'oung Pao*, 1912, p. 375. The text is in the *T'u šu tsi č'eṅ* and *Či wu miṅ ši t'u k'ao* (*l. c.*). I cannot trace it in the edition of the *Yu yaṅ tsa tsu* in the *Tsin tai pi šu* or *Pai hai.*

in the northern part of China (北方多種之),— a statement repeated in the K'ai-pao pen ts'ao. The Yu yan tsa tsu, which is well informed on the cultivated plants of Western and Central Asia, does not contain the tradition relating to Čan K'ien, but, on the other hand, does not speak of the tree as a novel introduction, nor does it explain its name. It begins by saying that "the kernel of the walnut is styled 'toad' ha-mo 蝦蟇."[1]

Mon Šen 孟詵, who in the second half of the seventh century wrote the Ši liao pen ts'ao,[2] warns people from excessive indulgence in walnuts as being injurious to health.[3] The T'ai p'in hwan yü ki 太平寰宇記, by Yo Ši 樂史 (published during the period T'ai-p'in, A.D. 976–981), mentions the walnut as being cultivated in the prefecture of Fun-sian 鳳翔 in Šen-si Province, and in Kian čou 絳州 in Šan-si Province.[4]

According to the Pen ts'ao kan mu, the term hu t'ao first appears in the Pen ts'ao of the K'ai-pao period (968–976) of the Sung dynasty, written by Ma Či 馬志; that is to say, the plant or its fruit was then officially sanctioned and received into the pharmacopœia for the first time. We have seen that it was certainly known prior to that date. K'ou Tsun-ši 寇宗奭, in his Pen ts'ao yen i 本草衍義 of 1116,[5] has a notice on the medicinal application of the fruit.

It is possible also to trace in general the route which the walnut has taken in its migration into China. It entered from Turkistan into Kan-su Province, as stated by Su Sun (see above, p. 258), and gradually spread first into Šen-si, and thence into the eastern provinces, but always remained restricted to the northern part of the country. Su Sun expressly says that walnuts do not occur in the south, but only in the north, being plentiful in Šen-si and Lo-yan (Ho-nan Province), while those grown in K'ai-fun (Pien čou 卞州) were not of good quality. In the south only a wild-growing variety was known, which is discussed below. Wan Ši-mou 王世懋, a native of Kian-su, who died in 1591, states in his Kwo su 果疏, a treatise on garden-fruits, that "the walnut is a northern fruit (pei kwo 北果), and thrives in mountains; that it is but rarely planted in the south, yet can be cultivated there."[6] Almost

---

[1] This definition is ascribed to the Ts'ao mu tse 草木子 in the Ko či kin yüan (Ch. 76, p. 5); that work was written by Ye Tse-k'i 葉子奇 in 1378 (WYLIE, Notes on Chinese Literature, p. 168).

[2] BRETSCHNEIDER, Bot. Sin., pt. I, p. 45.

[3] T'an Sun pai k'un leu t'ie, Ch. 99, p. 12.

[4] T'ai p'in hwan yü ki, Ch. 30, p. 4; Ch. 47, p. 4 (ed. of Kin-lin šu kü, 1882).

[5] Ch. 18, p. 6 b (ed. of Lu Sin-yüan).

[6] Also J. DE LOUREIRO (Flora cochinchinensis, p. 702) states that the habitat of Juglans regia is only in the northern provinces of China.

all the district and prefectural gazetteers of Šen-si Province enumerate the walnut in the lists of products. The "Gazetteer of Šan-tuṅ"[1] mentions walnuts for the prefectures of Ts'i-nan, Yen-čou, and Ts'in-čou, the last-named being the best. The Gazetteer of the District of Tuṅ-ṅo 東阿[2] in the prefecture of Tai-ṅan in Šan-tuṅ reports an abundance of walnuts in the river-valleys. An allusion to oil-production from walnuts is found in the "Gazetteer of Lu-nan," where it is said, "Of all the fruits growing in abundance, there is none comparable to the walnut. What is left on the markets is sufficient to supply the needs for lamp-oil."[3] Also under the heading "oil," walnut-oil is mentioned as a product of this district.[4]

Juglans regia, in its cultivated state, has been traced by our botanists in Šan-tuṅ, Kiaṅ-su, Hu-pei, Yün-nan, and Se-č'wan.[5] Wilson nowhere saw trees that could be declared spontaneous, and considers it highly improbable that *Juglans regia* is indigenous to China. His opinion is certainly upheld by the results of historical research.

A wild species (*Juglans mandshurica* or *cathayensis* Dode) occurs in Manchuria and the Amur region, Či-li, Hu-pei, Se-č'wan, and Yün-nan.[6] This species is a characteristic tree of the Amur and Usuri valleys.[7] It is known to the Golde under the name *kočoa* or *košoa*, to the Manāgir as *korčo*, to the Gilyak as *tiv-alys*. The Golde word is of ancient date, for we meet it in the ancient language of the Jurči, Jučen, or Niüči in the form *xušu*[8] and in Manchu as *xôsixa*. The great antiquity of this word is pointed out by the allied Mongol word *xusiga*. The whole series originally applies to the wild and indigenous species,

---

[1] *Šan tuṅ t'uṅ či*, Ch. 9, p. 15.

[2] Ch. 2, p. 32 (1829).

[3] Quotation from *Lu-nan či* 潞南志, in the *Šaṅ čou tsuṅ či* 商州總志 (General Gazetteer of Šaṅ-čou), 1744, Ch. 8, p. 3.

[4] *Ibid.*, Ch. 8, p. 9. Oil was formerly obtained from walnuts in France both for use at table and for varnishing and burning in lamps, also as a medicine supposed to possess vermifuge properties (AINSLIE, Materia Indica, Vol. I, p. 464).

[5] See particularly C. S. SARGENT, Plantae Wilsonianae, Vol. III, pp. 184–185 (1916). J. ANDERSON (Report on the Expedition to Western Yunan, p. 93, Calcutta, 1871) mentions walnuts as product of Yün-nan. According to the *Tien hai yü heṅ či* (Ch. 10, p. 1 b; above, p. 228), the best walnuts with thin shells grow on the Yaṅ-pi or Yaṅ-p'ei River 漾濞江 of Yün-nan.

[6] FORBES and HEMSLEY, *Journal of the Linnean Society*, Botany, Vol. XXVI, p. 493; SARGENT, *op. cit.*, pp. 185 *et seq.* J. DE LOUREIRO (Flora cochinchinensis, p. 702), writing in 1788, has a species *Juglans camirium* (Annamese *deầu lai*) "habitat agrestis cultaque in Cochinchina;" and a *Juglans catappa* (Annamese *cây mo cua*) "habitat in sylvis Cochinchinae montanis."

[7] GRUM-GRŽIMAILO, Description of the Amur Province (in Russian), p. 313.

[8] W. GRUBE, Schrift und Sprache der Jučen, p. 93.

Juglans mandshurica. Manchu *xôsixa* designates the tree, while its fruit is called *xôwalama* or *xôwalame usixa* (-*ixa* being a frequent termination in the names of plants and fruits). The cultivated walnut is styled *mase*.[1] One of the earliest explorers of the Amur territory, the Cossack chieftain Poyarkov, who reached the Amur in 1644, reported that walnuts and hazel-nuts were cultivated by the Daur or Dahur on the Dseya and Amur.[2]

The same species is known to the aboriginal tribes of Yün-nan. The Pa-yi and Šan style its fruit *twai*;[3] the Nyi Lo-lo, *se-mi-ma*; the Ahi Lo-lo, *sa-mi.* The Čuṅ-kia of Kwei-čou call it *dsao*; the Ya-č'io Miao, *či* or *ši*; the Hwa Miao, *klaeo*; while other Miao tribes have the Chinese loan-word *he-dao.*[4]

The wild walnut has not remained unknown to the Chinese, and it is curious that it is designated *šan hu t'ao* 山 胡 桃, the term *šan* ("mountain") referring to wild-growing plants. The "wild Iranian peach" is a sort of linguistic anomaly. It is demonstrated by this term that the wild indigenous species was discovered and named by the Chinese only in times posterior to the introduction of the cultivated variety; and that the latter, being introduced from abroad, was not derived from the wild-growing species. The case is identical with that of the wild alfalfas and vines. Č'en Hao-tse, who wrote a treatise on flowers in 1688,[5] determines the difference between the cultivated and wild varieties thus: the former has a thin shell, abundant meat, and is easy to break;[6] the latter has a thick and hard shell, which must be cracked with a hammer, and occurs in Yen and Ts'i (Či-li and Šan-tuṅ). This observa-

---

[1] K'ien-luṅ's Polyglot Dictionary, Ch. 28, p. 55.

[2] L. v. SCHRENCK, Reisen und Forschungen im Amur-Lande, Vol. III, p. 160.

[3] F. W. K. MÜLLER, *T'oung Pao*, Vol. III, 1892, p. 26.

[4] S. R. CLARKE, Tribes in South-West China, p. 312.

[5] *Hwa kiṅ*, Ch. 3, p. 49 b.

[6] According to the *Či wu miṅ ši t'u k'ao* (Ch. 31, p. 3 b), the walnuts with thin shells grow only in the prefecture of Yuṅ-p'iṅ 永 平 in Či-li, being styled *lu žaṅ ho t'ao* 露 穰 核 桃 In Č'aṅ-li, which belongs to this prefecture, these nuts have been observed by F. N. MEYER (Agricultural Explorations in the Orchards of China, p. 51), who states, "Some trees produce small hard-shelled nuts of poor flavor, while others bear fine large nuts, with a really fine flavor, and having shells so thin that they can be cracked with the fingers like peanuts. Between these extremes one finds many gradations in hardness of shell, size, and flavor." "In England the walnut presents considerable differences, in the shape of the fruit, in the thickness of the husk, and in the thinness of the shell; this latter quality has given rise to a variety called the thin-shelled, which is valuable, but suffers from the attacks of titmice" (DARWIN, Variation of Animals and Plants under Domestication, Vol. I, p. 445). A variety of walnut with thin shells grows on the Greek Island Paros (T. v. HELDREICH, Nutzpflanzen Griechenlands, p. 59).

tion is quite to the point; the shell of the walnut gradually became more refined under the influence of cultivation.

The earliest texts alluding to the wild walnut are not older than the T'ang period. The *Pei hu lu* 北戶錄, written by Twan Kuṅ-lu 段公路 about A.D. 875,[1] contains the following text concerning a wild walnut growing in the mountains of southern China:—

"The wild walnut has a thick shell and a flat bottom 底平. In appearance it resembles the areca-nut. As to size, it is as large as a bundle of betel-leaves.[2] As to taste, it comes near the walnuts of Yin-p'iṅ[3] and Lo-yu, but is different from these, inasmuch as it has a fragrance like apricot extract. This fragrance, however, does not last long, but will soon vanish. The *Kwaṅ ǒi* says that the walnuts of Yin-p'iṅ have brittle shells, and that, when quickly pinched, the back of the kernel will break. Liu Ši-luṅ 柳世隆, in his *Sie lo yu yüan* 詩樂遊苑, remarks, with reference to the term *hu t'ao*, that the Hu take to flight like rams,[4] and that walnuts therefore are prophets of auspicious omens. Čeṅ K'ien 鄭虔[5] says that the wild walnut has no glumelle; it can be made into a seal by grinding off the nut for this purpose. Judging from these data, it may be stated that this is not the walnut occurring in the mountains of the south."[6]

The *Liṅ piao lu i* 嶺表錄異, by Liu Sün 劉恂 of the T'ang period,[7] who lived under the reign of the Emperor Čao Tsuṅ (A.D. 889–904), contains the following information on a wild walnut:—

"The slanting or glandular walnut (*p'ien ho t'ao* 偏核桃) is produced in the country Čan-pi 占畢.[8] Its kernel cannot be eaten. The

---

[1] Cf. PELLIOT, *Bull. de l'Ecole française*, Vol. IX, p. 223.

[2] *Fu-liu*, usually written 扶留, is first mentioned in the *Wu lu ti li ǒi* 吳錄地理志 by Čaṅ Pu 張勃 of the third or beginning of the fourth century (see *Ts'i min yao šu*, Ch. 10, p. 32). It refers to *Piper betle* (BRETSCHNEIDER, *Chinese Recorder*, Vol. III, 1871, p. 264; C. IMBAULT-HUART, Le bétel, *T'oung Pao*, Vol. V, 1894, p. 313). The Chinese name is a transcription corresponding to Old Annamese *blâu;* Mĩ-sŏn, Uy-lô, and Hung *plu;* Khmer *m-luw*, Stieṅ *m-lu*, Bahnar *bö-lou*, Kha *b-lu* ("betel").

[3] See above, p. 264.

[4] A jocular interpretation by punning *t'ao* 桃 upon *t'ao* 逃 (both in the same tone).

[5] Author of the lost *Hu pen ts'ao* 胡本草 (BRETSCHNEIDER, Bot. Sin., pt. I, p. 45). He appears to have been the first who drew attention to the wild walnut. His work is repeatedly quoted in the *Pei hu lu*.

[6] *Pei hu lu*, Ch. 3, p. 4 b (ed. of Lu Sin-yüan).

[7] Ch. B, p. 5 (ed. of *Wu yin tien*).

[8] The two characters are wrongly inverted in the text of the work. In the text of the *Pei hu lu* that follows, the name of this country is given in the form Čan-pei 占卑. From the mention of the Malayan Po-se in the same text, it follows that

Hu 胡 people gather these nuts in abundance, and send them to the Chinese officials, designating them as curiosities 珍異. As to their shape, they are thin and pointed; the head is slanting like a sparrow's beak. If broken and eaten, the kernel has a bitter taste resembling that of the pine-seeds of Sin-ra 新羅松子.[1] Being hot by nature, they are employed as medicine, and do not differ from the kernels of northern China."

The *Pei hu lu*[2] likewise mentions the same variety of glandular walnut (*p'ien ho-t'ao*) as growing in the country Čan-pei 占卑, shaped like the crescent of the moon, gathered and eaten by the Po-se,[3] having a very fine fragrance, stronger than the peach-kernels of China, but of the same effect in the healing of disease.

The species here described may be identical with *Juglans cathayensis*, called the Chinese butternut, usually a bush, but in moist woods forming a tree from twelve to fifteen metres tall; but I do not know that this plant occurs in any Malayan region. With reference to Čan-pi, however, it may be identical with the fruit of *Canarium commune* (family *Burseraceae*), called in Malayan *kanari*, in Javanese *kenari*. J. CRAWFURD,[4] who was not yet able to identify this tree, offers the following remarks: "Of all the productions of the Archipelago the one which yields the finest edible oil is the *kanari*. This is a large handsome tree, which yields a nut of an oblong shape nearly of the size of a walnut. The kernel is as delicate as that of a filbert, and abounds in oil. This

Čan-pi is a Malayan territory probably to be located on Sumatra. For this reason I am inclined to think that Čan-pi 占畢 is identical with Čan-pei 詹卑; that is, Jambi, the capital of eastern Sumatra (HIRTH and ROCKHILL, Chau Ju-kua, pp. 65, 66; see further GROENEVELDT, Notes on the Malay Archipelago, pp. 188, 196; and GERINI, Researches on Ptolemy's Geography, p. 565; *Liň wai tai ta*, Ch. 2, p. 12). From a phonetic point of view, however, the transcription 占畢, made in the T'ang period, represents the ancient sounds *čan-pit, and would presuppose an original of the form *čambit, čambir, or jambir, whereas 卑 is without a final consonant. The country Čan-pei is first mentioned under the year A.D. 852 (大中 sixth year), when Wu-sie-ho 勿邪葛 and six men from there came to the Chinese Court with a tribute of local products (*T'ai p'iň hwan yü ki*, Ch. 177, p. 15 b). A second embassy is on record in 871 (PELLIOT, *Bull. de l'Ecole française*, Vol. IV, p. 347).

[1] *Pinus koraiensis* Sieb. et Zucc. (J. MATSUMURA, Shokubutsu mei-i, pp. 266–267, ed. 1915), in Japanese *čōsen-matsu* ("Korean pine"); see also STUART, Chinese Materia Medica, p. 333. Sin-ra (Japanese Šin-ra, Širaki) is the name of the ancient kingdom of Silla, in the northern part of Korea.

[2] Ch. 3, p. 5 (ed. of Lu Sin-yüan).

[3] 波斯 certainly is here not Persia, for the *Pei hu lu* deals with the products of Kwaň-tuň, Annam, and the countries south of China (PELLIOT, *Bull. de l'Ecole française*, Vol. IX, p. 223). See below, p. 468. The *Pei hu lu* has presumably served as the source for the text of the *Liň piao lu i*, quoted above.

[4] History of the Indian Archipelago, Vol. I, p. 383.

is one of the most useful trees of the countries where it grows. The nuts are either smoked and dried for use, or the oil is expressed from them in their recent state. The oil is used for all culinary purposes, and is more palatable and finer than that of the coconut. The kernels, mixed up with a little sago meal, are made into cakes and eaten as bread. The *kanari* is a native of the same country with the sago tree, and is not found to the westward. In Celebes and Java it has been introduced in modern times through the medium of traffic."

The *Yu yan tsa tsu*[1] speaks of a *man hu t'ao* 蔓胡桃 as "growing in the kingdom of Nan-čao 南詔 in Yün-nan; it is as large as a flat conch, and has two shells of equal size; its taste is like that of the cultivated walnut. It is styled also 'creeper in the land of the Man' (*Man čun t'en-tse* 蠻中藤子)." It will be remembered that Twan Č'en-ši, the author of this work, describes also the cultivated walnut (p. 264).

The *T'ai p'in yü lan* contains another text attributed to the *Lin piao lu i* relating to a wild walnut, which, however, is not extant in the edition of this work published in the collection *Wu yin tien* in 1775. This text is as follows: "The large walnut has a thick and firm shell. It is larger than that of the areca-nut.[2] It has much meat, but little glumelle. It does not resemble the nuts found in northern China. It must be broken with an axe or hammer. The shell, when evenly smoothed over the bottom, is occasionally made into a seal, for the crooked structure of the shell (*ko* 隔) resembles the seal characters."[3]

In the *Lin wai tai ta* 嶺外代答,[4] written by Čou K'ü-fei 周去非 in 1178, mention is made, among the plants of southern China and Tonking, of a "stone walnut (*ši hu t'ao* 石胡桃), which is like stone, has hardly any meat, and tastes like the walnut of the north." Again, a wild species is involved here. I have not found the term *ši hu t'ao* in any other author.

The various names employed by the T'ang writers for the wild

---

[1] Ch. 19, p. 9 b (ed. of *Tsin tai pi šu*); or Ch. 19, p. 9 a (ed. of *Pai hai*).

[2] This sentence, as well as the first, agrees with the definition given by the *Pei hu lu* with reference to a wild walnut (above, p. 268).

[3] *T'ai p'in yü lan*, Ch. 971, p. 8 b. The same text is cited by the *Pen ts'ao kan mu* and the *Ko či kin yüan* (Ch. 76, p. 5 b), which offer the reading *šan hu t'ao* 山 胡桃 ("wild walnut") instead of "large walnut." The *Kwan k'ün fan p'u* (Ch. 58, p. 26) also has arranged this text under the general heading "wild walnut." The *Pen ts'ao kan mu* opens it with the sentence, "In the southern regions there is a wild walnut." The restriction to South China follows also from the text as given in the *T'ai p'in yü lan*.

[4] Ch. 8, p. 10 b (ed. of *Či pu tsu čai ts'un šu*).

varieties (*p'ien hu t'ao, šan hu t'ao, man hu t'ao, ta hu t'ao*), combined with the fact that two authors describe both the varieties *p'ien* and *šan*, raise the question whether this nomenclature does not refer to different plants, and whether, aside from the wild walnut, other nuts may not also be included in this group. In this respect it is of interest to note that the hickory, recently discovered in Če-kiaṅ by F. N. MEYER, and determined by SARGENT[1] under the name *Carya cathayensis*, is said by Meyer to be called *shan-gho-to* in the colloquial language; and this evidently is identical with our *šan hu t'ao*. This certainly does not mean that this term refers exclusively to the hickory, but only that locally the hickory falls also within the category of *šan hu t'ao*. The distribution of the hickory over China is not yet known, and the descriptions we have of *šan hu t'ao* do not refer to Če-kiaṅ.

In the *P'an šan či* 盤山志, a description of the P'an mountains,[2] the term *šan ho t'ao* is given as a synonyme for the bark of *Catalpa bungei* (*ts'iu p'i* 楸皮), which is gathered on this mountain for medicinal purposes,— presumably because the structure of this bark bears some superficial resemblance to that of a walnut. Wild walnuts, further, are mentioned as growing on Mount Si fu žuṅ 西芙蓉山, forming part of the Ma-ku Mountains 麻姑山 situated in Fu-čou 撫州 in the prefecture of Kien-č'aṅ 建昌府, Kiaṅ-si Province.[3]

While the cultivated walnut was known in China during the fourth century under the Tsin dynasty, the wild species indigenous to southern China was brought to the attention of scholars only several centuries later, toward the close of the T'ang period. This case furnishes an excellent object-lesson, in that it reveals the fallacies to which botanists and others are only too frequently subject in drawing conclusions from mere botanical evidence as to cultivated plants. The favorite argumentation is, that if, in a certain region, a wild and a corresponding cultivated species co-exist, the cultivated species is simply supposed to have been derived from the wild congener. This is a deceptive conclusion. The walnut (as well as the vine) of China offers a

---

[1] Plantae Wilsonianae, Vol. III, p. 187.

[2] Ch. 15, p. 2 b, of the edition published in 1755 by order of K'ien-luṅ. The P'an šan is situated three or four days' journey east of Peking, in the province of Či-li, the summit being crowned by an interesting Buddhist temple, and there being an imperial travelling-station at its foot. It was visited by me in September, 1901. F. N. MEYER (Agricultural Explorations in the Orchards of China, p. 52) says that in the Pangshan district east of Peking one may still find a few specimens of the real wild walnut growing in ravines among large bowlders in the mountains.

[3] *Ma-ku šan či* (Ch. 3, p. 6 b), written by members of the family Hwaṅ 黃, and published in 1866 by the Tuṅ t'ien šu wu 洞天書屋. These mountains contain thirty-six caves dedicated to the Taoist goddess Ma-ku.

specific case apt to teach just the opposite: a wild walnut (probably in several species) is indigenous to China, nevertheless the species cultivated in this area did not spring from domestic material, but from seeds imported from Iranian and Tibetan regions of Central Asia. The botanical dogma has been hurled against many deductions of Hehn: botanists proclaimed that vine, fig, laurel, and myrtle have been indigenous to Greece and Italy in a wild state since time immemorial; likewise pomegranate, cypress, and plantain on the Aegean Islands and in Greece; hence it was inferred that also the cultivations of these plants must have been indigenous, and could not have been introduced from the Orient, as insisted on by Hehn. This is nothing but a sophism: the botanists still owe us the proof that the cultivated species were really derived from indigenous stock. A species may indeed be indigenous to a certain locality; and yet, as brought about by historical inter-relations of the peoples, the same or a similar species in the cultivated state may have been introduced from an outside quarter. It is only by painstaking historical research that the history of cultivated plants can be exactly determined. ENGLER (above, p. 258) doubts the occurrence of the wild walnut in China, because a cultivated species was introduced there from Tibet! It is plain now where such logic will lead us. Wilson deserves a place of honor among botanists, for, after close study of the subject in China, he recognized that "it is highly improbable that *Juglans regia* is indigenous to China."

With reference to the walnut, conditions are the same in China as in the Mediterranean region: there also *Juglans regia* grows spontaneously; still better, cultivated varieties reached the Greeks from Persia; the Greeks handed these on to the Romans; the Romans transplanted them to Gallia and Germania. *Juglans regia* occupies an extensive natural area throughout the temperate zone, stretching from the Mediterranean through Iran and the Himalaya as far as southern China and the Chinese maritime provinces. Despite this natural distribution, the fact remains that Iran has been the home and the centre of the best-cultivated varieties, and has transmitted these to Greece, to India, to Central Asia, and to China.

Dr. T. TANAKA has been good enough to furnish the following information, extracted from Japanese literature, in regard to the walnut.

"Translation of the notice on *ko-tō* (*kurumi*), 'walnut,' from a Japanese herbal *Yamato honzō* 大和本草, by Kaibara Ekken 貝原 益軒 (Ch. 10, p. 23), published in 1709.

"*Kurumi* 胡桃 (*kotō*). There are three sorts of walnut. The first is called *oni-gurumi* 鬼胡桃 ('devil walnut'). It is round in shape, and has a thick, hard skin (shell), difficult to break; it has very little meat. In the *Honzō* (*Pen ts'ao*, usually referring to the *Pen ts'ao kan mu*) it is called 山 胡 桃 (*yama-gurumi*, *san hu t'ao*). It is customary to open the shell by first baking it a little while in a bed of charcoal, and suddenly plunging it in water to cool off; then it is taken out of the fire, the shell is struck at the joint so that it is crushed, and the meat can be easily removed. The second variety is called *hime-gurumi* 姫 ク ル ミ ('demoiselle walnut'), and has a thin shell which is somewhat flat in form; it is very easily broken when struck with an iron hammer at the joint. It has plenty of meat, is rich in oil, and has a better taste than the one mentioned before. The names 'devil' and 'demoiselle' are derived from the appearance of the nuts, the one being rough and ugly, while the other is beautiful.

"The third variety, which is believed to have come from Korea, has a thin shell, easily cracked, with very little meat, but of the best quality. Moṅ Šen 孟 詵 (author of the *Si liao pen ts'ao* 食 療 本 草, second half of the seventh century) says, 'The walnut, when eaten, increases the appetite, stimulates the blood-circulation, and makes one appear glossy and elegant. It may be considered as a good medicine of high merit.' For further details refer to the prescriptions of the *Pen ts'ao*.

"Translation of the notice on walnut from the *Honzō kōmoku keimō* (Ch. 25, pp. 26–27) by Ono Ranzan; revised edition by Iguči Bōši of 1847 (first edition 1804).

"*kotō*, *kurimi* (walnut, *Juglans regia* L., var. *sinensis* Cas., ex MATSUMURA, Shokubutsu Mei-i, ed. 1915, Vol. I, p. 189).

"Japanese names: *tō-kurimi* ('Chinese walnut'); *čōsen-kurimi* ('Korean walnut').

"Chinese synonymes: *kaku-kwa* (*Jibutsu imei*); *činsō kyohō* (ibid.); *inpei činkwa* (ibid.); *kokaku* (*Jibutsu konšu*); *kenša* (ibid.); *tōšūši* (*Kunmō jikwai*).

"Names for kernels: *kama* (*Rōya taisui-hen*).

"Other names for *šan hu t'ao*: *sankakutō* (*Hokuto-roku*); *banzai-ši* (*Jonan Hoši*); *šū* (*Kummō jikwai*).

"The real walnut originated in Korea, and is not commonly planted in Japan.

"The leaves are larger than those of *onigurumi* (giant walnut, *Juglans sieboldiana* Maxim., ex Matsumura, *l.c.*). The shells are also larger, measuring more than 1 *sun* (1.193 inches) in length, and having more striations on the surface. The kernels are also larger, and have more folds.

"The variety commonly planted in our country is *onigurumi*, the abbreviated name of which is *kurumi;* local names are *ogurumi* (Province of Kaga), *okkoromi* (eastern provinces), and so on. This giant walnut grows to a large tree. Its leaves are much like those of the lacquer-tree (*Rhus vernificera* DC.) and a little larger; they have finely serrated margins. Its new leaves come out in the spring. It flowers in the autumn.

"The flower-clusters resemble chestnut-catkins, but are much larger, ranging in length from six to seven *sun;* they are yellowish white and pendulous. A single flower is very small, like that of a chestnut. The fruit is peach-shaped and green, but turns black when ripe. The shells are very hard and thick, and can be opened by being put on the fire for a little while; then insert a knife in the slit or fissure between the shells, which thus break. The kernels are good for human food, and are also used for feeding little birds.

"One species called *hime-gurumi* ('demoiselle walnut,' *Juglans cordiformis* Maxim., ex Matsumura, *l.c.*), or *me-gurumi* ('female walnut,' from the province of Kaga), has thin shells with fewer furrows, and the kernels can easily be taken out. Under the heading *šūkai* (*či-kie*, explanatory information in the *Pen ts'ao*), this kind of walnut is described as 'a walnut produced in Činšo (Č'en-ts'aṅ, a place in Fuṅsiaṅ fu, Šen-si, China) with thin shells and many surfaces,' so we call it *činsō-gurumi* (*č'en-ts'aṅ hu-t'ao*).[1] This variety is considered the best of all *yama-gurumi* (*šan hu t'ao*, wild walnuts), because no other variety has such saddle-shaped kernels entirely removable from the shells.

"A species called *karasu-gurumi* ('crow walnut') is a product of the province of Ečigo; it has a shell that opens by itself when ripe, and looks like a crow's bill when opened, whence it is called 'crow walnut.'

"Another variety from Ōsio-mura village of the Aidzu district is called *gonroku-gurumi* ('Gonroku's walnut'); it has a very small shell capable of being used as *ojime* ('string-fastener of a pouch'). This name is taken from the personal name of a man called Anazawa Gonroku, in whose garden this variety originated. It is said that the same kind has been found in the province of Kai.

"A variety found at Noširo, province of Ušū (Uzen and Ugo), is much larger in size, and has thinner shells, easily crushed by hand, so that the kernels may be taken out without using any tools. The name of this variety is therefore *teuči-gurumi* ('hand-crushed walnut')."

The most interesting point in these Japanese notes is presented by

---

[1] Compare above, p. 264.

the tradition tracing the cultivated walnut of Japan to Korea. The Koreans again have a tradition that walnuts reached them from China about fifteen hundred years ago in the days of the Silla Kingdom.[1] The Korean names for the fruit are derived from the Chinese: *ho do* being the equivalent of *hu t'ao, kan do* corresponding to *k'ian t'ao,* and *ha do* to *ho t'ao.* The Geography of the Ming Dynasty states that walnuts are a product of Korea.[2]

---

[1] *Korea Review*, Vol. II, 1902, p. 394.
[2] *Ta Min i t'un či*, Ch. 89 p. 4 b.

THE POMEGRANATE

5. A. DE CANDOLLE[1] sums up the result of his painstaking investigation of the diffusion of the pomegranate (*Punica granatum*, the sole genus with two species only within the family *Punicaceae*) as follows: "To conclude, botanical, historical, and philological data agree in showing that the modern species is a native of Persia and some adjacent countries. Its cultivation began in prehistoric time, and its early extension, first toward the west and afterwards into China, has caused its naturalization in cases which may give rise to errors as to its true origin, for they are frequent, ancient, and enduring." In fact, the pomegranate occurs spontaneously in Iran on stony ground, more particularly in the mountains of Persian Kurdistan, Baluchistan, and Afghanistan. I am in full accord with A. de Candolle's opinion, which, as will be seen, is signally corroborated by the investigation that follows, and am not in the least disturbed by A. ENGLER's view[2] that the pomegranate occurs wild in Greece and on the islands of the Grecian Archipelago, and that, accordingly, it is indigenous in anterior Asia and part of the Balkan Peninsula, while its propagation in Italy and Spain presumably followed its cultivation in historical times. First, as stated also by G. BUSCHAN,[3] these alleged wild trees of Greece are not spontaneous, but have reverted from cultivation to a wild state.[4] Second, be this as it may, all ancient Greek accounts concerning the pomegranate relate exclusively to the cultivated, in no case to the wild species; and it is a gratuitous speculation of O. SCHRADER,[5] who follows suit with Engler, that the Greek word ῥοά was originally applied to the indigenous wild species, and subsequently transferred to the cultivated one. As will be shown hereafter, the Greek term is a loan-word. The naturalization of the fruit in the Mediterranean basin is, as A. DE CANDOLLE justly terms it, an extension of the original

---

[1] Origin of Cultivated Plants, p. 240.

[2] In Hehn's Kulturpflanzen, p. 246 (8th ed.).

[3] Vorgeschichtliche Botanik, p. 159.

[4] I am unable, however, to share Buschan's view that the wild specimens of Iran and north-western India also belong to this class; that area is too extensive to allow of so narrow an interpretation. In this case, Buschan is prejudiced in order to establish his own hypothesis of an indigenous origin of the tree in Arabia (see below).

[5] In Hehn's Kulturpflanzen, p. 247.

276

area; and Hehn is quite right in dating its cultivation on the part of the Greeks to a time after the Homeric epoch, and deriving it from Asia Minor.

G. Buschan[1] holds that Europe is out of the question as to the indigenous occurrence of the pomegranate, and with regard to *Punica protopunica*, discovered by Balfour on the Island of Socotra, proposes Arabia felix as the home of the tree; but he fails to explain the diffusion of the tree from this alleged centre. He opposes Loret's conclusions with reference to Egypt, where he believes that the tree was naturalized from the time of the Eighteenth Dynasty; but he overlooks the principal point made by Loret, namely, that the Egyptian name is a Semitic loan-word.[2] Buschan's theory conflicts with all historical facts, and has not been accepted by any one.

The pomegranate-tree is supposed to be mentioned in the Avesta under the name *haδānaēpata*,[3] the wood serving as fuel, and the juice being employed in sacrificial libations; but this interpretation is solely given by the present Parsī of India and Yezd, and is not certain. The fruit, however, is mentioned in Pahlavi literature (above, p. 193).

There are numerous allusions to the pomegranate of Persia on the part of Mohammedan authors and European travellers, and it would be of little avail to cite all these testimonies on a subject which is perfectly well known. Suffice it to refer to the *Fārs Nāmah*[4] and to give the following extract from A. Olearius:[5]—

"Pomegranate-trees, almond-trees, and fig-trees grow there without any ordering or cultivation, especially in the Province of Kilan, where you have whole forests of them. The wild pomegranates, which you find almost every where, especially at Karabag, are sharp or sowrith.

---

[1] Vorgeschichtliche Botanik, p. 159.

[2] This fact was simultaneously and independently found by an American Egyptologist, Ch. E. Moldenke (Über die in altägyptischen Texten erwähnten Bäume, p. 115, doctor dissertation of Strassburg, Leipzig, 1887); so that Loret (Flore pharaonique, p. 76) said, "Moldenke est arrivé presque en même temps que moi, et par des moyens différents, ce qui donne une entière certitude à notre découverte commune, à la détermination du nom égyptien de la grenade." See also C. Joret, Plantes dans l'antiquité, Vol. I, p. 117. Buschan's book appeared in 1895; nevertheless he used Loret's work in the first edition of 1887, instead of the second of 1892, which is thoroughly revised and enlarged.

[3] For instance, Yasna, 62, 9; 68, 1. Cf. also A. V. W. Jackson, Persia Past and Present, p. 369.

[4] G. Le Strange, Description of the Province of Fars in Persia, p. 38 (London, 1912). See also d'Herbelot, Bibliothèque orientale, Vol. III, p. 188; and F. Spiegel, Eranische Altertumskunde, Vol. I, p. 252.

[5] Voyages of the Ambassadors to the Great Duke of Muscovy, and the King of Persia (1633-39), p. 232 (London, 1669).

They take out of them the seed, which they call *Nardan*, wherewith they drive a great trade, and the Persians make use of it in their sawces, whereto it gives a colour, and a picquant tast, having been steep'd in water, and strain'd through a cloath. Sometimes they boyl the juyce of these Pomegranates, and keep it to give a colour to the rice, which they serve up at their entertainments, and it gives it withall a tast which is not unpleasant. . . . The best pomegranates grow in Jescht, and at Caswin, but the biggest, in Karabag."

Mirza Haidar mentions a kind of pomegranate peculiar to Baluristan (Kafiristan), sweet, pure, and full-flavored, its seeds being white and very transparent.[1]

"Grapes, melons, apples, and pomegranates, all fruits, indeed, are good in Samarkand."[2] The pomegranates of Khojand were renowned for their excellence.[3] The Emperor Jahāngīr mentions in his Memoirs the sweet pomegranates of Yazd and the subacid ones of Farrāh, and says of the former that they are celebrated all over the world.[4] J. CRAWFURD[5] remarks, "The only good pomegranates which, indeed, I have ever met with are those brought into upper India by the caravans from eastern Persia."

The *Yu yan tsa tsu*[6] states that the pomegranates of Egypt 勿斯離 (Wu-se-li, *Mwir-si-li, Mirsir)[7] in the country of the Arabs (Ta-ši, *Ta-džik) weigh up to five and six catties.

Also in regard to the pomegranate we meet the tradition that its introduction into China is due to General Čaṅ K'ien. In the same manner as in the case of the walnut, this notion looms up only in post-Han authors. It is first recorded by Lu Ki 陸璣, who lived under the Western Tsin dynasty (A.D. 265–313), in his work *Yü ti yün šu* 與弟雲書. This text has been handed down in the *Ts'i min yao šu* of Kia Se-niu of the sixth century.[8] There it is said that Čaṅ K'ien, while an envoy of the Han in foreign countries for eighteen years, obtained *t'u-lin* 塗林, this term being identical with *ṅan-ši-liu* 安石榴. This tradition is repeated in the *Po wu či*[9] of Čaṅ Hwa and in the

[1] ELIAS and ROSS, Tarikh-i-Rashidi, p. 386.

[2] A. S. BEVERIDGE, Memoirs of Babur, p. 77.

[3] *Ibid.*, p. 8. They are also extolled by Ye-lu Č'u-ts'ai (BRETSCHNEIDER, Mediæval Researches, Vol. I, p. 19).

[4] H. M. ELLIOT, History of India as told by Its Own Historians, Vol. VI, p. 348.

[5] History of the Indian Archipelago, Vol. I, p. 433.

[6] 續集 Ch. 10, p. 4 b (ed. of *Tsin tai pi šu*).

[7] Old Persian Mudrāya, Hebrew Mizraim, Syriac Mezroye.

[8] Ch. 4, p. 14 b (new ed., 1896).

[9] See above, p. 258.

Tu i či 獨異志, written by Li Yu 李尤 (or Li Yūan 元) of the T'ang dynasty. Another formal testimony certifying to the acceptance of this creed at that period comes from Fuṅ Yen 封演 of the T'ang in his *Fuṅ ši wen kien ki* 封氏聞見記,[1] who states that Čaṅ K'ien obtained in the Western Countries the seeds of *ši-liu* 石榴 and alfalfa (*mu-su*), and that at present these are to be found everywhere in China. Under the Sung this tradition is repeated by Kao Č'eṅ 高承.[2] Č'en Hao-tse, in his *Hwa kiṅ*,[3] published in 1688, states it as a cold-blooded fact that the seeds of the pomegranate came from the country Ṅan-si or An-si (Parthia), and that Čaṅ K'ien brought them back. There is nothing to this effect in Čaṅ K'ien's biography, nor is the pomegranate mentioned in the Annals of the Han.[4] The exact time of its introduction cannot be ascertained, but the tree is on record no earlier than the third and fourth centuries A.D.[5]

Li Ši-čen ascribes the term *ṅan-ši-liu* to the *Pie lu* 別錄, but he cites no text from this ancient work, so that the case is not clear.[6] The earliest author whom he quotes regarding the subject is T'ao Huṅ-kiṅ (A.D. 452–536), who says, "The pomegranate, particularly as regards its blossoms, is charming, hence the people plant the tree in large numbers. It is also esteemed, because it comes from abroad. There are two varieties, the sweet and the sour one, only the root of the latter being used by physicians." According to the *Ts'i min yao šu*, Ko Huṅ 葛洪 of the fourth century, in his *Pao p'u tse* 抱朴子, speaks of the occurrence of bitter *liu* 苦榴 on stony mountains. These, indeed,

---

[1] Ch. 7, p. 1 b (ed. of *Ki fu ts'uṅ šu*).

[2] *Ši wu ki yüan* 事物紀原 (ed. of *Si yin hüan ts'uṅ šu*), Ch. 10, p. 34 b.

[3] Ch. 3, p. 37, edition of 1783; see above, p. 259.

[4] The Čaṅ-K'ien legend is repeated without criticism by BRETSCHNEIDER (Bot. Sin., pt. 1, p. 25; pt. 3, No. 280), so that A. DE CANDOLLE (Origin of Cultivated Plants, p. 238) was led to the erroneous statement that the pomegranate was introduced into China from Samarkand by Čaṅ K'ien, a century and a half before the Christian era. The same is asserted by F. P. SMITH (Contributions towards the Materia Medica of China, p. 176), G. A. STUART (Chinese Materia Medica, p. 361), and HIRTH (*T'oung Pao*, Vol. VI, 1895, p. 439).

[5] It is mentioned in the *Kin kwei yao lio* (Ch. c, p. 27) of the second century A.D., "Pomegranates must not be eaten in large quantity, for they injure man's lungs." As stated (p. 205), this may be an interpolation in the original text.

[6] The *Pie lu* is not quoted to this effect in the *Čeṅ lei pen ts'ao* (Ch. 22, p. 39), but the *Či wu miṅ ši t'u k'ao* (Ch. 15, p. 102; and 32, p. 36 b) gives two different extracts from this work relating to our fruit. In one, its real or alleged medical properties are expounded; in the other, different varieties are enumerated, while not a word is said about foreign origin. I am convinced that in this form these two texts were not contained in the *Pie lu*. The question is of no consequence, as the work itself is lost, and cannot be dated exactly. All that can be said with certainty is that it existed prior to the time of T'ao Huṅ-kiṅ.

are the particular places where the pomegranate thrives. Su Suṅ of the Sung period states that the pomegranate was originally grown in the Western Countries (Si yü 西域), and that it now occurs everywhere; but neither he nor any other author makes a positive statement as to the time and exact place of origin. The Yao siṅ lun, Pen ts'ao ši i, and Pen ts'ao yen i[1] give merely a botanical notice, but nothing of historical interest.

The pomegranate (ši-liu) is mentioned in the "Poem on the Capital of Wu" 吳都賦 by Tso Se 左思, who lived in the third century under the Wu dynasty (A.D. 222–280). P'an Yo 潘岳, a poet of the fourth century A.D., says, "Pomegranates are the most singular trees of the empire and famous fruits of the Nine Provinces.[2] A thousand seed-cases are enclosed by the same membrane, and what looks like a single seed in fact is ten."

The Tsin Luṅ ṅan k'i kü ču 晉隆安起居注 ("Annotations on the Conditions of the period Luṅ-ṅan [A.D. 397–402] of the Tsin Dynasty") contains the following note:[3] "The pomegranates (ṅan ši liu) of the district Lin-yüan 臨沅 in Wu-liṅ 武陵[4] are as large as cups; they are not sour to the taste. Each branch bears six fruits."

Lu Hui 陸翽 of the Tsin dynasty, in his Ye čuṅ ki 鄴中記,[5] states that in the park of Ši Hu 石虎 there were pomegranates with seeds as large as cups, and they were not sour. Ši Hu or Ši Ki-luṅ 石季龍 ruled from A.D. 335 to 349, under the appellation T'ai Tsu 大祖 of the Hou Čao dynasty, as "regent celestial king" (kü-še t'ien waṅ), and shifted the capital to Ye 鄴, the present district of Lin-čaṅ 臨漳, in the prefecture of Čaṅ-te 彰德 in Ho-nan.[6]

The pomegranate is mentioned in the Ku kin ču 古今注,[7] written by Ts'ui Pao 崔豹 during the middle of the fourth century, with reference to the pumelo 柑 (Citrus grandis), the fruit of which is compared in shape with the pomegranate. The Ts'i min yao šu (l.c.) gives rules for the planting of pomegranates.

---

[1] Ch. 18, p. 7 (ed. of Lu Sin-yüan); the other texts see in Čeṅ lei pen ts'ao, l. c.

[2] 九州, the ancient division of China under the Emperor Yü.

[3] T'ai p'iṅ yü lan, Ch. 970, p. 4 b. Regarding the department of records styled k'i kü ču, see The Diamond, p. 35. In the Yüan kien lei han (Ch. 402, p. 2) the same text is credited to the Suṅ šu.

[4] In Hu-nan Province.

[5] Ed. of Wu yiṅ tien, p. 12.

[6] Regarding his history, see L. WIEGER, Textes historiques, pp. 1095–1100. BRETSCHNEIDER's (Bot. Sin., pt. 1, p. 211) note, that, besides the Ye čuṅ ki of Lu Hui, there is another work of the same name by Ši Hu, is erroneous; Ši Hu is simply the "hero" of the Ye čuṅ ki.

[7] Ch. c, p. 1 (ed. of Han Wei ts'uṅ šu or Ki fu ts'uṅ šu). Cf. also below, p. 283.

The Annals of the Liu Sung Dynasty, A.D. 420–477 (*Suṅ šu*), contain the following account: "At the close of the period *Yüan-kia* 元嘉 (A.D. 424–453), when T'ai Wu (A.D. 424–452) 太武 of the Wei dynasty conquered the city Ku 鼓城,[1] he issued orders to search for sugarcane and pomegranates (*ṅan ši liu*). Čaṅ Č'aṅ 張暢 said that pomegranates (*ši-liu*) come from Ye." This is the same locality as mentioned above.

The *Siaṅ kwo ki* 襄國記[2] reports that in the district of Luṅ-kaṅ 龍岡縣[3] there are good pomegranates (*ši liu*). These various examples illustrate that in the beginning the tree was considered as peculiar to certain localities, and that accordingly a gradual dissemination must have taken place. Apparently no ancient Chinese author is informed as to the locality from which the tree originally came, nor as to the how and when of the transplantation.

The *Kwaṅ či* 廣志, written by Kwo Yi-kuṅ 郭義恭 prior to A.D. 527, as quoted in the *Ts'i min yao šu*, discriminates between two varieties of pomegranate (*ṅan ši liu*), a sweet and a sour one, in the same manner as T'ao Huṅ-kiṅ.[4] This distinction is already made by Theophrastus.[5] As stated above, there was also a bitter variety.[6]

It is likewise a fact of great interest that we have an isolated instance of the occurrence of a pomegranate-tree that reverted to the wild state. The *Lü šan ki* 廬山記[7] contains this notice: "On the summit of the Hiaṅ-lu fuṅ 香爐峯 ('Censer-Top') there is a huge rock on which several people can sit. There grows a wild pomegranate (*šan ši-liu* 山石榴) drooping from the rock. In the third month it produces blossoms. In color these resemble the [cultivated] pomegranate, but they

---

[1] Modern Čeṅ-tiṅ fu in Či-li Province.

[2] Thus in *T'ai p'iṅ yü lan*, Ch. 970, p. 5 b; the *Ts'i min yao šu* (Ch. 4, p. 14) ascribes the same text to the *Kiṅ k'ou ki* 京口記.

[3] At present the district which forms the prefectural city of Šun-te in Či-li Province.

[4] Above, p. 279.

[5] Historia plantarum, II. II, 7.

[6] Pliny (XIII, 113) distinguishes five varieties,—dulcia, acria, mixta, acida, vinosa.

[7] *T'ai p'iṅ yü lan*, Ch. 970, p. 5. The Lü Mountain is situated in Kiaṅ-si Province, twenty-five *li* south of Kiu-kiaṅ. A work under the title *Lü šan ki* was written by Č'en Liṅ-kü 陳令擧 in the eleventh century (WYLIE, Notes on Chinese Literature, p. 55); but, as the *T'ai p'iṅ yü lan* was published in A.D. 983, the question here must be of an older work of the same title. In fact, there is a *Lü šan ki* by Kiṅ Ši 景式 of the Hou Čou dynasty; and the *Yüan kien lei han* (Ch. 402, p. 2) ascribes the same text to the *Čou Kiṅ Ši Lü šan ki*. The John Crerar Library of Chicago (No. 156) possesses a *Lü šan siao či* in 24 chapters, written by Ts'ai Yiṅ 蔡瀛 and published in 1824.

are smaller and pale red. When they open, they display a purple calyx of bright and attractive hues." A poem of Li Te-yü 李德裕 (787–849) opens with the words, "In front of the hut where I live there is a wild pomegranate."[1]

Fa Hien 法顯, the celebrated Buddhist traveller, tells in his *Fu kwo ki* 佛國記 ("Memoirs of Buddhist Kingdoms"), written about A.D. 420, that, while travelling on the upper Indus, the flora differed from that of the land of Han, excepting only the bamboo, pomegranate, and sugar-cane.[2] This passage shows that Fa Hien was familiar with that tree in China. Hüan Tsaṅ observed in the seventh century that pomegranates were grown everywhere in India.[3] Soleiman (or whoever may be the author of this text), writing in A.D. 851, emphasizes the abundance of the fruit in India.[4] Ibn Baṭūṭa says that the pomegranates of India bear fruit twice a year, and emphasizes their fertility on the Maldive Islands.[5] Seedless pomegranates came to the household of the Emperor Akbar from Kabul.[6]

The pomegranate occurred in Fu-nan (Camboja), according to the *Nan Ts'i šu* or History of the Southern Ts'i (A.D. 479–501), compiled by Siao Tse-hien in the beginning of the sixth century.[7] It is mentioned again by Čou Ta-kwan of the Yüan dynasty, in his book on the "Customs of Camboja."[8] In Haṅ-čou, large and white pomegranates were styled *yü liu* 玉榴 ("jade" *liu*), while the red ones were regarded as inferior or of second quality.[9]

The following ancient terms for the pomegranate, accordingly, are on record:—

(1) 塗林 *t'u-lin*, *du-lim. Aside from the *Po wu či*, this term is used by the Emperor Yüan of the Liang dynasty in a eulogy of the fruit.[10] HIRTH[11] identified this word with an alleged Indian *darim;* and, according to him, Čaṅ K'ien must have brought the Indian name to

---

[1] *Li wei kuṅ pie tsi*, Ch. 2, p. 8 (*Ki fu ts'uṅ šu, t'ao* 10).

[2] Cf. J. LEGGE, A Record of Buddhistic Kingdoms, p. 24.

[3] *Ta T'aṅ si yü ki*, Ch. 2, p. 8 b (S. BEAL, Buddhist Records of the Western World, Vol. I, p. 88).

[4] M. REINAUD, Relation des voyages, Vol. I, p. 57.

[5] DEFRÉMERY and SANGUINETTI, Voyages d'Ibn Batoutah, Vol. III, p. 129.

[6] H. BLOCHMANN, Ain I Akbari, Vol. I, p. 65.

[7] PELLIOT, Le Fou-nan, *Bull. de l'Ecole française*, Vol. III, p. 262.

[8] PELLIOT, *ibid.*, Vol. II, p. 168.

[9] *Moṅ liaṅ lu* 夢粱錄 by Wu Tse-mu 吳自牧 of the Sung (Ch. 18, p. 5 b; ed. of *Či pu tsu čai ts'uṅ šu*).

[10] *Yüan kien lei han*, Ch. 402, p. 3 b. Further, in the lost *Hu pen ts'ao*, as follows from a quotation in a note to the *Pei hu lu* (Ch. 3, p. 12).

[11] *T'oung Pao*, Vol. VI, 1895, p. 439.

China. How this would have been possible, is not explained by him. The Sanskrit term for the pomegranate (and this is evidently what Hirth hinted at) is *dāḍima* or *dālima*, also *dāḍimva*, which has passed into Malayan as *delima*.[1] It is obvious that the Chinese transcription bears some relation to this word; but it is equally obvious that the Chinese form cannot be fully explained from it, as it leads only to *du-lim, not, however, to *dalim*. There are two possibilities: the Chinese transcription might be based either on an Indian vernacular or Apabhraṁça form of a type like *dulim, *duḍim,[2] or on a word of the same form belonging to some Iranian dialect. The difficulty of the problem is enhanced by the fact that no ancient Iranian word for the fruit is known to us.[3] It appears certain, however, that no Sanskrit word is intended in the Chinese transcription, otherwise we should meet the latter in the Sanskrit-Chinese glossaries. The fact remains that these, above all the *Fan yi miṅ yi tsi*, do not contain the word *t'u-lin;* and, as far as I know, Chinese Buddhist literature offers no allusion to the pomegranate. Nor do the Chinese say, as is usually stated by them in such cases, that the word is of Sanskrit origin; the only positive information given is that it came along with General Čaṅ K'ien, which is to say that the Chinese were under the impression that it hailed from some of the Iranian regions visited by him. *Dulim, dulima, or *durim, durima, accordingly, must have been a designation of the pomegranate in some Iranian language.

(2) 丹若 *tan-žo*, *dan-zak, dan-yak, dan-n'iak. This word appears in the *Ku kin ču*[4] and in the *Yu yaṅ tsa tsu*.[5] Apparently it represents a transcription, but it is not stated from which language it is derived. In my estimation, the foundation is an Iranian word still unknown to us, but congeners of which we glean from Persian *dānak* ("small grain"),

---

[1] J. CRAWFURD (History of the Indian Archipelago, Vol. I, p. 433) derives this word from the Malayan numeral five, with reference to the five cells into which the fruit is divided. This, of course, is a mere popular etymology. There is no doubt that the fruit was introduced into the Archipelago from India; it occurs there only cultivated, and is of inferior quality. On the Philippines it was only introduced by the Spaniards (A. DE MORGA, Philippine Islands, p. 275, ed. of Hakluyt Society).

[2] The vernacular forms known to me have the vowel *a;* for instance, Hindustānī *darim*, Bengalī *ḍālim, dālim* or *dārim;* Newārī, *dhāḍe*. The modern Indo-Aryan languages have also adopted the Persian word *anār*.

[3] In my opinion, the Sanskrit word is an Iranian loan-word, as is also Sanskrit *karaka*, given as a synonyme for the pomegranate in the Amarakoṣa. The earliest mention of *dāḍima* occurs in the Bower Manuscript; the word is absent in Vedic literature.

[4] At least it is thus stated in cyclopædias; but the editions of the work, as reprinted in the *Han Wei ts'uṅ šu* and *Ki fu ts'uṅ šu*, do not contain this term.

[5] Ch. 18, p. 3 b (ed. of *Pai hai*).

dāna ("grain, berry, stone of a fruit, seed of grain or fruit"), *dāngū* ("kind of grain"), Šina *danu* ("pomegranate");[1] Sanskrit *dhanika, dhanyāka*, or *dhanīyaka* ("coriander"; properly "grains"). The notion conveyed by this series is the same as that underlying Latin *granatum*, from *granum* ("grain"); cf. Anglo-Saxon *cornæppel* and English *pomegranate* ("apple made up of grains").

(3) 安石榴 *ṅan ši liu* or 石榴 *ši liu*. This transcription is generally taken in the sense "the plant *liu* of the countries Ṅan and Ši, or of the country Ṅan-ši." This view is expressed in the *Po wu či*, which, as stated, also refers to the Čaṅ-K'ien legend, and to the term *t'u-lin*, and continues that this was the seed of the *liu* of the countries Ṅan and Ši; hence, on the return of Čaṅ K'ien to China, the name *ṅan-ši-liu* was adopted.[2] Bretschneider intimates that Ṅan and Ši were little realms dependent on K'aṅ at the time of the Han. Under the T'ang, the name Ṅan referred to Bukhāra, and Ši to Tašḵend; but it is hardly credible that these two geographical names (one does not see for what reason) should have been combined into one, in order to designate the place of provenience of the pomegranate. It is preferable to assume that 安石 *ṅan ši*, *an-sek, an-sak, ar-sak*, represents a single name and answers to Arsak, the name of the Parthian dynasty, being on a par with 安息 *ṅan-si*, *Ar-sik, and 安西 *ṅan-si*, *Ar-sai. In fact, 安石 is the best possible of these transcriptions. We should expect, of course, to receive from the Chinese a specific and interesting story as to how and when this curious name, which is unique in their botanical nomenclature, was transmitted;[3] but nothing of the kind appears to be on record, or the record, if it existed, seems to have been lost. It is manifest that also the plant-name *liu* (*riu, r'u) presents the transcription of an Iranian word, and that the name in its entirety was adopted by the Chinese from an Iranian community outside of Parthia, which had received the tree or shrub from a Parthian region, and therefore styled it "Parthian pomegranate." It is not likely that the tree was transplanted to China directly from Parthia; we have to assume rather that the transplantation was a gradual process, in which the

---

[1] W. LEITNER, Races and Languages of Dardistan, p. 17.

[2] It is not correct, as asserted by BRETSCHNEIDER (*Chinese Recorder*, 1871, p. 222), to say that this definition emanates from Li Ši-čen, who, in fact, quotes only the *Po wu či*, and presents no definition of his own except that the word *liu* means 瘤 *liu* ("goitre"); this, of course, is not to be taken seriously. In Jehol, a variety of pomegranate is styled *hai* 海 *liu* (O. FRANKE, Beschreibung des Jehol-Gebietes, p. 75); this means literally, "*liu* from the sea," and signifies as much as "foreign *liu*."

[3] Cf. *ṅan-si hiaṅ* 安息香 ("Parthian incense") as designation for styrax benzoin (p. 464).

Iranian colonies outside of Iran proper, those of Sogdiana and Turkistan, played a prominent part. We know the Sogdian word for the pomegranate, which is written *n'r'kh*, and the reading of which has been reconstructed by R. GAUTHIOT[1] in the form *nārāk(a), developed from *anār-āka. This we meet again in Persian *anār*, which was adopted in the same form by the Mongols, while the Uigur had it as *nara*. At all events, however, it becomes necessary to restore, on the basis of the Chinese transcription, an ancient *riu, *ru, of some Iranian dialect. This lost Iranian word, in my opinion, presents also the foundation of Greek ῥόα or ῥοιά,—the origin of which has been hitherto unexplained or incorrectly explained,[2]—and the Semitic names, Hebrew *rimmōn*, Arabic *rummān*, Amharic *rūmān*, Syriac *rūmōnō*, Aramaic *rummāna*, from which Egyptian *arhmāni* or *anhmānī* (Coptic *erman* or *herman*) is derived.[3]

(4) 若 榴 *žo-liu*, *zak (yak, n'iak)-liu (riu). This hybrid compound, formed of elements contained in 2 and 3, is found in the dictionary *Kwaṅ ya* 廣 雅, written by Čaṅ Yi 張 揖 about A.D. 265.[4] It is also employed by the poet P'an Yo of the fourth century, mentioned above.[5] Eventually also this transcription might ultimately be traced to an Iranian prototype. Japanese *zakuro* is based on this Chinese form.[6]

While the direct historical evidence is lacking, the Chinese names of the tree point clearly to Iranian languages. Moreover, the tree itself is looked upon by the Chinese as a foreign product, and its first introduction into China appears to have taken place in the latter part of the third century A.D.

In my opinion, the pomegranate-tree was transplanted to India,

---

[1] Essai sur le vocalisme du sogdien, p. 49. Cf. also Armenian *nrneni* for the tree and *nurn* for the fruit.

[2] The etymologies of the Greek word enumerated by SCHRADER (in Hehn, Kulturpflanzen, p. 247) are so inane and far-fetched that they do not merit discussion. It is not necessary, of course, to hold that an immediate transmission of the Persian word took place, but we must look to a gradual propagation and to missing links by way of Asia Minor. According to W. MUSS-ARNOLT (*Transactions Am. Phil. Assoc.*, Vol. XXIII, 1892, p. 110), the Cyprian form ῥυδία forbids all connection with the Hebrew. It is not proved, however, that this dialectic word has any connection with ῥόα; it may very well be an independent local development.

[3] V. LORET, Flore pharaonique, p. 76. Portuguese *roma*, *romeira*, from the Arabic; Anglo-Saxon *rēad-æppel*.

[4] This is the date given by WATTERS (Essays on the Chinese Language, p. 38). BRETSCHNEIDER (Bot. Sin., pt. 1, p. 164) fixes the date at about 227–240.

[5] *T'aṅ lei han*, Ch. 183, p. 9.

[6] Written also 楉 榴. E. KAEMPFER (Amoenitates exoticae, p. 800) already mentions this term as *dsjakurjo*, vulgo *sakuro*, with the remark, "Rara est hoc coelo et fructu ingrato."

likewise from Iranian regions, presumably in the first centuries of our era. The tree is not mentioned in Vedic, Pāli, or early Sanskrit literature; and the word *dālima, dāḍima*, etc., is traceable to Iranian *dulim(a), which we have to reconstruct on the basis of the Chinese transcription. The Tibetans appear to have received the tree from Nepal, as shown by their ancient term *bal-poi seu-šiṅ* ("*seu* tree of Nepal").[1] From India the fruit spread to the Malayan Archipelago and Camboja. Both Čam *dalim* and Khmer *tātim*[2] are based on the Sanskrit word. The variety of pomegranate in the kingdom of Nan-čao in Yün-nan, with a skin as thin as paper, indicated in the *Yu yaṅ tsa tsu*,[3] may also have come from India. J. ANDERSON[4] mentions pomegranates as products of Yün-nan.

Pomegranate-wine was known throughout the anterior Orient at an early date. It is pointed out under the name *āsis* in Cant. VIII, 2 (Vulgata: *mustum*) and in the Egyptian texts under the name *šedeh-it*.[5] Dioscorides[6] speaks of pomegranate-wine (ῥοίτης οἶνος). Ye-lu Č'u-ts'ai, in his *Si yu lu* (account of his journey to Persia, 1219–24), speaking of the pomegranates of Khojand, which are "as large as two fists and of a sour-sweet taste," says that the juice of three or five fruits is pressed out into a vessel and makes an excellent beverage.[7] In the country Tun-sün 頓遜 (Tenasserim) there is a wine-tree resembling the pomegranate; the juice of its flowers is gathered and placed in jars, whereupon after several days it turns into good wine.[8] The inhabitants of Hai-nan made use of pomegranate-flowers in fermenting their wine.[9] I have not found any references to pomegranate-wine prepared by the Chinese, nor is it known to me that they actually make such wine.

It is known that the pomegranate, because of its exuberant seeds, is regarded in China as an emblem alluding to numerous progeny; it has become an anti-race-suicide symbol. The oldest intimation of this symbolism looms up in the *Pei ši* 北史, where it is told that two pomegranates were presented to King Nan-te 安德 of Ts'i 齊 on the occasion

---

[1] This matter has been discussed by me in *T'oung Pao*, 1916, pp. 408–410. In Lo-lo we have *sa-bu-se* in the A-hi dialect and *se-bu-se* in Nyi. *Sa* or *se* means "grain" (corresponding to Tibetan *sa* in *sa-bon*, "seed"). The last element *se* signifies "tree." The fruit is *se-bu-ma* (*ma*, "fruit").

[2] AYMONIER and CABATON, Dictionnaire čam-français, p. 220.

[3] Ch. 18, p. 3 b.

[4] Report on the Expedition to Western Yunan, p. 93 (Calcutta, 1871).

[5] V. LORET, Flore pharaonique, pp. 77, 78.

[6] v, 34.

[7] BRETSCHNEIDER, Mediæval Researches, Vol. I, p. 19.

[8] *Liaṅ šu*, Ch. 54, p. 3.

[9] HIRTH, Chau Ju-kua, p. 177.

of his marriage to the daughter of Li Tsu-šou 李祖收. The latter explained that the pomegranate encloses many seeds, and implies the wish for many sons and grandsons. Thus the fruit is still a favorite marriage gift or plays a rôle in the marriage feast.[1] The same is the case in modern Greece. Among the Arabs, the bride, when dismounting before the tent of the bridegroom, receives a pomegranate, which she smashes on the threshold, and then flings the seeds into the interior of the tent.[2] The Arabs would have a man like the pomegranate,—bitter-sweet, mild and affectionate with his friends in security, but tempered with a just anger if the time call him to be a defender in his own or in his neighbor's cause.[3]

---

[1] See, for instance, H. Doré, Recherches sur les superstitions en Chine, pt. 1 Vol. II, p. 479.

[2] A. Musil, Arabia Petraea, Vol. III, p. 191.

[3] C. M. Doughty, Travels in Arabia Deserta, Vol. I, p. 564.

SESAME AND FLAX

6. In A. DE CANDOLLE's book[1] we read, "Chinese works seem to show that sesame was not introduced into China before the Christian era. The first certain mention of it occurs in a book of the fifth or sixth century, entitled *Ts'i min yao šu*. Before this there is confusion between the name of this plant and that of flax, of which the seed also yields an oil, and which is not very ancient in China." Bretschneider is cited as the source for this information. It was first stated by the latter that, according to the *Pen ts'ao, hu ma* 胡麻 (*Sesamum orientale*) was brought by Čaṅ K'ien from Ta-yüan.[2] In his "Botanicon Sinicum"[3] he asserts positively that *hu ma*, or foreign hemp, is a plant introduced from western Asia in the second century B.C.[4] The same dogma is propounded by STUART.[5]

All that there is to this theory amounts to this. T'ao Huṅ-kiṅ (A.D. 451–536) is credited in the *Pen ts'ao kaṅ mu*[6] with the statement that "*hu ma* 胡麻 ('hemp of the Hu') originally grew in Ta-yüan (Fergana) 本生大宛,[7] and that it hence received the name *hu ma* ('Iranian hemp')." He makes no reference to Čaṅ K'ien or to the time when the introduction must have taken place; and to every one familiar with Chinese records the passage must evoke suspicion through its lack of precision and chronological and other circumstantial evidence. The records regarding Ta-yüan do not mention *hu ma*, nor does this term ever occur in the Annals. Now, T'ao Huṅ-kiṅ was a Taoist adept, a drug-hunter and alchemist, an immortality fiend; he never crossed the boundaries of his country, and certainly had no special information concerning Ta-yüan. He simply drew on his imagination by arguing, that, because *mu-su* (alfalfa) and grape sprang

---

[1] Origin of Cultivated Plants, p. 420.

[2] *Chinese Recorder*, 1871, p. 222; adopted by HIRTH, *T'oung Pao*, Vol. VI, 1895, p. 439, and maintained again in *Journal Am. Or. Soc.*, 1917, p. 92.

[3] Pt. II, p. 206.

[4] *Ibid.*, p. 204, he says, however, that the *Pen ts'ao* does not speak of flax, and that its introduction must be of more recent date. This conflicts with his statement above.

[5] Chinese Materia Medica, p. 404.

[6] Ch. 22, p. 1. Likewise in the earlier *Čeṅ lei pen ts'ao*, Ch. 24, p. 1 b.

[7] This tradition is reproduced without any reference in the *Pen ts'ao yen i* of 1116 (Ch. 20, p. 1, ed. of Lu Sin-yüan).

288

from Ta-yüan (that is, a Hu country), *hu ma* also, being a Hu plant, must likewise have emanated from that quarter. Such vagaries cannot be accepted as history. All that can be inferred from the passage in question is that T'ao Huṅ-kiṅ may have been familiar with *hu ma*. Li Ši-čen, quoting the *Moṅ k'i pi t'an* 夢溪筆談 by Šen Kwa 沈括[1] of the eleventh century, says, "In times of old there was in China only 'great hemp' *ta ma* 大麻 (*Cannabis sativa*) growing in abundance. The envoy of the Han, Čaṅ K'ien, was the first to obtain the seeds of oil-hemp 油麻[2] from Ta-yüan; hence the name *hu ma* in distinction from the Chinese species *ta ma*." The Čaṅ-K'ien tradition is further voiced in the *T'uṅ či* of Čeṅ Tsiao (1108–62) of the Sung.[3] The *T'ai p'iṅ yü lan*,[4] published in A.D. 983, quotes a *Pen ts'ao kiṅ* of unknown date as saying that Čaṅ K'ien obtained from abroad *hu ma* and *hu tou*.[5] This legend, accordingly, appears to have arisen under the Sung (A.D. 960–1278); that is, over a millennium after Čaṅ K'ien's lifetime. And then there are thinking scholars who would make us accept such stuff as the real history of the Han dynasty!

In the T'ang period this legend was wholly unknown: the *T'aṅ Pen ts'ao* does not allude to any introduction of *hu ma*, nor does this work speak of Čaṅ K'ien in this connection.

A serious book like the *T'u kiṅ pen ts'ao* of Su Suṅ, which for the first time has also introduced the name *yu ma* ("oil hemp"), says only that the plant originally grew in the territory of the Hu, that in appearance it is like hemp, and that hence it receives the name *hu ma*.

Unfortunately it is only too true that the Chinese confound *Sesamum indicum* (family *Pedaliaceae*) and *Linum usitatissimum* (family *Linaceae*) in the single term *hu ma* ("Iranian hemp"); the only apparent reason for this is the fact that the seeds of both plants yield an oil which is put to the same medicinal use. The two are totally different plants, nor do they have any relation to hemp. Philologically, the case is somewhat analogous to that of *hu tou* (p. 305). It is most probable that the two are but naturalized in China and introduced from Iranian regions, for both plants are typically ancient West-Asiatic cultivations. The alleged wild sesame of China[6] is doubtless an escape from cultivation.

---

[1] This is the author wrongly called "Ch'en Ts'ung-chung" by BRETSCHNEIDER (Bot. Sin., pt. II, p. 377). Ts'un-čuṅ 存中 is his *hao*.

[2] A synonyme of *hu ma*.

[3] Ch. 75, p. 33.

[4] Ch. 841, p. 6 b.

[5] See below, p. 305.

[6] FORBES and HEMSLEY, *Journal Linnean Soc.*, Vol. XXVI, p. 236.

Herodotus[1] emphasizes that the only oil used by the Babylonians is made from sesame. Sesame is also mentioned among their products by the Babylonian priest Berosus (fourth century B.C.).[2]

Aelius Gallus, a member of the Equestrian order, carried the Roman arms into Arabia, and brought back from his expedition the report that the Nomades (nomads) live on milk and the flesh of wild animals, and that the other peoples, like the Indians, express a wine from palms and oil from sesame.[3] According to Pliny, sesame comes from India, where they make an oil from it, the color of the seeds being white.[4] Both the seeds and the oil were largely employed in Roman pharmacology.[5] Megasthenes[6] mentions the cultivation of sesame in India. It likewise occurs in the Atharva Veda and in the Institutes of Manu (Sanskrit *tila*).[7] A. DE CANDOLLE's view[8] that it was introduced into India from the Sunda Isles in prehistoric times, is untenable. This theory is based on a purely linguistic argument: "Rumphius gives three names for the sesame in these islands, very different one from the other, and from the Sanskrit word, which supports the theory of a more ancient existence in the archipelago than on the continent." This alleged evidence proves nothing whatever for the history of the plant, but is merely a fact of language.[9] There can now be no doubt that from a botanical viewpoint the home of the genus is in tropical Africa, where twelve species occur, while there are only two in India.[10]

In the *Fan yi min yi tsi*,[11] a Sanskrit synonyme of "sesame" is given as 阿提目多伽 a-t'i-mu-to-k'ie, *a-di-muk-ta-g'a, i.e., Sanskrit *adhimuktaka*, which is identified with *kü-šen* (see below) and *hu-ma*. An old gloss explains the term as "the foreign flower of pious thoughtfulness" (*šan se i hwa* 善思夷華), an example of which is the lighting of a lamp fed with the oil of three flowers (sandal, soma, and campaka [*Michelia champaca*]) and the placing of this lamp on the altar of the

---

[1] I, 193.

[2] MÜLLER, Fragmenta historiae graecae, Vol. II, p. 496. Regarding Egypt, see V. LORET, Flore pharaonique, p. 57.

[3] Pliny, VI, 28, §161.

[4] Sesama ab Indis venit. Ex ea et oleum faciunt; colos eius candidus (XVIII, 22, §96).

[5] Pliny, XXII, 64, §132.

[6] Strabo, XV. I, 13.

[7] JORET, Plantes dans l'antiquité, Vol. II, p. 269.

[8] Origin of Cultivated Plants, p. 422.

[9] The Malayan languages possess a common name for *Sesamum indicum:* Javanese and Malayan *leña*, Batak *loña*, Čam *loñö* or *lañö;* Khmer *loño*.

[10] A. ENGLER, Pflanzenfamilien, Vol. IV, pt. 3 b, p. 262.

[11] Ch. 8, p. 6 (see above, p. 254).

Triratna.[1] From the application of *adhimuktaka* it becomes self-evident also that sesame-oil must be included in this series. The frequent mention of this oil for sacred lamps is familiar to all readers of the Buddhist Jātaka. The above Sanskrit-Chinese Dictionary adds the following comment: "This plant is in appearance like the 'great hemp' (*Cannabis sativa*). It has red flowers and green leaves. Its seeds can be made into oil; also they yield an aromatic. According to the *Tsuṅ kiṅ yin nie lun* 宗鏡引攝論, sesame (*kü-šeṅ*) is originally charcoal, and, while for a long time buried in the soil, will change into sesame. In the western countries (India) it is customary in anointing the body with fragrant oil to use first aromatic flowers and then to take sesame-seeds. These are gathered and soaked till thoroughly bright; afterwards they proceed to press the oil out of the sesame, which henceforth becomes fragrant."

Of greater importance for our purpose is the antiquity of sesame in Iran. According to Herodotus[2], it was cultivated by the Chorasmians, Hyrcanians, Parthians, Sarangians, and Thamanæans. In Persia sesame-oil was known at least from the time of the first Achæmenides.[3] G. WATT[4] even looks to Persia and Central Asia as the home of the species; he suggests that it was probably first cultivated somewhere between the Euphrates valley and Bukhara south to Afghanistan and upper India, and was very likely diffused into India proper and the Archipelago, before it found its way to Egypt and Europe.

Sesamum indicum (var. *subindivisum* Dl.) is cultivated in Russian Turkistan and occupies there the first place among the oil-producing plants. It thrives in the warmest parts of the valley of Fergana, and does not go beyond an elevation of two thousand five hundred feet. It is chiefly cultivated in the districts of Namanga and Andijan, though not in large quantity.[5] Its Persian name is *kunjut*.

While there is no doubt that this species was introduced into China from Iranian regions, the time as to when this introduction took place remains obscure. First, there is no historical and dependable record of this event; second, the confusion brought about by the Chinese in treating this subject is almost hopeless. Take the earliest notice of *hu ma* cited by the *Pen ts'ao* and occurring in the *Pie lu:* "*Hu ma* is also called *kü-šeṅ* 巨勝. It grows on the rivers and in the marshes of

----

[1] Cf. EITEL, Handbook of Chinese Buddhism, p. 4.

[2] III, 117.

[3] JORET, *op. cit.*, Vol. II, p. 71. Sesame is mentioned in Pahlavi literature (above, p. 193).

[4] Gingelly or Sesame Oil, p. 11 (Handbooks of Commercial Products, No. 21).

[5] S. KORŽINSKI, Vegetation of Turkistan (in Russian), p. 50.

Šaṅ-taṅ 上黨 (south-eastern portion of Šan-si), and is gathered in the autumn. What is called ts'iṅ žaṅ 青蘘 are the sprouts of the kü-šeṅ. They grow in the river-valleys of Čuṅ-yüan 中原 (Ho-nan)." Nothing is said here about a foreign introduction or a cultivation; on the contrary, the question evidently is of an indigenous wild swamp-plant, possibly *Mulgedium sibiriacum*.[1] Both *Sesamum* and *Linum* are thoroughly out of the question, for they grow in dry loam, and sesame especially in sandy soil. Thus suspicion is ripe that the terms *hu ma* and *kü-šeṅ* originally applied to an autochthonous plant of Šan-si and Ho-nan, and that *hu ma* in this case moves on the same line as the term *hu šeṅ* in the *Li sao* (p. 195). This suspicion is increased by the fact that *hu ma* occurs in a passage ascribed to Hwai-nan-tse, who died in 122 B.C., and cited in the *T'ai p'iṅ yü lan*.[2] Moreover, the *Wu ši* (or *p'u*) *pen ts'ao*, written in the first half of the third century by Wu P'u 吳普, in describing *hu ma*, alludes to the mythical Emperor Šen-nuṅ and to Lei kuṅ 雷公, a sage employed by the Emperor Hwaṅ in his efforts to perfect the art of healing.

The meaning of *kü-šeṅ* is "the great superior one." The later authors regard the term as a variety of *Sesamum*, but give varying definitions of it: thus, T'ao Huṅ-kiṅ states that the kind with a square stem is called *kü-šeṅ* (possibly *Mulgedium*), that with a round stem *hu ma*. Su Kuṅ of the T'ang says that the plant with capsules (*kio* 角) of eight ridges or angles (*pa leṅ* 八棱) is called *kü-šeṅ;* that with quadrangular capsules, *hu ma*. The latter definition would refer to *Sesamum indicum*, the capsule of which is oblong quadrangular, two-valved and two-celled, each cell containing numerous oily seeds.

Moṅ Šen 孟詵, in his *Ši liao pen t'sao* (written in the second half of the seventh century), observes that "the plants cultivated in fertile soil produce octangular capsules, while those planted in mountainous fields have the capsules quadrangular, the distinction arising from the difference of soil conditions, whereas the virtues of the two varieties are identical. Again, Lei Hiao 雷斆 of the fifth century asserts that *kü-šeṅ* is genuine, when it has seven ridges or angles, a red color, and a sour taste, but that it is erroneous to style *hu ma* the octangular capsules with two pointed ends, black in color, and furnishing a black oil. There is no doubt that in these varying descriptions entirely different plants are visualized. Kao Č'eṅ of the Sung, in his *Ši wu ki yüan*,[3]

---

[1] STUART, Chinese Materia Medica, p. 269. This identification, however, is uncertain.

[2] Ch. 989, p. 6 b.

[3] Ch. 10, p. 29 b (see above, p. 279).

admits that it is unknown what the *hu ma* spoken of in the *Pen-ts'ao* literature really is.

I have also prepared a translation of Li Ši-čen's text on the subject, which Bretschneider refrained from translating; but, as there are several difficult botanical points which I am unable to elucidate, I prefer to leave this subject to a competent botanist. In substance Li Ši-čen understands by *hu ma* the sesame, as follows from his use of the modern term *či ma* 脂麻. He says that there are two crops, an early and a late one,[1] with black, white, or red seeds; but how he can state that the stems are all square is unintelligible. The criticism of the statements of his predecessors occupies much space, but I do not see that it enlightens us much. The best way out of this difficulty seems to me Stuart's suggestion that the Chinese account confounds *Sesamum, Linum,* and *Mulgedium.* The Japanese naturalist Ono Ranzan[2] is of the same opinion. He says that there is no variety of sesame with red seed, as asserted by Li Ši-čen (save that the black seeds of sesame are reddish in the immature stage), and infers that this is a species of *Linum* which always produces red seeds exclusively. Ono also states that there is a close correlation between the color of the seeds and the angles of the capsules: a white variety will always produce two or four-angled capsules, while hexangular and octangular capsules invariably contain only black seeds. Whether or in how far this is correct I do not know. The confusion of *Sesamum* and *Linum* arose from the common name *hu ma,* but unfortunately proves that the Chinese botanists, or rather pharmacists, were bookworms to a much higher degree than observers; for it is almost beyond comprehension how such radically distinct plants can be confounded by any one who has even once seen them. In view of this disconsolate situation, the historian can only beg to be excused.

7. It is a point of great culture-historical interest that the Chinese have never utilized the flax-fibre in the manufacture of textiles, but that hemp has always occupied this place from the time of their earliest antiquity.[3] This is one of the points of fundamental diversity between East-Asiatic and Mediterranean civilizations,— there hemp, and here flax, as material for clothing. There are, further, two important facts to be considered in this connection,— first, that the Aryans

---

[1] In S. COULING's Encyclopædia Sinica (p. 504) it is stated that in China there is only one crop, but late and early varieties exist.

[2] *Honzō kōmoku keimō,* Ch. 18, p. 2.

[3] In a subsequent study on the plants and agriculture of the Indo-Chinese, I hope to demonstrate that the Indo-Chinese nations, especially the Chinese and Tibetans, possess a common designation for "hemp," and that hemp has been cultivated by them in a prehistoric age. There also the history of hemp will be discussed.

(Iranians and Indo-Aryans) possess an identical word for "hemp" (Avestan *bangha*, Sanskrit *bhanga*), while the European languages have a distinct designation, which is presumably a loan-word pointing to Finno-Ugrian and Turkish; and, second, that there is a common Old-Turkish word for "hemp" of the type *kändir*, which stands in some relation to the Finno-Ugrian appellations.[1] It is most likely that the Scythians brought hemp from Asia to Europe.[2] On the other hand, it is well known what vital importance flax and linen claimed in the life of the Egyptians and the classical peoples.[3] Flax is the typically European, hemp the typically Asiatic textile. Surely *Linum usitatissimum* was known in ancient Iran and India. It was and is still wild in the districts included between the Persian Gulf, the Caspian Sea, and the Black Sea.[4] It was probably introduced into India from Iran, but neither in India nor in Iran was the fibre ever used for garments: the plant was only cultivated as a source of linseed and linseed-oil.[5] Only a relatively modern utilization of flax-fibres for weaving is known from a single locality in Persia,—Kāzirūn, in the province of Fars. This account dates from the beginning of the fourteenth century, and the detailed description given of the process testifies to its novelty and exceptional character.[6] This exception confirms the rule. The naturalization of *Linum* in China, of course, is far earlier than the fourteenth century. As regards the utilization of *Linum*, the Chinese fall in line with Iranians and Indo-Aryans; and it is from Iranians that they received the plant. The case is a clear index of the fact that the Chinese never were in direct contact with the Mediterranean culture-area, and that even such cultivated plants of this area as reached them were not transmitted from there directly, but solely through the medium of Iranians. The case is further apt to illustrate how superficial, from the viewpoint of technical culture, the influence of the Greeks on the Orient must have been since Alexander's campaign, as an industry like flax-weaving was not promoted by them, although the material was offered there by nature.

For botanical reasons it is possible that *Linum usitatissimum* was introduced into China from Fergana. There it is still cultivated, and only for the exclusive purpose of obtaining oil from the seeds.[7] As has

---

[1] Z. GOMBOCZ, Bulgarisch-türkische Lehnwörter, p. 92.

[2] Cf. for the present, A. DE CANDOLLE, Origin of Cultivated Plants, p. 148.

[3] Pliny, XIX, 1-3; H. BLÜMNER, Technologie, Vol. I, 2d ed., p. 191.

[4] A. DE CANDOLLE, Origin of Cultivated Plants, p. 130.

[5] See the interesting discussion of WATT, Commercial Products of India, p. 721.

[6] G. LE STRANGE, Description of the Province of Fars in Persia, p. 55.

[7] S. KORŽINSKI, Vegetation of Turkistan (in Russian), p. 51.

been pointed out, the plant is indigenous also in northern Persia, and must have been cultivated there from ancient times, although we have no information on this point from either native documents or Greek authors.[1]

BRETSCHNEIDER[2] says that "flax was unknown to the ancient Chinese; it is nowadays cultivated in the mountains of northern China (probably also in other parts) and in southern Mongolia, but only for the oil of its seeds, not for its fibres; the Chinese call it *hu ma* ('foreign hemp'); the *Pen ts'ao* does not speak of it; its introduction must be of more recent date." This is erroneous. The *Pen ts'ao* includes this species under the ambiguous term *hu ma;* and, although the date of the introduction cannot be ascertained, the event seems to have taken place in the first centuries of our era.

At present, the designation *hu ma* appears to refer solely to flax. A. HENRY[3] states under this heading, "This is flax (*Linum usitatissimum*), which is cultivated in San-si, Mongolia, and the mountainous parts of Hu-pei and Se-č'wan. In the last two provinces, from personal observation, flax would seem to be entirely cultivated for the seeds, which are a common article in Chinese drug-shops, and are used locally for their oil, utilized for cooking and lighting purposes." In another paper,[4] the same author states that *Linum usitatissimum* is called at Yi-č'an, Se-č'wan, *šan či ma* 山脂[5]麻 ("mountain sap-hemp"), and that it is cultivated in the mountains of the Patun district, not for the fibre, but for the oil which the seed yields.

Chinese *hu ma* has passed into Mongol as *xuma* (*khuma*) with the meaning "sesame,"[6] and into Japanese as *goma*, used only in the sense of *Sesamum indicum*,[7] while *Linum usitatissimum* is in Japanese *ama* or *ičinen-ama*.[8]

Yao Min-hwi 姚明煇, in his book on Mongolia (*Mon-ku či*),[9] mentions *hu ma* among the products of that country. There are several wild-growing species of *Linum* in northern China and Japan,— *ya ma*

---

[1] JORET, Plantes dans l'antiquité, Vol. II, p. 69.

[2] Bot. Sin., pt. II, p. 204.

[3] Chinese Jute, p. 6 (publication of the Chinese Maritime Customs, Shanghai, 1891).

[4] Chinese Names of Plants, p. 239 (*Journal China Branch Royal As. Soc.*, Vol. XXII, 1887).

[5] The popular writing 芝, according to the *Pen ts'ao kan mu*, is incorrect.

[6] KOVALEVSKI, Dictionnaire mongol, p. 934.

[7] MATSUMURA, No. 2924.

[8] *Ibid.*, No. 1839.

[9] Ch. 3, p. 41 (Shanghai, 1907).

亞麻 (Japanese *nume-goma* or *aka-goma*), *Linum perenne*, and Japanese *matsuba-ninjin* or *matsuba-nadešiko*, *Linum possarioides*.[1] FORBES and HEMSLEY,[2] moreover, enumerate *Linum nutans* for Kan-su, and *L. stelleroides* for Či-li, Šan-tuṅ, Manchuria, and the Korean Archipelago. In northern China, *Linum sativum* (*Šan-si hu ma* 山西胡麻) is cultivated for the oil of its seeds.[3]

---

[1] MATSUMURA, Nos. 1837, 1838; STUART, Chinese Materia Medica, p. 242.

[2] *Journal Linnean Soc.*, Vol. XXIII, p. 95.

[3] This species is figured and described in the *Či wu miṅ ši t'u k'ao*.

THE CORIANDER

8. The *Po wu či*, faithful to its tendencies regarding other Iranian plants, generously permits General Čan K'ien to have also brought back from his journey the coriander, *hu swi* 胡荽 (*Coriandrum sativum*).[1] Li Ši-čen, and likewise K'an-hi's Dictionary, repeat this statement without reference to the *Po wu či*;[2] and of course the credulous community of the Changkienides has religiously sworn to this dogma.[3] Needless to say that nothing of the kind is contained in the General's biography or in the Han Annals.[4] The first indubitable mention of the plant is not earlier than the beginning of the sixth century A.D.; that is, about six centuries after the General's death, and this makes some difference to the historian.[5] The first *Pen ts'ao* giving the name *hu-swi* is the *Ši liao pen ts'ao*, written by Mon Šen in the seventh century, followed by the *Pen ts'ao ši i* of Č'en Ts'an-k'i in the first half of the eighth century. None of these authors makes any observation on foreign introduction. In the literature on agriculture, the cultivation of the coriander is first described in the *Ts'i min yao šu* of the sixth century, where, however, nothing is said about the origin of the plant from abroad.

An interesting reference to the plant occurs in the Buddhist dictionary *Yi ts'ie kin yin i* (*l.c.*), where several variations for writing

---

[1] This passage is not a modern interpolation, but is of ancient date, as it is cited in the *Yi ts'ie kin yin i*, Ch. 24, p. 2 (regarding this work, see above, p. 258). Whether it was contained in the original edition of the *Po wu či*, remains doubtful.

[2] Under 葫 ("garlic") *K'an-hi* cites the dictionary *T'an yün*, published by Sun Mien in A.D. 750, as saying that the coriander is due to Čan K'ien.

[3] BRETSCHNEIDER, *Chinese Recorder*, 1871, p. 221, where the term *hu-swi* is wrongly identified with parsley, and Bot. Sin., pt. I, p. 25; HIRTH, *T'oung Pao*, Vol. VI, 1895, p. 439.

[4] The coriander is mentioned in several passages of the *Kin kwei yao lio* by the physician Čan Čun-kin of the second century A.D.; but, as stated above (p. 205), there is no guaranty that these passages belonged to the original edition of the work. "To eat pork together with raw coriander rots away the navel" (Ch. C, p. 23 b). "In the fourth and eighth months do not eat coriander, for it injures the intellect" (*ibid.*, p. 28). "Coriander eaten for a long time makes man very forgetful; a patient must not eat coriander or *hwan-hwa ts'ai* 黃花菜 (*Lampsana apogonoides*)," *ibid.*, p. 29.

[5] An incidental reference to *hu swi* is made in the *Pen ts'ao kan mu* in the description of the plant *küan er* (see BRETSCHNEIDER, Bot. Sin., pt. II, No. 438), and ascribed to Lu Ki, who lived in the latter part of the third century A.D. In my opinion, this reading is merely due to a misprint, as there is preserved no description of the *hu-swi* by Lu Ki.

the character *swi* are given, also the synonymes *hiaṅ ts'ai* 香菜 ("fragrant vegetable") and *hiaṅ sün* 香荽.[1] In Kiaṅ-nan the plant was styled *hu swi* 胡荽, also *hu ki* 葫蔧, the pronunciation of the latter character being explained by 祇 *k'i*, *gi. The coriander belongs to the five vegetables of strong odor (p. 303) forbidden to the geomancers and Taoist monks.[2]

I have searched in vain for any notes on the plant that might elucidate its history or introduction; but such do not seem to exist, not even in the various *Pen ts'ao*. As regards the Annals, I found only a single mention in the *Wu Tai Ši*,[3] where the coriander is enumerated among the plants cultivated by the Uigur. In tracing its foreign origin, we are thrown back solely on the linguistic evidence.

The coriander was known in Iran: it is mentioned in the Būndahišn.[4] Its medical properties are discussed in detail by Abu Mansur in his Persian pharmacopœia.[5] SCHLIMMER[6] observes, "Se cultive presque partout en Perse comme plante potagère; les indigènes le croient antiaphrodisiaque et plus spécialement anéantissant les érections." It occurs also in Fergana.[7] It was highly appreciated by the Arabs in their pharmacopœia, as shown by the long extract devoted to it by Ibn al-Baiṭār.[8] In India it is cultivated during the cold season. The Sanskrit names which have been given on p. 284, mean simply "grain," and are merely attributes,[9] not proper designations of the plant, for which in fact there is no genuine Sanskrit word. As will be seen below, Sanskrit *kustumburu* is of Iranian origin; and there is no doubt in my mind that the plant came to India from Iran, in the same manner as it appears to have spread from Iran to China.

胡荽 or 蔧 *hu-swi*, *ko(go)-swi (su), appears to be the transcription of an Iranian form *koswi, košwi, gošwi. Cf. Middle Persian *gošniz*;

---

[1] Two dictionaries, the *Tse yüan* 字苑 and *Yün lio* 韻略, are quoted in this text, but their date is not known to me. As stated in the *Pen ts' ao ši i* and *Ši wu ki yüan* (Ch. 10, p. 30; above, p. 279), the change from *hu swi* to *hiaṅ swi* was dictated by a taboo imposed by Ši Lo 石勒 (A.D. 273-333), who was himself a Hu (cf. below, p. 300); but we have no contemporaneous account to this effect, and the attempt at explanation is surely retrospective.

[2] *Pen ts'ao kaṅ mu*, Ch. 26, p. 6 b; and STUART, Chinese Materia Medica, p. 28.

[3] Ch. 74, p. 4.

[4] Above, p. 192.

[5] ACHUNDOW, Abu Mansur, p. 112.

[6] Terminologie, p. 156.

[7] S. KORŽINSKI, Vegetation of Turkistan (in Russian), p. 51.

[8] L. LECLERC, Traité des simples, Vol. III, pp. 170-174.

[9] Such are also the synonymes *sūkṣmapatra*, *tīkṣṇapatra*, *tīkṣṇaphala* ("with leaves or fruits of sharp taste").

New Persian *kišnīz, kušnīz,* and *gišnīz,* also *šūnīz;*[1] Kurd *ksnis* or *kišniš;* Turkish *kišniš;* Russian *kišnêts;* Aramaic *kusbarta* and *kusbar* (Hebrew *gad,* Punic γοίδ, are unconnected), Arabic *kozbera* or *kɔsberet;* Sanskrit *kustumburu* and *kustumbarī;* Middle and Modern Greek κουσбαράs[2] and κισυνήτζι.

According to the *Hui k'ian či,* the coriander is called in Turkistan (that is, in Turkī) *yuṅ-ma-su* 永 麻 素.

It is commonly said that the coriander is indigenous to the Mediterranean and Caucasian regions (others say southern Europe, the Levant, etc.), but it is shown by the preceding notes that Iran should be included in this definition. I do not mean to say, however, that Iran is the exclusive and original home of the plant. Its antiquity in Egypt and in Palestine cannot be called into doubt. It has been traced in tombs of the twenty-second dynasty (960–800 B.C.),[3] and Pliny[4] states that the Egyptian coriander is the best. In Iran the cultivation seems to have been developed to a high degree; and the Iranian product was propagated in all directions,— in China, India, anterior Asia, and Russia.

The Tibetan name for the coriander, *u-su,* may be connected with or derived from Chinese *hu-sui.* L. A. WADDELL[5] saw the plant cultivated in a valley near Lhasa. It is also cultivated in Siam.[6]

Coriander was well known in Britain prior to the Norman Conquest, and was often employed in ancient Welsh and English medicine and cookery.[7] Its Anglo-Saxon name is *cellendre, coliandre,* going back to Greek *koriándron, koríannon.*

---

[1] Another Persian word is *bûghunj.* According to STEINGASS (Persian Dictionary), *tälki* or *tälgi* denotes a "wild coriander."

[2] The second element of the Arabic, Sanskrit, and Greek words seems to bear some relation to Coptic *beršiu, berešu* (V. LORET, Flore pharaonique, p. 72). In Greece, coriander is still cultivated, but only sparsely, near Theben, Corinth, and Cyparissia (TH. v. HELDREICH, Nutzpflanzen Griechenlands, p. 41).

[3] V. LORET, *op. cit.,* p. 72; F. WOENIG, Pflanzen im alten Aegypten, p. 225.

[4] XX, 20, §82.

[5] Lhasa, p. 316.

[6] PALLEGOIX, Description du royaume thai, Vol. I, p. 126.

[7] FLÜCKIGER and HANBURY, Pharmacographia, p. 329.

THE CUCUMBER

9. Another dogma of the Changkienomaniacs is that the renowned General should have also blessed his countrymen with the introduction of the cucumber (*Cucumis sativus*), styled *hu kwa* 胡瓜 ("Iranian melon") or *hwaṅ kwa* 黃瓜 ("yellow melon").[1] The sole document on which this opinion is based is presented by the recent work of Li Ši-čen,[2] who hazards this bold statement without reference to any older authority. Indeed, such an earlier source does not exist: this bit of history is concocted *ad hoc*, and merely suggested by the name *hu kwa*. Any plants formed with the attribute *hu* were ultimately palmed off on the old General as the easiest way out of a difficult problem, and as a comfortable means of saving further thought.

Li Ši-čen falls back upon two texts only of the T'ang period,— the *Pen ts'ao ši i*, which states that the people of the north, in order to avoid the name of Ši Lo 石勒 (A.D. 273–333), who was of Hu descent, tabooed the term *hu kwa*, and replaced it by *hwaṅ kwa;*[3] and the *Ši i lu* 拾遺錄 by Tu Pao 杜寶, who refers this taboo to the year 608 (fourth year of the period Ta-ye of the Sui dynasty).[4] If this information be correct, we gain a chronological clew as to the *terminus a quo:* the cucumber appears to have been in China prior to the sixth century A.D. Its cultivation is alluded to in the *Ts'i min yao šu* from the beginning of the sixth century, provided this is not an interpolation of later times.[5]

According to ENGLER,[6] the home of the cucumber would most prob-

---

[1] BRETSCHNEIDER, *Chinese Recorder*, 1871, p. 21 (accordingly adopted by DE CANDOLLE, Origin of Cultivated Plants, p. 266); STUART, Chinese Materia Medica, p. 135. In Japanese, the cucumber is *ki-uri*.

[2] *Pen ts'ao kaṅ mu*, Ch. 28, p. 5 b.

[3] A number of other plant-names was hit by this taboo (cf. above, p. 298): thus the plant *lo-lo* 羅勒 (*Ocimum basilicum*), which bears the same character as Ši Lo's personal name, as already indicated in the *Ts'i min yao šu* (see also *Ši wu ki yüan*, Ch. 10, p. 30 b; *Či wu miṅ ši t'u k'ao*, Ch. 5, p. 34; and *Pen ts'ao kaṅ mu*, Ch. 26, p. 22 b). He is said to have also changed the name of the myrobalan *ho-li-lo* (below, p. 378) into *ho-tse* 訶子. There is room for doubt, however, whether any of these plants existed in the China of his time; the taboo explanations may be makeshifts of later periods.

[4] This is the *Ta ye ši i lu* (Records relative to the Ta-ye period, 605–618), mentioned by BRETSCHNEIDER (Bot. Sin., pt. 1, p. 195). The *Pen ts'ao kaṅ mu* (Ch. 22, p. 1) quotes the same work again on the taboo of the term *hu ma* (p. 288), which in 608 was changed into *kiao ma* 交麻.

[5] Cf. *Či wu miṅ ši t'u k'ao*, Ch. 5, p. 43.

[6] In Hehn, Kulturpflanzen, p. 323.

300

ably be in India; and WATT[1] observes, "There seems to be no doubt that one at least of the original homes of the cucumber was in North India, and its cultivation can be traced to the most ancient classic times of Asia." DE CANDOLLE[2] traces the home of the plant to northwestern India. I am not yet convinced of the correctness of this theory, as the historical evidence in favor of India, as usual in such cases, is weak;[3] and the cultivation of the cucumber in Egypt and among the Semites is doubtless of ancient date.[4] At any rate, this *Cucurbitacea* belongs to the Egypto-West-Asiatic culture-sphere, and is not indigenous to China. There is, however, no trace of evidence for the gratuitous speculation that its introduction is due to General Čan K'ien. The theory that it was transmitted from Iranian territory is probable, but there is thus far no historical document to support it. The only trace of evidence thereof appears from the attribute Hu.

Abu Mansur mentions the cucumber under the name *qittā*, adding the Arabic-Persian *xiyār* and *kawanda* in the language of Khorasan.[5] The word *xiyār* has been adopted into Osmanli and into Hindustānī in the form *xīrā*. Persian *xāwuš* or *xāwaš* denotes a cucumber kept for seed; it means literally "ox-eye" (*gāv-aš; Avestan aši, Middle Persian aš, Sanskrit akṣi, "eye"), corresponding to Sanskrit *gavākṣī* ("a kind of cucumber"). A Pahlavi word for "cucumber" is *vātraṅ*, which developed into New Persian *bādraṅ, bālaṅ,* or *vāraṅ* (Afghan *bādraṅ*).[6]

---

[1] Commercial Products of India, p. 439. In Sanskrit the cucumber is *trapuṣa*.

[2] *Op. cit.,* p. 265.

[3] Such a positive assertion as that of de Candolle, that the cucumber was cultivated in India for at least three thousand years, cannot be accepted by any serious historian.

[4] V. LORET, Flore pharaonique, p. 75; C. JORET, Plantes dans l'antiquité, Vol. I, p. 61.

[5] ACHUNDOW, Abu Mansur, p. 106.

[6] This series is said to mean also "citron." The proper Persian word for the latter fruit is *turunj* (Afghan *turanj*, Balūči *trunj*). The origin of this word, as far as I know, has not yet been correctly explained, not even by HÜBSCHMANN (Armen. Gram., p. 266). VULLERS (Lexicon persico-latinum, Vol. I, p. 439) tentatively suggests derivation from Sanskrit *suraṅga*, which is surely impossible. The real source is presented by Sanskrit *mātuluṅga* ("citron," *Citrus medica*).

CHIVE, ONION, AND SHALLOT

10. Although a number of alliaceous plants are indigenous to China,[1] there is one species, the chive (*Allium scorodoprasum;* French *rocambole*), to which, as already indicated by its name *hu swan* 胡蒜 or *hu* 葫 ("garlic of the Hu, Iranian garlic"), a foreign origin is ascribed by the Chinese. Again, the worn-out tradition that also this introduction is due to Čaṅ K'ien, is of late origin, and is first met with in the spurious work *Po wu či*, and then in the dictionary *T'aṅ yün* of the middle of the eighth century.[2] Even Li Ši-čen[3] says no more than that "people of the Han dynasty obtained the *hu swan* from Central Asia." It seems difficult, however, to eradicate a long-established prejudice or an error even from the minds of scholars. In 1915 I endeavored to rectify it, especially with reference to the wrong opinion expressed by Hirth in 1895, that garlic in general must have been introduced into China for the first time by Čaṅ K'ien. Nevertheless the same misconception is repeated by him in 1917,[4] while a glance at the Botanicon Sinicum[5] would have convinced him that at least four species of *Allium* are of a prehistoric antiquity in China. The first mention of this Central-Asiatic or Iranian species of *Allium* is made by T'ao Huṅ-kiṅ (A.D. 451–536), provided the statement attributed to him in the *Čeṅ lei pen ts'ao* and *Pen ts'ao kaṅ mu* really emanates from him.[6] When the new *Allium* was introduced, the necessity was felt of distinguishing it from the old, indigenous *Allium sativum*, that was designated by the plain root-word *swan*. The former, accordingly, was characterized as *ta swan* 大蒜 ("large *Allium*"); the latter, as *siao* 小 *swan* ("small *Allium*"). This distinction is said to have first been recorded by T'ao Huṅ-kiṅ. Also the *Ku kin ču* is credited with the mention of *hu swan*; this, however, is not the older *Ku kin ču* by Ts'ui Pao of the fourth century, but, as expressly stated in the *Pen ts'ao*, the later re-edition by Fu Hou

---

[1] Cf. *T'oung Pao*, 1915, pp. 96–99.

[2] BRETSCHNEIDER, Bot. Sin., pt. III, No. 244.

[3] *Pen ts'ao kaṅ mu*, Ch. 26, p. 6 b.

[4] *Journal Am. Or. Soc.*, Vol. XXXVII, p. 92.

[5] Pt. II, Nos. 1–4, 63, 357–360, and III, Nos. 240–243.

[6] The *Kin kwei yao lio* (Ch. c, p. 24 b) of the second century A.D. mentions *hu swan*, but this in all probability is a later interpolation (above, p. 205).

302

伏候 of the tenth century. However, this text is now inserted in the older *Ku kin ču*,[1] which teems with interpolations.

Ta swan is mentioned also as the first among the five vegetables of strong odor tabooed for the Buddhist clergy, the so-called *wu hun* 五葷.[2] This series occurs in the Brahmajāla-sūtra, translated in A.D. 406 by Kumārajīva.[3] If the term *ta swan* was contained in the original edition of this work, we should have good evidence for carrying the date of the chive into the Eastern Tsin dynasty (A.D. 317–419).

11. There is another cultivated species of *Allium* (probably *A. fistulosum*) derived from the West. This is first mentioned by Sun Se-miao 孫思邈,[4] in his *Ts'ien kin ši či* 千金食治 (written in the beginning of the seventh century), under the name *hu ts'uṅ* 葫蔥, because the root of this plant resembles the *hu swan* 葫蒜. It was usually styled *swan-ts'uṅ* 蒜蔥 or *hu* 胡 *ts'uṅ* (the latter designation in the *K'ai pao pen ts'ao* of the Sung). In the *Yin šan čeṅ yao* (p. 236), written in 1331 under the Yüan, it is called *hui-hui ts'uṅ* 回回蔥 ("Mohammedan onion").[5] This does not mean, however, that it was only introduced by Mohammedans; but this is simply one of the many favorite alterations of ancient names, as they were in vogue during the Mongol epoch. This *Allium* was cultivated in Se-č'wan under the T'ang, as stated by Moṅ Šen 孟詵 in his *Ši liao pen ts'ao*, written in the second half of the seventh century. Particulars in regard to the introduction are not on record.

12. There is a third species of *Allium*, which reached China under the T'ang, and which, on excellent evidence, may be attributed to Persia. In A.D. 647 the Emperor T'ai Tsuṅ solicited from all his tributary nations their choicest vegetable products,[6] and their response to the imperial call secured a number of vegetables hitherto unknown in China. One of these is described as follows: "*Hun-t'i* onion 渾提蔥 resembles in appearance the onion (*ts'uṅ*, *Allium fistulosum*), but is whiter and more bitter. On account of its smell, it serves as a remedy.

---

[1] Ch. c, p. 3 b.

[2] This subject is treated in the *Pen ts'ao kaṅ mu* (Ch. 26, p. 6 b) under the article *swan*, and summed up by STUART (Chinese Materia Medica, p. 28). See, further, DE GROOT, Le Code du Mahāyāna en Chine, p. 42, where the five plant-names are unfortunately translated wrongly (*hiṅ-k'ü*, "asafœtida" [see p. 361], is given an alleged literal translation as "le lys d'eau montant"!), and CHAVANNES and PELLIOT, Traité manichéen, pp. 233–235.

[3] BUNYIU NANJIO, Catalogue of the Buddhist Tripiṭaka, No. 1087.

[4] Cf. below, p. 306.

[5] *Pen ts'ao kaṅ mu*, Ch. 26, p. 5.

[6] We shall come back to this important event in dealing with the history of the spinach.

In its appearance it is like *lan-liṅ-tuṅ* 蘭凌冬,[1] but greener. When dried and powdered, it tastes like cinnamon and pepper. The root is capable of relieving colds."[2] The *Fuṅ ši wen kien ki*[3] adds that *hun-t'i* came from the Western Countries (*Si yü*).

Hun-t'i is a transcription answering to ·ancient *gwun-de, and corresponds to Middle Persian *gandena*, New Persian *gändänä*, Hindī *gandanā*, Bengālī *gundīna* (Sanskrit *mleccha-kanda*, "bulb of the barbarians"), possibly the shallot (*Allium ascalonicum;* French *échalotte, ciboule*) or *A. porrum*, which occurs in western Asia and Persia, but not in China.[4]

Among the vegetables of India, Hüan Tsaṅ[5] mentions 葷陀 *hun-t'o* (*hun-da) *ts'ai*. JULIEN left this term untranslated; BEAL did not know, either, what to make of it, and added in parentheses *kaṇḍu* with an interrogation-mark. WATTERS[6] explained it as "*kunda* (properly the olibanum-tree)." This is absurd, as the question is of a vegetable cultivated for food, while the olibanum is a wild tree offering no food. Moreover, *hun* cannot answer to *kun;* and the Sanskrit word is not *kunda*, but *kundu* or *kunduru*. The mode of writing, *hun*, possibly is intended to allude to a species of *Allium*. Hüan Tsaṅ certainly transcribed a Sanskrit word, but a Sanskrit plant-name of the form *hunda* or *gunda* is not known. Perhaps his prototype is related to the Iranian word previously discussed.

---

[1] The parallel text in the *Ts'e fu yüan kwei* (Ch. 970, p. 12) writes only *liṅ-tuṅ*. This plant is unidentified.

[2] *T'aṅ hui yao*, Ch. 100, p. 3 b; and Ch. 200, p. 14 b.

[3] Ch. 7, p. 1 b (above, p. 232).

[4] A. DE CANDOLLE, Origin of Cultivated Plants, pp. 68–71; LECLERC, Traité des simples, Vol. III, pp. 69–71; ACHUNDOW, Abu Mansur, pp. 113, 258. Other Persian names are *tärä* and *kawar*. They correspond to Greek πράσον, Turkish *prāsa*, Arabic *kurāt*. The question as to whether the species *ascalonicum* or *porrum* should be understood by the Persian term *gändänä*, I have to leave in suspense and to refer to the decision of competent botanists. SCHLIMMER (Terminologie, p. 21) identifies Persian *gändänä* with *Allium porrum;* while, according to him, *A. ascalonicum* should be *musir* in Persian. VULLERS (Lexicon persico-latinum, Vol. II, p. 1036) translates the word by "porrum." On the other hand, STUART (Chinese Materia Medica, p. 25), following F. P. Smith, has labelled Chinese *hiai* 薤, an *Allium* anciently indigenous to China, as *A. ascalonicum*. If this be correct, the Chinese would certainly have recognized the identity of the foreign *hun-t'i* with *hiai*, provided both should represent the same species, *ascalonicum*. Maybe also the two were identical species, but differentiated by cultivation.

[5] *Ta T'aṅ si yü ki*, Ch. 2, p. 8 b.

[6] On Yuan Chwang's Travels, Vol. I, p. 178.

GARDEN PEA AND BROAD BEAN

13. Among the many species of pulse cultivated by the Chinese, there are at least two to which a foreign origin must be assigned. Both are comprised under the generic term *hu tou* 胡豆 ("bean of the Hu," or "Iranian bean"), but each has also its specific nomenclature. It is generally known that, on account of the bewildering number of species and variations and the great antiquity of their cultivation, the history of beans is fraught with graver difficulties than that of any other group of plants.

The common or garden pea (*Pisum sativum*) is usually styled *wan tou* 豌豆 (Japanese *širo-endō*), more rarely *ts'iṅ siao tou* 青小豆 ("green small pulse"), *ts'iṅ pan tou* 青斑豆 ("green streaked pulse"), and *ma lei* 麻累. A term 畢豆 *pi tou*, *pit (pir) tou, is regarded as characteristic of the T'ang period; while such names as *hu tou, žuṅ šu* 戎菽 ("pulse of the Žuṅ"),[1] and *hui-hu tou* 回鶻豆 ("pulse of the Uigur;" in the *Yin šan čeṅ yao* of the Mongol period changed also into *hui-hui tou* 回回豆, "Mohammedan pulse") are apt to bespeak the foreign origin of the plant.[2] Any document alluding to the event of the introduction, however, does not appear to exist in Chinese records. The term *hu tou* occurs in the present editions of the *Ku kin ču*,[3] *hu-ša* 虎沙 being given as its synonyme, and described as "resembling the *li tou* 貍豆, but larger, the fruit of the size of a child's fist and eatable." The term *li tou* is doubtfully identified with *Mucuna capitata;*[4] but the species of the *Ku kin ču* defies exact identification; and, as is well known, this book, in its present form, is very far from being able to claim absolute credence or authenticity. Also the *Kwaṅ či*, written prior to A.D. 527, contains the term *hu tou;*[5] but this name, unfortunately, is ambiguous. Li Ši-čen acquiesces in the general statement that the pea has come from the Hu and Žuṅ or from the Western Hu (Iranians); he cites, however, a few texts, which, if they be authentic, would permit us to

---

[1] This term is ambiguous, for originally it applies to the soy-bean (*Glycine hispida*), which is indigenous to China.

[2] Cf. *Pen ts'ao kaṅ mu*, Ch. 24, p. 7; and *Kwaṅ k'ün faṅ p'u*, Ch. 4, p. 11. The list of the names for the pea given by BRETSCHNEIDER (*Chinese Recorder*, 1871, p. 223) is rather incomplete.

[3] Ch. B, p. 1 b.

[4] STUART, Chinese Materia Medica, p. 269. The word *li* is also written 黧.

[5] *T'ai p'iṅ yü lan*, Ch. 841, p. 6 b.

305

fix approximately the date as to when the pea became known to the Chinese. Thus he quotes the *Ts'ien kin fan* 千金方 of the Taoist adept Sun Se-miao 孫思邈,[1] of the beginning of the seventh century, as mentioning the term *hu tou* with the synonymes *ts'in siao tou* and *ma-lei*. The *Ye čun ki*[2] of the fourth century A.D. is credited with the statement that, when Ši Hu tabooed the word *hu* 胡, the term *hu tou* was altered into *kwo tou* 國豆 ("bean of the country," "national bean"). According to Li Ši-čen, these passages allude to the pea, for anciently the term *hu tou* was in general use instead of *wan tou*. He further refers to the *T'an ši* 唐史 as saying that the *pi tou* comes from the Western Žun and the land of the Uigur, and to the dictionary *Kwan ya* by Čan Yi (third century A.D.) as containing the terms *pi tou*, *wan tou*, and *liu tou* 留豆. It would be difficult to vouchsafe for the fact that these were really embodied in the *editio princeps* of that work; yet it would not be impossible, after all, that, like the walnut and the pomegranate, so also the pea made its appearance on Chinese soil during the fourth century A.D. There can be no doubt of the fact that it was cultivated in China under the T'ang, and even under the Sui (A.D. 590–617). In the account of Liu-kiu (Formosa) it is stated that the soil of the island is advantageous for the cultivation of *hu tou*.[3] Wu K'i-tsün[4] contradicts Li Ši-čen's opinion, stating that the terms *hu tou* and *wan tou* apply to different species.

None of the Chinese names can be regarded as the transcription of an Iranian word. Pulse played a predominant part in the nutrition of Iranian peoples. The country Ši (Tashkend) had all sorts of pulse.[5] Abu Mansur discusses the pea under the Persian name *xullär* and the Arabic *julban*.[6] Other Persian words for the pea are *nujüd* and *gergeru* or *xereghan*.[7]

A wild plant indigenous to China is likewise styled *hu tou*. It is first disclosed by Č'en Ts'an-k'i of the T'ang period, in his *Pen ts'ao ši i*, as growing wild everywhere in rice-fields, its sprouts resembling the bean. In the *Či wu min ši t'u k'ao*[8] we meet illustrations of two wild

---

[1] Regarding this author, see WYLIE, Notes on Chinese Literature, pp. 97, 99; BRETSCHNEIDER, Bot. Sin., pt. 1, p. 43; L. WIEGER, Taoisme, le canon, pp. 142, 143, 182; PELLIOT, *Bull. de l'Ecole française*, Vol. IX, pp. 435–438.

[2] See above, p. 280.

[3] *Sui šu*, Ch. 81, p. 5 b.

[4] *Či wu min ši t'u k'ao*, Ch. 2, p. 150.

[5] *T'ai p'in hwan yü ki*, Ch. 186, p. 7 b.

[6] ACHUNDOW, Abu Mansur, pp. 41, 223.

[7] The latter is given by SCHLIMMER (Terminologie, p. 464).

[8] Ch. 2, pp. 11, 15.

plants. One is termed *hui-hui tou* ("Mohammedan bean"), first mentioned in the *Kiu hwaṅ pen ts'ao* of the fourteenth century, called also *na-ho tou* 那合豆, the bean being roasted and eaten. The other, named *hu tou*, is identified with the wild *hu tou* of Č'en Ts'aṅ-k'i; and Wu K'i-tsün, author of the *Či wu miṅ ši t'u k'ao*, adds the remark, "What is now called *hu tou* grows wild, and is not the *hu tou* [that is, the pea] of ancient times."

14. On the other hand, the term *hu tou* 胡豆 refers also to *Faba sativa* (*F. vulgaris*, the vetch or common bean), according to BRET-SCHNEIDER,[1] "one of the cultivated plants introduced from western Asia into China, in the second century B.C., by the famous general Chang K'ien." This is an anachronism and a wild statement, which he has not even supported by any Chinese text.[2] The history of the species in China is lost, or was never recorded. The supposition that it was introduced from Iran is probable. It is mentioned under the name *pag* (*gāvirs*) in the Būndahišn as the chief of small-seeded grains.[3] Abu Mansur has it under the Persian name *bāqilā* or *bāqlā*.[4] Its cultivation in Egypt is of ancient date.[5]

15. *Ts'an tou* 蠶豆 ("silkworm bean," so called because in its shape it resembles an old silkworm), Japanese *soramame*, the kidney-bean or horse-bean (*Vicia faba*), is also erroneously counted by BRET-SCHNEIDER[6] among the Čaṅ-K'ien plants, without any evidence being produced. It is likewise called *hu tou* 胡豆, but no historical documents touching on the introduction of this species are on record. It is not mentioned in T'ang or Sung literature, and seems to have been introduced not earlier than the Yüan period (1260–1367). It is spoken of in the *Nuṅ šu* 農書 ("Book on Agriculture") of Waṅ Čeṅ 王禎 of that period, and in the *Kiu hwaṅ pen ts'ao* 救荒本草 of the early

---

[1] Bot. Sin., pt. II, No. 29.

[2] The only text to this effect that I know of is the *Pen ts'ao kiṅ*, quoted in the *T'ai p'iṅ yü lan* (Ch. 841, p. 6 b), which ascribes to Čaṅ K'ien the introduction of sesame and *hu tou;* but which species is meant (*Pisum sativum, Faba sativa*, or *Vicia faba*) cannot be guessed. The work in question certainly is not the *Pen ts'ao kiṅ* of Šen-nuṅ, but it must have existed prior to A.D. 983, the date of the publication of the *T'ai p'iṅ yü lan.*

[3] WEST, Pahlavi Texts, Vol. I, p. 90.

[4] ACHUNDOW, Abu Mansur, p. 20.

[5] V. LORET, Flore pharaonique, p. 94.

[6] *Chinese Recorder*, 1871, p. 221 (thus again reiterated by DE CANDOLLE, Origin of Cultivated Plants, p. 318). The *Kwaṅ k'ün faṅ p'u* (Ch. 4, p. 12 b) refers the above text from the *T'ai p'iṅ yü lan* to this species, but also to the pea. This confusion is hopeless.

Ming,[1] which states that "now it occurs everywhere." Li Ši-čen says that it is cultivated in southern China and to a larger extent in Se-č'wan. Wan Ši-mou 王世懋, who died in 1591, in his *Hio pu tsa šu* 學圃雜疏, a work on horticulture in one chapter,[2] mentions an especially large and excellent variety of this bean from Yün-nan. This is also referred to in the old edition of the Gazetteer of Yün-nan Province (*Kiu Yün-nan t'un či*) and in the Gazetteer of the Prefecture of Mun-hwa in Yün-nan, where the synonyme *nan tou* 南豆 ("southern bean") is added, as the flower turns its face toward the south. The New-Persian name of the plant is *bāgelā*.[3]

---

[1] *Či wu min ši t'u k'ao*, Ch. 2, p. 142. BRETSCHNEIDER (Bot. Sin., pt. 1, p. 52) has recognized *Vicia faba* among the illustrations of this work.

[2] Cf. the Imperial Catalogue, Ch. 116, p. 37 b.

[3] SCHLIMMER, Terminologie, p. 562. Arabic *bāqilā*. Finally, the *Fan yi min yi tsi* (section 27) offers a Sanskrit term 勿 伽 *wu-kia*, *mwut-g'a, translated by *hu tou* and explained as "a green bean." The corresponding Sanskrit word is *mudga* (*Phaseolus mungo*), which the Tibetans have rendered as *mon sran rdeu*, the term Mon alluding to the origin from northern India or Himalayan regions (*Mém. Soc. finno-ougrienne*, Vol. XI, p. 96). The Persians have borrowed the Indian word in the form *mung*, which is based on the Indian vernacular *munga* or *mungu* (as in Singhalese; Pali *mugga*). *Phaseolus mungo* is peculiar to India, and is mentioned in Vedic literature (MACDONELL and KEITH, Vedic Index, Vol. II, p. 166).

SAFFRON AND TURMERIC

16. Saffron is prepared from the deep orange-colored stigmas, with a portion of the style, of the flowers of *Crocus sativus* (family *Irideae*). The dried stigmas are nearly 3 cm long, dark red, and aromatic, about twenty thousand of them making a pound, or a grain containing the stigmas and styles of nine flowers. It is a small plant with a fleshy bulb-like corm and grassy leaves with a beautiful purple flower blossoming in the autumn. As a dye, condiment, perfume, and medicine, saffron has always been highly prized, and has played an important part in the history of commerce. It has been cultivated in western Asia from remote ages, so much so that it is unknown in a wild state. It was always an expensive article, restricted mostly to the use of kings and the upper classes, and therefore subject to adulteration and substitutes.[1] In India it is adulterated with safflower (*Carthamus tinctorius*), which yields a coloring-agent of the same deep-orange color, and in Oriental records these products are frequently confused. Still greater confusion prevails between *Crocus* and *Curcuma* (a genus of *Zingiberaceae*), plants with perennial root-stocks, the dried tubers of which yield the turmeric of commerce, largely used in the composition of curry-powder and as a yellow dye. It appears also that the flowers of *Memecylon tinctorium* were substituted for saffron as early as the seventh century. The matter as a subject of historical research is therefore somewhat complex.

Orientalists have added to the confusion of Orientals, chiefly being led astray by the application of our botanical term *Curcuma*, which is derived from an Oriental word originally relating to *Crocus*, but also confounded by the Arabs with our *Curcuma*. It cannot be too strongly emphasized that Sanskrit *kuṅkuma* strictly denotes *Crocus sativus*, but never our *Curcuma* or turmeric (which is Sanskrit *haridrā*),[2] and

---

[1] Pliny already knew that there is nothing so much adulterated as saffron (adulteratur nihil aeque.—XXI, 17, §31). E. WIEDEMANN (*Sitzber. Phys.-med. Soz. Erl.*, 1914, pp. 182, 197) has dealt with the adulteration of saffron from Arabic sources. According to WATT (Commercial Products of India, p. 430), it is too expensive to be extensively employed in India, but is in request at princely marriages, and for the caste markings of the wealthy.

[2] This is not superfluous to add, in view of the wrong definition of *kuṅkuma* given by EITEL (Handbook of Chinese Buddhism, p. 80). Sanskrit *kāvera* ("saffron") and *kāverī* ("turmeric") do not present a confusion of names, as the two words are derived from the name of the trading-place Kāvera, Chaveris of Ptolemy and Caber of Cosmas (see MacCRINDLE, Christian Topography of Cosmas, p. 367).

309

that our genus *Curcuma* has nothing whatever to do with *Crocus* or saffron.

As regards Chinese knowledge of saffron, we must distinguish two long periods,—first, from the third century to the T'ang dynasty inclusive, in which the Chinese received some information about the plant and its product, and occasionally tribute-gifts of it; and, second, the Mongol period (1260–1367), when saffron as a product was actually imported into China by Mohammedan peoples and commonly used. This second period is here considered first.

Of no foreign product are the notions of the Chinese vaguer than of saffron. This is chiefly accounted for by the fact that *Crocus sativus* was hardly ever transplanted into their country,[1] and that, although the early Buddhist travellers to India caught a glimpse of the plant in Kashmir, their knowledge of it always remained rather imperfect. First of all, they confounded saffron with safflower (*Carthamus tinctori-us*), as the products of both plants were colloquially styled "red flower" (*huṅ hwa* 紅花). Li Ši-čen[2] annotates, "The foreign (*fan* 番) or Tibetan red flower [saffron] comes from Tibet (Si-fan), the places of the Mohammedans, and from Arabia (T'ien-faṅ 天方). It is the *huṅ-lan* [*Carthamus*] of those localities. At the time of the Yüan (1260–1367) it was used as an ingredient in food-stuffs. According to the *Po wu či* of Čaṅ Hwa, Čaṅ K'ien obtained the seeds of the *huṅ-lan* [*Carthamus*] in the Western Countries (Si yü), which is the same species as that in question [saffron], although, of course, there is some difference caused by the different climatic conditions." It is hence erroneous to state, as asserted by F. P. SMITH,[3] that "the story of Čaṅ K'ien is repeated for the saffron as well as for the safflower;" and it is due to the utmost confusion that STUART[4] writes, "According to the *Pen-ts'ao*, *Crocus* was brought from Arabia by Čaṅ K'ien at the same time that he brought the safflower and other Western plants and drugs." Čaṅ K'ien in Arabia! The *Po wu či* speaks merely of safflower (*Carthamus*), not of saffron (*Crocus*),—two absolutely distinct plants, which even belong to different families; and there is no Chinese text whatever that would link the saffron with Čaṅ K'ien. In fact, the Chinese have nothing to say re-

---

[1] It is curious that the Armenian historian Moses of Khorene, who wrote about the middle of the fifth century, attributes to China musk, saffron, and cotton (YULE, Cathay, Vol. I, p. 93). Cotton was then not manufactured in China; likewise is saffron cultivation out of the question for the China of that period.

[2] *Pen ts'ao kaṅ mu*, Ch. 15, p. 14 b.

[3] Contributions towards the Materia Medica of China, p. 189.

[4] Chinese Materia Medica, p. 131.

garding the introduction or cultivation of saffron.[1] The confusion of Li Ši-čen is simply due to an association of the two plants known as "red flower." Safflower is thus designated in the *Ts'i min yao šu*, further by Li Čuň 李中 of the T'ang and in the *Suň ši*, where the *yen-či* red flower is stated to have been sent as tribute by the prefecture of Hiň-yüan 興元 in Šen-si.[2]

The fact that Li Ši-čen in the above passage was thinking of saffron becomes evident from two foreign words added to his nomenclature of the product: namely, 泊夫藍 *ki-fu-lan* and 撤法郎 *sa-fa-tsi*. The first character in the former transcription is a misprint for 咱 *tsa* (*tsap, dzap); the last character in the latter form must be emendated into 郎 *laň*.[3] *Tsa-fu-lan* and *sa-fa-laň* (Japanese *safuran*, Siamese *faran*), as was recognized long ago, represent transcriptions of Arabic *za'ferān* or *za'farān*, which, on its part, has resulted in our "saf-

---

[1] BRETSCHNEIDER (*Chinese Recorder*, 1871, p. 222) asserts that saffron is not cultivated in Peking, but that it is known that it is extensively cultivated in other parts of China. I know nothing about this, and have never seen or heard of any saffron cultivation in China, nor is any Chinese account to that effect known to me. *Crocus sativus* is not listed in the great work of F. B. FORBES and W. B. HEMSLEY (An Enumeration of All the Plants known from China Proper, comprising Vols. 23, 26, and 36 of the *Journal of the Linnean Society*), the most comprehensive systematic botany of China. ENGLER (in Hehn, Kulturpflanzen, p. 270) says that *Crocus* is cultivated in China. WATT (Dictionary, Vol. II, p. 593) speaks of Chinese saffron imported into India. It is of especial interest that Marco Polo did not find saffron in China, but he reports that in the province of Fu-kien they have "a kind of fruit, resembling saffron, and which serves the purpose of saffron just as well" (YULE, Marco Polo, Vol. II, p. 225). It may be, as suggested by Yule after Flückiger, that this is *Gardenia florida*, the fruits of which are indeed used in China for dyeing-purposes, producing a beautiful yellow color. On the other hand, the *Pen ts'ao kaň mu ši i* (Ch. 4, p. 14 b) contains the description of a "native saffron" (*t'u huň hwa* 土 紅花, in opposition to the "Tibetan red flower" or genuine saffron) after the Continued Gazetteer of Fu-kien 福建續志, as follows: "As regards the native saffron, the largest specimens are seven or eight feet high. The leaves are like those of the *p'i-p'a* 枇杷 (*Eriobotrya japonica*), but smaller and without hair. In the autumn it produces a white flower like a grain of maize (*su-mi* 粟米, *Zea mays*). It grows in Fu-čou and Nan-ňen-čou 南恩州 [now Yaň-kiaň 陽江 in Kwaň-tuň] in the mountain wilderness. That of Fu-čou makes a fine creeper, resembling the *fu-yuň* (*Hibiscus mutabilis*), green above and white below, the root being like that of the *ko* 葛 (*Pachyrhizus thunbergianus*). It is employed in the pharmacopœia, being finely chopped for this purpose and soaked overnight in water in which rice has been scoured; then it is soaked for another night in pure water and pounded: thus it is ready for prescriptions." This species has not been identified, but may well be Marco Polo's pseudo-saffron of Fu-kien.

[2] *T'u šu tsi č'eň*, XX, Ch. 158.

[3] Cf. WATTERS, Essays on the Chinese Language, p. 348. This transcription, however, does not prove, as intimated by Watters, that "this product was first imported into China from Persia direct or at least obtained immediately from Persian traders." The word *zafarān* is an Arabic loan-word in Persian, and may have been brought to China by Arabic traders as well.

fron."[1] It is borne out by the very form of these transcriptions that
they cannot be older than the Mongol period when the final consonants
had disappeared. Under the T'ang we should have *dzap-fu-lam and
*sat-fap-laṅ. This conclusion agrees with Li Ši-čen's testimony that
saffron was mixed with food at the time of the Yüan,— an Indo-Persian
custom. Indeed, it seems as if not until then was it imported and used
in China; at least, we have no earlier document to this effect.

Saffron is not cultivated in Tibet. There is no *Crocus tibetan us*, as
tentatively introduced by PERROT and HURRIER[2] on the basis of the
Chinese term "Tibetan red flower." This only means that saffron is
exported from Tibet to China, chiefly to Peking; but Tibet does not
produce any saffron, and imports it solely from Kashmir. STUART[3]
says that "*Ts'aṅ huṅ hwa* 藏紅花 ('Red flower from Tsaṅ,' that is,
Central Tibet) is given by some foreign writers as another name for
saffron, but this has not been found mentioned by any Chinese writer."
In fact, that term is given in the *Pen ts'ao kaṅ mu ši i*[4] and the *Či wu
miṅ ši t'u k'ao* of 1848,[5] where it is said to come from Tibet (Si-tsaṅ)
and to be the equivalent of the *Fan huṅ hwa* of the *Pen ts'ao kaṅ mu.*
Ts'aṅ hwa is still a colloquial name for saffron in Peking; it is also called
simply *huṅ hwa* ("red flower").[6] By Tibetans in Peking I heard it
designated *gur-kum*, *ša-ka-ma*, and *dri-bzaṅ* ("of good fragrance").
Saffron is looked upon by the Chinese as the most valuable drug sent
by Tibet, *ts'aṅ hiaṅ* ("Tibetan incense") ranking next.

Li Ši-čen[7] holds that there are two *yü-kin* 鬱金,—the *yü-kin* aromatic,
the flowers of which only are used; and the *yü-kin* the root of which is
employed. The former is the saffron (*Crocus sativus*); the latter, a
Curcuma. As will be seen, however, there are at least three *yü-kin*.

Of the genus *Curcuma*, there are several species in China and
Indo-China,—*C. leucorrhiza* (*yü-kin*), *C. longa* (*kiaṅ hwaṅ* 姜 or 薑黃,

---

[1] The Arabs first brought saffron to Spain; and from Arabic *za'farān* are derived
Spanish *azafran*, Portuguese *açafrão* or *azafrão*, Indo-Portuguese *safrão*, Italian
zafferano, French *safran*, Rumanian *sofrán*. The same Arabic root ('*aṣfur*, "yellow")
has supplied also those Romance words that correspond to our safflow, safflower
(*Carthamus tinctorius*), like Spanish *azafranillo*, *alazor*, Portuguese *açafroa*, Italian
asforo, French *safran*; Old Armenian *zavhran*, New Armenian *zafran*; Russian
safran; Uigur *sakparan*.

[2] Mat. méd. et pharmacopée sino-annamites, p. 94.

[3] Chinese Materia Medica, p. 132.

[4] Ch. 4, p. 14 b.

[5] Ch. 4, p. 35 b.

[6] It should be borne in mind that this name is merely a modern colloquialism,
but *huṅ hwa*, when occurring in ancient texts, is not "saffron," but "safflower"
(*Carthamus tinctorius*); see below, p. 324.

[7] *Pen ts'ao kaṅ mu*, Ch. 14, p. 18.

"ginger-yellow"), *C. pallida, C. petiolata, C. zedoaria*. Which particular species was anciently known in China, is difficult to decide; but it appears that at least one species was utilized in times of antiquity. *Curcuma longa* and *C. leucorrhiza* are described not earlier than the T'ang period, and the probability is that either they were introduced from the West; or, if on good botanical evidence it can be demonstrated that these species are autochthonous,[1] we are compelled to assume that superior cultivated varieties were imported in the T'ang era. In regard to *yü-kin* (*C. leucorrhiza*), Su Kuṅ of the seventh century observes that it grows in Šu (Se-č'wan) and Si-žuṅ, and that the Hu call it 馬蔋 *ma-šu*, *mo-džut (dzut),[2] while he states with reference to *kiaṅ-hwaṅ* (*C. longa*) that the Žuṅ 戎人 call it 蔋 *šu*, *džut (dzut, dzur); he also insists on the close resemblance of the two species. Likewise Č'en Ts'aṅ-k'i, who wrote in the first part of the eighth century, states concerning *kiaṅ-hwaṅ* that the kind coming from the Western Barbarians (Si Fan) is similar to *yü-kin* and *šu yao* 蔋藥.[3] Su Suṅ of the Sung remarks that *yü-kin* now occurs in all districts of Kwaṅ-tuṅ and Kwaṅ-si, but does not equal that of Se-č'wan, where it had previously existed. K'ou Tsuṅ-ši[4] states that *yü-kin* is not aromatic, and that in his time it was used for the dyeing of woman's clothes. Li Ši-čen reminds us of the fact that *yü-kin* was a product of the Hellenistic Orient (Ta Ts'in): this is stated in the *Wei lio* of the third century,[5] and the *Liaṅ šu*[6] enumerates *yü-kin* among the articles traded from Ta Ts'in to western India.[7]

The preceding observations, in connection with the foreign names

---

[1] According to LOUREIRO (Flora Cochin-Chinensis, p. 9), *Curcuma longa* grows wild in Indo-China.

[2] This foreign name has not been pointed out by Bretschneider or Stuart or any previous author.

[3] This term is referred (whether correctly, I do not know) to *Kæmpferia pundurata* (STUART, Chinese Materia Medica, p. 227). Another name for this plant is 蓬莪茂 *p'uṅ-no šu* (not *mou*), *buṅ-ṅa. Now, Ta Miṅ states that the *Curcuma* growing on Hai-nan is 蓬莪蔋 *p'uṅ-no šu*, while that growing in Kiaṅ-nan is *kiaṅ-hwaṅ* (*Curcuma longa*). *Kæmpferia* belongs to the same order as *Curcuma*, —*Scitamineae*. According to Ma Či of the Sung, this plant grows in Si-žuṅ and in all districts of Kwaṅ-nan; it is poisonous, and the people of the West first test it on sheep: if these refuse to eat it, it is discarded. Chinese *p'uṅ-ṅo*, *buṅ-ṅa, looks like a transcription of Tibetan *boṅ-ṅa*, which, however, applies to aconite.

[4] *Pen ts'ao yen i*, Ch. 10, p. 3.

[5] *San kwo či*, Ch. 30, p. 13.

[6] Ch. 78, p. 7.

[7] The question whether in this case *Curcuma* or *Crocus* is meant, cannot be decided; both products were known in western Asia. Č'en Ts'aṅ-k'i holds that the *yü-kin* of Ta Ts'in was safflower (see below).

šu and *ma-šu,* are sufficient to raise serious doubts of the indigenous character of *Curcuma;* and for my part, I am strongly inclined to believe that at least two species of this genus were first introduced into Se-č'wan by way of Central Asia. This certainly would not exclude the possibility that other species of this genus, or even other varieties of the imported species, pre-existed in China long before that time; and this is even probable, in view of the fact that a fragrant plant *yü* 鬱, which was mixed with sacrificial wine, is mentioned in the ancient *Čou li,* the State Ceremonial of the Čou Dynasty, and in the *Li ki.* The commentators, with a few exceptions, agree on the point that this ancient *yü* was a *yü-kin;* that is, a *Curcuma.*[1]

In India, *Curcuma longa* is extensively cultivated all over the country, and probably so from ancient times. The plant (Sanskrit *haridrā*) is already listed in the Bower Manuscript. From India the rhizome is exported to Tibet, where it is known as *yuṅ-ba* or *skyer-pa,* the latter name originally applying to the barberry, the wood and root of which, like *Curcuma,* yield a yellow dye.

Ibn al-Baiṭār understands by *kurkum* the genus *Curcuma,* not *Crocus,* as is obvious from his definition that it is the great species of the tinctorial roots. These roots come from India, being styled *hard* in Persian; this is derived from Sanskrit *haridrā* (*Curcuma longa*). Ibn Hassan, however, observes that the people of Basra bestow on *hard* the name *kurkum,* which is the designation of saffron, and to which it is assimilated; but then he goes on to confound saffron with the root of *wars,* which is a *Memecylon* (see below).[2] Turmeric is called in Persian *zird-čūbe* or *darzard* ("yellow wood"). According to GARCIA DA ORTA, it was much exported from India to Arabia and Persia; and there was unanimous opinion that it did not grow in Persia, Arabia, or Turkey, but that all comes from India.[3]

The name *yü-kin,* or with the addition *hiaṅ* ("aromatic"),[4] is frequently referred in ancient documents to two different plants of Indian and Iranian countries,— *Memecylon tinctorium* and *Crocus sativus,* the

---

[1] Cf. BRETSCHNEIDER, Bot. Sin., pt. II, No. 408.

[2] LECLERC, Traité des simples, Vol. III, p. 167.

[3] C. MARKHAM, Colloquies, p. 163.

[4] As a matter of principle, the term *yü-kin hiaṅ* strictly refers to saffron. It is this term which BRETSCHNEIDER (Bot. Sin., pt. II, No. 408) was unable to identify, and of which STUART (Chinese Materia Medica, p. 140) was compelled to admit, "The plant is not yet identified, but is probably not *Curcuma.*" The latter remark is to the point. The descriptions we have of *yü-kin⟨hiaṅ,* and which are given below, exclude any idea of a *Curcuma.* The modern Japanese botanists apply the term *yü-kin hiaṅ* (Japanese *ukkonkō*) to *Tulipa gesneriana,* a flower of Japan (MATSUMURA, No. 3193).

latter possibly confounded again with *Curcuma*.[1] It is curious that in the entire *Pen-ts'ao* literature the fact has been overlooked that under the same name there is also preserved the ancient description of a tree. This fact has escaped all European writers, with the sole exception of PALLADIUS. In his admirable Chinese-Russian Dictionary[2] he gives the following explanation of the term *yü-kin:* "Designation of a tree in Ki-pin; yellow blossoms, which are gathered, and when they begin to wither, are pressed, the sap being mixed with other odorous substances; it is found likewise in Ta Ts'in, the blossoms being like those of saffron, and is utilized in the coloration of wine."

A description of this tree *yü-kin* is given in the Buddhist dictionary *Yi ts'ie kiṅ yin i*[3] of A.D. 649 as follows: "This is the name of a tree, the habitat of which is in the country Ki-pin 罽賓 (Kashmir). Its flowers are of yellow color. The trees are planted from the flowers. One waits till they are faded; the sap is then pressed out of them and mixed with other substances. It serves as an aromatic. The grains of the flowers also are odoriferous, and are likewise employed as aromatics."

I am inclined to identify this tree with *Memecylon tinctorium, M. edule,* or *M. capitellatum* (*Melastomaceae*), a very common, small tree or large shrub in the east and south of India, Ceylon, Tenasserim, and the Andamans. The leaves are employed in southern India for dyeing a "delicate yellow lake." The flowers produce an evanescent yellow.[4] In restricting the habitat of the tree to Kashmir, Hüan Yiṅ is doubtless influenced by the notion that saffron (*yü-kin*) was an exclusive product of Kashmir (see below).

The same tree is described by Abu Mansur under the name *wars* as a saffron-like plant of yellow color and fragrant, and employed by Arabic women for dyeing garments.[5] The ancients were not acquainted

---

[1] A third identification has been given by BRETSCHNEIDER (*Chinese Recorder,* 1871, p. 222), who thought that probably the sumbul (*Sumbulus moschatus*) is meant. This is a mistaken botanical name, but he evidently had in mind the so-called muskroot of *Euryangium* or *Ferula sumbul,* of musk-like odor and acrid taste. The only basis for this identification might be sought in the fact that one of the synonymes given for *yü-kin hiaṅ* in the *Pen ts'ao* is *ts'ao še hiaṅ* 草麝香 ("vegetable musk"); this name itself, however, is not explained. Saffron, of course, has no musk odor; and the term *ts'ao še hiaṅ* surely does not relate to saffron, but is smuggled in here by mistake. The *Tien hai yü heṅ či* (Ch. 3, p. 1 b, see above, p. 228) also equates *yü-kin hiaṅ* with *ts'ao še hiaṅ,* adding that the root is like ginger and colors wine yellow. This would decidedly hint at a *Curcuma.*

[2] Vol. II, p. 202.

[3] Ch. 24, p. 8 (cf. Beginnings of Porcelain, p. 115; and above, p. 258).

[4] WATT, Dictionary of the Economic Products of India, Vol. V, p. 227.

[5] ACHUNDOW, Abu Mansur, p. 145.

with this dye. Abu Hanīfa has a long discourse on it.[1] Ibn Hassan knew the root of *wars*, and confounded it with saffron.[2] Ibn al-Baiṭar offers a lengthy notice of it.[3] Two species are distinguished,— one from Ethiopia, black, and of inferior quality; and another from India, of a brilliant red, yielding a dye of a pure yellow. A variety called *bārida* dyes red. It is cultivated in Yemen. Also the association with *Curcuma* and *Crocus* is indicated. Išak Ibn Amrān remarks, "It is said that *wars* represents roots of *Curcuma*, which come from China and Yemen"; and Ibn Massa el-Basri says, "It is a substance of a brilliant red which resembles pounded saffron." This explains why the Chinese included it in the term *yü-kin*. LECLERC also has identified the *wars* of the Arabs with *Memecylon tinctorium*, and adds, "L'ouars n'est pas le produit exclusif de l'Arabie. On le rencontre abondamment dans l'Inde, notamment aux environs de Pondichéry qui en a envoyé en Europe, aux dernières expositions. Il s'appelle *kana* dans le pays."[4] The *Yamato honzō* speaks of *yü-kin* as a dye-stuff coming from Siam; this seems to be also *Memecylon*.

The fact that the Chinese included the product of *Memecylon* in the term *yü-kin* appears to indicate that this cheap coloring-matter was substituted in trade for the precious saffron.

While the Chinese writers on botany and pharmacology have overlooked *yü-kin* as the name of a tree, they have clearly recognized that the term principally serves for the designation of the saffron, the product of the *Crocus sativus*. This fact is well borne out by the descriptions and names of the plant, as well as by other evidence.

The account given of Central India in the Annals of the Liang Dynasty[5] expressly states that *yü-kin* is produced solely in Kashmir (Ki-pin), that its flower is perfectly yellow and fine, resembling the flower *fu-yuṅ* (*Hibiscus mutabilis*). Kashmir was always the classical land famed for the cultivation of saffron, which was (and is) thence exported to India, Tibet, Mongolia, and China. In Kashmir, Uḍḍiyāna,

---

[1] ACHUNDOW, Abu Mansur, p. 272.

[2] LECLERC, Traité des simples, Vol. III, p. 167.

[3] *Ibid.*, p. 409.

[4] Arabic *wars* has also been identified with *Flemingia congesta* (WATT, Dictionary, Vol. III, p. 400) and *Mallotus philippinensis* (*ibid.*, Vol. V, p. 114). The whole subject is much confused, particularly by FLÜCKIGER and HANBURY (Pharmacographia, p. 573; cf. also G. JACOB, Beduinenleben, p. 15, and Arab. Geographen, p. 166), but this is not the place to discuss it. The Chinese description of the *yü-kin* tree does not correspond to any of these plants.

[5] *Liaṅ šu*, Ch. 54, p. 7 b. This work was compiled by Yao Se-lien in the first half of the seventh century from documents of the Liang dynasty, which ruled from A.D. 502 to 556.

and Jāguḍa (Zābulistān) it was observed by the famous pilgrim Hüan Tsaṅ in the seventh century.[1] The Buddhist traveller Yi Tsiṅ (671–695) attributes it to northern India.[2]

The earliest description of the plant is preserved in the *Nan čou i wu či*, written by Wan Čen in the third century A.D.,[3] who says, "The habitat of *yü-kin* is in the country Ki-pin (Kashmir), where it is cultivated by men, first of all, for the purpose of being offered to the Buddha. After a few days the flower fades away, and then it is utilized on account of its color, which is uniformly yellow. It resembles the *fu-yuṅ* (*Hibiscus*) and a young lotus (*Nelumbium speciosum*), and can render wine aromatic." This characteristic is fairly correct, and unequivocally applies to the *Crocus*, which indeed has the appearance of a liliaceous plant, and therefore belongs to the family *Irideae* and to the order *Liliiflorae*. The observation in regard to the short duration of the flowers is to the point.[4]

In A.D. 647 the country Kia-p'i 伽ㆍ仳 in India offered to the Court *yü-kin hiaṅ*, which is described on this occasion as follows: "Its leaves are like those of the *mai-men-tuṅ* 麥門冬 (*Ophiopogon spicatus*). It blooms in the ninth month. In appearance it is similar to *fu-yuṅ* (*Hibiscus mutabilis*). It is purple-blue 紫碧 in color. Its odor may be perceived at a distance of several tens of paces. It flowers, but does not bear fruit. In order to propagate it, the root must be taken."[5]

---

[1] S. JULIEN, Mémoires sur les contrées occidentales, Vol. I, pp. 40, 131; Vol. II, p. 187 (story of the Saffron-Stūpa, *ibid.*, Vol. I, p. 474; or S. BEAL, Buddhist Records, Vol. II, p. 125); W. W. ROCKHILL, Life of the Buddha, p. 169; S. LÉVI, *Journal asiatique*, 1915, I, pp. 83–85.

[2] TAKAKUSU's translation, p. 128; he adds erroneously, "species of *Curcuma*."

[3] *Pen ts'ao kaṅ mu*, Ch. 14, p. 22.

[4] Compare Pliny's (XXI, 17, §34) description of *Crocus*: "Floret vergiliarum occasu paucis diebus folioque florem expellit. Viret bruma et colligitur; siccatur umbra, melius etiam hiberna."

[5] *T'aṅ hui yao*, Ch. 200, pp. 14 a-b. This text was adopted by the *Pen ts'ao kaṅ mu* (Ch. 14, p. 22), which quotes it from the T'ang Annals. Li Si-čen comments that this description agrees with that of the *Nan čou i wu či*, except in the colors of the flower, which may be explained by assuming that there are several varieties; in this he is quite correct. The flower, indeed, occurs in a great variation of colors,—purple, yellow, white, and others. W. WOODVILLE (Medical Botany, Vol. IV, p. 763) gives the following description of *Crocus*: "The root is bulbous, perennial: the flower appears after the leaves, rising very little above the ground upon a slender succulent tube: the leaves rise higher than the flower, are linear, simple, radical, of a rich green colour, with a white line running in the centre, and all at the base inclosed along with the tube of the flower in a membranous sheath. The flower is large, of a bluish purple, or lilac colour: the corolla consists of six petals, which are nearly elliptical, equal, and turned inwards at the edges. The filaments are three, short, tapering, and support long erect yellow antherae. The germen is roundish, from

The last clause means that the plant is propagated from bulbs. There is a much earlier tribute-gift of saffron on record. In A.D. 519, King Jayavarman of Fu-nan (Camboja) offered saffron with storax and other aromatics to the Chinese Court.[1] Accordingly we have to assume that in the sixth century saffron was traded from India to Camboja. In fact we know from the T'ang Annals that India, in her trade with Camboja and the anterior Orient, exported to these countries diamonds, sandal-wood, and saffron.[2] The T'ang Annals, further, mention saffron as a product of India, Kashmir, Uḍḍiyāna, Jāguḍa, and Baltistan.[3] In A.D. 719 the king of Ṅan (Bukhārā) presented thirty pounds of saffron to the Chinese Emperor.[4]

Li Ši-čen has added to his notice of *yü-kin hiaṅ* a Sanskrit name 茶矩摩 *č'a-kü-mo*, *dža-gu-ma, which he reveals from the Suvarṇaprabhāsa-sūtra.[5] This term is likewise given, with the translation *yü-kin*, in the Chinese-Sanskrit Dictionary *Fan yi miṅ yi tsi*.[6] This name has been discussed by me and identified with Sanskrit *jāguḍa* through the medium of a vernacular form *jāguma, the ending *-ma* corresponding to that of Tibetan *ša-ka-ma*.[7]

A singular position is taken by Č'en Ts'aṅ-k'i, who reports, "*Yü-kin* aromatic grows in the country Ta Ts'in. It flowers in the second or third month, and has the appearance of the *huṅ-lan* (safflower, *Carthamus tinctorius*).[8] In the fourth or fifth month the flowers are gathered and make an aromatic." This, of course, cannot refer to the saffron which blooms in September or October. Č'en Ts'aṅ-k'i has created confusion, and has led astray Li Ši-čen, who wrongly enumerates *huṅ-lan hwa* among the synonymes of *yü-kin hiaṅ*.

The inhabitants of Ku-lin (Quilon) 故臨 rubbed their bodies with

which issues a slender style, terminated by three long convoluted stigmata, of a deep yellow colour. The capsule is roundish, three-lobed, three-celled, three-valved, and contains several round seeds. It flowers in September and October."

[1] According to the *Liaṅ šu;* cf. PELLIOT, *Bull. de l'Ecole française*, Vol. III, p. 270.

[2] *T'aṅ šu*, Ch. 221 A, p. 10 b.

[3] *Kiu T'aṅ šu*, Ch. 221 B, p. 6; 198, pp. 8 b, 9; *T'aṅ šu*, Ch. 221 A, p. 10 b; cf. CHAVANNES (Documents sur les Tou-kiue occidentaux, pp. 128, 150, 160, 166), whose identification with *Curcuma longa* is not correct.

[4] CHAVANNES, *ibid.*, p. 203.

[5] The passage in which Li Ši-čen cites this term demonstrates clearly that he discriminated well between *Crocus* and *Curcuma;* for he adds that "*č'a-kü-mo* is the aromatic of the *yü-kin* flower (*Crocus*), but that, while it is identical in name with the *yü-kin* root (*Curcuma*) utilized at the present time, the two plants are different."

[6] Ch. 8, p. 10 b.

[7] *T'oung Pao*, 1916, p. 458.

[8] See below, p. 324.

yü-kin after every bath, with the intention of making it resemble the
"gold body" of a Buddha.[1] Certainly they did not smear their bodies
with "turmeric,"[2] which is used only as a dye-stuff, but with saffron.
Annamese mothers rub the bodies of their infants with saffron-powder
as a tonic to their skin.[3]

The *Ain-i Akbari*, written 1597 in Persian by Abul Fazl 'Allami
(1551–1602), gives detailed information on the saffron cultivation in
Kashmir,[4] from which the following extract may be quoted: "In the
village of Pāmpūr, one of the dependencies of Vīhī (in Kashmir), there
are fields of saffron to the extent of ten or twelve thousand *bighas*, a
sight that would enchant the most fastidious. At the close of the
month of March and during all April, which is the season of cultivation,
the land is plowed up and rendered soft, and each portion is prepared
with the spade for planting, and the saffron bulbs are hard in the ground.
In a month's time they sprout, and at the close of September, it is at
its full growth, shooting up somewhat over a span. The stalk is white,
and when it has sprouted to the height of a finger, one bud after another
begins to flower till there are eight flowers. It has six lilac-tinted petals.
Usually among six filaments, three are yellow and three ruddy. The
last three yield the saffron. [There are three stamens and three stigmas
in each flower, the latter yielding the saffron.] When the flowers are
past, leaves appear upon the stalk. Once planted it will flower for six
years in succession. The first year, the yield is small: in the second as
thirty to ten. In the third year it reaches its highest point, and the
bulbs are dug up. If left in the same soil, they gradually deteriorate,
but if taken up, they may be profitably transplanted."

The Emperor Jahāngīr was deeply impressed by the saffron plantations of Kashmir, and left the following notes in his Memoirs:[5]—

"As the saffron was in blossom, his Majesty left the city to go to
Pāmpūr, which is the only place in Kashmir where it flourishes. Every
parterre, every field, was, as far as the eye could reach, covered with
flowers. The stem inclines toward the ground. The flower has five
petals of a violet color, and three stigmas producing saffron are found
within it, and that is the purest saffron. In an ordinary year, 400

---

[1] *Lin wai tai ta*, Ch. 2, p. 13.

[2] HIRTH, Chau Ju-kua, p. 91.

[3] PERROT and HURRIER, Mat. méd. et pharmacopée sino-annamites, p. 94.
Cf. also MARCO POLO's observation (YULE's edition, Vol. II, p. 286) that the faces
of stuffed monkeys on Java are daubed with saffron, in order to give them a manlike
appearance.

[4] Translation of H. BLOCHMANN, Vol. I, p. 84; Vol. II, p. 357.

[5] H. M. ELLIOT, History of India as told by Its Own Historians, Vol. VI, p. 375

maunds, or 3200 Khurāsānī *maunds*, are produced. Half belongs to
the Government, half to the cultivators, and a *sīr* sells for ten rupees;
but the price sometimes varies a little. It is the established custom to
weigh the flowers, and give them to the manufacturers, who take them
home and extract the saffron from them, and upon giving the extract,
which amounts to about one-fourth weight of the flower, to the public
officers, they receive in return an equal weight of salt, in lieu of money
wages."

The ancient Chinese attribute saffron not only to Kashmir, but also
to Sasanian Persia. The *Čou šu*[1] enumerates *yü-kin* among the products
of Po-se (Persia); so does the *Sui šu*.[2] In fact, *Crocus* occurs in Persia
spontaneously, and its cultivation must date from an early period.
Aeschylus alludes to the saffron-yellow footgear of King Darius.[3]
Saffron is mentioned in Pahlavi literature (above, p.193). The plant is
well attested for Derbend, Ispahan, and Transoxania in the tenth
century by Istaxri and Edrisi.[4] Yāqūt mentions saffron as the principal
production of Rud-Derawer in the province Jebal, the ancient Media,
whence it was largely exported.[5] Abu Mansur describes it under the
Arabic name *zafarān*.[6] The Armenian consumers esteem most highly
the saffron of Khorasan, which, however, is marketed in such small
quantities that the Persians themselves must fill the demand with
exportations from the Caucasus.[7] According to SCHLIMMER,[8] part of
the Persian saffron comes from Baku in Russia, another part is cultivated in Persia in the district of Kain, but in quantity insufficient to
fill the demand. In two places,—Rudzabar (identical with the above
Rud-Derawer), a mountainous tract near Hamadan, and Mount
Derbend, where saffron cultivation had been indicated by previous
writers,—he was unable to find a trace of it.

It is most probable that it was from Persia that the saffron-plant
was propagated to Kashmir. A reminiscence of this event is preserved
in the Sanskrit term *vāhlīka*, a synonyme of "saffron," which means
"originating from the Pahlava."[9] The Buddhists have a legend to the

---

[1] Ch. 50, p. 6.

[2] Ch. 83, p. 7 b; also *Wei šu*, Ch. 102, p. 5 b.

[3] HEHN, Kulturpflanzen, p. 264.

[4] A. JAUBERT, Géographie, pp. 168, 192.

[5] B. DE MEYNARD, Dictionnaire géogr. de la Perse, p. 267. See also G. FER-
RAND, Textes relatifs à l'Extrême-Orient, Vol. II, pp. 618, 622.

[6] ACHUNDOW, Abu Mansur, p. 76.

[7] E. SEIDEL, Mechithar, p. 151. CHARDIN (Voyages en Perse, Vol. II, p. 14)
even says that the saffron of Persia is the best of the world.

[8] Terminologie, p. 165.

[9] Cf. *T'oung Pao*, 1916, p. 459.

effect that Madhyāntika, the first apostle of Buddha's word in Kashmir, planted the saffron there.[1] If nothing else, this shows at least that the plant was regarded as an introduction. The share of the Persians in the distribution of the product is vividly demonstrated by the Tibetan word for "saffron," *kur-kum, gur-kum, gur-gum,* which is directly traceable to Persian *kurkum* or *karkam,* but not to Sanskrit *kuṅkuma.*[2] The Tibetans carried the word to Mongolia, and it is still heard among the Kalmuk on the Wolga. By some, the Persian word (Pahlavi *kulkem*) is traced to Semitic, Assyrian *karkuma,* Hebrew *karkōm,* Arabic *kurkum;* while others regard the Semitic origin as doubtful.[3] It is beyond the scope of this notice to deal with the history of saffron in the west and Europe, on which so much has been written.[4]

From the preceding investigation it follows that the word *yü-kin* 鬱金, owing to its multiplicity of meaning, offers some difficulty to the translator of Chinese texts. The general rule may be laid down that *yü-kin,* whenever it hints at a plant or product of China, denotes a species of *Curcuma;* but that, when used with reference to India, Indo-China, and Iran, the greater probability is in favor of *Crocus.* The term *yü-kin hian* ("yü-kin aromatic"), with reference to foreign countries, almost invariably appears to refer to the latter plant, which indeed served as an aromatic; while the same term, as will be seen below, with reference to China, again denotes *Curcuma.* The question may now be raised, What is the origin of the word *yü-kin?* And what was its original meaning? In 1886 HIRTH[5] identified *yü-kin* with Persian *karkam* ("saffron"), and restated this opinion in 1911,[6] by falling back on an ancient pronunciation *hat-kam. Phonetically this is not very convincing, as the Chinese would hardly have employed an initial *h* for

---

[1] SCHIEFNER, Tāranātha, p. 13; cf. also J. PRZYLUSKI, *Journal asiatique,* 1914 II, p. 537.

[2] *T'oung Pao,* 1916, p. 474. Cf. also Sogdian *kurkumba* and Tokharian *kurkama.*

[3] HORN, Grundriss der iranischen Philologie, Vol. I, pt. 2, p. 6. Besides *kurkum,* there are Persian *kākbān* and *kāfīša,* which denote "saffron in the flower." Old Armenian *k'rk'um* is regarded as a loan from Syriac *kurkemā* (HÜBSCHMANN, Armen. Gram., p. 320).

[4] In regard to saffron among the Arabs, see LECLERC, Traité des simples, Vol. II, pp. 208-210. In general cf. J. BECKMANN, Beyträge zur Geschichte der Erfindungen, 1784, Vol. II, pp. 79-91 (also in English translation); FLÜCKIGER and HANBURY, Pharmacographia, pp. 663-669; A. DE CANDOLLE, Géographie botanique, p. 857, and Origin of Cultivated Plants, p. 166; HEHN, Kulturpflanzen (8th ed.), pp. 264-270; WATT, Dictionary, Vol. II, p. 592; W. HEYD, Histoire du commerce du levant, Vol. II, p. 668, etc.

[5] *Journal China Branch Roy. As. Soc.,* Vol. XXI, p. 221.

[6] Chau Ju-kua, p. 91.

the reproduction of a foreign *k;* but the character *yü* in transcriptions usually answers to *ut, ud. The whole theory, however, is exposed to much graver objections. The Chinese themselves do not admit that *yü-kin* represents a foreign word; nowhere do they say that *yü-kin* is Persian, Sanskrit, or anything of the sort; on the contrary, they regard it as an element of their own language. Moreover, if *yü-kin* should originally designate the saffron, how, then, did it happen that this alleged Persian word was transferred to the genus *Curcuma,* some species of which are even indigenous to China, and which, at any rate, has been acclimated there for a long period? The case, indeed, is not simple, and requires closer study. Let us see what the Chinese have to say concerning the word *yü-kin.* PELLIOT[1] has already clearly, though briefly, outlined the general situation by calling attention to the fact that as early as the beginning of the second century, *yü-kin* is mentioned in the dictionary *Šwo wen* as the name of an odoriferous plant, offered as tribute by the people of Yü, the present Yü-lin in Kwaṅ-si Province; hence he inferred that the sense of the word should be "gold of Yü," in allusion to the yellow color of the product. We read in the *Šwi kiṅ ču* 水經注[2] as follows: "The district Kwei-lin 桂林郡 of the Ts'in dynasty had its name changed into the Yü-lin district 鬱林郡 in the sixth year of the period Yüan-tiṅ (111 B.C.) of the Emperor Wu of the Han dynasty. Waṅ Maṅ made it into the Yü-p'iṅ district 鬱平. Yiṅ Šao 應邵 [second century A.D.], in his work *Ti li fuṅ su ki* 地理風俗記, says, 'The *Čou li* speaks of the *yü žen* 鬱人 ('officials in charge of the plant *yü*'), who have charge of the jars serving for libations; whenever libations are necessary for sacrifices or for the reception of guests, they attend to the blending of the plant *yü* with the odoriferous wine *č'aṅ,* pour it into the sacred vases, and arrange them in their place.'[3] *Yü* is a fragrant plant. Flowers of manifold plants are boiled and mixed with wine fermented by means of black millet as an offering to the spirits: this is regarded by some as what is now called *yü-kin hiaṅ* 鬱金香 (*Curcuma*); while others contend that it was brought as tribute by the people of Yü, thus connecting the name of the plant with that of the clan and district of Yü." The latter is the explanation

---

[1] *Bull. de l'Ecole française,* Vol. III, p. 270.

[2] This work is a commentary to the *Šwi kiṅ,* a canonical book on water-courses, supposed to have been written by Saṅ K'in under the Later Han dynasty, but it was elaborated rather in the third century. The commentary is due to Li Tao-yüan of the Hou Wei period, who died in A.D. 527 (his biography is in *Wei šu,* Ch. 89; *Pei ši,* Ch. 27). Regarding the various editions of the work, see PELLIOT, *Bull. de l'Ecole française,* Vol. VI, p. 364, note 4.

[3] Cf. BIOT, *Le Tcheou-li,* Vol. I, p. 465.

favored by the *Šwo wen*.[1] Both explanations are reasonable, but only one of the two can be correct.[2] My own opinion is this: *yü* is an ancient Chinese name for an indigenous Chinese aromatic plant; whether *Curcuma* or another genus, can no longer be decided with certainty.[3] The term *yü-kin* means literally "gold of the *yü* plant," "gold" referring to the yellow rhizome,[4] *yü* to the total plant-character; the concrete significance, accordingly, is "*yü*-rhizome" or "*yü*-root." I do not believe, however, that *yü-kin* is derived from the district or clan of Yü; for this is impossible to assume, since *yü* as the name of a plant existed prior to the name of that district. This is clearly evidenced by the text of the *Šwi kiṅ ču:* for it was only in 111 B.C. that the name Yü-lin ("Grove of the *Yü* Plant") came into existence, being then substituted for the earlier Kwei-lin ("Grove of *Cinnamomum cassia*"). It is the plant, consequently, which lent its name to the district, not the district which named the plant. As in so many cases, the Chinese confound cause and effect. The reason why the name of this district was altered into Yü-lin is now also obvious. It must have been renowned under the Han for the wealth of its *yü-kin* plants, which was less conspicuous under the Ts'in, when the cassia predominated there. At any rate, *yü-kin* is a perfectly authentic and legitimate constituent of the Chinese language, and not a foreign word. It denotes an indigenous *Curcuma*; while under the T'ang, as we have seen, additional species of this genus may have been introduced from abroad. The word *yü-kin* then underwent a psychological treatment similar to *yen-či:* as *yen-či*, "safflower," was transformed to any cosmetic or rouge, so *yü-kin* "turmeric," was grafted on any dyes producing similar tinges of yellow. Thus it was applied to the saffron of Kashmir and Persia.

---

[1] The early edition of this work did not contain the form *yü-kin*, but merely the plain, ancient *yü*. Solely the *Fan yi miṅ yi tsi* (Ch. 8, p. 10 b) attributes (I believe, erroneously) the term *yü-kin* to the *Šwo wen*.

[2] Li Ši-čen says that the district Yü-lin of the Han period comprises the territory of the present *čou* 州 of Sün 潯, Liu 柳, Yuṅ 邕 , and Pin 賓 of Kwaṅ-si and Kwei-čou, and that, according to the *Ta Miṅ i t'uṅ či*, only the district of Lo-č'eṅ 羅城 in Liu-čou fu (Kwaṅ-si) produces *yü-kin hiaṅ*, which is that here spoken of (that is, *Crocus*), while in fact *Curcuma* must be understood.

[3] There is also the opinion that the ancient *yü* must be a plant similar to *lan* 蘭, an orchidaceous plant (see the *P'i ya* of Lu Tien and the *T'uṅ či* of Čeṅ Tsiao).

[4] PALLEGOIX (Description du royaume Thai ou Siam, Vol. I, p. 126) says, "Le curcuma est une racine bulbeuse et charnue, d'un beau jaune d'or."

SAFFLOWER

17. A. DE CANDOLLE,[1] while maintaining that the cultivation of safflower[2] (*Carthamus tinctorius*) is of ancient date both in Egypt and India, asserts on Bretschneider's authority that the Chinese received it only in the second century B.C., when Čaṅ K'ien brought it back from Bactriana. The same myth is repeated by STUART.[3] The biography of the general and the Han Annals contain nothing to this effect. Only the *Po wu či* enumerates *hwaṅ lan* 黃藍 in its series of Čaṅ-K'ien plants, adding that it can be used as a cosmetic (*yen-či* 燕支).[4] The *Ku kin ču*, while admitting the introduction of the plant from the West, makes no reference to the General. The *Ts'i min yao šu* discusses the method of cultivating the flower, but is silent as to its introduction. The fact of this introduction cannot be doubted, but it is hardly older than the third or fourth century A.D. under the Tsin dynasty. The introduction of safflower drew the attention of the Chinese to an indigenous wild plant (*Basella rubra*) which yielded a similar dye and cosmetic, and both plants and their products were combined or confounded under the common name *yen-či*.

Basella rubra, a climbing plant of the family *Basellaceae*, is largely cultivated in China (as well as in India) on account of its berries, which contain a red juice used as a rouge by women and as a purple dye for making seal-impressions. This dye was the prerogative of the highest

---

[1] Origin of Cultivated Plants, p. 164.

[2] Regarding the history of this word, see YULE, Hobson-Jobson, p. 779.

[3] Chinese Materia Medica, p. 94. It is likewise an erroneous statement of Stuart that Tibet was regarded by the Chinese as the natural habitat of this plant. This is due to a confusion with the term *Si-ts'aṅ huṅ hwa* ("red flower of Tibet"), which refers to the saffron, and is so called because in modern times saffron is imported into China from Kashmir by way of Tibet (see p. 312). Neither *Carthamus* nor saffron is grown in the latter country.

[4] Some editions of the *Po wu či* add, "At present it has also been planted in the land of Wei 魏 (China)," which might convey the impression that it had only been introduced during the third century A.D., the lifetime of Čaṅ Hwa, author of that work. In the commentary to the *Pei hu lu* (Ch. 3, p. 12), the *Po wu či* is quoted as saying, "The safflower (*huṅ hwa* 紅花, 'red flower') has its habitat in Persia, Su-le (Kashgar), and Ho-lu 河臝. Now that of Liaṅ-han 梁漢 is of prime quality, a tribute of twenty thousand catties being annually sent to the Bureau of Weaving and Dyeing." The term *huṅ hwa* in the written language does not refer to "saffron," but to "safflower." Java produced the latter (Javanese *kasumba*), not saffron, as translated by HIRTH (Chau Ju-kua, p. 78). The Čaṅ-K'ien story is repeated in the *Hwa kiṅ* of 1688 (Ch. 5, p. 24 b).

boards of the capital, the prefects of Šun-t'ien and Mukden, and all provincial governors.[1] Under the name *lo k'wei* 落葵 it is mentioned by T'ao Huṅ-kiṅ (A.D. 451–536), who refers to its cultivation, to the employment of the leaves as a condiment, and to the use of the berries as a cosmetic.[2] This probably came into use after the introduction of safflower. The *Ku kin ču*,[3] written by Ts'ui Pao in the middle of the fourth century, states, "The leaves of *yen-či* 燕支 resemble those of the thistle (*ki* 薊) and the *p'u-kuṅ* 蒲公 (*Taraxacum officinalis*). Its habitat is in the Western Countries 西方, where the natives avail themselves of the plant for dyeing, and designate it *yen-či* 燕支, while the Chinese call it *huṅ-lan* (紅藍 'red indigo,' *Carthamus tinctorius*); and the powder obtained from it, and used for painting the face, is styled *yen-či fen* 粉. [At present, because people value a deep-red color 絳, they speak of the *yen-či* flower which dyes; the *yen-či* flower, however, is not the dye-plant *yen-či*, but has its own name, *huṅ-lan* (*Carthamus tinctorius*). Of old, the color intermediate between *či* 赤 and white is termed *huṅ* 紅, and this is what is now styled *huṅ-lan*.]"[4] It would follow from this text that *Basella* was at an early date confounded with *Carthamus*, but that originally the term *yen-či* related to *Carthamus* only.

The *Pei hu lu*[5] contains the following information in regard to the *yen-či* flower: "There is a wild flower growing abundantly in the rugged mountains of Twan-čou 端州.[6] Its leaves resemble those of the *lan* 藍 (*Indigofera*); its flowers, those of the *liao* 蓼 (*Polygonum*, probably *P. tinctorium*). The blossoms 穗, when pulled out, are from two to three inches long, and yield a green-white pigment. It blooms in the first month. The natives gather the bursting seeds while still in their shells, in order to sell them. They are utilized in the preparation of a cosmetic 燕支粉, and particularly also for dyeing pongee and other silks. Its red is not inferior to that of the *lan* flower. Si Ts'o-č'i

---

P. HOANG, Mélanges sur l'administration, pp. 80–81.

[1] RETSCHNEIDER, Bot. Sin., pt. II, No. 148; pt. III, No. 258.

[3] Ch. c, p. 5 (ed. of *Han Wei ts'uṅ šu*). In regard to the historicity of this work, the critical remarks of the Imperial Catalogue (cf. WYLIE, Notes on Chinese Literature, p. 159) must be kept in mind. Cf. also above, p. 242.

[4] The passage enclosed in brackets, though now incorporated in the text of the *Ku kin ču*, is without any doubt later commentatorial wisdom. This is formally corroborated by the *Pei hu lu* (Ch. 3, p. 12), which omits all this in quoting the relevant text of the *Ku kin ču*.

[5] Ch. 3, p. 11 (see above, p. 268).

[6] Name of the prefecture of Čao-k'iṅ 肇慶 in Kwaṅ-tuṅ Province. This wild flower is *Basella rubra*.

習鑿齒, in his *Yü sie ši čuň šu* 與謝侍中書, says,[1] 'These are *huň-lan (Carthamus)*:[2] did you know these previously, Sir, or not? The people of the north gather these flowers, and dye materials a red-yellow by rubbing their surface with it. The fresh blossoms are made into a cosmetic.[3] Women, when dressing, use this pigment, it being the fashion to apply only a piece the size of a small bean. When distributed evenly, the paint is pleasing, as long as it is fresh. In my youth I observed this cosmetic again and again; and to-day I have for the first time beheld the *huň-lan* flower. Afterwards I shall raise its seeds for your benefit, Sir. The Hiuň-nu styled a wife *yen-či* 閼氏,[4] a word just as pleasing as *yen-či* 烟支 ('cosmetic'). The characters 閼 and 烟 have the same sound *yen;* the character 氏 has the sound 支 *či*. I expect you knew this before, Sir, or you may read it up in the Han Annals.' Čeň K'ien 鄭虔[5] says that a cosmetic may be prepared from pomegranate flowers."[6]

The curious word *yen-či* has stirred the imagination of Chinese scholars. It is not only correlated with the Hiuň-nu word *yen-či*, as was first proposed by Si Ts'o-č'i, but is also connected with a Yen-či mountain. Lo Yüan, in his *Er ya i*, remarks that the Hiuň-nu had a Yen-či mountain, and goes on to cite a song from the *Si ho kiu ši* 西河舊事,[7] which says, "If we lose our K'i-lien mountain 祁連山,[8] we cause our herds to diminish in number; if we lose our Yen-či mountain, we cause our women to go without paint."[9] The *Pei pien pei tui* 北邊備對, a work of the Sung period, states, "The *yen-či* 焉支 of the Yen-či mountain 焉支山 is the *yen-či* 燕脂 of the present time. This moun-

---

[1] This author is stated to have lived under the Tsin dynasty (A.D. 265–419) in the *T'u šu tsi č'eň*, XX, Ch. 158, where this passage is quoted; but his book is there entitled *Yü yen waň šu* 與燕王書. The same passage is inserted in the *Er ya i* of Lo Yüan 羅顥 of the twelfth century, where the title is identical with that given above.

[2] In the text of the *T'u šu:* "At the foot of the mountain there are *huň lan.*"

[3] *Carthamus* was already employed for the same purposes in ancient Egypt.

[4] This is the Hiuň-nu word for a royal consort, handed down in the Han Annals (*Ts'ien Han šu*, Ch. 94 A, p. 5). See my Language of the Yüe-chi, p. 10.

[5] Author of the lost *Hu pen ts'ao* (above, p. 268).

[6] Then follow a valueless anecdote anent a princess of the T'ang dynasty preparing a cosmetic, and the passage of the *Ku kin ču* given above.

[7] Mentioned in the T'ang literature, but seems to date from an earlier period (BRETSCHNEIDER, Bot. Sin., pt. 1, p. 190).

[8] A mountain-range south-west of Kan čou in Kan-su (*Ši ki*, Ch. 123, p. 4). The word *k'i-lien* belongs to the language of the Hiuň-nu and means "heaven." In my opinion, it is related to Manchu *kulun*, which has the same meaning. The interpretations given by WATTERS (Essays, p. 362) and SHIRATORI (Sprache der Hiung-nu, p. 8) are not correct.

[9] The same text is quoted in the commentary to the *Pei hu lu* (Ch. 3, p. 11 b).

tain produces *huṅ-lan* (*Carthamus*) which yields *yen-či* ('cosmetic')."
All this, of course, is pure fantasy inspired by the homophony of the two
words *yen-či* ("cosmetic") and Hiuṅ-nu *yen-či* ("royal consort").
Another etymology propounded by Fu Hou 伏侯 in his *Čuṅ hwa ku
kin ču* 中華古今注 (tenth century) is no more fortunate: he explains
that *yen-či* is produced in the country Yen 燕, and is hence styled 臙脂
yen-či ("sap of Yen"). Yen was one of the small feudal states at the
time of the Čou dynasty. This is likewise a philological afterthought,
for there is no ancient historical record to the effect that the state of
Yen should have produced (exclusively or pre-eminently) *Basella* or
Carthamus. It is perfectly certain that *yen-či* is not Chinese, but the
transcription of a foreign word: this appears clearly from the ancient
form 燕支, which yields no meaning whatever; 支, as is well known,
being a favorite character in the rendering of foreign words. This is
further corroborated by the vacillating modes of writing the word,
to which Li Ši-čen adds 䉤荍,[1] while he rejects as erroneous 臙肢
and 胭支, and justly so. Unfortunately we are not informed as to the
country or language from which the word was adopted: the *Ku kin
ču* avails itself only of the vague term Si faṅ ("Western Countries"),
where *Carthamus* was called *yen-či*; but in no language known to me is
there any such name for the designation of this plant or its product.
The Sanskrit name for safflower is *kusumbha;* and if the plant had come
from India, Chinese writers would certainly not have failed to express
this clearly. The supposition therefore remains that it was introduced
from some Iranian region, and that *yen-či* represents a word from an
old Iranian dialect now extinct, or an Iranian word somehow still
unknown. The New-Persian name for the plant is *gāwdžīla;* in Arabic
it is *qurtum*.[2]

Li Ši-čen distinguishes four kinds of *yen-či:* (1) From *Carthamus
tinctorius*, the juice of the flowers of which is made into a rouge (the
information is chiefly drawn from the *Ku kin ču*, as cited above).
(2) From *Basella rubra*, as described in the *Pei hu lu*. (3) From the
šan-liu 山榴 flower [unidentified, perhaps a wild pomegranate: above,
p. 281], described in the *Hu pen ts'ao*. (4) From the tree producing
gum lac (*tse-kuṅ* 紫鉚),[3] this product being styled 胡燕脂 *hu yen-či*
("foreign cosmetic") and described in the *Nan hai yao p'u* 南海藥譜
of Li Sün 李珣.[4] "At present," Li Ši-čen continues, "the southerners

---

[1] Formed with the classifier 155, "red."

[2] ACHUNDOW, Abu Mansur, p. 105.

[3] See below, p. 476.

[4] He lived in the second half of the eighth century.

make abundant use of *tse-kuṅ* cosmetic, which is commonly called *tse-kuṅ*. In general, all these substances may be used as remedies in blood diseases.[1] Also the juice from the seeds of *lo k'wei* 落葵 (*Basella rubra*) may be taken, and, mixed evenly with powder, may be applied to the face. Also this is styled *hu yen-či*." Now it becomes clear why *Basella rubra*, a plant indigenous to China, is termed *hu yen-či* in the *T'uṅ či* of Čeṅ Tsiao and by Ma Či of the tenth century: this name originally referred to the cosmetic furnished by *Butea frondosa* or other trees on which the lac-insect lives,[2]—trees growing in Indo-China, the Archipelago, and India. This product, accordingly, was foreign, and hence styled "foreign cosmetic" or "cosmetic of the barbarians" (*hu yen-či*). Since *Basella* was used in the same manner, that name was ultimately transferred also to the cosmetic furnished by this indigenous plant.

What is not stated by Li Ši-čen is that *yen-či* is also used with reference to *Mirabilis jalapa*, because from the flowers of this plant is derived a red coloring-matter often substituted for carthamine.[3] It is obvious that the term *yen-či* has no botanical value, and for many centuries has simply had the meaning "cosmetic."

Fan Č'eṅ-ta (1126–93), in his *Kwei hai yü heṅ či*,[4] mentions a *yen-či* 臙脂 tree, strong and fine, with a color like *yen-či* (that is, red), good for making arrowheads, and growing in Yuṅ čou, also in the caves of this department, and in the districts of Kwei-lin, in Kwaṅ-si Province. A. HENRY[5] gives for Yi-č'aṅ in Se-č'wan a plant-name *yen-či ma* 煙脂 麻 ("cosmetic hemp"), identified with *Patrinia villosa*.

---

[1] On account of the red color of the berries.

[2] See p. 478.

[3] STUART, Chinese Materia Medica, p. 264; MATSUMURA, No. 2040; PERROT and HURRIER, Matière médicale et pharmacopée sino-annamites, p. 116, where *lo-k'wei* is erroneously given as Chinese name of the plant.

[4] Ed. of *Či pu tsu čai ts'uṅ šu*, p. 28 b.

[5] Chinese Names of Plants, p. 239 (*Journal China Branch Roy. As. Soc.*, Vol. XXII, 1887).

JASMINE

18. The *Nan fan ts'ao mu čwan* 南方草木狀, the oldest Chinese work devoted to the botany of southern China, attributed to Ki Han 稽含, a minister of the Emperor Hwei 惠 (A.D. 290–309), contains the following notice:[1]—

"The *ye-si-min* 耶悉茗 flower and the *mo-li* 末利 flower (*Jasminum officinale*, family *Oleaceae*) were brought over from western countries by Hu people 胡人, and have been planted in Kwan-tun (Nan hai 南海). The southerners are fond of their fragrant odor, and therefore cultivate them . . . The *mo-li* flower resembles the white variety of *ts'ian-mi* 薔薇 (*Cnidium monnieri*), and its odor exceeds that of the *ye-si-min.*"

In another passage of the same work[2] it is stated that the *či-kia* 指甲 flower (*Lawsonia alba*),[3] *ye-si-min*, and *mo-li* were introduced by Hu people from the country Ta Ts'in; that is, the Hellenistic Orient.

The plant *ye-si-min* has been identified with *Jasminum officinale;* the plant *mo-li*, with *Jasminum sambac*. Both species are now cultivated in China on account of the fragrancy of the flowers and the oil that they yield.[4]

The passage of the *Nan fan ts'ao mu čwan*, first disclosed by BRET-SCHNEIDER,[5] has given rise to various misunderstandings. HIRTH[6] remarked, "This foreign name, which is now common to all European languages, is said to be derived from Arabic-Persian *jāsamīn* [read *yāsmīn*], and the occurrence of the word in a Chinese record written about A.D. 300 shows that it must have been in early use." WATTERS[7] regarded *yāsmīn* as "one of the earliest Arabian words to be found in Chinese literature." It seems never to have occurred to these authors

---

[1] Ch. A, p. 2 (ed. of *Han Wei ts'un šu*).

[2] Ch. B, p. 3.

[3] See below, p. 334.

[4] The sambac is a favored flower of the Chinese. In Peking there are special gardeners who cultivate it exclusively. Every day in summer, the flower-buds are gathered before sunrise (without branches or leaves) and sold for the purpose of perfuming tea and snuff, and to adorn the head-dress of Chinese ladies. *Jasminum officinale* is not cultivated in Peking (BRETSCHNEIDER, *Chinese Recorder*, Vol. III, 1871, p. 225).

[5] *Chinese Recorder*, Vol. III, p. 225.

[6] China and the Roman Orient, p. 270.

[7] Essays on the Chinese Language, p. 354.

329

that at this early date we know nothing about an Arabic or Persian language; and this *rapprochement* is wrong, even in view of the Chinese work itself, which distinctly says that both *ye-si-miṅ* and *mo-li* were introduced from Ta Ts'in, the Hellenistic Orient. PELLIOT[1] observes that the authenticity of the Chinese book has never been called into doubt, but expresses surprise at the fact that jasmine figures there under its Arabic name. But Arabic is surely excluded from the languages of Ta Ts'in. Moreover, thanks to the researches of L. AUROUSSEAU,[2] we now know that the *Nan faṅ ts'ao mu čwaṅ* is impaired by inter-polations. The passage in question may therefore be a later addition, and, at all events, cannot be enlisted to prove that prior to the year 300 there were people from western Asia in Canton.[3] Still less is it credible that, as asserted in the Chinese work, the *Nan yüe hiṅ ki* 南越行記 ascribed to Lu Kia 陸賈, who lived in the third and second centuries B.C., should have alluded to the two species of *Jasminum*.[4] In fact, this author is made to say only that in the territory of Nan Yüe the five cereals have no taste and the flowers have no odor, and merely that these flowers are particularly fragrant. Their names are not given, and it is Ki Han who refers them to *ye-si-miṅ* and *mo-li*. It is out of the question that at the time of Lu Kia these two foreign plants should have been introduced over the maritime route into southern China; Lu Kia, if he has written this passage, may have as well had two other flowers in mind.

The fact must not be overlooked, either, that the alleged introduction from Ta Ts'in is not contained in the historical texts relative to that country, nor is it confirmed by any other coeval or subsequent source.

The *Pei hu lu*[5] mentions the flower under the names *ye-si-mi* 耶悉弭 and white *mo-li* 白末利花 as having been transplanted to China by Persians, like the *p'i-ši-ša* or gold-coin flower.[6] The *Yu yaṅ tsa tsu* has furnished a brief description of the plant,[7] stating that its habitat is in Fu-lin and in Po-se (Persia). The *Pen ts'ao kaṅ mu, Kwaṅ k'üṅ faṅ p'u*,[8] and *Hwa kiṅ*[9] state that the habitat of jasmine (*mo-li*) was

---

[1] *Bull. de l'Ecole française*, Vol. II, p. 146.

[2] See above, p. 263.

[3] HIRTH, Chau Ju-kua, p. 6, note 1.

[4] This point is discussed neither by Bretschneider nor by Hirth, who do not at all mention this reference.

[5] Ch. 3, p. 16 (see above, p. 268).

[6] See below, p. 335.

[7] Translated by HIRTH, *Journal Am. Or. Soc.*, Vol. XXX, 1910, p. 22.

[8] Ch. 22, p. 8 b.

[9] Ch. 4, p. 9.

originally in Persia, and that it was thence transplanted into Kwaṅ-tuṅ. The first-named work adds that it is now (sixteenth century) cultivated in Yün-nan and Kwaṅ-tuṅ, but that it cannot stand cold, and is unsuited to the climate of China. The *Tan k'ien tsuṅ lu* 丹鉛 總錄 of Yaṅ Šen 楊愼 (1488–1559) is cited to the effect that "the name *nai* 柰 used in the north of China is identical with what is termed in the Tsin Annals 晉書 *tsan nai hwa* 簪 ('hair-pin') 柰花.[1] As regards this flower, it entered China a long time ago."

Accordingly we meet in Chinese records the following names for jasmine:[2]—

(1) 耶悉茗 *ye-si-miṅ*, * ya-sit(siδ)-miṅ, = Pahlavi *yāsmīn*, New Persian *yāsamīn*, *yāsmīn*, *yāsmūn*, Arabic *yasmin*, or 野悉蜜 *ye-si-mi*, *ya-sit-mit (in *Yu yaṅ tsa tsu*) = Middle Persian *yāsmīr (?).[3] Judging from this philological evidence, the statement of the *Yu yaṅ tsa tsu*, and Li Ši-čen's opinion that the original habitat of the plant was in Persia, it seems preferable to think that it was really introduced from that country into China. The data of the *Nan faṅ ts'ao mu čwaṅ* are open to grave suspicion; but he who is ready to accept them is compelled to argue, that, on the one hand, the Persian term was extant in western Asia at least in the third century A.D., and that, on the other hand, the Indian word *mallikā* (see No. 2) had reached Ta Ts'in about the same time. Either suggestion would be possible, but is not confirmed by any West-Asiatic sources.[4] The evidence presented by the Chinese work is isolated; and its authority is not weighty enough, the relation of the modern text to the original issue of about A.D. 300 is too obscure, to derive from it such a far-reaching conclusion. The Persian-Arabic word has become the property of the entire world: all European languages have adopted it, and the Arabs diffused it along the east coast of Africa (Swahīlī *yasmini*, Madagasy *dzasimini*).

(2) 末利 or 茉莉 *mo-li*,[5] **mwat(mwal)-li* = *malli*, transcription of

---

[1] This is the night-blooming jasmine (*Nyctanthes arbor tristis*), the musk-flower of India (STUART, Chinese Materia Medica, p. 287).

[2] There are numerous varieties of *Jasminum*,—about 49 to 70 in India, about 39 in the Archipelago, and about 15 in China and Japan.

[3] From the Persian loan-word in Armenian, *yasmik*, HÜBSCHMANN (Armen. Gram., p. 198) justly infers a Pahlavi *yāsmīk, beside *yāsmīn*. Thus also *yāsmīt or *yāsmīr may have existed in Pahlavi.

[4] It is noteworthy also that neither Dioscorides nor Galenus was acquainted with jasmine.

[5] For the expression of the element *li* are used various other characters which may be seen in the *Kwaṅ k'ün faṅ p'u* (Ch. 22, p. 8 b); they are of no importance for the phonetic side of the case.

Sanskrit *mallikā* (*Jasminum sambac*), Tibetan *mal-li-ka*, Siamese *ma-li*,[1] Khmer *māly* or *mlih*, Čam *molih*. Malayan *melati* is derived from Sanskrit *mālatī*, which refers to *Jasminum grandiflorum*. Mongol *melirge* is independent. Hirth's identification with Syriac *molo*[2] must be rejected.

(3) 散沫 *san-mo*, *san-mwat (Fukien *mwak*). This word is given in the *Nan fan ts'ao mu čwan*[3] as a synonyme of *Lawsonia alba*, furnishing the henna; but a confusion has here arisen, for the transcription does not answer to any foreign name of *Lawsonia*, but apparently corresponds to Arabic *zanbaq* ("jasmine"), from which the botanical term *sambac* is derived. It is out of the question that this word was known to Ki Han: it is clearly an interpolation in his text.

(4) 鬘華 *man hwa* ("*man* flower") occurs in Buddhist literature, and is apparently an abridgment of Sanskrit *sumanā* (*Jasminum grandiflorum*), which has been adopted into Persian as *suman* or *saman*.

Jasminum officinale occurs in Kashmir, Kabul, Afghanistan, and Persia; in the latter country also in the wild state.

Jasmine is discussed in Pahlavi literature (above, p. 192) and in the Persian pharmacopœia of Abu Mansur.[4] Č'an Te noticed the flower in the region of Samarkand.[5] It grows abundantly in the province of Fars in Persia.[6]

Oil of jasmine is a famous product among Arabs and Persians, being styled in Arabic *duhn az-zanbaq*. Its manufacture is briefly described in Ibn al-Baiṭār's compilation.[7] According to Istaxrī, there is in the province of Dārābejird in Persia an oil of jasmine that is to be found nowhere else. Sābūr and Šīrāz were renowned for the same product.[8]

The oil of jasmine manufactured in the West is mentioned in the *Yu yaṅ tsa tsu* as a tonic. It was imported into China during the Sung period, as we learn from the *Wei lio* 緯畧,[9] written by Kao Se-sun 高似孫, who lived toward the end of the twelfth and in the beginning of the thirteenth century. Here it is stated, "The *ye-si-min* flower is a flower of the western countries, snow-white in color. The Hu 胡 (Iranians or foreigners) bring it to Kiao-čou and Canton, and every one

[1] Pallegoix, Description du royaume Thai, Vol. I, p. 147.

[2] *Journal Am. Or. Soc.*, Vol. XXX, 1910, p. 23.

[3] Ch. B, p. 3. See below, p. 334.

[4] Achundow, Abu Mansur, p. 147.

[5] Bretschneider, Mediæval Researches, Vol. I, p. 131.

[6] G. Le Strange, Description of the Province of Fars, p. 51.

[7] L. Leclerc, Traité des simples, Vol. II, p. 111.

[8] P. Schwarz, Iran, pp. 52, 94, 97, 165.

[9] Ch. 9, p. 9.

is fond of its fragrance and plants this flower. According to the *Kwaṅ tou t'u kiṅ* 廣州圖經 ('Gazetteer of Kwaṅ-tuṅ Province'), oil of jasmine is imported on ships; for the Hu gather the flowers to press from them oil, which is beneficial for leprosy 麻風.[1] When this fatty substance is rubbed on the palm of the hand, the odor penetrates through the back of the hand."

---

[1] According to the Arabs, it is useful as a preventive of paralysis and epilepsy (LECLERC, *l. c.*).

HENNA

19. It is well known that the leaves of *Lawsonia alba* or *L. inermis*, grown all over southern China, are extensively used by women and children as a finger-nail dye, and are therefore styled *či kia hwa* 指甲 花 ("finger-nail flower").[1] This flower is mentioned in the *San fu hwaṅ t'u*,[2] of unknown authorship and date, as having been transplanted from Nan Yüe (South China) into the Fu-li Palace at the time of the Han Emperor Wu (140–87 B.C.). This is doubtless an anachronism or a subsequent interpolation in the text of that book. The earliest datable reference to this plant is again contained in the *Nan faṅ ts'ao mu čwaṅ* by Ki Han,[3] by whom it is described as a tree from five to six feet in height, with tender and weak branches and leaves like those of the young elm-tree 楡 (*Ulmus campestris*), the flowers being snow-white like *ye-si-miṅ* and *mo-li*, but different in odor. As stated above (p. 329), this work goes on to say that these three plants were introduced by Hu people from Ta Ts'in, and cultivated in Kwaṅ-tuṅ.[4] The question arises again whether this passage was embodied in the original edition. It is somewhat suspicious, chiefly for the reason that Ki Han adds the synonyme *san-mo*, which, as we have seen, in fact relates to jasmine.

The *Pei hu lu*,[5] written about A.D. 875 by Twan Kuṅ-lu, contains the following text under the heading *či kia hwa:* "The finger-nail flower is fine and white and of intense fragrance. The barbarians 番人 now plant it. Its name has not yet been explained. There are, further, the jasmine and the white *mo-li*. All these were transplanted to China by the Persians (Po-se). This is likewise the case with the *p'i-ši-ša* 毗尸 沙 (or 'gold coin') flower (*Inula chinensis*). Originally it was only produced abroad, but in the second year of the period Ta-t'uṅ 大同 (A.D. 536 of the Liang dynasty) it came to China for the first time (始來中土)." In the *Yu yaṅ tsa tsu*,[6] written about fifteen years earlier, we read, "The gold-coin flower 金錢花, it is said, was originally produced abroad. In the second year of the period Ta-t'uṅ of the

---

[1] Cf. *Notes and Queries on China and Japan*, Vol. I, 1867, pp. 40–41. STUART, Chinese Materia Medica, p. 232.

[2] Ch. 3, p. 9 b (see above, p. 263).

[3] Ch. B, p. 3 (ed. of *Han Wei ts'uṅ šu*).

[4] Cf. also HIRTH, China and the Roman Orient, p. 268.

[5] Ch. 3, p. 16 (see above, p. 268).

[6] Ch. 19, p. 10 b.

334

Liang (A.D. 536) it came to China. At the time of the Liang dynasty, people of Kiṅ čou 荆州 used to gamble in their houses at backgammon with gold coins. When the supply of coins was exhausted, they resorted to gold-coin flowers. Hence Yü Huṅ 魚弘 said, 'He who obtains flowers makes money.' " The same work likewise contains the following note:[1] "P'i-ši-ša 毗尸沙 is a synonyme for the gold-coin flower,[2] which was originally produced abroad, and came to China in the first year of the period Ta-t'uṅ of the Liang (A.D. 535)." The gold-coin flower visualized by Twan Kuṅ-lu and Twan Č'eṅ-ši assuredly cannot be *Inula chinensis*, which is a common, wild plant in northern China, and which is already mentioned in the *Pie lu* and by T'ao Huṅ-kiṅ.[3] It is patent that this flower introduced under the Liang must have been a different species. The only method of solving the problem would be to determine the prototype of *p'i-ši-ša*, which is apparently the transcription of a foreign word. It is not stated to which language it belongs; but, judging from appearances, it is Sanskrit, and should be traceable to a form like *viṣīṣa (or *viçeṣa). Such a Sanskrit plant-name is not to be found, however. Possibly the word is not Sanskrit.[4]

The *Pei hu lu*, accordingly, conceives the finger-nail flower as an introduction due to the Persians, but does not allude to its product, the henna. I fail to find any allusion to henna in other books of the T'ang period. I am under the impression that the use of this cosmetic did not come into existence in China before the Sung epoch, and that the practice was then introduced (or possibly only re-introduced) by Mohammedans, and was at first restricted to these. It is known that also the leaves of *Impatiens balsamina* (*fuṅ sien* 鳳仙) mixed with alum are now used as a finger-nail dye, being therefore styled *žan či kia ts'ao* 染指甲草 ("plant dyeing finger-nails"),[5]—a term first appearing in the *Kiu hwaṅ pen ts'ao*, published early in the Ming period. The earliest source that mentions the practice is the *Kwei sin tsa ši* 癸辛

---

[1] Ch. 19, p. 10 a.

[2] The addition of 中 before *kin* in the edition of *Pai hai* surely rests on an error.

[3] Cf. also BRETSCHNEIDER, Bot. Sin., pt. III, p. 158.

[4] The new Chinese Botanical Dictionary (p. 913) identifies the gold-coin flower with *Inula britannica*. In Buddhist lexicography it is identified with Sanskrit *jāti* (*Jasminum grandiflorum;* cf. EITEL, Handbook, p. 52). The same word means also "kind, class"; so does likewise *viçeṣa*, and the compound *jāti-viçeṣa* denotes the specific characters of a plant (HOERNLE, Bower Manuscript, p. 273). It is therefore possible that this term was taken by the Buddhists in the sense of "species of *Jasminum*," and that finally *viçeṣa* was retained as the name of the flower.

[5] STUART, Chinese Materia Medica, p. 215; *Pen ts'ao kaṅ mu*, Ch. 17 B, p. 12 b.

雜識[1] by Čou Mi 周密 (1230–1320), who makes the following observation: "As regards the red variety of the *fuṅ sien* flower (*Impatiens
balsamina*), the leaves are used, being pounded in a mortar and mixed
with a little alum.[2] The finger-nails must first be thoroughly cleaned,
and then this paste is applied to them. During the night a piece of
silk is wrapped around them, and the dyeing takes effect. This process
is repeated three or five times. The color resembles that of the *yen-či*
(*Basella rubrum*). Even by washing it does not come off, and keeps
for fully ten days. At present many Mohammedan women are fond
of using this cosmetic for dyeing their hands, and also apply it to cats
and dogs for their amusement." The *Pen ts'ao kaṅ mu* quotes only the
last clause of this text. From what Čou Mi says, it does not appear
that the custom was of ancient date; on the contrary, it does not seem
to be older than the Sung period.

None of the early *Pen ts'ao* makes mention of *Lawsonia*. It first
appears in the *Pen ts'ao kaṅ mu*. All that Li Ši-čen is able to note
amounts to this: that there are two varieties, a yellow and a white one,
which bloom during the summer months; that its odor resembles that
of *mu-si* 木犀 (*Osmanthus fragrans*); and that it can be used for dyeing
the finger-nails, being superior in this respect to the *fuṅ sien* flower
(*Impatiens balsamina*). Čeṅ Kaṅ-čuṅ 鄭剛中, an author of the Sung
period, mentions the plant under the name *i hiaṅ hwa* 異香花 ("flower
of peculiar fragrance").

It has generally been believed hitherto that the use of henna and
the introduction of *Lawsonia* into China are of ancient date; but, in
fact, the evidence is extremely weak. In my opinion, as far as the employment of henna is concerned, we have to go down as far as the
Sung period. It is noteworthy also that no foreign name of ancient date,
either for the plant or its product, is on record. F. P. SMITH and STUART
parade the term 海蒳 *hai-na* (Arabic *hinnā*) without giving a reference.
The very form of this transcription shows that it is of recent date: in
fact, it occurs as late as the sixteenth century in the *Pen ts'ao kaṅ mu*,[3]
then in the *K'ün faṅ p'u* of 1630[4] and the *Nuṅ čeṅ ts'üan šu* 農政全書,
published in 1619 by Sü Kwaṅ-k'i 徐光啟, the friend and supporter
of the Jesuits. It also occurs in the *Hwa kiṅ* of 1688.[5]

It is well known what extensive use of henna (Arabic *hinnā*, hence

---

[1] 續集上, p. 17 (ed. of *Pai hai*).

[2] In this manner the dye is also prepared at present.

[3] Ch. 17 B, p. 12 b.

[4] *Kwaṅ k'ün faṅ p'u*, Ch. 26, p. 4 b. The passages of the first edition are
especially indicated.

[5] Ch. 5, p. 23 b.

Malayan *inei*) has been made in the west from ancient times. The Egyptians stained their hands red with the leaves of the plant[1] (Egyptian *puqer*, Coptic *kuper* or *khuper*, Hebrew *kopher*, Greek κύπρος). All Mohammedan peoples have adopted this custom; and they even dye their hair with henna, also the manes, tails, and hoofs of horses.[2] The species of western Asia is identical with that of China, which is spontaneous also in Baluchistan and in southern Persia.[3] Ancient Persia played a prominent rôle as mediator in the propagation of the plant.[4] "They [the Persians] have also a custom of painting their hands, and, above all, their nails, with a red color, inclining to yellowish or orange, much near the color that our tanners nails are of. There are those who also paint their feet. This is so necessary an ornament in their married women, that this kind of paint is brought up, and distributed among those that are invited to their wedding dinners. They therewith paint also the bodies of such as dye maids, that when they appear before the Angels Examinants, they may be found more neat and handsome. This color is made of the herb, which they call *Chinne*, which hath leaves like those of liquorice, or rather those of myrtle. It grows in the Province of Erak, and it is dry'd, and beaten, small as flower, and there is put thereto a little of the juyce of sour pomegranate, or citron, or sometimes only fair water; and therewith they color their hands. And if they would have them to be of a darker color, they rub them afterwards with wall-nut leaves. This color will not be got off in fifteen days, though they wash their hands several times a day."[5] It

---

[1] V. LORET, Flore pharaonique, p. 80; WŒNIG, Pflanzen im alten Aegypten, p. 349.

[2] L. LECLERC, Traité des simples, Vol. I, p. 469; G. JACOB, Studien in arabischen Geographen, p. 172; A. v. KREMER, Culturgeschichte des Orients unter den Chalifen, Vol. II, p. 325.

[3] C. JORET, Plantes dans l'antiquité, Vol. II, p. 47.

[4] SCHWEINFURTH, Z. Ethnologie, Vol. XXIII, 1891, p. 658.

[5] A. OLEARIUS, Voyages of the Ambassadors to the Great Duke of Muscovy and the King of Persia (1633–39), p. 234 (London, 1669). I add the very exact description of the process given by SCHLIMMER (Terminologie, p. 343): "C'est avec la poudre fine des feuilles sèches de cette plante, largement cultivée dans le midi de la Perse, que les indigènes se colorent les cheveux, la barbe et les ongles en rouge-orange. La poudre, formée en pâte avec de l'eau plus ou moins chaude, est appliquée sur les cheveux et les ongles et y reste pendant une ou deux heures, ayant soin de la tenir constamment humide en empêchant l'évaporation de son eau; après quoi la partie est lavée soigneusement; l'effet de l'application du henna est de donner une couleur rouge-orange aux cheveux et aux ongles. Pour transformer cette couleur rougeâtre en noir luisant, on enduit pendant deux ou trois autres heures les cheveux ou la barbe d'une seconde pâte formée de feuilles pulvérisées finement d'une espèce d'indigofère, cultivée sur une large échelle dans la province de Kerman. Ces manipulations se pratiquent d'ordinaire au bain persan, où la chaleur humide diminue

seems more likely that the plant was transmitted to China from Persia than from western Asia, but the accounts of the Chinese in this case are too vague and deficient to enable us to reach a positive conclusion.

In India, *Lawsonia alba* is said to be wild on the Coromandel coast. It is now cultivated throughout India. The use of henna as a cosmetic is universal among Mohammedan women, and to a greater or lesser extent among Hindu also; but that it dates "from very ancient times," as stated by WATT,[1] seems doubtful to me. There is no ancient Sanskrit term for the plant or the cosmetic (*mendhī* or *mendhikā* is Neo-Sanskrit), and it would be more probable that its use is due to Mohammedan influence. JORET[2] holds that the tree, although it is perhaps indigenous, may have been planted only since the Mohammedan invasion.[3]

FRANÇOIS PYRARD, who travelled from 1601 to 1610, reports the henna-furnishing plant on the Maldives, where it is styled *innapa* (=*hinā-fai*, "henna-leaf"). "The leaves are bruised," he remarks, "and rubbed on their hands and feet to make them red, which they esteem a great beauty. This color does not yield to any washing, nor until the nails grow, or a fresh skin comes over the flesh, and then (that is, at the end of five or six months) they rub them again."[4]

---

singulièrement la durée de l'opération." While the Persians dye the whole of their hands as far as the wrist, also the soles of their feet, the Turks more commonly only tinge the nails; both use it for the hair.

[1] Commercial Products of India, p. 707.

[2] Plantes dans l'antiquité, Vol. II, p. 273.

[3] Cf. also D. HOOPER, Oil of Lawsonia alba, *Journal As. Soc. Bengal*, Vol. IV, 1908, p. 35.

[4] Voyage of F. Pyrard, ed. by A. GRAY, Vol. II, p. 361 (Hakluyt Society). The first edition of this work appeared in Paris, 1611.

THE BALSAM-POPLAR

20. Under the term *hu t'un* (Japanese *kotō*) 胡桐 ("t'ung tree of the Hu, Iranian *Paulownia imperialis;*" that is, *Populus balsamifera*), the Annals of the Former Han Dynasty mention a wild-growing tree as characteristic of the flora of the Lob-nor region; for it is said to be plentiful in the kingdom of Šan-šan 鄯善.[1] It is self-evident from the nomenclature that this was a species new to the Chinese, who discovered it in their advance through Turkistan in the second century B.C., but that the genus was somewhat familiar to them. The commentator Moṅ K'aṅ states on this occasion that the *hu t'un* tree resembles the mulberry (*Morus alba*), but has numerous crooked branches. A more elaborate annotation is furnished by Yen Ši-ku (A.D. 579–645), who comments, "The *hu t'un* tree resembles the *t'un* 桐 (*Paulownia imperialis*), but not the mulberry; hence the name *hu t'un* is bestowed upon it. This tree is punctured by insects, whereupon flows down a juice, that is commonly termed *hu t'un lei* 胡桐涙 ('*hu-t'un* tears'), because it is said to resemble human tears.[2] When this substance penetrates earth or stone, it coagulates into a solid mass, somewhat on the order of rock salt, called *wu-t'un kien* 梧桐鹼 ('natron of the *wu-t'un* tree,' *Sterculia platanifolia*). It serves for soldering metal, and is now used by all workmen."[3]

The *T'un tien* 通典, written by Tu Yu 杜佑 between the years 766 and 801, says that "the country Lou 樓[4] among the Si Žuṅ 西戎 produces an abundance of tamarisks 檉柳 (*Tamarix chinensis*), *hu t'un*, and *pai ts'ao* 白草 ('white herb or grass'),[5] the latter being eaten by

---

[1] *Ts'ien Han šu*, Ch. 96 A, p. 3 b. Cf. A. WYLIE, *Journal Anthropological Institute*, Vol. X, 1881, p. 25.

[2] Pliny (XII, 18, § 33) speaks of a thorny shrub in Ariana on the borders of India, valuable for its tears, resembling the myrrh, but difficult of access on account of the adhering thorns (Contermina Indis gens Ariana appellatur, cui spina lacrima pretiosa murrae simili, difficili accessu propter aculeos adnexos). It is not known what plant is to be understood by the Plinian text; but the analogy of the "tears" with the above Chinese term is noteworthy.

[3] This text has been adopted by the *T'ai p'in hwan yü ki* (Ch. 181, p. 4) in describing the products of Lou-lan.

[4] Abbreviated for Lou-lan 樓蘭, the original name of the kingdom of Šan-šan.

[5] This is repeated from the Han Annals, which add also rushes. The "white grass" is explained by Yen Ši-ku as "resembling the grass *yu* 茅 (*Setaria viridis*), but finer and without awns; when dried, it assumes a white color, and serves as fodder for cattle and horses."

cattle and horses. The *hu t'uṅ* looks as if it were corroded by insects. A resin flows down and comes out of this tree, which is popularly called '*hu-t'uṅ* tears'. It can be used for soldering gold (or metal) and silver. In the colloquial language, they say also *lü* 律 instead of *lei*, which is faulty."[1]

The *T'aṅ pen ts'ao*[2] is credited with this statement: "*Hu t'uṅ lei* is an important remedy for the teeth. At present this word is the name of a place west of Aksu. The tree is full of small holes. One can travel for several days and see nothing but *hu t'uṅ* trees in the forests. The leaves resemble those of the *t'uṅ* (*Paulownia*). The resin which is like glue flows out of the roots."

The *Liṅ piao lu i*[3] states positively that *hu t'uṅ lei* is produced in Persia, being the sap of the *hu t'uṅ* tree, and adds that there are also "stone tears," *ši lei* 石淚, which are collected from stones.

Su Kuṅ, the reviser of the *Pen ts'ao* of the T'ang, makes this observation:[4] "*Hu t'uṅ lei* is produced in the plains and marshes as well as in the mountains and valleys lying to the west of Su-čou 肅州. In its shape it resembles yellow vitriol (*hwaṅ fan* 黃礬),[5] but is far more solid. The worm-eaten trees are styled *hu t'uṅ* trees. When their sap filters into earth and stones, it forms a soil-made product like natron. This tree is high and large, its bark and leaves resembling those of the white poplar and the green *t'uṅ* 青桐. It belongs to the family of mulberries, and is hence called *hu t'uṅ* tree. Its wood is good for making implements."

Han Pao-šeṅ 韓保昇, who edited the *Šu pen ts'ao* 蜀本草 about the middle of the tenth century, states, "The tree occurs west of Liaṅ-čou 涼州 (in Kan-su). In the beginning it resembles a willow; when it has grown, it resembles a mulberry and the *t'uṅ*. Its sap sinks into the soil, and is similar to earth and stone. It is used as a dye like the ginger-stone (*kiaṅ ši* 薑石).[6] It is extremely salty and bitter. It is dissolved by the application of water, and then becomes like alum shale or saltpetre. It is collected during the winter months."

Ta Miṅ 大明, who wrote a *Pen ts'ao* about A.D. 970, says with reference to this tree, "There are two kinds,— a tree-sap which is not employed in the pharmacopœia, and a stone-sap collected on the

---

[1] Cf. *Čeṅ lei pen ts'ao*, Ch. 13, p. 33.

[2] As quoted in the *Či wu miṅ ši t'u k'ao*, Ch. 35, p. 8 b.

[3] Ch. B, p. 7 a (see above, p. 268).

[4] *Čeṅ lei pen ts'ao*, l.c.

[5] F. DE MÉLY, Lapidaire chinois, p. 149.

[6] A variety of stalactite (see F. DE MÉLY, Lapidaire chinois, p. 94; GEERTS, Produits, p. 343; *Čeṅ lei pen ts'ao*, Ch. 5, p. 32).

surface of stones; this one only is utilized as a medicine. It resembles in appearance small pieces of stone, and those colored like loess take the first place. The latter are employed as a remedy for toothache." Su Suṅ, in his *T'u kiṅ pen ts'ao*, remarks that it then occurred among the Western Barbarians (Si Fan), and was traded by merchants. He adds that it was seldom used in the recipes of former times, but that it is now utilized for toothache and regarded as an important remedy in families.

Li Ši-čen[1] refers to the chapter on the Western Countries (*Si yü čwan*) in the Han Annals, stating that the tree was plentiful in the country Kü-ši 車師 (Turfan). No such statement is made in the Annals of the Han with regard to this country, but, as we have seen, only with reference to Šan-šan.[2] He then gives a brief résumé of the matter, setting down the two varieties of "tree-tears" and "stone-tears."

The Ming Geography mentions *hu t'uṅ lei* as a product of Hami. The *Kwaṅ yü ki*[3] notices it as a product of the Chikin Mongols between Su-čou and Ša-čou. The *Si yü wen kien lu*,[4] written in 1777, states in regard to this tree that it is only good as fuel on account of its crooked growth: hence the natives of Turkistan merely call it *odon* or *otun*, which means "wood, fuel" in Turkish.[5] The tree itself is termed in Turkī *tograk*.

The *Hui k'iaṅ či*[6] likewise describes the *hu t'uṅ* tree of Hami, saying that the Mohammedans use its wood as fuel, but that some with ornamental designs is carved into cases for writing-brushes and into saddles.

BRETSCHNEIDER[7] has identified this tree with *Populus euphratica*, the wood of which is used as fuel in Turkistan. It is not known, however, that this tree produces a resin, such as is described by the Chinese. Moreover, this species is distributed through northern China;[8] while all Chinese records, both ancient and modern, speak of the *hu t'uṅ*

---

[1] *Pen ts'ao kaṅ mu*, Ch. 34, p. 22.

[2] There is a passage in the *Šwi kiṅ ču* where the *hu t'uṅ* is mentioned, and may be referred to Kü-ši (CHAVANNES, *T'oung Pao*, 1905, p. 569).

[3] Above, p. 251.

[4] Ch. 7, p. 9 (WYLIE, Notes on Chinese Literature, p. 64).

[5] This passage has already been translated correctly by W. SCHOTT (*Abh. Berl. Ak.*, 1842, p. 370). It was not quite comprehended by BRETSCHNEIDER (Mediæval Researches, Vol. II, p. 179), who writes, "The characters *hu t'ung* here are intended to render a foreign word which means 'fuel'."

[6] Above, p. 230.

[7] Mediæval Researches, Vol. II, p. 179.

[8] FORBES and HEMSLEY, *Journal Linnean Society*, Vol. XXVI, p. 536.

exclusively as a tree peculiar to Turkistan and Persia. The correct identification of the tree is *Populus balsamifera,* var. *genuina* Wesm.[1] The easternmost boundary of this tree is presented by the hills of Kumbum east of the Kukunōr, which geographically is part of Central Asia. The same species occurs also in Siberia and North America; it is called *liard* by the French of Canada. It is met with, further, wild and cultivated, in the inner ranges of the north-western Himalaya, from Kunawar, altitude 8000 to 13000 feet, westwards. In western Tibet it is found up to 14000 feet.[2] The buds contain a balsam-resin which is considered antiscorbutic and diuretic, and was formerly imported into Europe under the name *baume facot* and *tacamahaca*[3] *communis* (or *vulgaris*). WATT says that he can find no account of this exudation being utilized in India. It appears from the Chinese records that the tree must have been known to the Iranians of Central Asia and Persia, and we shall not fail in assuming that these were also the discoverers of the medical properties of the balsam. It is quite credible that it was efficacious in alleviating pain caused by carious teeth, as it would form an air-tight coating around them.

---

[1] MATSUMURA, Shokubutsu mei-i, No. 2518.

[2] G. WATT, Dictionary of the Economic Products of India, Vol. VI, p. 325.

[3] The *tacamahaca* (a word of American-Indian origin) was first described by NICOLOSO DE MONARDES (Dos libros el uno que trata de todas las cosas que traen de nuestras Indias Occidentales, Sevilla, 1569): "Assi mismo traen de nueva España otro genero de Goma, o resina, que llaman los Indios Tacamahaca. Y este mismo nombre dieron nuestros Españoles. Es resina sacada por incision de un Arbol grande como Alamo, que es muy oloroso, echa el fruto colorado como simiente de Peonia. Desta Resina o goma, usan mucho los Indios en sus enfermedades, mayormente en hinchazones, en qualquiera parte del cuerpo que se engendran, por que las ressuelue madura, y deshaze marauillosamente," etc. A copy of this very scarce work is in the Edward E. Ayer collection of the Newberry Library, Chicago; likewise the continuation Segunda parte del libro, de las cosas que se traen de nuestras Indias Occidentales (Sevilla, 1571).

MANNA

21. The word "manna," of Semitic origin (Hebrew *mān*, Arabic *mann*), has been transmitted to us through the medium of Greek μάννα in the translation of the Septuaginta and the New Testament. Manna is a saccharine product discharged from the bark or leaves of a number of plants under certain conditions, either through the puncture of insects or by making incisions in the trunk and branches. Thus there are mannas of various nature and origin. The best-known manna is the exudation of *Fraxinus ornus* (or *Ornus europaea*), the so-called manna-ash, occurring in the Mediterranean region and Asia Minor.[1] The chief constituent of manna is manna-sugar or mannite, which occurs in many other plants besides *Fraxinus*.

The Annals of the Sui Dynasty ascribe to the region of Kao-č'aṅ 高 昌 (Turfan) a plant, styled *yaṅ ts'e* 羊 刺 ("sheep-thorn"), the upper part of which produces honey of very excellent taste.[2]

Č'en Ts'aṅ-k'i, who wrote in the first part of the eighth century, states that in the sand of Kiao-ho 交 河 (Yarkhoto) there is a plant with hair on its top, and that in this hair honey is produced; it is styled by the Hu (Iranians) 結 孜 (= 勃) 羅 *k'ie-p'o-lo*, *k'it(k'ir)-bwuδ-la.*[3] The first element apparently corresponds to Persian *xār* ("thorn") or the dialectic form *γār;*[4] the second, to Persian *burra* or *bura* ("lamb"),[5] so that the Chinese term *yaṅ ts'e* presents itself as a literal rendering of the Persian (or rather a Middle-Persian or Sogdian) expression. In New Persian the term *xar-i-šutur* ("camel-thorn") is used, and, according to Aitchison, also *xar-i-buzi* ("goat's thorn").[6]

It is noteworthy that the Chinese have preserved a Middle-Persian word for "manna," which has not yet been traced in an Iranian source. The plant (*Hedysarum alhagi*), widely diffused over all the arid lowlands

---

[1] Cf. the excellent investigation of D. Hanbury, Science Papers, pp. 355–368.

[2] *Sui šu*, Ch. 83, p. 3 b. The same text is also found in the *Wei šu* and *Pei ši;* in the *T'ai p'iṅ hwan yü ki* (Ch. 180, p. 11 b) it is placed among the products of Kü-ši 車 師 in Turfan.

[3] Stuart (Chinese Materia Medica, p. 258) erroneously writes the first character 給. He has not been able to identify the plant in question.

[4] P. Horn, Grundriss der iranischen Philologie, Vol. I, pt. 2, p. 70.

[5] In dialects of northern Persia also *varre*, *varra*, and *werk* (J. de Morgan, Mission en Perse, Vol. V, p. 208).

[6] Cf. D. Hooper, *Journal As. Soc. Bengal*, Vol. V, 1909, p. 33.

343

of Persia, furnishes manna only in certain districts. Wherever it fails to yield this product, it serves as pasture to the camels (hence its name "thorn of camels"), and, according to the express assurance of SCHLIMMER,[1] also to the sheep and goats. "Les indigènes des contrées de la Perse, où se fait la récolte de *teren-djebin*, me disent que les pasteurs sont obligés par les institutions communales de s'éloigner avec leurs troupeaux des plaines où la plante mannifère abonde, parce que les moutons et chèvres ne manqueraient de faire avorter la récolte." In regard to a related species (*Hedysarum semenowi*), S. KORŽINSKI[2] states that it is particularly relished by the sheep which fatten on it.

The *Lian se kun tse ki* 梁四公子記[3] is cited in the *Pen ts'ao kan mu* as follows: "In Kao-č'an there is manna (*ts'e mi* 刺蜜). Mr. Kie 杰 公 says, In the town Nan-p'in 南平[4] 城 the plant *yan ts'e* is devoid of leaves, its honey is white in color and sweet of taste. The leaves of the plant *yan ts'e* in Salt City (Yen č'en 鹽城) are large, its honey is dark 青 in color, and its taste is indifferent. Kao-č'an is the same as Kiao-ho, and is situated in the land of the Western Barbarians (Si Fan 西番);[5] at present it forms a large department (*ta čou* 大州)."

Wan Yen-te, who was sent on a mission to Turfan in A.D. 981, mentions the plant and its sweet manna in his narrative.[6]

Čou K'ü-fei, who wrote the *Lin wai tai ta* in 1178, describes the "genuine manna (sweet dew)" 眞甘露 of Mosul (勿斯離 Wu-se-li) as follows:[7] "This country has a number of famous mountains. When the autumn-dew falls, it hardens under the influence of the sun-rays into a substance of the appearance of sugar and hoar-frost, which is gathered and consumed. It has purifying, cooling, sweet, and nutritious qualities, and is known as genuine manna."[8]

Wan Ta-yüan 汪大淵, in his *Tao i či lio* 島夷志畧 of 1349,[9] has

---

[1] Terminologie, p. 357.

[2] Vegetation of Turkistan (in Russian), p. 77.

[3] The work of Čan Yüe (A.D. 667–730); see The Diamond, this volume, p. 6.

[4] Other texts write 乎 *hu*.

[5] This term, which in general denotes Tibet, but certainly cannot refer to Tibet in this connection, has evidently misled STUART (Chinese Materia Medica, p. 258) into saying that the substance is spoken of as coming from Tangut.

[6] Cf. W. SCHOTT, Zur Uigurenfrage II, p. 47 (*Abh. Berl. Akad.*, 1875).

[7] Ch. 3, p. 3 b (ed. of *Či pu tsu čai ts'un šu*). Regarding the term *kan lu*, which also translates Sanskrit *amṛita*, see CHAVANNES and PELLIOT, Traité manichéen, p. 155.

[8] The same text with a few insignificant changes has been copied by Čao Žu-kwa (HIRTH's translation, p. 140).

[9] Regarding this work, cf. PELLIOT, *Bull. de l'Ecole française*, Vol. IV, p. 255.

the following note regarding manna (*kan lu*) in Ma-k'o-se-li:[1] "Every year during the eighth and ninth months it rains manna, when the people make a pool to collect it. At sunrise it will condense like water-drops, and then it is dried. Its flavor is like that of crystallized sugar. They also store it in jars, mixing it with hot water, and this beverage serves as a remedy for malaria. There is an old saying that this is the country of the Amṛitarāja-tathāgata 甘 露 王 如 來."[2]

Li Ši-čen, after quoting the texts of Č'en Ts'an-k'i, the *Pei ši*, etc.,[3] arrives at the conclusion that these data refer to the same honey-bearing plant, but that it is unknown what plant is to be understood by the term *yan ts'e*.

The Turkī name for this plant is *yantaq*, and the sweet resin accumulating on it is styled *yantaq šäkärī* ("*yantaq* sugar").[4]

The modern Persian name for the manna is *tär-ängubīn* (Arabic *terenjobīn;* hence Spanish *tereniabin*); and the plant which exudates the sweet substance, as stated, is styled *xar-i-šutur* ("camel-thorn"). The manna suddenly appears toward the close of the summer during the night, and must be gathered during the early hours of the morning. It is eaten in its natural state, or is utilized for syrup (*šire*) in Central Asia or in the sugar-factories of Meshed and Yezd in Persia.[5] The Persian word became known to the Chinese from Samarkand in the transcription *ta-lan-ku-pin* 達 郎 古 賓.[6] The product is described under the title *kan lu* 甘 露 ("sweet dew") as being derived from a small plant, one to two feet high, growing densely, the leaves being fine like those of an *Indigofera* (*lan*). The autumn dew hardens on the surface of the stems, and this product has a taste like sugar. It is gathered and boiled into sweetmeats. Under the same name, *kan-lu*, the *Kwan yü ki*[7] describes a small plant of Samarkand, on the leaves of which accumulates in the autumn a dew as sweet in taste as honey, the leaves resem-

---

[1] Unidentified. It can hardly be identified with Mosul, as intimated by ROCKHILL.

[2] ROCKHILL, *T'oung Pao*, 1915, p. 622. This Buddhist term has crept in here owing to the fact that *kan lu* ("sweet dew") serves as rendering of Sanskrit *amṛita* ("the nectar of the gods") and as designation for manna.

[3] Also the *Yu yan tsa tsu*, but this passage refers to India and to a different plant, and is therefore treated below in its proper setting.

[4] A. v. LE COQ, Sprichwörter und Lieder aus Turfan, p. 99. If the supposition of B. MUNKACSI (*Keleti szemle*, Vol. XI, 1910, p. 353) be correct, that Hungarian *gyanta* (*gyánta, jánta, gyenta*, "resin") and *gyantár* ("varnish") may be Turkish loan-words, the above Turkī name would refer to the resinous character of the plant.

[5] VÁMBÉRY, Skizzen aus Mittelasien, p. 189.

[6] *Ta Min i t'un či*, Ch. 89, p. 23.

[7] Ch. 24, p. 26, of the edition printed in 1744; this passage is not contained in the original edition of 1600 (cf. above, p. 251, regarding the various editions).

bling those of an Indigofera (*lan*); and in the same work[1] this plant is referred to Qarā-Khoja 火 州 under the name *yań ts'e*. Also the Ming Annals[2] contain the same reference. The plant in question has been identified by D. HANBURY with the camel-thorn (*Alhagi camelorum*), a small spiny plant of the family *Leguminosae*, growing in Iran and Turkistan.[3]

In the fourteenth century, ODORIC of Pordenone found near the city Huz in Persia manna of better quality and in greater abundance than in any part of the world.[4] The Persian-Arabic manna was made known in Europe during the sixteenth century by the traveller and naturalist PIERRE BELON DU MONS (1518–64),[5] who has this account: "Les Caloieres auoyêt de la Māne liquide recueillie en leurs montagnes, qu'ils appellent *Tereniabin*, a la differēnce de la dure: Car ce que les autheurs Arabes ont appellé *Tereniabin*, est gardée en pots de terre comme miel, et la portent vendre au Caire: qui est ce qu' Hippocrates nomma miel de Cedre, et les autres Grecs ont nommé Rosée du mont Liban: qui est differente à la Manne blanche seiche. Celle que nous auons en France, apportée de Brianson, recueillie dessus les Meleses à la sommité des plus hautes montagnes, est dure, differente à la susdicte. Parquoy estant la Manne de deux sortes, lon en trouve au Caire de l'vne et de l'autre es boutiques des marchands, exposée en vente. L'vne est appellée Manne, et est dure: l'autre Tereniabin, et est liquide: et pource qu'en auons fait plus long discours au liure des arbres tousiours verds, n'en dirons autre chose en ce lieu." The Briançon manna mentioned by Belon is collected from the larch-trees (*Pinus larix*) of southern France.[6] GARCIA DA ORTA[7] described several kinds of manna, one brought to Ormuz from the country of the Uzbeg under the name *xirquest* or *xircast*, "which means the milk of a tree called *quest*, for *xir* [read *šīr*] is milk in the Persian language, so that it is the dew that falls

---

[1] Ch. 24, p. 6, of the original edition; and Ch. 24, p. 30 b, of the edition of 1744.

[2] Ch. 329 (cf. BRETSCHNEIDER, Mediæval Researches, Vol. II, p. 192).

[3] The plant is said to occur also in India (Sanskrit *viçāladā* and *gāndhārī*; that is, from Gāndhāra), Arabia, and Egypt, but, curiously, in those countries does not produce a sugar-like secretion. Consequently it cannot be claimed as the plant which furnished the manna to the Israelites in the desert (see the Dictionnaire de la Bible by F. VIGOUROUX, Vol. I, col. 367). The manna of northern India became known to the Chinese in recent times (see *Lu č'ań kuń ši k'i* 盧 長 公 史 隙, p. 44, in *Ts'iń čao t'ań ts'uń šu*).

[4] YULE, Cathay, new ed., Vol. II, p. 109; CORDIER's edition of Odoric, p. 59.

[5] Les Observations de plusieurs singularitez, pp. 228–229 (Anvers, 1555).

[6] FLÜCKIGER and HANBURY, Pharmacographia, p. 416.

[7] C. MARKHAM, Colloquies, p. 280.

from these trees, or the gum that exudes from them.[1] The Portuguese corrupted the word to *siracost.*" The other kind he calls *tiriam-jabim* or *trumgibim* (Persian *tär-ängubīn*). "They say that it is found among the thistles and in small pieces, somewhat of a red color. It is said that they are obtained by shaking the thistles with a stick, and that they are larger than a coriander-seed when dried, the color, as I said, between red and vermilion. The vulgar hold that it is a fruit, but I believe that it is a gum or resin. They think this is more wholesome than the kind we have, and it is much used in Persia and Ormuz." "Another kind comes in large pieces mixed with leaves. This is like that of Calabria, and is worth more money, coming by way of Baçora, a city of renown in Persia. Another kind is sometimes seen in Goa, liquid in leather bottles, which is like coagulated white honey. They sent this to me from Ormuz, for it corrupts quickly in our land, but the glass flasks preserve it. I do not know anything more about this medicine." JOHN FRYER[2] speaks of the mellifluous dew a-nights turned into manna, which is white and granulated, and not inferior to the Calabrian. According to G. WATT,[3] *shirkhist* is the name for the white granular masses found in Persia on the shrub *Cotoneaster nummularia;* white *taranjabin* (=*tär-ängubīn*) is obtained from the camel-thorn (*Alhagi camelorum* and *A. maurorum*), growing in Persia, and consisting of a peculiar sugar called melezitose and cane-sugar. The former is chiefly brought from Herat, and is obtained also from *Atraphaxis spinosa* (*Polygonaceae*).[4]

It is thus demonstrated also from a philological and historical point of view that the *yan ts'e* and *k'ie-p'o-lo* of the Chinese represent the species *Alhagi camelorum.*

Another Persian name for manna is *xoškenjubīn,* which means "dry honey." An Arabic tradition explains it as a dew that falls on trees in the mountains of Persia; while another Arabic author says, "It is dry honey brought from the mountains of Persia. It has a detestable odor. It is warm and dry, warmer and dryer than honey. Its properties in general are more energetic than those of honey."[5] This product, called

---

[1] Garcia's etymology is only partially correct. The Persian word is *šīr-xešt,* which means "goat's milk." Hence Armenian *širixišd, širxešd, širaxušg,* or *širaxuš* (cf. E. SEIDEL, Mechithar, p. 210).

[2] New Account of East India and Persia, Vol. II, p. 201.

[3] *Agricultural Ledger,* 1900, No. 17, p. 188.

[4] See FLÜCKIGER and HANBURY, *op. cit.,* p. 415. According to SCHLIMMER (Terminologie, p. 357), this manna comes from Herat, Khorasan, and the district Lor-šehrestanek.

[5] L. LECLERC, Traité des simples, Vol. II, p. 32.

in India *guzangabin*, is collected from the tamarisk (*Tamarix gallica*, var. *mannifera* Ehrenb.) in the valleys of the Peninsula of Sinai and also in Persia.[1] In the latter country, the above name is likewise applied to a manna obtained from *Astragalus florulentus* and *A. adscendens* in the mountain-districts of Chahar-Mahal and Faraidan, and especially about the town of Khonsar, south-west of Ispahan. The best sorts of this manna, which are termed *gaz-alefi* or *gaz-khonsar* (from the province Khonsar), are obtained in August by shaking it from the branches, the little drops finally sticking together and forming a dirty, grayish-white, tough mass. According to SCHLIMMER,[2] the shrub on which this manna is formed is common everywhere, without yielding, however, the slightest trace of manna, which is solely obtained in the small province Khonsar or Khunsar. The cause for this phenomenon is sought in the existence there of the *Coccus mannifer* and in the absence of this insect in other parts of the country. Several Persian physicians of Ispahan, and some European authors, have attributed to the puncture of this insect the production of manna in Khonsar; and Schlimmer recommends transporting and acclimatizing the insect to those regions where *Tamarix* grows spontaneously.

It has been stated that the earliest allusion to tamarisk-manna is to be found in Herodotus,[3] who says in regard to the men of the city Callatebus in Asia Minor that they make honey out of wheat and the fruit of the tamarisk. The case, however, is different; Herodotus does not allude to the exudation of the tree.

STUART[4] states that tamarisk-manna is called *č'en žu* 檉乳. The tamarisk belongs to the flora of China, three species of it being known.[5] The Chinese, as far as I know, make no reference to a manna from any of these species; and the term pointed out by Stuart merely refers to the sap in the interior of the tree, which, according to the *Pen ts'ao*, is used in the Materia Medica. Čen Tsiao 鄭樵 of the Sung period, in his *T'un či* 通志,[6] simply defines *č'en žu* as "the sap in the wood or trunk of the tamarisk."[7]

---

[1] See particularly D. HOOPER, Tamarisk Manna, *Journal As. Soc. Bengal*, Vol. V, 1909, pp. 31–36.

[2] Terminologie, p. 359.

[3] VII, 31.

[4] Chinese Materia Medica, p. 259.

[5] BRETSCHNEIDER, Bot. Sin., pt. II, No. 527; *Pen ts'ao kan mu*, Ch. 35 B, p. 9.

[6] Ch. 76, p. 12.

[7] The Turkī name for the tamarisk is *yulgun*. In Persian it is styled *gaz* or *gazm* (Kurd *gazo* or *gezu*), the fruit *gazmāzak* or *gazmāzū* (*gaz basrah*, the manna of the tree); further, *balangmušt*, *balangmusk*, or *balanjmušk*, and Arabic-Persian *kizmāzaj*.

There is, further, an oak-manna collected from *Quercus vallonea* Kotschy and *Q. persica*. These trees are visited in the month of August by immense numbers of a small white Coccus, from the puncture of which a saccharine fluid exudes, and solidifies in little grains. The people go out before sunrise, and shake the grains of manna from the branches on to linen cloths spread out beneath the trees. The exudation is also collected by dipping into vessels of hot water the small branches on which it is formed, and evaporating the saccharine solution to a syrupy consistence, which in this state is used for sweetening food, or is mixed with flour to form a sort of cake.[1]

Aside from the afore-mentioned mannas, SCHLIMMER[2] describes two other varieties which I have not found in any other author. One he calls in Persian *šiker eighal* ("sugar *eighal*"), saying that it is produced by the puncture of a worm in the plant. This worm he has himself found in fresh specimens. This manna is brought to Teheran by the farmers of the Elburs, Lawistan, and Dimawend, but the plant occurs also in the environment of Teheran and other places. Although this manna almost lacks sweetness, it is a remarkable pectoral and alleviates obstinate coughs. The other is the manna of *Apocynum syriacum*, known in Persia as *šiker al-ošr* and imported from Yemen and Hedjaz. According to the Persian pharmacologists, it is the product of a nocturnal exudation solidified during the day, similar to small pieces of salt, either white, or gray, and even black. It is likewise employed medicinally.

Manna belonged to the food-products of the ancient Iranians, and has figured in their kitchen from olden times. When the great king sojourned in Media, he received daily for his table a hundred baskets full of manna, each weighing ten mines. It was utilized like honey for the sweetening of beverages.[3] I am inclined to think that the Iranians diffused this practice over Central Asia.

The *Yu yan tsa tsu* has a reference to manna of India, as follows: "In northern India there is a honey-plant growing in the form of a creeper with large leaves, without withering in the autumn and winter. While it receives hoar-frost and dew, it forms the honey." According to G. WATT,[4] some thirteen or fourteen plants in India are known to

---

[1] FLÜCKIGER and HANBURY, Pharmacographia, p. 416; HANBURY, Science Papers, p. 287; SCHLIMMER (Terminologie, p. 358) attributes the oak-manna to the mountains of Kurdistan in Persia.

[2] Terminologie, p. 359.

[3] C. JORET, Plantes dans l'antiquité, Vol. II, p. 93. Regarding manna in Persia, see also E. SEIDEL, Mechithar, p. 163.

[4] Commercial Products of India, p. 929.

yield, under the parasitic influence of insects or otherwise, a sweet fluid called "manna." This is regularly collected and, like honey, enters more largely than sugar into the pharmaceutical preparations of the Hindu.

The silicious concretion of crystalline form, found in the culms or joints of an Indian bamboo (*Bambusa arundinacea*) and known as tabashir, is styled in India also "bamboo manna," — decidedly a misnomer. On the other hand, a real manna has sometimes been discovered on the nodes of certain species of bamboo in India.[1] The subject of tabashir has nothing to do with manna, nor with Sino-Iranian relations; but, as the early history of this substance has not yet been correctly expounded, the following brief notes may not be unwelcome.[2] Specimens of tabashir, procured by me in China in 1902, are in the American Museum of Natural History in New York.[3]

We now know that tabashir is due to an ancient discovery made in India, and that at an early date it was traded to China and Egypt. In recent years the very name has been traced in the form *tabasis* (τάβασις) in a Greek papyrus, where it is said that the porous stone is brought down [to Alexandria] from [upper] Egypt: the articles of Indian commerce were shipped across the Red Sea to the Egyptian ports, and then freighted on the Nile downward to the Delta.[4] The Indian origin of the article is evidenced, above all, by the fact that the Greek term *tabasis* (of the same phonetic appearance as Persian *tabāšīr*) is connected with Sanskrit *tavak-kṣīrā* (or *tvak-kṣīrā; kṣīrā,* "vegetable juice"), and permits us to reconstruct a Prākrit form *tabašīra;* for the Greek importers or exporters naturally did not derive the word from Sanskrit, but from a vernacular idiom spoken somewhere on the west coast of India. Or, we have to assume that the Greeks received the word from the Persians, and the Persians from an Indian Prākrit.[5]

The Chinese, in like manner, at first imported the article from India, calling it "yellow of India" (*T'ien-ču hwan* 天竺黃). It is first mentioned under this designation as a product of India in the Materia Medica published in the period K'ai-pao (A.D. 968–976), the *K'ai pao*

---

[1] See G. WATT, *Agricultural Ledger*, 1900, No. 17, pp. 185–189.

[2] The latest writer on the subject, G. F. KUNZ (The Magic of Jewels and Charms, pp. 233–235, Philadelphia, 1915), has given only a few historical notes of mediæval origin.

[3] Cat. No. 70, 13834. This is incidentally mentioned here, as Dr. Kunz states that very little of the material has reached the United States.

[4] H. DIELS, Antike Technik, p. 123.

[5] The Persian *tabāšīr* is first described by Abu Mansur (ACHUNDOW, p. 95), and is still eaten as a delicacy by Persian women (*ibid.*, p. 247). In Armenian it is *dabašir*.

pen ts'ao; but at the same time we are informed that it was then obtained from all bamboos of China,[1] and that the Chinese, according to their habit, adulterated the product with scorched bones, the arrowroot from *Pachyrhizus angulatus,* and other stuff.[2] The *Pen ts'ao yen i* of 1116[3] explains the substance as a natural production in bamboo, yellow like loess. The name was soon changed into "bamboo-yellow" (*ču hwaṅ* 竹 黃) or "bamboo-grease" (*ču kao*).[4] It is noticeable that the Chinese do not classify tabashir among stones, but conceive it as a production of bamboo, while the Hindu regard it as a kind of pearl.

The earliest Arabic author who has described the substance is Abū Dulaf, who lived at the Court of the Samanides of Bokhāra, and travelled in Central Asia about A.D. 940. He says that the product comes from Mandūrapatan in northwestern India (Abulfeda and others state that Tāna on the island of Salsette, twenty miles from Bombay, was the chief place of production), and is exported from there into all countries of the world. It is produced by rushes, which, when they are dry and agitated by the wind, rub against one another; this motion develops heat and sets them afire. The blaze sometimes spreads over a surface of fifty parasangs, or even more. Tabashir is the product of these rushes.[5] Other Arabic authors cited by Ibn al-Baiṭār derive the substance from the Indian sugarcane, and let it come from all coasts of India; they dwell at length on its medicinal properties.[6] GARCIA DA ORTA (1563), who was familiar with the drug, also mentions the burning of the canes, and states it as certain that the reason they set fire to them is to reach the heart; but sometimes they do not follow this practice, as appears from many specimens which are untouched by fire. He justly says that the Arabic name (*tabašir,* in his Portuguese spelling *tabaxir*) is derived from the Persian, and means "milk or juice, or moisture." The ordinary price for the product in Persia and Arabia was its weight in silver. The canes, lofty and large like ash-trees,

---

[1] The *Čeṅ lei pen ts'ao* (Ch. 13, p. 48) cites the same text from a work *Lin hai či* 臨 海 誌, apparently an other work than the *Lin hai i wu či* mentioned by BRET-SCHNEIDER (Bot. Sin., pt. I, p. 169).

[2] The following assertion by STUART (Chinese Materia Medica, p. 64)is erroneous: "The Chinese did not probably derive the substance originally from India, but it is possible that the knowledge of its medicinal uses were derived from that country, where it has been held in high esteem from very early times." The knowledge of this product and the product itself first reached the Chinese from India, and naturally induced them to search for it in their own bamboos.

[3] Ch. 14, p. 4 b (ed. of Lu Sin-yūan).

[4] *Pen ts'ao kaṅ mu,* Ch. 37, p. 9.

[5] G. FERRAND, Textes relatifs à l'Extrême-Orient, p. 225.

[6] L. LECLERC, Traité des simples, Vol. II, pp. 399–401.

according to his statement, generate between the knots great humidity, like starch when it is much coagulated. The Indian carpenters, who work at these canes, find thick juice or pith, which they put on the lumbar region or reins, and in case of a headache on the forehead; it is used by Indian physicians against over-heating, external or internal, and for fevers and dysentery.[1] The most interesting of all accounts remains that of ODORIC OF PORDENONE (died in 1331), who, though he does not name the product and may partially confound it with bezoar, alludes to certain stones found in canes of Borneo, "which be such that if any man wear one of them upon his person he can never be hurt or wounded by iron in any shape, and so for the most part the men of that country do wear such stones upon them."[2]

J. A. DE MANDELSLO[3] gives the following notice of tabashir: "It is certain that on the coast of Malabar, Coromandel, Bisnagar, and near to Malacca, this sort of cane (called by the Javians *mambu* [bamboo]) produces a drug called *sacar mambus*, that is, sugar of mambu. The Arabians, the Persians, and the Moores call it *tabaxir*, which in their language signifies a white frozen liquor. These canes are as big as the body of a poplar, having straight branches, and leaves something longer than the olive-tree. They are divided into divers knots, wherein there is a certain white matter like starch, for which the Persians and Arabians give the weight in silver, for the use they make of it in physick, against burning feavers, and bloudy fluxes, but especially upon the first approaches of any disease."

---

[1] C. MARKHAM, Colloquies of Garcia da Orta, pp. 409-414. A list of Sanskrit synonymes for tabashir is given by R. SCHMIDT (ZDMG, Vol. LXV, 1911, p. 745).

[2] YULE, Cathay, new ed. by CORDIER, Vol. II, p. 161.

[3] Voyages and Travels, p. 120 (London, 1669).

ASAFŒTIDA

22. The riddles of asafœtida begin with the very name: there is no adequate explanation of our word *asa* or *assa*. The new Oxford English Dictionary ventures to derive it from Persian *āzā* or *aza*. This word, however, means nothing but "mastic," a product entirely different from what we understand by asafœtida (p. 252). In no Oriental language is there a word of the type *asa* or *aza* with reference to this product, so it could not have been handed on to Europe by an Oriental nation. KAEMPFER, who in 1687 studied the plant in Laristan, and was fairly familiar with Persian, said that he was ignorant of the origin of·the European name.[1] LITTRÉ, the renowned author of the Dictionnaire français, admits that the origin of *asa* is unknown, and wisely abstains from any theory.[2] The supposition has been advanced that *asa* was developed from the *laser* or *laserpitium* of Pliny (XIX, 5), the latter having thus been mutilated by the druggists of the middle ages. This etymology, first given by GARCIA DA ORTA,[3] has been indorsed by E. BORSZCZOW,[4] a Polish botanist, to whom we owe an excellent investigation of the asa-furnishing plants. Although this explanation remains as yet unsatisfactory, as the alleged development from *laser* to *asa* is merely inferred, but cannot actually be proved from mediæval documents,[5] it is better, at any rate, than the derivation from the Persian.

Asafœtida is a vegetable product consisting of resin, gum, and essential oil in varying proportions, the resin generally amounting to more than one-half, derived from different umbelliferous plants, as *Ferula narthex, alliacea, fœtida, persica,* and *scorodosma* (or *Scorodosma*

---

[1] Amoenitates exoticae, p. 539.

[2] The suggestion has also been made that *asa* may be derived from Greek *asi* (?) ("disgust") or from Persian *anguza* ("asafœtida"); thus at least it is said by F. STUHLMANN (Beiträge zur Kulturgeschichte Ostafrikas, p. 609). Neither is convincing. The former moves on the same high level as Li Ši-čen's explanation of *a-wei* ("The barbarians call out *a*, expressing by this exclamation their horror at the abominable odor of this resin").

[3] C. MARKHAM, Colloquies, p. 41. JOHN PARKINSON (Theatrum botanicum, p. 1569, London, 1640) says, "There is none of the ancient Authours either Greeke, Latine, or Arabian, that hath made any mention of *Asa*, either *dulcis* or *fœtida*, but was first depraved by the Druggists and Apothecaries in forraigne parts, that in stead of Laser said Asa, from whence ever since the name of *Asa* hath continued."

[4] *Mémoires de l'Acad. de St. Pétersbourg,* Vol. III, No. 8, 1860, p. 4.

[5] DUCANGE does not even list the word "asafœtida."

353

fœtidum).[1] It is generally used in India as a condiment, being especially eaten with pulse and rice. Wherever the plant grows, the fresh leaves are cooked and eaten as a green vegetable, especially by the natives of Bukhāra, who also consider as a delicacy the white under part of the stem when roasted and flavored with salt and butter. In the pharmacopœia it is used as a stimulant and antispasmodic.

Abu Mansur, the Persian Li Ši-čen of the tenth century, discriminates between two varieties of asafœtida (Persian *anguyān*, Arabic *anjudān*), a white and a black one, adding that there is a third kind called by the Romans *sesalius*. It renders food easily digestible, strengthens the stomach, and alleviates pain of the joints in hands and feet. Rubbed into the skin, it dispels swellings, especially if the milky juice of the plant is employed. The root macerated in vinegar strengthens and purifies the stomach, promotes digestion, and acts as an appetizer.[2]

The *Ferula* and *Scorodosma* furnishing asafœtida are typically Iranian plants. According to Abu Hanīfa,[3] asa grows in the sandy plains extending between Bost and the country Kīkān in northern Persia. Abu Mansur designates the leaves of the variety from Sarachs near Merw as the best. According to Istaxrī, asa was abundantly produced in the desert between the provinces Seistan and Makran; according to Edrīsi, in the environment of Kaleh Bust in Afghanistan. KAEMPFER observed the harvest of the plant in Laristan in 1687, and gives the following notice on its occurrence:[4] "Patria eius sola est Persia, non Media, Libya, Syria aut Cyrenaica regio. In Persia plantam hodie alunt saltem duorum locorum tractus, videlicet campi montesque circa Heraat, emporium provinciae Chorasaan, et jugum montium in provincia Laar, quod a flumine Cuur adusque urbem Congo secundum Persici sinus tractum extenditur, duobus, alibi tribus pluribusve parasangis a litore." Herat is a renowned place of production, presumably the exclusive centre of production at the present day, whence the product is shipped to India.

The exact geographical distribution has been well outlined by E. BORSZCZOW.[5] Aside from Persia proper, *Scorodosma* occurs also on the Oxus, on the Aral Sea, and in an isolated spot on the east coast of the Caspian Sea. Judging from Chinese accounts, plants yielding asa appear to have occurred also near Khotan (see below), Turfan, and

---

[1] The genus *Ferula* contains about sixty species.

[2] ACHUNDOW, Abu Mansur, p. 8.

[3] LECLERC, Traité des simples, Vol. I, p. 142.

[4] Amoenitates exoticae, p. 291.

[5] Ferulaceen der aralo-caspischen Wüste (*Mémoires de l'Acad. de St. Pétersbourg*, Vol. III, No. 8, 1860, p. 16).

Shahrokia.[1] We do not know, however, what species here come into question.

Čao Žu-kwa states that the home of asafœtida is in Mu-kŭ-lan 木 俱 蘭, in the country of the Ta-ši (Ta-džik, Arabs).[2] Mu-kŭ-lan is identical with Mekrān, the Gedrosia of the ancients, the Makā of the Old-Persian inscriptions. Alexander the Great crossed Gedrosia on his campaign to India, and we should expect that his scientific staff, which has left us so many valuable contributions to the flora of Iran and north-western India, might have also observed the plant furnishing asafœtida; in the floristic descriptions of the Alexander literature, however, nothing can be found that could be interpreted as referring to this species. H. BRETZL[3] has made a forcible attempt to identify a plant briefly described by Theophrastus,[4] with *Scorodosma fœtidum;* and A. HORT,[5] in his new edition and translation of Theophrastus, has followed him. The text runs thus: "There is another shrub [in Aria] as large as a cabbage, whose leaf is like that of the bay in size and shape. And if any animal should eat this, it is certain to die of it. Wherefore, wherever there were horses, they kept them under control" [that is, in Alexander's army]. This in no way fits the properties of *Ferula* or *Scorodosma,* which is non-poisonous, and does not hurt any animal. It is supposed also that the *laserpitium* or *silphion* and *laser* of Pliny[6] should, at least partially, relate to asafœtida; this, however, is rejected by some authors, and appears to me rather doubtful. GARCIA DA ORTA[7] has already denied any connection between that plant of the ancients and asa. L. LECLERC[8] has discussed at length this much-disputed question.

The first European author who made an exact report of asafœtida

---

[1] BRETSCHNEIDER, Mediæval Researches, Vol. II, pp. 193, 254. The interpretation of *lu-wei* ("rushes") as asafœtida in the *Si yu ki* (*ibid.,* Vol. I, p. 85) seems to me a forced and erroneous interpretation.

[2] HIRTH and ROCKHILL, Chao Ju-kua, p. 224.

[3] Botanische Forschungen des Alexanderzuges, p. 285.

[4] Histor. plant., IV. IV, 12.

[5] Vol. I, p. 321.

[6] XIX, 15. The Medic juice, called *silphion,* and mentioned as a product of Media by Strabo (XI. XIII, 7), might possibly allude to a product of the nature ot asafœtida, especially as it is said in another passage (XV. II, 10) that *silphion* grew in great abundance in the deserts of Bactriana, and promoted the digestion of the raw flesh on which Alexander's soldiers were forced to subsist there. According to others, the *silphion* of the ancients is *Thapsia garganica* (ENGLER, Pflanzenfamilien, Vol. III, pt. 8, p. 247). Regarding the Medic oil (oleum Medicum) see Ammianus Marcellinus, XXIII, 6.

[7] C. Markham, Colloquies, p. 44.

[8] Traité des simples, Vol. I, p. 144.

was GARCIA DA ORTA in 1563. However, living and studying in Goa, India, he did not learn from what plant the product was derived. On its use in India he comments as follows: "The thing most used throughout India, and in all parts of it, is that Assa-fetida, as well for medicine as in cookery. A great quantity is used, for every Gentio who is able to get the means of buying it will buy it to flavor his food. The rich eat much of it, both Banyans and all the Gentios of Cambay, and he who imitates Pythagoras. These flavor the vegetables they eat with it; first rubbing the pan with it, and then using it as seasoning with everything they eat. All the other Gentios who can get it, eat it, and laborers who, having nothing more to eat than bread and onions, can only eat it when they feel a great need for it. The Moors all eat it, but in smaller quantity and only as a medicine. A Portuguese merchant highly praised the pot-herb used by these Banyans who bring this Assa-fetida, and I wished to try it and see whether it pleased my taste, but as I do not know our spinach very well, it did not seem so palatable to me as it did to the Portuguese who spoke to me about it. There is a respected and discreet man in these parts, holding an office under the king, who eats Assa-fetida to give him an appetite for his dinner, and finds it very good, taking it in doses of two drachms. He says there is a slightly bitter taste, but that this is appetising like eating olives. This is before swallowing, and afterwards it gives the person who takes it much content. All the people in this country tell me that it is good to taste and to smell."

CHR. ACOSTA or DA COSTA[1] gives the following account: "Altiht, anjuden, Assa fetida, dulce y odorata medicina (de que entre los Doctores ha auido differencia y controuersia) es ona Goma, que del Coraçone traen a Ormuz, y de Ormuz a la India, y del Guzarate y del reyno Dely (tierra muy fria) la qual por la otra parte confina con el Coraçone, y con la region de Chiruan, como siente Auicena. Esta Goma es llamada de los Arabios Altiht, y Antit, y delos Indios Ingu, o Ingara. El arbol de adonde mana, se llama Anjuden, y otros le llaman Angeydan.

"La Assa se aplica para leuãtar el miembro viril, cosa muy vsada en aquellas partes: y no viene a proposito para la diminucion del coito, vsar del tal çumo de Regaliza. Y en las diuisiones pone Razis Altiht por medicina para las fiestas de Venus: y Assa dulcis no la pone Doctor Arabe, ni Griego, ni Latino, que sea de autoridad, porque Regaliza se llama en Arabio Cuz, y el çumo del cozido, y reduzido en forma de Arrope, le llaman los Arabes Robalçuz, y los Españoles corrompiendole

---

[1] Tractado de las drogas, y medicinas de las Indias orientales, p. 362 (Burgos, 1578).

el nombre le llaman Rabaçuz. De suerte que Robalçuz en Arabio, quiere
dezir çumo basto de Regaliza: porque Rob, es çumo basto, y Al, ar-
ticulo de genitiuo, de, y Cuz, regaliza, y todo junto significa çumo
basto de Regaliza: y assi no se puede llamar a este çumo Assa dulcis.
Los Indios la loan para el estomago, para facilitar el vientre, y para
consumir las ventosidadas. Tambien curan con esta medicina los
cauallos, que echan mucha ventosidad. En tanto tienen esta medicina
que le llama aquella gente, principalmente la de Bisnaguer, manjar
delos Dioses."

JOHN FRYER[1] relates, "In this country Assa Fœtida is gathered at
a place called Descoon;[2] some deliver it to be the juice of a cane or reed
inspissated; others, of a tree wounded: It differs much from the stinking stuff called *Hing*, it being of the Province of Carmania:[3] This latter
is that the Indians perfume themselves with, mixing it in all their pulse,
and make it up in wafers to correct the windiness of their food, which
they thunder up in belchings from the crudities created in their stomachs; never thinking themselves at ease without this *Theriac:* And this
is they cozen the Europeans with instead of *Assa Fœtida*, of which
it bears not only the smell, but color also, only it is more liquid."

J. A. DE MANDELSLO[4] reports as follows: "The Hingh, which our
drugsters and apothecaries call Assa fœtida, comes for the most part
from Persia, but that which the Province of Utrad produces in the Indies
is the best, and there is a great traffick driven in it all over Indosthan.
The plant which produces it is of two kinds; one grows like a bush, and
hath small leaves, like rice, and the other resembles a turnip-leaf, and
its greenness is like that of fig-tree leaves. It thrives best in stony and
dry places, and its gum begins to come forth towards the latter end
of summer, so that it must be gathered in autumn. The traffick of it
is so much the greater in those parts, upon this account, that the
Benjans of Guzuratta make use of it in all their sawces, and rub their

---

[1] New Account of East India and Persia, Vol. II, p. 195 (Hakluyt Soc., 1912).

[2] Kuh-i Dozgan, west of Kuristan.

[3] *Hing* is mentioned by FRYER (Vol. I, p. 286) as in use among the natives of
southern India, "to correct all distempers of the brain, as well as stomach," "a sort
of liquid Assa Fœtida, whereby they smell odiously." This is the product of *Ferula
alliacea*, collected near Yezd in Khorasan and in the province of Kerman, and
chiefly used by the natives of Bombay (FLÜCKIGER and HANBURY, Pharmacographia,
pp. 319–320; WATT, Commercial Products of India, p. 534). Fryer's distinction between *hing* and asafœtida shows well that there were different kinds and grades of
the article, derived from different plants. Thus there is no reason to wonder that
the Chinese Buddhist authors discriminate between *hiṅgu* and *a-wei* (CHAVANNES
and PELLIOT, Traité manichéen, p. 234); the *ĕ'ou ts'ai* ("stinking vegetable") is
probably also a variety of this product.

[4] Voyages and Travels, p. 67 (London, 1669).

pots and drinking vessels therewith, by which means they insensibly accustom themselves to that strong scent, which we in Europe are hardly able to endure."

The Chinese understand by the term *a-wei* products of two different plants. Neither Bretschneider nor Stuart has noted this. Li Ši-čen[1] states that "there are two kinds of *a-wei*,— one an herb, the other a tree. The former is produced in Turkistan (Si yü), and can be sun-dried or boiled: this is the kind discussed by Su Kuṅ. The latter is produced among the Southern Barbarians (Nan Fan), and it is the sap of the tree which is taken: this is the kind described by Li Sün, Su Suṅ, and Č'en Č'eṅ." Su Kuṅ of the T'ang period reports that "*a-wei* grows among the Western Barbarians (Si Fan) and in K'un-lun.[2] Sprouts, leaves, root, and stems strongly resemble the *pai či* 白 芷 (*Angelica anomala*). The root is pounded, and the sap extracted from it is dried in the sun and pressed into cakes. This is the first quality. Cut-up pieces of the root, properly dried, take the second rank. Its prominent characteristic is a rank odor, but it can also stop foul smells; indeed, it is a strange product. The Brahmans say that *hün-kü* (Sanskrit *hiṅgu*, see below) is the same as *a-wei*, and that the coagulated juice of the root is like glue; also that the root is sliced, dried in the sun, and malodorous. In the western countries (India) its consumption is forbidden.[3] Habitual enjoyment of it is said to do away with foul breath. The barbarians (戎 人) prize it as the Chinese do pepper." This, indeed, relates to the plant or plants yielding asa, and Li Ši-čen comments that its habitat is in Hwo čou (Qarā-Khoja) and Ša-lu-hai-ya (Shahrokia).[4] Curiously enough, such a typical Iranian plant is passed over with silence in the ancient historical texts relative to Sasanian Persia. The only mention of it in the pre-T'ang Annals occurs in the *Sui šu*[5] with reference to the country Ts'ao 漕 north of the Ts'uṅ-liṅ (identical with the Ki-pin of the Han), while the *T'ai p'iṅ hwan yü ki*[6] ascribes *a-wei* to Ki-pin.

The *Yu yaṅ tsa tsu*[7] contains the following account of the product:

---

[1] *Pen ts'ao kaṅ mu*, Ch. 34, p. 21.

[2] K'un-lun is given as place of production in the *Kwaṅ či*, written prior to A.D. 527, but there it is described as the product of a tree (see below).

[3] It was prohibited to the monks of the Mahāyāna (cf. S. Lévi, *Journal asiatique*, 1915, I, p. 87).

[4] BRETSCHNEIDER, Mediæval Researches, Vol. II, pp. 253, 254, also 193.

[5] Ch. 83, p. 8 (also in the *Pei ši*).

[6] Ch. 182, p. 12 b.

[7] Ch. 18, p. 8 b.

"*A-wei* is produced in Gazna 伽闍郍 (*Gia-ja-na);[1] that is, in northern India. In Gazna its name is *hiṅ-yü* (Sanskrit *hiṅgu*). Its habitat is also in Persia, where it is termed *a-yü-tsie* (see below). The tree grows to a height of eight and nine feet.[2] The bark is green and yellow. In the third month the tree forms leaves which resemble a rodent's ear. It does not flower, nor does it produce fruit. The branches, when cut, have a continuous flow of sap like syrup, which consolidates, and is styled *a-wei*. The monk from the country Fu-lin, Wan 彎 by name, and the monk from Magadha, T'i-p'o 提婆 (*De-bwa, Sanskrit Deva), agree in stating that the combination[3] of the sap with rice or beans, and powdered, forms what is called *a-wei*."[4]

Another description of *a-wei* by the Buddhist monk Hwei Ži 慧日, born in A.D. 680, has been made known by S. Lévi.[5] The Chinese pilgrim points out that the plant is lacking in China, and is not to be seen in other kingdoms except in the region of Khotan. The root is as large as a turnip and white; it smells like garlic, and the people of Khotan feed on this root. The Buddhist pilgrim Yi Tsiṅ, who travelled in A.D. 671–695, reports that *a-wei* is abundant in the western limit of India, and that all vegetables are mixed with it, clarified butter, oil, or any spice.[6]

Li Sün, who wrote in the second half of the eighth century, states that, "according to the *Kwaṅ ži*, *a-wei* grows in the country K'un-lun; it is a tree with a sap of the appearance of the resin of the peach-tree. That which is black in color does not keep; that of yellow color is the best. Along the Yangtse in Yün-nan is found also a variety like the one imported in ships, juicy, and in taste identical with the yellow brand, but not yellow in color." Su Suṅ of the Sung period remarks that there is *a-wei* only in Kwaṅ-čou (Kwaṅ-tuṅ), and that it is the coagulated sap of a tree, which does not agree with the statement of Su Kuṅ. Č'en Č'eṅ 陳承, a distinguished physician, who wrote the *Pen ts'ao*

---

[1] In the *Pen ts'ao kaṅ mu*, where the text is quoted from the *Hai yao pen ts'ao* of Li Sün, Persia is coupled with Gazna. Gazna is the capital of Jāguḍa, the Tsao-kü-č'a of Hüan Tsaṅ, the Zabulistan of the Arabs. Hüan Tsaṅ reported that asafœtida is abundant there (S. Julien, Mémoires sur les contrées occidentales, Vol. II, p. 187. Cf. S. Lévi, *Journal asiatique*, 1915, I, p. 83).

[2] Thus in the text of the *Pen ts'ao*; in the edition of *Pai hai:* eighty or ninety feet. In fact, the stems of *Ferula* reach an average height of from eight to ten feet.

[3] Instead of 如 of the text I read 和 with the *Pen ts'ao*.

[4] The translation of this passage by Hirth (Chau Ju-kua, p. 225) does not render the sense correctly. The two monks mean to say that the sap or resin is a condiment added to a dish of rice or beans, and that the whole mixture bears the name *a-wei*.

[5] *Journal asiatique*, 1915, I, p. 89.

[6] Takakusu, I-tsing, pp. 128, 137.

pie šwo about A.D. 1090, says, "*A-wei* is classed among trees. People of Kian-su and Če-kian have now planted it. The odor of the branches and leaves is the same, but they are tasteless and yield no sap." The above K'un-lun refers to the K'un-lun of the Southern Sea;[1] and Li Ši-čen comments that "this tree grows in Sumatra and Siam, and that it is not very high. The natives take a bamboo tube and stick it into the tree; the tube gradually becomes filled with the sap of the tree, and during the winter months they smash the tube and obtain the sap." Then he goes on to tell the curious tale of the sheep, in the same manner as Čao Žu-kwa.[2]

Čao Žu-kwa's notice that the resin is gathered and packed in skin bags is correct; for GARCIA DA ORTA[3] reports that the gum, obtained by making cuts in the tree, is kept in bullock's hides, first anointed with blood, and then mixed with wheat flour. It is more difficult to account for the tradition given by the Chinese author, that, in order to neutralize the poison of the plant, a sheep is tied to the base of the tree and shot with arrows, whereupon the poison filters into the sheep that is doomed to death, and its carcass forms the asafœtida. This bit of folk-lore was certainly transmitted by Indian, Persian, or Arabic navigators, but any corresponding Western tradition has not yet been traced. Hobeich Ibn el-Hacen, quoted by Ibn al-Baiṭār,[4] insists on the poisonous action of the plant, and says that the harvests succeed in Sind only when asa is packed in a cloth and suspended at the mouth of water-courses, where the odor spread by the harvest will kill water-dogs and worms. Here we likewise meet the notion that the poisonous properties of the plant are capable of killing animals, and the sheep of the Chinese tradition is obviously suggested by the simile of white sheep-fat and the white vegetable fat of asa. In reality, sheep and goats are fond of the plant and fatten on it.[5] The asa ascribed to the country Ts'en-t'an in the *Sun Ši*[6] was surely an imported article.

---

[1] Not to the K'un-lun mountains, as assumed by STUART (Chinese Materia Medica, p. 173).

[2] Needless to say, this Malayan asafœtida can have been but a substitute; but to what plant it refers, I am unable to say. The *Tun si yan k'ao* (Ch. 2, p. 18; 3, p. 6 b), published in 1618, mentions *a-wei* as product of Siam and Java. T'an Ts'ui 檀萃, in his *Tien hai yü hen či*, written in 1799 (Ch. 3, p. 4, ed. of *Wen yin lou yü ti ts'un šu*), states that the *a-wei* of Yün-nan is produced in Siam, being imported from Siam to Burma and brought from Burma up the Kin-ša kian.

[3] C. MARKHAM, Colloquies, p. 47.

[4] LECLERC, Traité des simples, Vol. I, p. 447.

[5] E. KAEMPFER, Amoenitates exoticae, p. 540; C. JORET, Plantes dans l'antiquité, Vol. II, p. 100.

[6] Ch. 490; cf. HIRTH, Chao Ju-kua, p. 127. I am not convinced that Ts'en-t'an is identical with Ts'en-pa or Zanguebar.

In regard to the modern employment of the article, S. W. Williams[1] writes, "It is brought from Bombay at the rate of $15 a picul, and ranks high in the Materia Medica of the Chinese physician; it is exhibited in cholera, in syphilitic complaints and worms, and often forms an ingredient in the pills advertised to cure opium-smokers." It is chiefly believed, however, to assist in the digestion of meat and to correct the poison of stale meats (ptomaine poisoning), mushrooms, and herbs.[2] In Annam it is carried in small bags as a preventive of cholera.[3]

The following ancient terms for asafœtida are on record:—

(1) Persian 阿虞截 a-yü-tsie, *a-ñü-zet = Middle Persian *anguzad; New Persian angūža, angužad, anguyān, anguwān, angudān, angištak (stem angu+žad = "gum"[4]); Armenian ankužad, anjidan, Old Armenian angužat, angžat; Arabic anjudān. Garcia gives anjuden or angeidan as name of the tree from which asa is extracted.

(2) Sanskrit 興瞿 hiṅ-kü, *hiṅ-gu; 形虞 hiṅ-yü, *hiṅ-ñü; 薰渠 hün-k'ü, *hūn-gü; corresponding to Sanskrit hiṅgu. In my opinion, the Sanskrit word is an ancient loan from Iranian.[5] Garcia gives imgo or imgara as Indian name, and forms with initial i appear in Indian vernaculars: cf. Telugu inguva; cf., further, Japanese ingu, Malayan angu (according to J. Bontius, who wrote in 1658, the Javanese and Malayans have also the word hin).

(3) 阿魏 a-wei, *a-ñwai; 央匱 (in the Nirvāṇa-sūtra) yaṅ-kwei, *aṅ-kwai, correspond to an Indian or Iranian vernacular form of the type *aṅkwa or *aṅkwai, that we meet in Tokharian B or Kuča aṅkwa.[6] This form is obviously based on Iranian angu, angwa.

(4) Mongol 哈昔泥 xa-si-ni (thus given as a Mongol term in the Pen ts'ao kaṅ mu after the Yin šan čeṅ yao of the Mongol period, written in 1331), corresponds to Persian kasnī, kisnī, or gisnī ("asafœtida"), derived from the name of Gazni or Gazna, the capital of Zābulistan, which, according to Hüan Tsaṅ, was the habitat of the plant. A Mongol word of this type is not listed in the Mongol dictionaries of Kovalevski and Golstunski, but doubtless existed in the age of the Yüan,

---

[1] Chinese Commercial Guide, p. 80.

[2] Stuart, Chinese Materia Medica, p. 174.

[3] Perrot and Hurrier, Mat. méd. et pharmacopée sino-annamites, p. 161.

[4] Cf. Sanskrit jatuka (literally, "gum, lac") = asafœtida. Hübschmann, Armen. Gram., p. 98.

[5] D'Herbelot (Bibliothèque orientale, Vol. I, p. 226; Vol. II, p. 327) derived the Persian word (written by him angiu, engiu, ingu; Arabic ingiu, ingudan) from Indian henk and hengu, ingu, for the reason that in India this drug is principally used; this certainly is not correct.

[6] Cf. T'oung Pao, 1915, pp. 274–275.

when the Mongols introduced the condiment into China under that name, while they styled the root 穩展 *yin-čan*. In modern Mongol, the name of the product is *šingun*, which is borrowed from the Tibetan word mentioned below.

In the Tibetan dialect of Ladākh, asafœtida is called *hiṅ* or *sip*.[1] The name *sip* or *sup* was reported by Falconer, who was the first to discover in 1838 *Ferula narthex* in western Tibet on the slopes of the mountains dividing Ladākh from Kashmir.[2] The word *sip*, however, is not generally Tibetan, but only of local value; in all probability, it is not of Tibetan origin. The common Tibetan word is *šiṅ-kun*, which differs from the Iranian and Indian terms, and which, in view of the fact that the plant occurs in Tibetan regions, may be a purely Tibetan formation.

Finally it may be mentioned that, according to Borszczow,[3] *Scorodosma* is generally known to the inhabitants of the Aralo-Caspian territory under the name *sasyk-karai* or *keurök-kurai*, which means as much as "malodorous rush." The Bukharans call it *sasyk-kawar* or simply *kawar*.

---

[1] Ramsay, Western Tibet, p. 7.

[2] *Transactions Linnean Soc.*, Vol. XX, pt. 1, 1846, pp. 285–291.

[3] *Op. cit.*, p. 25.

GALBANUM

23. There is only a single Chinese text relative to galbanum, which is contained in the *Yu yan tsa tsu*,[1] where it is said, "*P'i-ts'i* 醒² 齊 (*bit-dzi, bir-zi, bir-zai) is a product of the country Po-se (Persia). In Fu-lin it is styled 預 勃 梨 他 *han-p'o-li-t'a* (*xan-bwiδ-li-da).[3] The tree grows to a height of more than ten feet, with a circumference of over a foot. Its bark is green, thin, and extremely bright. The leaves resemble those of the asafœtida plant (*a-wei*), three of them growing at the end of a branch. It does not flower or bear fruit. In the western countries people are accustomed to cut the leaves in the eighth month; and they continue to do this more and more till the twelfth month. The new branches are thus very juicy and luxuriant; without the trimming process, they would infallibly fade away. In the seventh month the boughs are broken off, and there is a yellow sap of the appearance of honey and slightly fragrant, which is medicinally employed in curing disease."

Hirth has correctly identified the transcription *p'i-ts'i* with Persian *bīrzai*, which, however, like the other Po-se words in the *Yu yan tsa tsu*, must be regarded as Pahlavi or Middle Persian;[4] and the Fu-lin *han-p'o-li-t'a* he has equated with Aramaic *xelbānita*, the latter from Hebrew *xelbenāh*, one of the four ingredients of the sacred perfume (Exodus, xxx, 34–38). This is translated by the Septuaginta χαλβάνη and by the Vulgate *galbanum*. The substance is mentioned in three passages

---

[1] Ch. 18, p. 11 b.

[2] HIRTH, who is the first to have translated this text (*Journal Am. Or. Soc.* Vol. XXX, p. 21), writes this character with the phonetic element 爾, apparently in agreement with the edition of the *Tsin tai pi šu;* but this character is not authorized by K'an-hi, and it is difficult to see how it could have the phonetic value *p'i;* we should expect *ni*. The above character is that given by K'an-hi, who cites under it the passage in question. It is thus written also in the *Min hian p'u* 名 香 譜 by Ye T'in-kwei 葉 廷 珪 (p. 10, ed. of *Hian yen ts'un šu*) and in the *Pen ts'ao kan mu* (Ch. 33, p. 6), where the pronunciation is explained by 別 *biet. The editors of cyclopædias were apparently staggered by this character, and most of them have chosen the phonetic *man*, which is obviously erroneous. None of our Chinese dictionaries lists the character.

[3] The *Pen ts'ao kan mu* (*l. c.*) annotates that the first character should have the sound 奪 *to*, *dwat, which is not very probable.

[4] There are also the forms *pirzed, bārzed* (LECLERC, Traité des simples, Vol. I, p. 201), *berzed, barije,* and *bazrud;* in India *bireja, ganda-biroza.* Another Persian term given by SCHLIMMER (Terminologie, p. 294) is *wešā.*

363

by Theophrastus:[1] it is produced in Syria from a plant called πάναξ ("all-heal"); it is only the juice (ὀπός) which is called χαλβάνη, and which "was used in cases of miscarriage as well as for sprains and such-like troubles, also for the ears, and to strengthen the voice. The root was used in childbirth, and for flatulence in beasts of burden, further in making the iris-perfume (ἴρινον μύρον) because of its fragrance; but the seed is stronger than the root. It grows in Syria, and is cut at the time of wheat-harvest."[2]

Pliny says that galbanum grows on the mountain Amanus in Syria as the exudation from a kind of *ferula* of the same name as the resin, sometimes known as *stagonitis*.[3] Its medicinal employment is treated by him in detail.[4] DIOSCORIDES[5] explains it as the gum of a plant which has the form of a *ferula*, growing in Syria, and called by some *metopion*. Abu Mansur[6] discusses the drug under the Arabic name *quinna* and the Persian name *bārzäd*. During the middle ages galbanum was well known in Europe from the fourteenth century onward.[7]

The philological result is confirmed by the botanical evidence, although Twan Č'eṅ-ši's description, made from an oral report, not as an eye-witness, is naturally somewhat deficient; but it allows us to recognize the characteristics of a *Ferula*. It is perfectly correct that the leaves resemble those of the asafœtida *Ferula*, as a glance at the excellent plates in the monograph of BORSZCZOW (*op. cit.*) will convince one. It is likewise correct that the leaves grow at the ends of the twigs, and usually by threes. It is erroneous, however, that the tree does not flower or bear fruit.[8] The process of collecting the sap is briefly but well described. Nothing positive is known about the importation of galbanum into China, although W. AINSLIE[9] stated in 1826 that it was

---

[1] Histor. plant., IX. 1, 2; IX. VII, 2; IX. IX, 2. The term occurs also in the Greek papyri.

[2] Cf. the new edition and translation of Theophrastus by A. HORT (Vol. II, p. 261). I do not see how the term "balsam of Mecca" (*ibid.*, p. 219), which is a misnomer anyhow, can be employed in the translation of an ancient Greek author.

[3] Dat et galbanum Syria in eodem Amano monte e ferula, quae eiusdem nominis, resinae modo; stagonitim appellant (XII, 56, § 126).

[4] XXIV, 13.

[5] III, 87 (cf. LECLERC, Traité des simples, Vol. III, p. 115).

[6] ACHUNDOW, Abu Mansur, p. 108.

[7] See, for instance, K. v. MEGENBERG, Buch der Natur (written in 1349–50), ed. F. Pfeiffer, p. 367; FLÜCKIGER and HANBURY, Pharmacographia, p. 321.

[8] The fruits are already mentioned by Theophrastus (Hist. plant., IX. IX, 2) as remedies.

[9] Materia Indica, Vol. I, p. 143.

sent from Bombay to China, and STUART[1] regards this as entirely probable; but this is merely a supposition unsupported by any tangible data: no modern name is known under which the article might come. The three names given for galbanum in the English-Chinese Standard Dictionary are all wrong: the first, *a-yü*, refers to asafœtida (see above, p. 361);[2] the second, 楓, denotes *Liquidambar orientalis;* and the third, *pai suṅ hiaṅ* ("white pine aromatic"), relates to *Pinus bungeana.* The *Pen ts'ao kaṅmu*[3] has the notice on *p'i-ts'i* as an appendix to "manna." Li Ši-čen, accordingly, did not know the nature of the product. He is content to cite the text of the *Yu yaṅ tsa tsu* and to define the medical properties of the substance after Č'en Ts'aṅ-k'i of the T'ang. Only under the T'ang was galbanum known in China.

The trees from which the product is obtained are usually identified with *Ferula galbaniflua* and *F. rubricaulis* or *erubescens*, both natives of Persia. The Syrian product used by the Hebrews and the ancients was apparently derived from a different though kindred species. *F. rubricaulis*, said by the botanist Buhse to be called in Persian *khassuih*,[4] is diffused all over northern Persia and in the Daēna Mountains in the southern part of the country; it is frequent in the Demawend and on the slopes of the Alwend near Hamadan.[5] No incisions are made in the plant: the sap flowing out of the lower part of the stalks and from the base of the leaves is simply collected. The gum is amber-yellow, of not disagreeable, strongly aromatic odor, and soon softens between the fingers. Its taste is slightly bitter. Only in the vicinity of Hamadan, where the plant is exuberant, has the collecting of galbanum developed into an industry.

SCHLIMMER[6] distinguishes two kinds,— a brown and a white-yellowish galbanum. The former (Persian *barzed* or *barije*), the product of *Ferula galbaniflua*, is found near De Gerdon in the mountains Sa-utepolagh between Teheran and Gezwin, in the valleys of Lars (Elburs), Khereghan, and Sawe, where the villagers gather it under the name *balubu*. The latter kind is the product of *Dorema anchezi* Boiss., en-

---

[1] Chinese Materia Medica, p. 181.

[2] This is the name given for galbanum by F. P. SMITH (Contributions towards the Materia Medica, p. 100), but it is mere guesswork.

[3] Ch. 33, p. 6.

[4] Evidently identical with what WATT (Commercial Products of India, p. 535) writes *khassnib*, explaining it as a kind of galbanum from Shīrāz. LOEW (Aram. Pflanzennamen, p. 163) makes *kassnih* of this word. The word intended is apparently the *kasnī* mentioned above (p. 361).

[5] BORSZCZOW, *op. cit.*, p. 35.

[6] Terminologie, p. 295.

countered by Buhse in the low mountains near Reshm (white galbanum). Galbanum is also called *kilyānī* in Persian.

Borszczow has discovered in the Aralo-Caspian region another species of *Ferula*, named by him *F. schair* from the native word *šair* (=Persian *šīr*, "milk-juice") for this plant. The juice of this species has the same properties as galbanum; also the plant has the same odor.

Abu Mansur[1] mentions a *Ferula* under the name *sakbīnaj* (Arabic form, Persian *sakbīna*), which his translator, the Persian physician Achundow, has identified with the Sagapenum resin of *Ferula persica*, said to be similar to galbanum and to be gathered in the mountains of Luristan. According to FLÜCKIGER and HANBURY,[2] the botanical origin of Sagapenum is unknown; but there is no doubt that this word (σαγάπηνον in Dioscorides, III, 95, and Galenus; *sacopenium* in Pliny, XII, 56), in mediæval pharmacy often written *serapinum*, is derived from the Persian word.

The galbanum employed in India is imported from Persia to Bombay. WATT[3] distinguishes three kinds known in commerce,—Levant, Persian solid, and Persian liquid. The first comes from Shiraz, the second has an odor of turpentine, and the third is the *gaoshir* or *jawāshir;* the latter being a yellow or greenish semi-fluid resin, generally mixed with the stems, flowers, and fruits of the plant. It is obtained from the stem, which, when injured, yields an orange-yellow gummy fluid. Generally, however, the galbanum of commerce forms round, agglutinated tears, about the size of peas, orange-brown outside, yellowish-white or bluish-green inside. The odor is not disagreeable, like that of asafœtida, and the taste is bitter.

Galbanum consists of about 65 per cent resin, 20 per cent gum, and from 3 to 7 per cent volatile oil.

---

[1] ACHUNDOW, Abu Mansur, p. 84.
[2] Pharmacographia, p. 342.
[3] Commercial Products of India, p. 535.

OAK–GALLS

24. Oak-galls (French *noix de galles*, Portuguese *galhas*) are globular excrescences caused by the gall-wasp (*Cynips quercus folii*) puncturing the twigs, leaves, and buds, and depositing its ova in several species of oak (chiefly *Quercus lusitanica* var. *infectoria*), to be found in Asia Minor, Armenia, Syria, and Persia. In times of antiquity, galls were employed for technical and medicinal purposes. In consequence of their large percentage (up to 60 per cent) of tannic or Gallo-tannic acid, they served for tanning, still further for the dyeing of wool and the manufacture of ink.[1] Both Theophrastus[2] and Dioscorides[3] mention galls under the name κηκίς. Abu Mansur describes galls under the Arabic name *afs*.[4]

The greater part of the galls found in Indian bazars come from Persia, being brought by Arab merchants.[5] The Sanskrit name *mājūphala* (*phala*, "fruit") is plainly a loan-word from the Persian *māzū*.

In Chinese records, oak-galls are for the first time mentioned under the term *wu-ši-tse* 無食子 as products of Sasanian Persia.[6] They first became known in China under the T'ang from Persia, being introduced in the Materia Medica of the T'ang Dynasty (*T'aṅ pen ts'ao*). The *T'aṅ pen ču* 唐本注 states that they grow in sandy deserts,[7] and that the tree is like the tamarisk (*č'eṅ* 檉). A commentary, cited as *kin ču* 今注, adds that they are produced in Persia, while the *Čeṅ lei pen ts'ao*[8] says that they grow in the country of the Western Žuṅ (Iranians). The *Yu yaṅ tsa tsu*[9] gives a description of the plant as follows: "*Wu-ši-tse* 無石子 are produced in the country Po-se (Persia),

---

[1] BLÜMNER, Technologie, Vol. I, 2d ed., pp. 251, 268.

[2] Hist. plant., III. VIII, 6.

[3] I, 146 (cf. LECLERC, Traité des simples, Vol. II, p. 457). See also Pliny, XIII, 63; XVI, 26; XXIV, 109.

[4] ACHUNDOW, Abu Mansur, p. 98.

[5] W. AINSLIE, Materia Indica, Vol. I, p. 145; WATT, Commercial Products of India, p. 911.

[6] Sui šu, Ch. 83, p. 7 b.

[7] According to another reading, "in sandy deserts of the Western Žuṅ" (that is, Iranians).

[8] Ch. 14, p. 20.

[9] Ch. 18, p. 9.

367

where they are styled 摩賊 *mo-tsei*, *mwa-džak.[1] The tree grows to
a height of from six to seven feet,[2] with a circumference of from eight to
nine feet. The leaves resemble those of the peach, but are more oblong.
It blossoms in the third month, the flowers being white, and their
heart reddish. The seeds are round like pills, green in the beginning,
but when ripe turning to yellow-white. Those punctured by insects
and perforated are good for the preparation of leather; those without
holes are used as medicine. This tree alternately produces galls one
year and acorns (跋屢子 *pa-lü tse*, *bwaδ-lu; Middle Persian *ballu,
barru [see below], New Persian *balut*), the size of a finger and three
inches long, the next."[3] The latter notion is not a Chinese fancy, but
the reproduction of a Persian belief.[4]

The Geography of the Ming (*Ta Min i t'un či*) states that galls are
produced in the country of the Arabs (Ta-ši) and all barbarians, and
that the tree is like the camphor-tree (*Laurus camphora*), the fruits
like the Chinese wild chestnuts (*mao-li* 茅栗).

The Chinese transcriptions of the Iranian name do not "all represent Persian *māzū*," as reiterated by Hirth after Watters, but reproduce older Middle-Persian forms. In fact, none of the Chinese render-
ings can be the equivalent of *māzū*.

(1) 摩賊 (*Yu yan tsa tsu*) *mo-tsei*, *mwa-džak (dzak, zak), answers
to a Middle Persian *madžak (madzak or mazak).

(2) 墨石 *mo-ši*, *mak-zak, = Middle Persian *maxzak.

(3) 無石 *wu-ši*, *mwu-zak, = Middle Persian *muzak.

(4) 沒石 *mu-ši*, *mut-zak, = Middle Persian *muzak. Compare
with these various forms Tamil *mācakai*, Telugu *mācikai*, and the
magican of Barbosa.

(5) 摩茶[5] *mo-t'u*, *mwa-du, = Middle Persian *madu.

沙沒律 *ša-mu-lü* (in Čao Žu-kwa), *ša-mut-lwut, answers to Iranian

---

[1] Instead of *tsei*, some editions write 澤 *tsö* (*dzak, džak), which is phonetically
the same.

[2] The text has 丈, which should be corrected into 尺, for the tree seldom rises
higher than six feet.

[3] The text of the following last clause is corrupted, and varies in the different
editions; it yields no acceptable sense. HIRTH's translation (Chao Ju-kua, p. 215)
is not intelligible to me. WATTERS (Essays on the Chinese Language, p. 349) is
certainly wrong in saying that "the Chinese do not seem to know even yet the
origin of these natural products" (oak-galls); this is plainly refuted by the above
description. The *T'u šu tsi č'en* (XX, Ch. 310) and *Či wu min ši t'u k'ao* (Ch. 35,
p. 21) even have a tolerably good sketch of the tree, showing galls on the leaves.

[4] E. SEIDEL, Mechithar, p. 127.

[5] The character 茶 *č'a* in Čao Žu-kwa, and thus adopted by HIRTH (p. 215), is
an error.

šah-baluṭ ("the edible chestnut," *Castanea vulgaris*), which appears in the Būndahišn (above, p. 193), as correctly identified by Hirth; but 蒲蘆 *p'u-lu* and *pa-lü* of the *Yu yan tsa tsu* (see above) would indicate that the Chinese heard *bulu* and *balu* without a final *t*, and such forms may have existed in Middle-Persian dialects. In fact, we have this type in the dialect of the Kurd in the form *berru*, and in certain Kurd dialects *barü* and *barru*.[1]

---

[1] Cf. J. DE MORGAN, Mission scientifique en Perse, Vol. V, p. 133. The Iranian term means literally "acorn of the Shah, royal acorn," somehow a certain analogy to Greek Διὸς βάλανος ("acorn of Zeus"). The origin of Greek καστάναιον or κάστανον is sought in Armenian *kask* ("chestnut") and *kaskeni* ("chestnut-tree"; see SCHRADER in Hehn, Kulturpflanzen, p. 402). According to the Armenian Geography of Moses of Khorene, the tree flourished in the Old-Armenian province Duruperan (Daron); according to Galenus, near Sardes in Asia Minor; according to Daūd, on Cyprus; according to Abu Mansur, also in Syria; while, according to the same author, Persia imported chestnuts from Adherbeijan and Arran; according to Schlimmer, from Russia (E. SEIDEL, Mechithar, p. 152). It is striking that the Chinese did not see the identity of the Iranian term with their *li* 栗, the common chestnut, several varieties of which grow in China.

INDIGO

25. As indicated by our word "indigo" (from Latin *indicum*), this dye-stuff took its origin from India. The indigo-plant (*Indigofera tinctoria*), introduced into Persia from India, is discussed by Abu Mansur under the name *nīl* or *līla*. The leaves are said to strengthen the hair. The hair, if previously dyed with henna, becomes brilliant black from the pounded leaves of the plant. Another species, *I. linifolia*, is still used in Persia for dyeing beard and hair black.[1] The Persian words are derived from Sanskrit *nīla*, as is likewise Arabic *nīlej*.[2] Also *nili hindi* ("Indian indigo") occurs in Persian. GARCIA DA ORTA has handed down a form *anil*,[3] and in Spanish the plant is called *añil* (Portuguese and Italian *anil*).[4] It may be permissible to assume that indigo was first introduced into Sasanian Persia under the reign of Khosrau I Anōšarwān (A.D. 531–579); for Masūdī, who wrote about A.D. 943, reports that this king received from India the book *Kalīla wa Dimna*, the game of chess, and the black dye-stuff for the hair, called the Indian.[5]

Under the designation *ts'iṅ tai* 青黛 ("blue cosmetic for painting the eyebrows") the Chinese became acquainted with the true indigo and the Iranian practice mentioned above. The term is first on record as a product of Ts'ao 漕 (Jāguḍa)[6] and Kü-lan 俱蘭 in the vicinity of Tokharestan;[7] during the T'ang period, the women of Fergana did not employ lead-powder, but daubed their eyebrows with *ts'iṅ tai*.[8] Ma Či of the tenth century says that "*ts'iṅ tai* came from the country Po-se (Persia), but that now in T'ai-yüan, Lu-liṅ, Nan-k'aṅ, and other

---

[1] ACHUNDOW, Abu Mansur, pp. 144, 271. SCHLIMMER (Terminologie, p. 395) gives *ringi rīš* and *wesme* as Persian words for indigo-leaves.

[2] LECLERC, Traité des simples, Vol. III, p. 384.

[3] C. MARKHAM, Colloquies, p. 51. The form *anil* is also employed by F. PYRARD (Vol. II, p. 359, ed. of Hakluyt Society), who says that indigo is found only in the kingdom of Cambaye and Surat.

[4] ROEDIGER and POTT (Z. f. Kunde d. Morg., Vol. VII, p. 125) regard this prefix *a* as the Semitic article (Arabic *al-nīl*, *an-nīl*).

[5] BARBIER DE MEYNARD and PAVET DE COURTEILLE, Les Prairies d'or, Vol. II, p. 203.

[6] *Sui šu*, Ch. 83, p. 8 (see above, p. 317).

[7] *T'ai p'iṅ hwan yü ki*, Ch. 186, p. 12. It was also found in Ki-pin (*ibid.*, Ch. 182, p. 12 b).

[8] *Ibid.*, Ch. 181, p. 13 b.

370

places, a dye-stuff of similar virtues is made from *tien* 澱 (the indigenous *Polygonum tinctorium*)."[1] Li Ši-čen holds the opinion that the Persian *ts'iṅ tai* was the foreign *lan-tien* 藍靛 (*Indigofera tinctoria*). It must not be forgotten that the genus *Indigofera* comprises some three hundred species, and that it is therefore impossible to hope for exact identifications in Oriental records. Says G. WATT[2] on this point, "Species of *Indigofera* are distributed throughout the tropical regions of the globe (both in the Old and New Worlds) with Africa as their headquarters. And in addition to the Indigoferas several widely different plants yield the self-same substance chemically. Hence, for many ages, the dye prepared from these has borne a synonymous name in most tongues, and to such an extent has this been the case that it is impossible to say for certain whether the *nīla* of the classic authors of India denoted the self-same plant which yields the dye of that name in modern commerce." "Indigo," therefore, is a generalized commercial label for a blue dye-stuff, but without botanical value. Thus also Chinese indigo is yielded by distinct plants in different parts of China.[3]

It is singular that the Chinese at one time imported indigo from Persia, where it was doubtless derived from India, and do not refer to India as the principal indigo-producing country. An interesting article on the term *ts'iṅ tai* has been written by HIRTH.[4]

---

[1] *Pen ts'ao kaṅ mu*, Ch. 16, p. 25 b.
[2] Commercial Products of India, p. 663.
[3] BRETSCHNEIDER, Bot. Sin., pt. II, p. 212.
[4] Chinesische Studien, pp. 243–258.

RICE

26. While rice is at present a common article of food of the Persian people, being particularly enjoyed as pilau,[1] it was entirely unknown in the days of Iranian antiquity. No word for "rice" appears in the Avesta.[2] Herodotus[3] mentions only wheat as the staple food of the Persians at the time of Cambyses. This negative evidence is signally confirmed by the Chinese annals, which positively state that there is no rice or millet in Sasanian Persia;[4] and on this point Chinese testimony carries weight, since the Chinese as a rice-eating nation were always anxious to ascertain whether rice was grown and consumed by foreign peoples. Indeed, the first question a travelling Chinese will ask on arrival at a new place will invariably refer to rice, its qualities and valuations. This is conspicuous in the memoirs of Čaṅ K'ien, the first Chinese who travelled extensively across Iranian territory, and carefully noted the cultivation of rice in Fergana (Ta-yūan), further for Parthia (An-si), and T'iao-či (Chaldæa). The two last-named countries, however, he did not visit himself, but reported what he had heard about them. In the Sasanian epoch, Chinese records tell us that rice was plentiful in Kuča, Kašgar (Su-lek), Khotan, and Ts'ao (Jăguḍa) north of the Ts'uṅ-liṅ;[5] also in Ši (Tashkend).[6] On the other hand, Aristobulus, a companion of Alexander on his expedition in Asia and author of an Alexander biography written after 285 B.C., states that rice grows in Bactriana, Babylonia, Susis, and in lower Syria;[7] and Diodorus[8] likewise emphasizes the abundance of rice in Susi-

---

[1] *T'oung Pao*, 1916, p. 481.

[2] Modi, in Spiegel Memorial Volume, p. XXXVII.

[3] III, 22.

[4] *Wei šu*, Ch. 102, pp. 5 b–6 a; *Čou šu*, Ch. 50, p. 6. Tabari (translation of Nöldeke, p. 244) mentions rice among the crops taxed by Khusrau I (A.D. 531–578); but this is surely an interpolation, as in the following list of taxes rice is not mentioned, while all other crops are. Another point to be considered is that in Arabic manuscripts, when the diacritical marks are omitted, the word *birinj* may be read as well *naranj*, which means "orange" (cf. Ouseley, Oriental Geography of Ebn Haukal, p. 221).

[5] *Sui šu*, Ch. 83, pp. 5 b, 7 b.

[6] *T'ai p'iṅ hwan yü ki*, Ch. 186, p. 7 b.

[7] Strabo, XV. 1, 18.

[8] XIX, 13.

372

ana. From these data Hehn[1] infers that under the rule of the Persians, and possibly in consequence of their rule, rice-cultivation advanced from the Indus to the Euphrates, and that from there came also the Greek name ὄρυζα. This rice-cultivation, however, can have been but sporadic and along the outskirts of Iran; it did not affect Persia as a whole. The Chinese verdict of "no rice" in Sasanian Persia appears to me conclusive, and it further seems to me that only from the Arabic period did the cultivation of rice become more general in Persia. This conclusion is in harmony with the account of Hwi Čao 慧超, a traveller in the beginning of the eighth century, who reports in regard to the people of Mohammedan Persia that they subsist only on pastry and meat, but have also rice, which is ground and made into cakes.[2] This conveys the impression that rice then was not a staple food, but merely a side-issue of minor importance. Yāqūt mentions rice for the provinces Khuzistān and Sabur.[3] Abu Mansur, whose work is largely based on Arabic sources, is the first Persian author to discuss fully the subject of rice.[4] Solely a New-Persian word for "rice" is known, namely birinǰ or gurinǰ (Armenian and Ossetic brinǰ), which is usually regarded as a loan-word from Sanskrit vrīhi; Afghan vriže (with Greek ὄρυζα, βρίζα) is still nearer to the latter. In view of the historical situation, the reconstruction of an Avestan *verenja[5] or an Iranian *vrinji,[6] and the theory of an originally Aryan word for "rice," seem to me inadmissible.

---

[1] Kulturpflanzen, p. 505.

[2] Hirth, Journal Am. Or. Soc., Vol. XXXIII, 1913, pp. 202, 204, 207.

[3] B. de Meynard, Dictionnaire géographique de la Perse, pp. 217, 294.

[4] Achundow, Abu Mansur, p. 5. J. Schiltberger (1396–1427), in his Bondage and Travels (p. 44, ed. of Hakluyt Society, 1879) speaks of the "rich country called Gilan, where rice and cotton alone is grown."

[5] P. Horn, Neupersische Etymologie, No. 208.

[6] H. Hübschmann, Persische Studien, p. 27.

PEPPER

27. The pepper-plant (*hu tsiao*, Japanese *košō*, 胡椒, *Piper nigrum*)
deserves mention in this connection only inasmuch as it is listed among
the products of Sasanian Persia.[1] Ibn Haukal says that pepper, sandal,
and various kinds of drugs, were shipped from Sīrāf in Persia to all
quarters of the world.[2] Pepper must have been introduced into Persia
from India, which is the home of the shrub.[3] It is already enumerated
among the plants of India in the Annals of the Han Dynasty.[4] The
Yu yan tsa tsu[5] refers it more specifically to Magadha,[6] pointing out
its Sanskrit name *marica* or *marīca* in the transcription 昧㽮支 *mei-
li-či*.[7] The term *hu tsiao* shows that not all plants whose names have
the prefix *hu* are of Iranian origin: in this case *hu* distinctly alludes
to India.[8] *Tsiao* is a general designation for spice-plants, principally
belonging to the genus *Zanthoxylon*. Li Ši-čen[9] observes that the black
pepper received its name only for the reason that it is bitter of taste
and resembles the *tsiao*, but that the pepper-fruit in fact is not a *tsiao*.
It is interesting to note that the authors of the various *Pen ts'ao* seem
to have lost sight of the fact of the Indian origin of the plant, and do
not even refer to the Han Annals. Su Kun states that *hu tsiao* grows
among the Si Žun, which plainly shows that he took the word *hu* in
the sense of peoples of Central Asia or Iranians, and substituted for it

[1] *Sui šu*, Ch. 83, p. 7 b; *Čou šu*, Ch. 50, p. 6; and *Wei šu*, Ch. 102, p. 6. According
to Hirth (Chau Ju-kua, p. 223), this would mean that pepper was brought to China
by Persian traders from India. I am unable to see this point. The texts in question
simply give a list of products to be found in Persia, and say nothing about exportation of any kind.

[2] W. Ouseley, Oriental Geography of Ebn Haukal, p. 133. Regarding the former importance of Sīrāf, which "in old times was a great city, very populous and
full of merchandise, being the port of call for caravans and ships," see G. Le Strange,
Description of the Province of Fars, pp. 41–43.

[3] In New Persian, pepper is called *pilpil* (Arabicized *filfil*, *fulful*), from the
Sanskrit *pippalī*.

[4] *Hou Han šu*, Ch. 118, p. 5 b.

[5] Ch. 18, p. 11.

[6] Cf. Sanskrit *māgadha* as an epithet of pepper.

[7] In fact, this form presupposes a vernacular type *merici.

[8] *Hu tsiao* certainly does not mean "Western Barbarians (Tartar) pepper,"
as conceived by Watters (Essays on the Chinese Language, p. 441). What had
the "Tartars" to do with pepper? The Uigur adopted simply the Sanskrit word in
the form *murč*.

[9] *Pen ts'ao kan mu*, Ch. 32, p. 3 b.

374

its synonyme Si Žuṅ; at least, it appears certain that the latter term bears no reference to India. Li Ši-čen gives as localities where the plant is cultivated, "all countries of the Southern Barbarians (Nan Fan), Kiao-či (Annam), Yün-nan, and Hai-nan."

Another point of interest is that in the *T'aṅ pen ts'ao* of Su Kuṅ appears a species called *šan hu tsiao* 山 胡 椒 or wild pepper, described as resembling the cultivated species, of black color, with a grain the size of a black bean, acrid taste, great heat, and non-poisonous. This plant-name has been identified with *Lindera glauca* by A. Henry,[1] who says that the fruit is eaten by the peasants of Yi-č'aṅ, Se-č'wan. The same author offers a *ye hu-tsiao* ("wild pepper"), being *Zanthoxylum setosum*.

Piper longum or *Chavica roxburghii*, Chinese 蓽 茇 or 撥 *pi-po*, *pit-pat(pal), from Sanskrit *pippalī*, is likewise attributed to Sasanian Persia.[2] This pepper must have been also imported into Iran from India, for it is a native of the hotter parts of India from Nepal eastward to Assam, the Khasia hills and Bengal, westward to Bombay, and southward to Travancore, Ceylon, and Malacca.[3] It is therefore surprising to read in the *Pen ts'ao* of the T'ang that *pi-po* grows in the country Po-se: this cannot be Persia, but refers solely to the Malayan Po-se. For the rest, the Chinese were very well aware of the Indian origin of the plant, as particularly shown by the adoption of the Sanskrit name. It is first mentioned in the *Nan faṅ ts'ao mu čwaṅ*, unless it be there one of the interpolations in which this work abounds, but it is mixed up with the betel-pepper (*Chavica betel*).

---

[1] Chinese Names of Plants, No. 45.

[2] *Čou šu*, Ch. 50, p. 6.

[3] Watt, Commercial Products of India, p. 891.

SUGAR

28. The sugar-cane (*Saccharum officinarum*) is a typically Indian or rather Southeast-Asiatic, and merely a secondary Iranian cultivation, but its history in Iran is of sufficient importance to devote here a few lines to this subject. The Sui Annals[1] attribute hard sugar (*ši-mi* 石蜜, literally, "stone honey") and *pan-mi* 半蜜 ("half honey") to Sasanian Persia and to Ts'ao (Jāguḍa). It is not known what kind of sugar is to be understood by the latter term.[2] Before the advent of sugar, honey was the universal ingredient for sweetening food-stuffs, and thus the ancients conceived the sugar of India as a kind of honey obtained from canes without the agency of bees.[3] The term *ši-mi* first appears in the *Nan fan ts'ao mu čwan*,[4] which contains the first description of the sugar-cane, and refers it to Kiao-či (Tonking); according to this work, the natives of this country designate sugar as *ši-mi*, which accordingly may be the literal rendering of a Kiao-či term. In A.D. 285 Fu-nan (Camboja) sent *ču-čö* 諸蔗 ("sugar-cane") as tribute to China.[5]

It seems that under the T'ang sugar was also imported from Persia to China; for Moṅ Šen, who wrote the *Ši liao pen ts'ao* in the second half of the seventh century, says that the sugar coming from Po-se (Persia) to Se-č'wan is excellent. Su Kuṅ, the reviser of the *T'aṅ pen ts'ao* of about A.D. 650, extols the sugar coming from the Si Žuṅ, which may likewise allude to Iranian regions. Exact data as to the introduction and dissemination of the sugar-cane in Persia are not available. E. O. v. Lippmann[6] has developed an elaborate theory to the effect that

---

[1] *Sui šu*, Ch. 83, p. 7 b.

[2] It is only contained in the *Sui šu*, not in the *Wei šu* (Ch. 102, p. 5 b), which has merely *ši-mi*. The sugar-cane was also grown in Su-le (Kashgar): *T'ai p'iṅ hwan yü ki*, Ch. 181, p. 12 b.

[3] Pliny, XII, 17.

[4] Ch. 1, p. 4.

[5] This word apparently comes from a language spoken in Indo-China; it is already ascribed to the dictionary *Šwo wen*. Subsequently it was replaced by *kan* 甘 ("sweet") *čö* or *kan* 芋 *čö*, presumably also the transcription of a foreign word. The *Nan Ts'i šu* mentions *ču-čö* as a product of Fu-nan (cf. Pelliot, *Bull. de l'Ecole française*, Vol. III, p. 262). In Č'i-t'u 赤土 (Siam) a wine of yellow color and fine aroma was prepared from sugar and mixed with the root of a Cucurbitacea (*Sui šu*, Ch. 82, p. 2 b).

[6] *Geschichte des Zuckers*, p. 93 (Leipzig, 1890); and *Abhandlungen*, Vol. I, p. 263. According to the same author, the Persians were the inventors of sugar-refining; but this is purely hypothetical.

376

the Christians of the city Gundēšāpūr, which was in connection with India and cultivated Indian medicine, should have propagated the cane and promoted the sugar-industry. This is no more than an ingenious speculation, which, however, is not substantiated by any documents. The facts in the case are merely, that according to the Armenian historian Moses of Khorene, who wrote in the second half of the fifth century, sugar-cane was cultivated in Elymais near Gundēšāpūr, and that later Arabic writers, like Ibn Haukal, Muqaddasī, and Yāqūt, mention the cultivation of the cane and the manufacture of sugar in certain parts of Persia. The above Chinese notice is of some importance in showing that sugar was known under the Sasanians in the sixth century. The Arabs, as is well known, took a profound interest in the sugar-industry after the conquest of Persia (A.D. 640), and disseminated the cane to Palestine, Syria, Egypt, etc. The Chinese owe nothing to the Persians as regards the technique of sugar-production. In A.D. 647 the Emperor T'ai Tsuň was anxious to learn its secrets, and sent a mission to Magadha in India to study there the process of boiling sugar, and this method was adopted by the sugar-cane growers of Yaň-čou. The color and taste of this product then were superior to that of India.[1] The art of refining sugar was taught the Chinese as late as the Mongol period by men from Cairo.[2]

---

[1] *T'aň hui yao*, Ch. 100, p. 21.

[2] YULE, Marco Polo, Vol. II, pp. 226, 230. The latest writer on the subject of sugar in Persia is P. SCHWARZ (*Der Islam*, Vol. VI, 1915, pp. 269–279), whose researches are restricted to the province of Ahwāz. In opposition to C. Ritter, who regarded Sīrāf on the Persian Gulf as the place whither the sugar-cane was first transplanted from India, he assigns this rôle to Hormuz; the first mention of refined sugar he finds in an Arabic poet of the seventh century. Lippmann's work is not known to him.

MYROBALAN

29. The myrobalan *Terminalia chebula*, *ho-li-lo* 訶黎勒 (*ha-ri-lak, Japanese *kariroku*, Sanskrit *harītakī*, Tokharian *arirāk*, Tibetan *a-ru-ra*, Newārī *halala;* Persian *halīla*, Arabic *halīlaj* and *ihlīligāt*), was found in Persia.[1] The tree itself is indigenous to India, and the fruit was evidently imported from India into Persia.[2] This is confirmed by the fact that it is called in New Persian *halīla* (Old Armenian *halile*), or *halīla-i kabūli*, hinting at the provenience from Kābul.[3]

In the "Treatise on Wine," *Tsiu p'u* 酒譜,[4] written by Tou Kin 竇蓳 of the Sung, it is said, "In the country Po-se there is a congee made from the three myrobalans (*san-lo tsiaṅ* 三勒漿),[5] resembling wine, and styled *an-mo-lo* 庵摩勒 (*āmalaka, Phyllanthus emblica*) or *p'i-li-lo* 毗梨勒 (*vibhītaka, Terminalia belerica*)." The source of this statement is not given. If Po-se in this case refers to Persia, it would go to show that the three myrobalans were known there.

On the other hand, there is quite a different explanation of the term *san-lo tsiaṅ*. According to Ma Či, who wrote in the tenth century, this is the designation for a wine obtained from a flower of sweet flavor, growing in the countries of the West and gathered by the Hu. The name of the flower is 陀得 *t'o-te*, *da-tik.[6] In this case the term *san-lo* may represent a transcription; it answers to ancient *sam-lak, sam-rak.

---

[1] *Sui šu*, Ch. 83, p. 7 b; *Čou šu*, Ch. 50, p. 6.

[2] Cf. *T'oung Pao*, 1915, pp. 275–276. *Ho-li-lo* were products of A-lo-yi-lo 阿羅偌羅 in the north of Uḍḍiyāna (*T'ai p'iṅ hwan yü ki*, Ch. 186, p. 12 b).

[3] Cf. G. FERRAND, Textes relatifs à l'Extrême-Orient, p. 227.

[4] Ed. of *T'aṅ Suṅ ts'uṅ šu*, p. 20.

[5] The *san lo* are the three plants the names of which terminate in *lo*,—*ho-li-lo* (*Terminalia chebula*), *p'i-li-lo* (*T. belerica*, Sanskrit *vibhītaka*, Persian *balīla*), and *a-mo-lo* or *an-mo-lo* (*Phyllanthus emblica*, Sanskrit *āmalaka*, Persian *amola*).

[6] The text is in the *T'u šu tsi č'eṅ*, XX, Ch. 182, *tsa hwa ts'ao pu, hui k'ao 2*, p. 13 b. I cannot trace it in the *Pen ts'ao kaṅ mu*.

378

THE "GOLD PEACH"

30. A fruit called yellow peach (*hwan t'ao* 黃桃) or gold peach (*kin t'ao* 金桃), of the size of a goose-egg, was introduced into China under the reign of the Emperor T'ai Tsuṅ of the T'ang (A.D. 629–649), being presented by the country K'aṅ 康 (Sogdiana).[1] This introduction is assigned to the year 647 in the *T'aṅ hui yao*,[2] where it is said that Sogdiana offered to the Court the yellow peach, being of the size of a goose-egg and golden in color, and hence styled also "gold peach." A somewhat earlier date for the introduction of this fruit is on record in the *Ts'e fu yüan kwei*,[3] which has the notice that in A.D. 625 (under the Emperor Kao Tsu) Sogdiana presented gold peaches (*kin t'ao*) and silver peaches (*yin t'ao*), and that by imperial order they were planted in the gardens. This fruit is not mentioned in the *Pen-ts'ao* literature; it is not known what kind of fruit it was. Maybe it was a peculiar variety of peach.

FU–TSE

31. *Fu-tse* 附子 is enumerated among the products of Sasanian Persia in the *Sui šu*.[4] *Pai* 白 *fu-tse* is attributed to the country Ts'ao (Jaguḍa) north of the Ts'uṅ-liṅ,[5] and to Ki-pin.[6]

In the form 付子 *fu-tse*, it occurs in a prescription written on a wooden tablet of the Han period, found in Turkistan.[7] *Fu-tse* 莤子 is identified with *Aconitum fischeri*, cultivated on a large scale in Čaṅ-miṅ hien in the prefecture of Lu-ṅan, Se-č'wan.[8] It is not known, however, that this species occurs in Persia.

Yi Tsiṅ calls attention to the fact that the medicinal herbs of India are not the same as those of China, and enumerates tubers of aconite together with *fu-tse* among the best drugs of China, and which are never found in India.[9]

---

[1] *Fuṅ ši wen kien ki*, Ch. 7, p. 1 b (ed. of *Ki fu ts'uṅ šu*).
[2] Ch. 200, p. 14; also *T'ai p'iṅ hwan yü ki*, Ch. 183, p. 3.
[3] Ch. 970, p. 8 b.
[4] Ch. 83, p. 7 b; also *Čou šu*, Ch. 50, p. 6.
[5] *Sui šu, ibid.*, p. 8 a.
[6] *T'ai p'iṅ hwan yü ki*, Ch. 182, p. 12 b.
[7] CHAVANNES, Documents de l'époque des Han, p. 115, No. 530.
[8] STUART, Chinese Materia Medica, p. 10.
[9] TAKAKUSU, Record of the Buddhist Religion, p. 148.

379

BRASSICA

32. Of the two species of mustard, *Brassica* or *Sinapis juncea* and *S. alba*, the former has always been a native of China (*kiai* 芥). The latter, however, was imported as late as the T'ang period. It is first mentioned by Su Kuṅ in the *Pen ts'ao* of the T'ang (about A.D. 650) as coming from the Western Žuṅ (Si Žuṅ),[1] a term which, as noted, frequently refers to Iranian regions. In the *Šu pen ts'ao* 蜀本草, published about the middle of the tenth century by Han Pao-šeṅ 韓保昇, we find the term 胡芥 *hu kiai* ("mustard of the Hu"). Č'en Ts'aṅ-k'i of the T'ang states that it grows in T'ai-yüan and Ho-tuṅ 河東 (Šan-si), without referring to the foreign origin. Li Ši-čen[2] annotates that this cultivation comes from the Hu and Žuṅ and abounds in Šu (Se-č'wan), hence the names *hu kiai* and *šu kiai* ("mustard of Se-č'wan"), while the common designation is *pai kiai* ("white mustard"). This state of affairs plainly reveals the fact that the plant was conveyed to China over the land-route of Central Asia, while no allusion is made to an oversea transplantation. As shown by me on a previous occasion,[3] the Si-hia word *si-na* ("mustard") appears to be related to Greek *sinapi*, and was probably carried into the Si-hia kingdom by Nestorian missionaries, who, we are informed by Marco Polo, were settled there. The same species was likewise foreign to the Tibetans, as is evidenced by their designation "white turnip" (*yuṅs-kar*). In India it is not indigenous, either: WATT[4] says that if met with at all, it occurs in gardens only within the temperate areas, or in upper India during the winter months; it is not a field crop.

This genus comprises nearly a hundred species, all natives of the north temperate zones, and most of them of ancient European cultivation (with an independent centre in China).

Abu Mansur[5] distinguishes under the Arabic name *karnab* five kinds of *Brassica*,— Nabathæan, *Brassica silvestris*, *B. marina*, *B. cypria*

---

[1] The same definition is given by T'aṅ Šen-wei in his *Čeṅ lei pen ts'ao* (Ch. 27, p. 15).

[2] *Pen ts'ao kaṅ mu*, Ch. 26, p. 12.

[3] *T'oung Pao*, 1915, p. 86.

[4] Commercial Products of India, p. 176.

[5] ACHUNDOW, Abu Mansur, p. 110.

380

(*qanbīt*) and Syrian from Mosul. He further mentions *Brassica rapa* under the name *šelgem* (Arabic *šaljam*).[1]

33. One of the synonymes of *yün-t'ai* 蕓薹 (*Brassica rapa*) is *hu ts'ai* 胡菜 ("vegetable of the Hu"). According to Li Ši-čen,[2] this term was first applied to this vegetable by Fu K'ien 服虔 of the second century A.D. in his *T'un su wen* 通俗文. If this information were correct, this would be the earliest example of the occurrence of the term Hu in connection with a cultivated plant; but this Hu does not relate to Iranians, for Hu Hia 胡洽, in his *Pai pin fan* 百病方, a medical work of the Sui period (A.D. 589–618), styles the plant *sai ts'ai* 塞菜, which, according to Li Ši-čen, has the same significance as *hu ts'ai*, and refers to 塞外 Sai-wai, the Country beyond the Passes, Mongolia. Some even believe that Yün-t'ai is a place-name in Mongolia, where this plant thrives, and that it received therefrom its name. Such localities abstracted from plant-names are usually afterthoughts and fictitious.[3] The term *yün-t'ai* occurs in the early work *Pie lu*.

SCHLIMMER[4] mentions *Brassica capitata* (Persian *kalam pīč*), *B. caulozapa* (*kalam gomri*), and *B. napus* or *rapa* (*šelgem*). I have already pointed out that the Persians were active in disseminating species of *Brassica* and *Raphanus* to Tibet, the Turks, and Mongolia.[5] Reference has been made above (p. 199) to the fact that *Brassica rapa* (*yün-t'ai*) was introduced into China from Turkish tribes of Mongolia under the Later Han dynasty, and it would be reasonable to conclude that these had previously received the cultivation from Iranians.[6] *Brassica rapa* is very generally cultivated in Persia and most parts of India during the dry season, from October until March.[7] *Yün-t'ai* is enumerated among the choice vegetables of the country 末禄 Mo-lu, *Mar-luk, in Arabia.[8]

The country of the Arabs produced the rape-turnip (*man-tsin* 蔓菁, *Brassica rapa-depressa*) with roots the size of a peck 斗, round, and of very sweet flavor.[9]

Yi Tsin, the Buddhist pilgrim of the seventh century, makes some comment on the difference between Indian and Chinese *Brassica* by saying,

---

[1] ACHUNDOW, Abu Mansur, p. 87.

[2] *Pen ts'ao kan mu*, Ch. 26, p. 9 b.

[3] Compare p. 401.

[4] Terminologie, p. 93.

[5] *T'oung Pao*, 1915, pp. 84, 87.

[6] The case would then be analogous to the history of the water-melon.

[7] W. ROXBURGH, Flora Indica, p. 497.

[8] *T'ai p'in hwan yü ki*, Ch. 186, p. 16 b.

[9] *Ibid.*, Ch. 186, p. 15 b.

"*Man-tsiṅ* occurs [in India] in sufficient quantity and in two varieties, one with white, the other with black seeds. In Chinese translation it is called mustard (*kie-tse* 芥子). As in all countries, oil is pressed from it for culinary purposes. When eating it as a vegetable, I found it not very different from the *man-tsiṅ* of China; but as regards the root, which is rather tough, it is not identical with our *man-tsiṅ*. The seeds are coarse, and again bear no relation to mustard-seeds. They are like those of *Hovenia dulcis* (*či-kü* 枳椇), transformed in their shape in consequence of the soil."[1]

---

[1] This sentence is entirely misunderstood by J. TAKAKUSU in his translation of Yi Tsiṅ's work (p. 44), where we read, "The change in the growth of this plant is considered to be something like the change of an orange-tree into a bramble when brought north of the Yangtse River." The text has: 其猶枳橘因地遷形. There is nothing here about an orange or a bramble or the Yangtse. The character 橘 is erroneously used for 椇, as is still the case in southern China (see STUART, Chinese Materia Medica, p. 209), and 枳椇 is a well-known botanical name for a rhamnaceous tree (not an orange), *Hovenia dulcis*. "Change of an orange-tree into a bramble" is nonsense in itself.

CUMMIN

34. Under the foreign term 蒔蘿 *ši-lo*, **ži-la*, the Chinese have not described the fennel (*Foeniculum vulgare*), as erroneously asserted by WATTERS[1] and STUART,[2] but cummin (*Cuminum cyminum*) and caraway (*Carum carui*). This is fundamentally proved by the prototype, Middle Persian *žīra* or *zīra*, Sanskrit *jīra*, of which *ši-lo* (**ži-la*) forms the regular transcription.[3] In India, *jīra* refers to both cummin and caraway.[4] Although *Cuminum* is more or less cultivated in most provinces of India, except Bengal and Assam, there is, according to WATT, fairly conclusive evidence that it is nowhere indigenous; but in several districts it would appear to be so far naturalized as to have been regarded as "wild," even by competent observers. No doubt, it was transmitted to India from Iran. Cummin was known to the ancient Persians, being mentioned in the inscription of Cyrus at Persepolis,[5] and at an early period penetrated from Iran to Egypt on the one hand, and to India on the other.[6]

Avicenna distinguishes four varieties of cummin (Arabic *kammūn*),[7] — that of Kirmān, which is black; that of Persia, which is yellow and more active than the others; that of Syria, and the Nabathæan.[8] Each variety is both spontaneous and cultivated. Abu Mansur regards that of Kirmān as the best, and styles it *zīre-i kirmān*.[9] This name, according to SCHLIMMER,[10] would refer to caraway, also called *zīre-i siah*,[11] while cummin is styled in Persian *zīre-i sebze* or *sefīd*. Caraway (*Carum*

---

[1] Essays on the Chinese Language, p. 440. He even adds "coriander," which is *hu swi* (p. 297).

[2] Chinese Materia Medica, p. 176. Fennel is *hwi hiaṅ* 茴 香, while a synonyme of cummin is *siao hwi hiaṅ* ("small fennel").

[3] In the same form, the word occurs in Tibetan, *zi-ra* (*T'oung Pao*, 1916, p. 475).

[4] G. WATT, Commercial Products of India, p. 442.

[5] JORET, Plantes dans l'antiquité, Vol. II, p. 66.

[6] *Ibid.*, p. 258.

[7] Hebrew *kammōn*, Assyrian *kamanu*, resulting in Greek κύμινον, Latin *cumīnum*, *cymīnum*, or *cimīnum*; Armenian *caman*; Persian *kamūn*.

[8] LECLERC, Traité des simples, Vol. III, p. 196.

[9] ACHUNDOW, Abu Mansur, pp. 112, 258.

[10] Terminologie, p. 112.

[11] In India, the Persian word *siah* refers to the black caraway (*Carum bulbocastanum*), which confirms Schlimmer's opinion. Also Avicenna's black cummin of Kirmān apparently represents this species. This plant is a native of Baluchistan, Afghanistan, Kashmir, and Lahūl, mainly occurring as a weed in cultivated land.

383

carui), however, is commonly termed in Persian *šāh-zīre* ("cummin of the Shah") or *zīre-i rūmī* ("Byzantine or Turkish cummin").[1]

While the philological evidence would speak in favor of a transmission of cummin from Persia to China, this point is not clearly brought out by our records. Č'en Ts'aṅ-k'i, who wrote in the first half of the eighth century, states that *ši-lo* grows in Fu-ši 佛誓 (Bhoja, Sumatra). Li Sün, in his *Hai yao pen ts'ao*, says after the *Kwaṅ čou ki* 廣州記 that the plant grows in the country Po-se;[2] and Su Suṅ of the Sung notes that in his time it occurred in Liṅ-nan (Kwaṅ-tuṅ) and adjoining regions. Now, the *Kwaṅ čou ki* is said to have been written under the Tsin dynasty (A.D. 265–420);[3] and, as will be shown below in detail, the Po-se of Li Sün almost invariably denotes, not Persia, but the Malayan Po-se. Again, it is Li Sün who does not avail himself of the Iranian form *ši-lo = žīra*, but of the Sanskrit form *jīraka*, possibly conveyed through the medium of the Malayan Po-se.

Li Ši-čen has entered under *ši-lo* another foreign word in the form 慈謀勒 *ts'e-mou-lo* (*dži-mu-lak), which he derived from the *K'ai pao pen ts'ao*, and which, in the same manner as *ši-lo*, he stamps as a foreign word. This transcription has hitherto defied identification,[4] because it is incorrectly recorded. It is met with correctly in the *Čen lei pen ts'ao*[5] in the form 慈勒 *ts'e-lo*, *dži-lak(rak), and this answers to Sanskrit *jīraka*. This form is handed down in the *Hai yao pen ts'ao*, written by Li Sün in the eighth century. Thus we have, on the one hand a Sanskrit form *jīraka*, conveyed by the Malayan Po-se to Kwaṅ-tuṅ in the T'ang period, and on the other hand the Iranian type *ši-lo = žīra*, which for phonetic reasons must likewise go back to the era of the T'ang, and which we should suppose had migrated overland to China. The latter point, for the time being, remains an hypothesis, which will perhaps be elucidated by the documents of Turkistan.

---

[1] Corresponding to Arabic *karāwyā*, the source of our word caraway.

[2] The *Čen lei pen ts'ao* (Ch. 13, p. 27 b) repeats this without citing a source.

[3] Cf. below, p. 475.

[4] STUART, Chinese Materia Medica, p. 176.

[5] Ch. 13, p. 17 b.

THE DATE–PALM

35. The Chinese records of the date-palm (*Phoenix dactylifera*) contain two points that are of interest to science: first, a contribution to the geographical distribution of the tree in ancient times; and, second, a temporary attempt at acclimating it in China. The tree is not indigenous there. It is for the first time in the T'ang period that we receive some information about it; but it is mentioned at an earlier date as a product of Sasanian Persia in both the *Wei šu* and *Sui šu*, under the name *ts'ien nien tsao* 千年棗 ("jujubes of thousand years," the jujube, *Zizyphus vulgaris*, being a native of China).[1] In the *Yu yan tsa tsu*,[2] the date is styled *Po-se tsao* 波斯棗 ("Persian jujube"), with the observation that its habitat is in Po-se (Persia), or that it comes from there.[3] The Persian name is then given in the form 窟莽 *k'u-man*, *k'ut(k'ur)-man*, which would correspond to a Middle Persian *xurman (*khurmang), Pāzand and New Persian *xurmā*, that was also adopted by Osmanli and Neo-Greek, χουρμᾶs ("date") and κουρμαδηά ("date-palm"), Albanian *korme*.[4] The *T'aṅ šu*[5] writes the same word 鶻莽 hu-maṅ, *guδ(gur)-maṅ, answering to a Middle-Persian form *gurmaṅ or *kurmaṅ. The New-Persian word is rendered 苦魯麻 *k'u-lu(ru)-ma* in the *Pen ts'ao kaṅ mu*;[6] this is the style of the Yüan transcriptions,[7]

---

[1] This name was bestowed upon the tree, not, as erroneously asserted by HIRTH (Chau Ju-kua, p. 210), "evidently on account of the stony hardness of the dates on reaching China," but, as stated in the *Pen ts'ao kaṅ mu* (Ch. 31, p. 8), owing to the long-enduring character of the tree 其 樹 性 耐 久 也. The same explanation holds good for the synonyme *wan sui tsao* ("jujube of ten thousand or numerous years"). Indeed, this palm lives to a great age, and trees of from one to two hundred years old continue to produce their annual crop.

[2] Ch. 18, p. 10.

[3] The same term, *Po-se tsao*, appears in a passage of the *Pei hu lu* (Ch. 2, p. 9 b), where the trunk and leaves of the sago-palm (*Sago rumphii*) are compared with those of the date.

[4] In Old Armenian of the fifth century we have the Iranian loan-word *armav*, and hence it is inferred that the *x* of Persian was subsequently prefixed (HÜBSCH-MANN, Persische Studien, p. 265; Armen. Gram., p. 111). The date of the Chinese transcriptions proves that the initial *x* existed in Pahlavi.

[5] Ch. 221 B, p. 13.

[6] Ch. 31, p. 21. It is interesting to note that Li Ši-čen endeavors to make out a distinction between *k'u-maṅ* and *k'u-lu-ma* by saying that the former denotes the tree, the latter the fruit; but both, in his opinion, are closely allied foreign words.

[7] The T'ang transcription, of course, is not "probably a distorted transcription of *khurma*," as asserted by BRETSCHNEIDER (Chinese Recorder, 1871, p. 266), but, on the contrary, is very exact.

385

and first occurs in the *Čo ken lu* 輟耕錄, published in 1366. The Persian word has also migrated into the modern Aryan languages of India, as well as into the Malayan group: Javanese *kurma;* Čam *kuramŏ;* Malayan, Dayak, and Sunda *korma;* Bugi and Makassar *koromma;* also into Khmer: *romŏ, lomŏ, amŏ.*

Following is the description of the tree given in the *Yu yan tsa tsu:* "It is thirty to forty feet in height,[1] and has a circumference of from five to six feet. The leaves resemble those of the *t'u t'en* 土藤 (a kind of rattan), and remain ever green. It blooms in the second month. The blossoms are shaped like those of the banana, and have a double bottom. They open gradually; and in the fissure are formed more than ten seed-cases, two inches long, yellow and white in color. When the kernel ripens, the seeds are black. In their appearance they resemble dried jujubes. They are good to eat and as sweet as candy."

Another foreign word for the date is handed down by Č'en Ts'an-k'i in his *Pen ts'ao ši i,* in the form 無漏 *wu-lou,* *bu-nu. He identifies this term with the "Persian jujube," which he says grows in Persia, and has the appearance of a jujube. Li Ši-čen annotates that the meaning of this word is not yet explained. Neither Bretschneider nor any one else has commented on this name. It is strikingly identical with the old Egyptian designation of the date, *bunnu.*[2] It is known that the Arabs have an infinite number of terms for the varieties of the date and the fruit in its various stages of growth, and it may be that they likewise adopted the Egyptian word and transmitted it to China. The common Arabic names are *nakhl* and *tamr* (Hebrew *tamar,* Syriac *temar*). On the other hand, the relation of *wu-lou* to the Egyptian word may be accidental, if we assume that *wu-lou* was originally the designation of *Cycas revoluta* (see below), and was only subsequently transferred to the date-palm.

The *Lin piao lu i*[3] by Liu Sūn contains the following interesting account:—

"In regard to the date ('Persian jujube'), this tree may be seen in the suburbs of Kwan-čou (Canton). The trunk of the tree is entirely without branches, is straight, and rises to a height of from thirty to forty feet. The crown of the tree spreads in all directions, and forms over ten branches. The leaves are like those of the 'sea coir-palm'

---

[1] It even grows to a height of sixty or eighty feet.

[2] V. LORET, Flore pharaonique, p. 34. I concur with Loret in the opinion that the Egyptian word is the foundation of Greek φοῖνιξ. The theory of HEHN (Kulturpflanzen, p. 273) and upheld by SCHRADER (ibid., p. 284), that the latter might denote the Phœnician tree, does not seem to me correct.

[3] Ch. B, p. 4 (see above, p. 268).

(*hai tsun* 海棕, *Chamaerops excelsa*).[1] The trees planted in Kwan-čou bear fruit once in three or five years. The fruits resemble the green jujube growing in the north, but are smaller. They turn from green to yellow. When the leaves have come out, the fruit is formed in clusters, each cluster generally bearing from three to twenty berries, which require careful handling. The foreign as well as the domestic kind is consumed in our country. In color it resembles that of granulated sugar. Shell and meat are soft and bright. Baked into cakes or steamed in water, they are savory. The kernel is widely different from that of the jujube of the north. The two ends are not pointed [as in the jujube], but doubly rolled up and round like a small piece of red kino 紫礦.[2] They must be carefully handled. When sown, no shoots sprout forth for a long time, so that one might suppose they would never mature."

The date is clearly described in this text; and we learn from it that the tree was cultivated in Kwan-tun, and its fruit was also imported during the T'ang period. As Liu Sün, author of that work, lived under the Emperor Čao Tsun (A.D. 889–904), this notice refers to the end of the ninth century.[3] A. DE CANDOLLE[4] states erroneously that the Chinese received the tree from Persia in the third century of our era.

In his note on the date, headed by the term *wu-lou tse*, Li Ši-čen[5] has produced a confusion of terms, and accordingly brought together

---

[1] In the text of this work, as cited in the *Pen ts'ao kan mu*, this clause is worded as follows: "The leaves are like those of the *tsun-lü* 棕櫚 (*Chamaerops excelsa*), and hence the people of that locality style the tree [the date] *hai tsun* ('sea,' that is, 'foreign coir-palm')." This would indeed appear more logical than the passage above, rendered after the edition of *Wu yin tien*, which, however, must be regarded as more authoritative. Not only in this extract, but also in several others, does the *Pen ts'ao kan mu* exhibit many discrepancies from the *Wu yin tien* edition; this subject should merit closer study. In the present case there is only one other point · worthy of special mention; and this is, that Li Ši-čen, in his section of nomenclature, gives the synonyme 番棗 *fan tsao* ("foreign jujube") with reference to the *Lin piao lu i*. This term, however, does not occur in the text of this work as transmitted by him, or in the *Wu yin tien* edition. The latter has added a saying of the Emperor Wen 文 of the Wei dynasty, which has nothing to do with the date, and in which is found the phrase 凡棗 *fan tsao* ("all jujubes"). In other editions, *fan* ("foreign") was perhaps substituted for this *fan*, so that the existence of the synonyme established by Li and adopted by Bretschneider appears to be very doubtful.

[2] See below, p. 478.

[3] It is singular that Bretschneider, who has given a rather uncritical digest of the subject from the *Pen ts'ao*, does not at all mention this transplantation of the tree. To my mind, this is the most interesting point to be noted. Whether date-palms are still grown in Kwan-tun, I am not prepared to say; but, as foreign authors do not mention the fact, I almost doubt it.

[4] *Origin of Cultivated Plants*, p. 303.

[5] *Pen ts'ao kan mu*, Ch. 31, p. 8.

a number of heterogeneous texts. BRETSCHNEIDER[1] has accepted all this in good faith and without criticism. It is hardly necessary to be a botanist in order to see that the texts of the *Nan fan ts'ao mu čwan* and *Čo ken lu*, alleged to refer to the date, bear no relation to this tree.[2] The *hai tsao* 海棗 described in the former work[3] may very well refer to *Cycas revoluta*.[4] The text of the other book, which Bretschneider does not quote by its title, and erroneously characterizes as "a writer of the Ming," speaks of six "gold fruit" (*kin kwo* 金果) trees growing in Č'en-tu, capital of Se-č'wan, and, according to an oral tradition, planted at the time of the Han. Then follows a description of the tree, the foreign name of which is given as *k'u-lu-ma* (see above), and which, according to Bretschneider, suits the date-palm quite well. It is hardly credible, however, that this tree could ever thrive in the climate of Se-č'wan, and Bretschneider himself admits that the fruit of *Salisburia adiantifolia* now bears also the name *kin kwo*. Thus, despite the fact that the Persian name for the date is added, the passage of the *Čo ken lu* is open to the suspicion of some misunderstanding.

Not only did the Chinese know that the date is a product of Persia, but they knew also that it was utilized as food by certain tribes of the

---

[1] *Chinese Recorder*, 1871, pp. 265–267.

[2] Bretschneider, it should be understood, was personally acquainted with only the flora of Peking and its environment; for the rest, his familiarity with Chinese plants was mere book-knowledge, and botany as a science was almost foreign to him. Research in the history of cultivated plants was in its very beginning in his days; and his methods relating to such subjects were not very profound, and were rather crude.

[3] Ch. B, p. 4. Also Wu K'i-tsūn, author of the *Či wu min ši t'u k'ao* (Ch. 17, p. 21), has identified the term *wu-lou-tse* with *hai tsao*.

[4] STUART, Chinese Materia Medica, p. 140; but Stuart falls into the other extreme by identifying with this species also the terms *Po-se tsao, ts'ien nien tsao*, etc., which without any doubt relate to the date. In Bretschneider's translation of the above text there is a curious misunderstanding. We read there, "In the year 285 A.D. Lin-yi offered to the Emperor Wu-ti a hundred trees of the *hai tsao*. The prince Li-sha told the Emperor that in his travels by sea he saw fruits of this tree, which were, without exaggeration, as large as a melon." The text reads, "In the fifth year of the period T'ai-k'an (A.D. 284), Lin-yi presented to the Court a hundred trees. Li Šao-kūn 李少君 (the well-known magician) said to the Emperor Wu of the Han, 'During my sea-voyages I met Nan-k'i Šen 安期生 (the magician of the Blest Islands), who ate jujubes of the size of a gourd, which is by no means an exaggeration.'" The two events are not interrelated; the second refers to the second century B.C. Neither, however, has anything to do with the date. The working of Chinese logic is visibly manifest: the sea-travels of Li Šao-kūn are combined with his fabulous jujube into the sea-jujube (*hai tsao*), and this imaginary product is associated with a real tree of that name. Li Ši-čen's example shows at what fancies the Chinese finally arrive through their wrong associations of ideas; and Bretschneider's example finally demonstrates that any Chinese data must first be taken under our microscope before being accepted by science.

East-African coast. The early texts relating to Ta Ts'in do not mention the palm; but at the end of the article Fu-lin (Syria), the *T'aṅ šu* speaks of two countries, 磨鄰 Mo-lin (*Mwa-lin, Mwa-rin) and 老勃薩 Lao-p'o-sa (*Lav-bwiδ-sar), as being situated 2000 *li* south-west of Fu-lin, and sheltering a dark-complexioned population. The land is barren, the people feed their horses on dried fish, and they themselves subsist on dates.[1] BRETSCHNEIDER[2] was quite right in seeking this locality in Africa, but it is impossible to accept his suggestion that "perhaps the Chinese names Mo-lin and Lao-p'o-sa are intended to express the country of the Moors (Mauritania) or Lybia." HIRTH[3] did not discuss this weak theory, and, while locating the countries in question along the west coast of the Red Sea, did not attempt to identify the transcriptions. According to Ma Twan-lin, the country Mo-lin is situated south-west of the country 秧薩羅 Yaṅ-sa-lo, which Hirth tentatively equated with Jerusalem. This is out of the question, as Yaṅ-sa-lo answers to an ancient Aṅ-saδ(sar)-la(ra).[4] Moreover, it is on record in the *T'ai p'iṅ hwan yü ki*[5] that Mo-lin is south-west of 勃薩羅 P'o-sa-lo (*Bwiδ-saδ-la), so that this name is clearly identical with that of Ma Twan-lin and the transcription of the T'ang Annals. In my opinion, the transcription *Mwa-lin is intended for the Malindi of Edrīsī or Mulanda of Yāqūt, now Malindi, south of the Equator, in Seyidieh Province of British East Africa. Edrīsī describes this place as a large city, the inhabitants of which live by hunting and fishing. They salt sea-fish for trade, and also exploit iron-mines, iron being the source of their wealth.[6] If this identification be correct, the geographical definition of the T'ang Annals (2000 *li* south-west of Fu-lin) is, of course, deficient; but we must not lose sight of the fact that these data rest on a hearsay report hailing from Fu-lin, and that, generally speaking, Chinese calculations of distances on sea-routes are not to be taken too seriously.[7] Under the Ming, the same country appears as 麻林 Ma-lin, the king of which sent an embassy to China in 1415 with a gift of

---

[1] In the transcription *hu-maṅ*, as given above, followed by the explanation that this is the "Persian jujube." The date is not a native of eastern Africa, nor does it thrive in the tropics, but it was doubtless introduced there by the Arabs (cf. F. STORBECK, *Mitt. Sem. Or. Spr.*, 1914, II, p. 158; A. ENGLER, Nutzpflanzen Ost-Afrikas, p. 12).

[2] Knowledge possessed by the Chinese of the Arabs, p. 25.

[3] China and the Roman Orient, p. 204.

[4] If Mo-lin was on the littoral of the Red Sea, it would certainly be an absurdity to define its location as south-west of Jerusalem.

[5] Ch. 184, p. 3.

[6] Dozy and DE GOEJE, Edrīsī's description de l'Afrique, p. 56 (Leiden, 1866).

[7] Cf. Chinese Clay Figures, pp. 80–81, note.

giraffes.[1] It likewise appears in the list of countries visited by Čeṅ Ho,[2] where Ma-lin and La-sa 刺撒 are named, the latter apparently being identical with the older Lao-p'o-sa.[3]

The Chinese knew, further, that the date thrives in the country of the Arabs (Ta-ši),[4] further, in Oman, Basra, and on the Coromandel Coast.[5] It is pointed out, further, for Aden and Ormuz.[6]

There is no doubt that the date-palm has existed in southern Persia from ancient times, chiefly on the littoral of the Persian Gulf and in Mekrān, Baluchistan. It is mentioned in several passages of the Būndahišn.[7] Its great antiquity in Babylonia also is uncontested (Assyrian *gišimmaru*).[8] Strabo[9] reports how Alexander's army was greatly distressed on its march through the barren Gedrosian desert. The supplies had to come from a distance, and were scanty and unfrequent, so much so that the army suffered greatly from hunger, the beasts of burden dropped, and the baggage was abandoned. The army was saved by the consumption of dates and the marrow of the palmtree.[10] Again he tells us that many persons were suffocated by eating unripe dates.[11] Philostratus speaks of a eunuch who received Apollonius of Tyana when he entered the Parthian kingdom, and offered him dates of amber color and of exceptional size.[12] In the Province of Fars, the date-palm is conspicuous almost everywhere.[13] In Babylon, Persian and Aramaic date-palms were distinguished, the former being held in greater esteem, as their meat perfectly detaches itself from the stone, while it partially adheres in the Aramaic date.[14] The same distinction

---

[1] *Ta Miṅ i l'uṅ či*, Ch. 90, p. 24.

[2] *Miṅ ši*, Ch. 304.

[3] It is not Ma-liṅ-la-sa, the name of a single country, as made out by GROENE-VELDT (Notes on the Malay Archipelago, p. 170).

[4] *T'ai p'iṅ hwan yü ki*, Ch. 186, p. 15 b.

[5] HIRTH, Chau Ju-kua, pp. 133, 137, 96.

[6] ROCKHILL, T'oung Pao, 1915, p. 609. The word *to-ša-pu*, not explained by him, represents Arabic *dūšāb* ("date-wine"); see LECLERC, Traité des simples, Vol. II, p. 49). NÖLDEKE (Persische Studien, II, p. 42) explains this word from *dūš* ("honey") and Persian *āb* ("water").

[7] Above, p. 193.

[8] Herodotus, I, 193; E. BONAVIA, Flora of the Assyrian Monuments, p. 3; HANDCOCK, Mesopotamian Archæology, pp. 12–13.

[9] XV, 2, § 7.

[10] Cf. Theophrastus, Histor. plant., IV. IV, 13.

[11] *Ibid.*, IV. IV, 5; and Pliny, XIII, 9.

[12] C. JORET, Plantes dans l'antiquité, Vol. II, p. 93.

[13] G. LE STRANGE, Description of the Province of Fars, pp. 31, 33, 35, 39, 40, etc.

[14] I. LOEW, Aramaeische Pflanzennamen, p. 112.

was made in the Sasanian empire: in the tax laws of Khosrau I (A.D. 531–578), four Persian date-palms were valued and taxed equally with six common ones.[1] As already remarked, the *Wei* and *Sui* Annals attribute the date to Sasanian Persia, and the date is mentioned in Pahlavi literature (above, p. 193). At present dates thrive in the low plains of Kerman and of the littoral of the Persian Gulf; but the crops are insufficient, so that a considerable importation from Bagdad takes place.[2]

A. DE CANDOLLE[3] asserts, "No Sanskrit name is known, whence it may be inferred that the plantations of the date-palm in western India are not very ancient. The Indian climate does not suit the species." There is the Sanskrit name *kharjūra* for *Phoenix sylvestris*, that already occurs in the Yajurveda.[4] This is the wild date or date-sugar palm, which is indigenous in many parts of India, being most abundant in Bengal, Bihar, on the Coromandel Coast, and in Gujarat. The edible date (*P. dactylifera*) is cultivated and self-sown in Sind and the southern Panjāb, particularly near Multan, Muzaffargarh, the Sind Sagar Doab, and in the Trans-Indus territory. It is also grown in the Deccan and Gujarat.[5] Its Hindī name is *khajūra*, Hindustānī *khajūr*, from Sanskrit *kharjūra*. It is also called *sindhi, seindi, sendri*, which names allude to its origin from Sind. Possibly Sanskrit *kharjūra* and Iranian *khurma(ṅ)*, at least as far as the first element is concerned, are anciently related.

---

[1] NÖLDEKE, Tabari, p. 245.

[2] SCHLIMMER, Terminologie, p. 175.

[3] Origin of Cultivated Plants, p. 303.

[4] MACDONELL and KEITH, Vedic Index, Vol. I, p. 215.

[5] G. WATT, Commercial Products of India, pp. 883, 885.

THE SPINACH

36. In regard to the spinach (*Spinacia oleracea*), BRETSCHNEIDER[1] stated that "it is said to come from Persia. The botanists consider western Asia as the native country of spinach, and derive the names *Spinacia, spinage, spinat, épinards*, from the spinous seeds; but as the Persian name is *esfinadsh*, our various names would seem more likely to be of Persian origin." The problem is not quite so simple, however. It is not stated straightforwardly in any Chinese source that the spinach comes from Persia; and the name "Persian vegetable" (*Po-se ts'ai*) is of recent origin, being first traceable in the *Pen ts'ao kan mu*, where Li Ši-čen himself ascribes it to a certain Fan Ši-yin 方士隱.

Strangely enough, we get also in this case a taste of the Čan-K'ien myth. At least, H. L. JOLY[2] asserts, "The *Chinese and Japanese Repository* says that Chang K'ien brought to China the spinach." The only Chinese work in which I am able to find this tradition is the *T'un či* 通志,[3] written by Čen Tsiao 鄭樵 of the Sung dynasty, who states in cold blood that Čan K'ien brought spinach over. Not even the *Pen ts'ao kan mu* dares repeat this fantasy. It is plainly devoid of any value, in view of the fact that spinach was unknown in the west as far back as the second century B.C. Indeed, it was unfamiliar to the Semites and to the ancients. It is a cultivation that comes to light only in mediæval times.

In perfect agreement with this state of affairs, spinach is not mentioned in China earlier than the T'ang period. As regards the literature on agriculture, the vegetable makes its first appearance in the *Čun šu šu* 種樹書, written toward the end of the eighth century.[4] Here it is stated that the spinach, *po-lin* 菠薐 (*pwa-lin), came from the country Po-lin 菠薐國 (*Pwa-lin, Palinga).

The first *Pen ts'ao* that speaks of the spinach is the *Čen lei pen ts'ao* written by T'an Šen-wei in A.D. 1108.[5] This Materia Medica describes altogether 1746 articles, compared with 1118 which are treated in the *Kia yu pu ču pen ts'ao* (published in the period Kia-yu, A.D. 1056–64), so that 628 new ones were added. These are expressly so designated in

---

[1] *Chinese Recorder*, 1871, p. 223.
[2] Legend in Japanese Art, p. 35.
[3] Ch. 75, p. 32 b.
[4] BRETSCHNEIDER, Bot. Sin., pt. 1, p. 79.
[5] Ch. 29, p. 14 b (print of 1587).

392

the table of contents preceding each chapter, and spinach ranks among these novelties. Judging from the description here given, it must have been a favorite vegetable in the Sung period. It is said to be particularly beneficial to the people in the north of China, who feed on meat and flour (chiefly in the form of vermicelli), while the southerners, who subsist on fish and turtles, cannot eat much of it, because their water food makes them cold, and spinach brings about the same effect.[1] The *Kia yü* (or *hwa*) *lu* 嘉語 (or 話) 錄 by Liu Yū-si 劉禹錫 (A.D. 772-842) is cited to the effect that "*po-liň* 波薐 was originally in the western countries, and that its seeds came thence to China[2] in the same manner as alfalfa and grapes were brought over by Čaň K'ien. Originally it was the country of Po-liň 頗陵, and an error arose in the course of the transmission of the word, which is not known to many at this time."

The first and only historical reference to the matter that we have occurs in the *T'aň hui yao*,[3] where it is on record, "At the time of the Emperor T'ai Tsuň (A.D. 627-649), in the twenty-first year of the period Čeň-kwan (A.D. 647), Ni-p'o-lo (Nepal) sent to the Court the vegetable *po-liň* 波稜, resembling the flower of the *huň-lan* 紅藍 (*Carthamus tinctorius*), the fruit being like that of the *tsi-li* 蒺藜 (*Tribulus terrestris*). Well cooked, it makes good eating, and is savory."[4]

This text represents not only the earliest datable mention of the vegetable in Chinese records, but in general the earliest reference to it that we thus far possess. This document shows that the plant then was a novelty not only to the Chinese, but presumably also to the people of Nepal; otherwise they would not have thought it worthy of being sent as a gift to China, which was made in response to a request of the

---

[1] JOHN GERARDE (The Herball or Generall Historie of Plantes, p. 260, London, 1597) remarks, "Spinach is evidently colde and moist, almost in the second degree, but rather moist. It is one of the potherbes whose substance is waterie."

[2] According to another reading, a Buddhist monk (*seň*) is said to have brought the seeds over, which sounds rather plausible. G. A. STUART remarks that the herb is extensively used by the monks in their lenten fare.

[3] Ch. 200, p. 14 b (also Ch. 100, p. 3 b). Cf. *Ts'e fu yüan kwei*, Ch. 970, p. 12, and *Pei hu lu*, Ch. 2, p. 19 b (ed. of Lu Sin-yüan).

[4] The *T'ai p'iň yü lan* (Ch. 980, p. 7) attributes this text to the T'ang Annals. It is not extant, however, in the account of Nepal inserted in the two *T'aň šu*, nor in the notice of Nepal in the *T'aň hui yao*. *Pen ts'ao kaň mu*, *T'u šu tsi č'eň*, and *Či wu miň ši t'u k'ao* (Ch. 5, p. 37) correctly cite the above text from the *T'aň hui yao*, with the only variant that the leaves of the *po-liň* resemble those of the *huň-lan*. The *Fuň ši wen kien ki* (Ch. 7, p. 1 b) by Fuň Yen of the ninth century (above, p. 232), referring to the same introduction, offers a singular name for the spinach in the form 波羅拔藻 *po-lo-pa-tsao*, *pa-la-bat-tsaw, or, if *tsao*, denoting several aquatic plants, does not form part of the transcription, *pa-la-bat(bar).

Emperor T'ai Tsuṅ that all tributary nations should present their choicest vegetable products. Yüan Wen 袁文, an author of the Sung period, in his work *Weṅ yu kien p'iṅ* 甕牖閒評,[1] states that the spinach (*po-liṅ*) comes from (or is produced in) the country Ni-p'o-lo (Nepal) in the Western Regions.[2] The *Kia yu pen ts'ao*, compiled in A.D. 1057, is the first Materia Medica that introduced the spinach into the pharmacopœia.[3]

The colloquial name is *po ts'ai* 菠菜 ("*po* vegetable"), *po* being abbreviated for *po-liṅ*. According to Waṅ Ši-mou 王世懋 (who died in 1591), in his *Kwa su su* 瓜蔬疏, the current name in northern China is *č'i ken ts'ai* 赤根菜 ("red-root vegetable"). The *Kwaṅ k'ün faṅ p'u* uses also the term *yiṅ-wu ts'ai* ("parrot vegetable"), named for the root, which is red, and believed to resemble a parrot. Aside from the term *Po-se ts'ai*, the *Pen ts'ao kaṅ mu ši i*[4] gives the synonymes *huṅ ts'ai* 紅菜 ("red vegetable") and *yaṅ* 洋 *ts'ai* ("foreign vegetable"). Another designation is *šan-hu ts'ai* ("coral vegetable").

A rather bad joke is perpetrated by the *Min šu* 閩書, a description of Fu-kien Province written at the end of the sixteenth or beginning of the seventeenth century, where the name *po-liṅ* is explained as 波棱 *po leṅ* ("waves and edges"), because the leaves are shaped like wave-patterns and have edges. There is nothing, of course, that the Chinese could not etymologize.[5]

There is no account in the traditions of the T'ang and Sung periods to the effect that the spinach was derived from Persia; and in view of the recent origin of the term "Persian vegetable," which is not even explained, we are tempted at the outset to dismiss the theory of a Persian origin. STUART[6] even goes so far as to say that, "as the Chinese have a tendency to attribute everything that comes from the south-west to Persia, we are not surprised to find this called *Po-se ts'ao*, 'Per-

---

[1] Ch. 4, p. 11 b (ed. of *Wu yiṅ tien*, 1775).

[2] 波稜出西域泥婆羅國. This could be translated also, "in the Western Regions and in the country Ni-p'o-lo."

[3] *Či wu miṅ ši t'u k'ao*, Ch. 4, p. 38 b.

[4] Ch. 8, p. 87 b.

[5] Of greater interest is the following fact recorded in the same book. The spinach in the north of China is styled "bamboo (*ču* 竹) *po-liṅ*," with long and bitter stems; that of Fu-kien is termed "stone (*ši* 石) *po-liṅ*," and has short and sweet stems.—The *Min šu*, in 154 chapters, was written by Ho K'iao-yüan 何喬遠 from Tsin-kiaṅ in Fu-kien; he obtained the degree of *tsin ši* in 1586 (cf. Cat. of the Imperial Library, Ch. 74, p. 19).

[6] Chinese Materia Medica, p. 417.

sian vegetable.' "[1] There is, however, another side to the case. In all probability, as shown by A. DE CANDOLLE,[2] it was Persia where the spinach was first raised as a vegetable; but the date given by him, "from the time of the Graeco-Roman civilization," is far too early.[3] A. de Candolle's statement that the Arabs did not carry the plant to Spain has already been rectified by L. LECLERC;[4] as his work is usually not in the hands of botanists or other students using de Candolle, this may aptly be pointed out here.

According to a treatise on agriculture (*Kitāb el-falāha*) written by Ibn al-Awwām of Spain toward the end of the eleventh century, spinach was cultivated in Spain at that time.[5] Ibn Haddjāj had then even written a special treatise on the cultivation of the vegetable, saying that it was sown at Sevilla in January. From Spain it spread to the rest of Europe. Additional evidence is afforded by the very name of the plant, which is of Persian origin, and was carried by the Arabs to Europe. The Persian designation is *aspanāh, aspanāj* or *asfināj;* Arabic *isfenāh* or *isbenāh.* Hence Mediæval Latin *spinachium* or *spinarium,*[6] Spanish

---

[1] The outcry of WATTERS (Essays on the Chinese Language, p. 347) against the looseness of the term Po-se, and his denunciation of the "Persian vegetable" as "an example of the loose way in which the word is used," are entirely out of place. It is utterly incorrect to say that "they have made it include, beside Persia itself, Syria, Turkey, and the Roman Empire, and sometimes they seem to use it as a sort of general designation for the abode of any barbarian people to the south-west of the Middle Kingdom." Po-se is a good transcription of Pārsa, the native designation of Persia, and strictly refers to Persia and to nought else. When F. P. Smith applied the name *po-ts'ai* to *Convolvulus reptans,* this was one of the numerous confusions and errors to which he fell victim. Likewise is it untrue, as asserted by Watters, that the term has been applied even to beet and carrot and other vegetables not indigenous in Persia. As on so many other points, Watters was badly informed on this subject also.

[2] Origin of Cultivated Plants, pp. 98–100.

[3] This conclusion, again, is the immediate outcome of Bretschneider's Chang-kienomania: for A. DE CANDOLLE says, "Bretschneider tells us that the Chinese name signifies 'herb of Persia,' and that Western vegetables were commonly introduced into China a century before the Christian era."

[4] Traité des simples, Vol. I, p. 61.

[5] L. LECLERC, Histoire de la médecine arabe, Vol. II, p. 112. The Arabic work has been translated into French by CLÉMENT-MULLET under the title Ibn al Awwam, le livre de l'agriculture (2 vols., Paris, 1864–67). De Candolle's erroneous theory that "the European cultivation must have come from the East about the fifteenth century," unfortunately still holds sway, and is perpetuated, for instance, in the last edition of the Encyclopædia Britannica.

[6] The earliest occurrence of this term quoted by DU CANGE refers to the year 1351, and is contained in the Transactio inter Abbatem et Monachos Crassenses. Spinach served the Christian monks of Europe as well as the Buddhists of China. O. SCHRADER (Reallexikon, p. 788) asserts that the vegetable is first mentioned by Albertus Magnus (1193–1280) under the name *spinachium,* but he fails to give a

espinaca, Portuguese *espinafre* or *espinacio*, Italian *spinace* or *spinaccio*, Provençal *espinarc*, Old French *espinoche* or *épinoche*, French *épinard*.[1] The Persian word was further adopted into Armenian *spanax* or *asbanax*, Turkish *spanâk* or *ispanâk*, Comanian *yspanac*, Middle Greek *spinakion*, Neo-Greek *spanaki(on)* or *spanakia* (plural). There are various spellings in older English, like *spynnage*, *spenege*, *spinnage*, *spinage*, etc. In English literature it is not mentioned earlier than the sixteenth century. W. Turner, in his "Herball" of 1568, speaks of "spinage or spinech as an herbe lately found and not long in use."

However, in the latter part of the sixteenth century, spinach was well known and generally eaten in England. D. Rembert Dodoens[2] describes it as a perfectly known subject, and so does John Gerarde,[3] who does not even intimate that it came but recently into use. The names employed by them are *Spanachea*, *Spinachia*, *Spinachæum olus*, *Hispanicum olus*, English *spinage* and *spinach*. John Parkinson[4] likewise gives a full description and recipes for the preparation of the vegetable.

The earliest Persian mention of the spinach, as far as I know, is made in the pharmacopœia of Abu Mansur.[5] The oldest source cited by Ibn al-Baiṭār (1197–1248)[6] on the subject is the "Book of Nabathæan Agriculture" (*Falāha nabaṭīya*), which pretends to be the Arabic translation of an ancient Nabathæan source, and is believed to be a forgery of the tenth century. This book speaks of the spinach as a known vegetable and as the most harmless of all vegetables; but the most interesting remark is that there is a wild species resembling the cultivated one, save that it is more slender and thinner, that the leaves are

---

specific reference. It is a gratuitous theory of his that the spinach must have been brought to Europe by the Crusaders; the Arabic importation into Spain has escaped him entirely.

[1] The former derivation of the word from "Spain" or from *spina* ("thorn"), in allusion to the prickly seeds, moves on the same high level as the performance of the *Min šu*. Littré cites Ménagier of the sixteenth century to the effect, "Les espinars sont ainsi appellés à cause de leur graine qui est espineuse, bien qu'il y en ait de ronde sans piqueron." In the Supplément, Littré points out the oriental origin of the word, as established by Devic.

[2] A Nievve Herball, or Historie of Plants, translated by H. Lyte, p. 556 (London, 1578).

[3] The Herball or Generall Historie of Plantes, p. 260 (London, 1597).

[4] Paradisus in sole paradisus terrestris, p. 496 (London, 1629).

[5] Achundow, Abu Mansur, p. 6.

[6] L. Leclerc, Traité des simples, Vol. I, p. 60.

more deeply divided, and that it rises less from the ground.[1] A. DE
CANDOLLE states that "spinach has not yet been found in a wild state,
unless it be a cultivated modification of *Spinacia tetandra* Steven, which
is wild to the south of the Caucasus, in Turkistan, in Persia, and in
Afghanistan, and which is used as a vegetable under the name of
šamum." The latter word is apparently a bad spelling or misreading
for Persian *šomīn* or *šūmīn* (Armenian *zomin* and *šomin*), another
designation for the spinach.

The spinach is not known in India except as an introduction by the
English. The agriculturists of India classify spinach among the English
vegetables.[2] The species *Spinacia tetrandra* Roxb., for which Rox-
BURGH[3] gives the common Persian and Arabic name for the spinach,
and of which he says that it is much cultivated in Bengal and the
adjoining provinces, being a pot-herb held in considerable estimation
by the natives, may possibly have been introduced by the Mohammedans. As a matter of fact, spinach is a vegetable of the temperate
zones and alien to tropical regions. A genuine Sanskrit word for the
spinach is unknown.[4] Nevertheless Chinese *po-liṅ*, *pwa-liṅ, must
represent the transcription of some Indian vernacular name. In Hindustānī we have *palak* as designation for the spinach, and *palaṅ* or
palak as name for *Beta vulgaris*, Puštu *pālak*,[5] apparently developed
from Sanskrit *pālaṅka, pālankya, palakyū, pālakyā,* to which our
dictionaries attribute the meaning "a kind of vegetable, a kind of
beet-root, *Beta bengalensis*"; in Bengālī *paluṅ*.[6] To render the coincidence with the Chinese form complete, there is also Sanskrit Pālakka

---

[1] Perhaps related to *Atriplex* L., the so-called wild spinach, chiefly cultivated
in France and eaten like spinach. The above description, of course, must
not be construed to mean that the cultivated spinach is derived from the
so-called wild spinach of the Nabathæans. The two plants may not be interrelated at all.

[2] N. G. MUKERJI, Handbook of Indian Agriculture, 2d ed., p. 300 (Calcutta,
1907); but it is incorrect to state that spinach originally came from northern Asia.
A. DE CANDOLLE (*op. cit.*, p. 99) has already observed, "Some popular works repeat
that spinach is a native of northern Asia, but there is nothing to confirm this supposition."

[3] Flora Indica, p. 718.

[4] A. BOROOAH, in his English-Sanskrit Dictionary, gives a word *çākaprabheda*
with this meaning, but this simply signifies "a kind of vegetable," and is accordingly an explanation.

[5] H. W. BELLEW, Report on the Yusufzais, p. 255 (Lahore, 1864).

[6] *Beta* is much cultivated by the natives of Bengal, the leaves being consumed
in stews (W. ROXBURGH, Flora Indica, p. 260). Another species, *Beta maritima*, is
also known as "wild spinach." It should be remembered that the genus *Beta* belongs
to the same family (*Chenopodiaceae*) as *Spinacia*.

or Pālaka[1] as the name of a country, which has evidently resulted in the assertion of Buddhist monks that the spinach must come from a country Palinga. The Nepalese, accordingly, applied a word relative to a native plant to the newly-introduced spinach, and, together with the product, handed this word on to China. The Tibetans never became acquainted with the plant; the word *spo ts'od*, given in the Polyglot Dictionary,[2] is artificially modelled after the Chinese term, *spo* (pronounced *po*) transcribing Chinese *po*, and *ts'od* meaning "vegetable."

Due regard being paid to all facts botanical and historical, we are compelled to admit that the spinach was introduced into Nepal from some Iranian region, and thence transmitted to China in A.D. 647. It must further be admitted that the Chinese designation "Persian vegetable," despite its comparatively recent date, cannot be wholly fictitious, but has some foundation in fact. Either in the Yūan or in the Ming period (more probably in the former) the Chinese seem to have learned the fact that Persia is the land of the spinach. I trust that a text to this effect will be discovered in the future. All available historical data point to the conclusion that the Persian cultivation can be but of comparatively recent origin, and is not older than the sixth century or so. The Chinese notice referring it to the seventh century is the oldest in existence. Then follow the Nabathæan Book of Agriculture of the tenth century and the Arabic introduction into Spain during the eleventh.

---

[1] The latter form is noted in the catalogue of the Mahāmayūrī, edited by S. Lévi (*Journal asiatique*, 1915, I, p. 42).

[2] Ch. 27, p. 19 b.

SUGAR BEET AND LETTUCE

37. In the preceding notes we observed that the name for a species of *Beta* was transferred to the spinach in India and still serves in China as designation for this vegetable. We have also a Sino-Iranian name for a *Beta*, 軍蓬, *kün-t'a*, *gwun-d'ar, which belonged to the choice vegetables of the country 末藤 Mo-lu, *Mar-luk, in Arabia.[1] The *Čen su wen* 證俗文[2] says that it is now erroneously called *ken ta ts'ai* 根大茶 or *ta ken ts'ai*, which is identical with *tien ts'ai* 甜茶 ("sweet vegetable"). STUART[3] gives the latter name together with 莙蓬 *kün-t'a*, identifying it with *Beta vulgaris*, the white sugar beet, which he says grows in China. Stuart, however, is mistaken in saying that this plant is not mentioned in the *Pen ts'ao*. It is noted both in the *Čen lei pen ts'ao*[4] and the *Pen ts'ao kan mu*,[5] the latter giving also the term *kün-t'a*, which is lacking in the former work. Li Ši-čen observes with reference to this term that its meaning is unexplained, a comment which usually betrays the foreign character of the word, but he fails to state the source from which he derived it. There is no doubt that this *kün-t'a* is merely a graphic variant of the above 軍蓬. The writing 莙 is as early as the T'ang period, and occurs in the *Yu yan tsa tsu*,[6] where the leaves of the *yu tien ts'ao* 油點草 ("herb with oily spots") are compared to those of the *kün-t'a*.[7] A description of the *kün-t'a* is not contained in that work, but from this incidental reference it must be inferred that the plant was well known in the latter half of the ninth century.

Beta vulgaris is called in New Persian *čugundur* or *čegonder*, and is mentioned by Abu Mansur.[8] The corresponding Arabic word is *silk*.[9] The Chinese transcription made in the T'ang period is apparently based on a Middle-Persian form of the type *gundar or *gundur. *Beta vulgaris* is a Mediterranean and West-Asiatic plant grown as far as the

---

[1] *T'ai p'in hwan yü ki*, Ch. 186, p. 16 b.

[2] Ch. 12, p. 3. This work was published in 1884 by Ho Yi-hin 郝懿行.

[3] Chinese Materia Medica, p. 68.

[4] Ch. 28, p. 9.

[5] Ch. 27, p. 1 b. Cf. also *Yamato honzō*, Ch. 5, p. 26.

[6] Ch. 9, p. 9 b.

[7] "On each leaf there are black spots opposite one another."

[8] ACHUNDOW, Abu Mansur, p. 81.

[9] LECLERC, Traité des simples, Vol. II, p. 274.

399

Caspian Sea and Persia. According to DE CANDOLLE,[1] its cultivation does not date from more than three or four centuries before our era. The Egyptian illustration brought forward by F. WOENIG[2] in favor of the assumption of an early cultivation in Egypt is not convincing to me.

It is therefore probable, although we have no record referring to the introduction, that *Beta vulgaris* was introduced into China in the T'ang period, perhaps by the Arabs, who themselves brought many Persian words and products to China. For this reason Chinese records sometimes credit Persian words to the Ta-ši (Arabs); for instance, the numbers on dice, which go as Ta-ši, but in fact are Persian.[3]

The real Chinese name of the plant is *tien ts'ai* 菾菜, the first character being explained in sound and meaning by 甜 *tien* ("sweet"). Li Ši-čen identifies *tien ts'ai* with *kün-t'a*. The earliest description of *tien ts'ai* comes from Su Kuň of the T'ang, who compares its leaves to those of *šen ma* 升麻 (*Actea spicata*, a ranunculaceous plant), adding that the southerners steam the sprouts and eat them, the dish being very fragrant and fine.[4] It is not stated, however, that *tien ts'ai* is an imported article.

38. Reference was made above to the memorable text of the *T'aň hui yao*, in which are enumerated the vegetable products of foreign countries sent to the Emperor T'ai Tsuň of the T'ang dynasty at his special request in A.D. 647. After mentioning the spinach of Nepal, the text continues thus:—

"Further, there was the *ts'o ts'ai* 酢菜 ('wine vegetable') with broad and long leaves.[5] It has a taste like a good wine and *k'u ts'ai* 苦菜 ('bitter vegetable,' lettuce, *Lactuca*), and in its appearance is like *kü* 苣,[6] but its leaves are longer and broader. Although it is somewhat bitter of taste, eating it for a long time is beneficial. *Hu k'in* 胡芹

---

[1] Origin of Cultivated Plants, p. 59; see also his Géographie botanique, p. 831

[2] Pflanzen im alten Aegypten, p. 218.

[3] See *T'oung Pao*, Vol. I, 1890, p. 95.

[4] A *tien ts'ai* mentioned by T'ao Huň-kiň, as quoted in the *Pen ts'ao kaň mu*, and made into a condiment *ča* 鮓 for cooking-purposes, is apparently a different vegetable.

[5] The corresponding text of the *Ts'e fu yüan kwei* (Ch. 970, p. 12) has the addition, "resembling the leaves of the *šen-hwo* 慎火." The text of the *Pei hu lu* (Ch. 2, p. 19 b) has, "resembling in its appearance the *šen-hwo*, but with leaves broader and longer." This tree, also called *kiň t'ien* 景天 (see *Yu yaň tsa tsu*, Ch. 19, p. 6), is believed to protect houses from fire; it is identified with *Sedum erythrostictum* or *Sempervivum tectorum* (BRETSCHNEIDER, Bot. Sin., pt. III, No. 205; STUART, Chinese Materia Medica, p. 401).

[6] A general term for plants like *Lactuca*, *Cichorium*, *Sonchus*.

resembles in its appearance the k'in 芹 ('celery,' *Apium graveolens*), and has a fragrant flavor."

Judging from the description, the vegetable *ts'o ts'ai* appears to have been a species of *Lactuca*, *Cichorium*, or *Sonchus*. These genera are closely allied, belonging to the family *Cichoraceae*, and are confounded by the Chinese under a large number of terms. A. DE CANDOLLE[1] supposed that lettuce (*Lactuca sativa*) was hardly known in China at an early date, as, according to Loureiro, Europeans had introduced it into Macao.[2] With reference to this passage, BRETSCHNEIDER[3] thinks that de Candolle "may be right, although the *Pen ts'ao* says nothing about the introduction; the *šen ts'ai* 生菜 (the common name of lettuce at Peking) or *pai-kü* 白苣 seems not to be mentioned earlier than by writers of the T'ang (618-906)." Again, DE CANDOLLE seized on this passage, and embodied it in his "Origin of Cultivated Plants" (p. 96). The problem, however, is not so simple. Bretschneider must have read the *Pen ts'ao* at that time rather superficially, for some species of *Lactuca* is directly designated there as being of foreign origin. Again, twenty-five years later, he wrote a notice on the same subject,[4] in which not a word is said about foreign introduction, and from which, on the contrary, it would appear that *Lactuca*, *Cichorium*, and *Sonchus*, have been indigenous to China from ancient times, as the bitter vegetable (*k'u ts'ai*) is already mentioned in the *Pen kin* and *Pie lu*. The terms *pai kü* 白苣 and *k'u kü* 苦苣 are supposed to represent *Cichorium endivia;* and *wo-kü* 萵苣, *Lactuca sativa*. In explanation of the latter name, Li Ši-čen cites the *Mo k'o hui si* 墨客揮犀 by P'eň Č'eň 彭乘, who wrote in the first half of the eleventh century, as saying that *wo ts'ai* 萵菜 ("*wo* vegetable") came from the country 喎 Kwa, and hence received its name.[5] The *Ts'in i lu* 清異錄, a work by T'ao Ku 陶穀 of the Sung period, says that "envoys from the country Kwa came to China, and at the request of the people distributed seeds of a vegetable; they were so generously rewarded that it was called *ts'ien kin ts'ai* 千金菜 ('vegetable of a thousand gold pieces'); now it is styled *wo-*

---

[1] Géographie botanique, p. 843.

[2] This certainly is a weak argument. The evidence, in fact, proves nothing. Europeans also introduce their own sugar and many other products of which China has a great plenty.

[3] *Chinese Recorder*, 1871, p. 223.

[4] Bot. Sin., pt. III, No. 257.

[5] I do not know how STUART (p. 229) gets at the definition "in the time of the Han dynasty." The same text is also contained in the *Sü po wu či* (Ch. 7, p. 1 b), written by Li Ši 李石 about the middle of the twelfth century.

kü."[1] These are vague and puerile anecdotes, without chronological specification. There is no country Kwa, which is merely distilled from the character 苐, and no such tradition appears in any historical text.[2] The term *wo-kü* was well known under the T'ang, being mentioned in the *Pen ts'ao ši i* of Č'en Ts'aṅ-k'i, who distinguishes a white and a purple variety, but is silent as to the point of introduction.[3] This author, however, as can be shown by numerous instances, had a keen sense of foreign plants and products, and never failed to indicate them as such. There is no evidence for the supposition that *Lactuca* was introduced into China from abroad. All there is to it amounts to this, that, as shown by the above passage of the *T'aṅ hui yao*, possibly superior varieties of the West were introduced.

In Persia, *Lactuca sativa* (Persian *kāhu*) occurs both wild and cultivated.[4] *Cichoreum* is *kasnī* in Persian, *hindubā* in Arabic and Osmanli.[5]

39. The *hu k'in*, mentioned in the above text of the *T'aṅ hui yao*, possibly represents the garden celery, *Apium graveolens* (Persian *kerefs* or *karafs*) (or possibly parsley, *Apium petroselinum*) of the west.[6] It appears to be a different plant from the *hu k'in* mentioned above (p. 196).

Hu k'in is likewise mentioned among the best vegetables of the country 末 祿 Mo-lu, *Mwat-luk, Mar-luk, in Arabia.[7]

In order to conclude the series of vegetables enumerated in the text of the *T'aṅ hui yao*, the following may be added here.

In A.D. 647 the king of Gandhāra (in north-western India) sent to the Chinese Court a vegetable styled *fu-t'u* 佛土菜 ("Buddha-land vegetable"), each stem possessing five leaves, with red flowers, a yellow pith, and purple stamens.[8]

---

[1] I have looked up the text of the *Ts'iṅ i lu*, which is reprinted in the *T'aṅ Suṅ ts'uṅ šu* and *Si yin hüan ts'uṅ šu*. The passage in question is in Ch. 2, p. 7 b, and printed in the same manner as in the *Pen ts'ao kaṅ mu*, save that the country is called Kao 高, not Kwa 苐 It is easy to see that these two characters could be confounded, and that only one of the two can be correct; but Kao does not help us any more than Kwa. Either name is fictitious as that of a country.

[2] We have had several other examples of alleged names of countries being distilled out of botanical names.

[3] K'ou Tsuṅ-ši is likewise; see his *Pen ts'ao yen i* (Ch. 19, p. 2).

[4] SCHLIMMER, Terminologie, p. 337.

[5] See ACHUNDOW, Abu Mansur, p. 146; E. SEIDEL, Mechithar, p. 134; LECLERC, Traité des simples, Vol. II, p. 28.

[6] Cf. ACHUNDOW, Abu Mansur, pp. 110, 257. Celery is cultivated only in a few gardens of Teheran, but it grows spontaneously and abundantly in the mountains of the Bakhtiaris (SCHLIMMER, Terminologie, p. 43).

[7] *T'ai p'iṅ hwan yü ki*, Ch. 186, p. 16 b.

[8] *T'aṅ hui yao*, Ch. 200, p. 4 b; and *T'aṅ šu*, Ch. 221 B, p. 7. The name of Gandhāra is abbreviated into *d'ar, but in the corresponding passage of the *T'aṅ hui yao* (Ch. 100, p. 3 b) and in the *Ts'e fu yüan kwei* (Ch. 970, p. 12) the name is written completely 健達 Kien-ta, *G'an-d'ar.

RICINUS

40. In regard to *Ricinus communis* (family *Euphorbiaceae*) the accounts of the Chinese are strikingly deficient and unsatisfactory. There can be no doubt that it is an introduced plant in China, as it occurs there only in the cultivated state, and is not mentioned earlier than the T'ang period (618–906) with an allusion to the Hu.[1] Su Kuṅ states in the *T'aṅ pen ts'ao*, "The leaves of this plant which is cultivated by man resemble those of the hemp (*Cannabis sativa*), being very large. The seeds look like cattle-ticks (*niu pei* 牛 蜱).[2] The stems of that kind which at present comes from the Hu[3] are red and over ten feet high. They are of the size of a *tsao kia* 皂 荚 (*Gleditschia sinensis*). The kernels are the part used, and they are excellent." It would seem from this report that two kinds of *Ricinus* are assumed, one presumably the white-stemmed variety known prior to Su Kuṅ's time, and the red-stemmed variety introduced in his age. Unfortunately we receive no information as to the exact date and provenience of the introduction.

The earliest mention of the plant is made by Herodotus,[4] who ascribes it to the Egyptians who live in the marshes and use the oil pressed from the seeds for anointing their bodies. He calls the plant *sillikyprion*,[5] and gives the Egyptian name as *kiki*.[6] In Hellas it grows spontaneously (αὐτόματα φύεται), but the Egyptians cultivate it along the banks of the rivers and by the sides of the lakes, where it produces fruit in abundance, which, however, is malodorous. This fruit is

---

[1] *Pen ts'ao kaṅ mu*, Ch. 17 A, p. 11. BRETSCHNEIDER (*Chinese Recorder*, 1871, p. 242) says that it cannot be decided from Chinese books whether *Ricinus* is indigenous to China or not, and that the plant is not mentioned before the T'ang. The allusion to the Hu escaped him.

[2] Hence the name 蓖 or 牌 麻 *pei ma* (only in the written language) for the plant (Peking colloquial *ta ma*, "great hemp"). This etymology has already been advanced by Su Suṅ of the Sung and confirmed by Li Ši-čen, who explains the insect as the "louse of cattle." This interpretation appears to be correct, for it represents a counterpart to Latin *ricinus*, which means a "tick": Nostri eam ricinum vocant a similitudine seminis (Pliny, xv, 7, § 25). The Chinese may have hit upon this simile independently, or, what is even more likely, received it with the plant from the West.

[3] This appears to be the foundation for STUART's statement (Chinese Materia Medica, p. 378) that the plant was introduced from "Tartary."

[4] II, 94.

[5] The common name was κρότων (Theophrastus, Hist. plant., I. x, 1), Latin *croton*.

[6] This word has not yet been traced in the hieroglyphic texts, but in Coptic. In the demotic documents *Ricinus* is *degam* (V. LORET, Flore pharaonique, p. 49).

403

gathered, and either pounded and pressed or roasted and boiled, and the oily fluid is collected. It is found to be unctuous and not inferior to olive-oil for burning in lamps, save that it emits a disagreeable odor. Seeds of Ricinus are known from Egyptian tombs, and the plant is still cultivated in Egypt. Pliny[1] states that it is not so long ago that the plant was introduced into Italy. A. DE CANDOLLE[2] traces its home to tropical Africa, and I agree with this view. Moreover, I hold that it was transplanted from Egypt to India, although, of course, we have no documentary proof to this effect. *Ricinus* does not belong to the plants which were equally known to the Iranians and Indo-Aryans. It is not mentioned in the Vedas or in the Laws of Manu.[3] The first datable references to it occur in the Bower Manuscript, where its oil and root are pointed out under the names *eraṇḍa, gandharva, rubūgaka,* and *vakṣaṇa.* Other names are *ruvu, ruvuka,* or *ruvūka, citraka, gandharva-hastaka, vyāghrapuccha* ("tiger's-tail"). The word *eraṇḍa* has become known to the Chinese in the form *i-lan* 伊蘭,[4] and was adopted into the language of Kuča (Tokharian B) in the form *hiraṇḍa.*[5] From India the plant seems to have spread to the Archipelago and Indo-China (Malayan, Sunda, and Javanese *jarak;* Khmer *lohoṅ;* Annamese *du du traṅ, kai-dua,* or *kai-du-du-tia;* Čam *tamṅön, lahauṅ, lahon*).[6] The Miao and the Lo-lo appear to be familiar with the plant: the former call it *zrwa-ño;*[7] the latter, *č'e-tu-ma* (that is, "fruit for the poisoning of dogs").[8]

In Iran the cultivation of *Ricinus* has assumed great importance, but no document informs us as to the time of its transplantation. It may be admitted, however, that it was well known there prior to our era.[9] The Persian name is *bedānjir, pandu, punde,* or *pendu;* in Arabic it is *xarva* or *xirva.*

---

[1] XV, 7, § 25.

[2] Origin of Cultivated Plants, p. 422.

[3] JORET, Plantes dans l'antiquité, Vol. II, p. 270.

[4] *Fan yi miṅ yi tsi,* section 24.

[5] S. LÉVI, *Journal asiatique,* 1911, II, p. 123.

[6] On the cultivation in Indo-China, see PERROT and HURRIER, Mat. méd. et pharmacopée sino-annamites, p. 107. Regarding the Archipelago, see A. DE CANDOLLE, *op. cit.,* p. 422; W. MARSDEN, History of Sumatra, p. 92; J. CRAWFURD, History of the Indian Archipelago, Vol. I, p. 382. The plant is reported wild from Sumatra and the Philippines, but the common Malayan name *jarak* hints at an historical distribution.

[7] F. M. SAVINA, Dictionnaire miao-tseu-français, pp. 205, 235.

[8] P. VIAL, Dictionnaire français-lolo, p. 290. Also the Arabs used *Ricinus* as a dog-poison (LECLERC, Traité des simples, Vol. II, p. 20).

[9] JORET, *op. cit.,* p. 72.

THE ALMOND

41. Iran was the centre from which the almond (*Amygdalus communis* or *Prunus amygdalus*) spread, on the one hand to Europe, and on the other to China, Tibet, and India. As to India, it is cultivated but occasionally in Kashmir and the Panjab, where its fruits are mediocre. It was doubtless imported there from Iran. The almond yields a gum which is still exported from Persia to Bombay, and thence re-exported to Europe.[1] The almond grows spontaneously in Afghanistan and farther to the north-east in the upper Zarafshan valley, and in the Chotkal mountains at an altitude of ·1000–1300 m, also in Aderbeidjan, Kurdistan, and Mesopotamia. According to SCHLIMMER,[2] *Amygdalus coparia* is very general on the high mountains, and its timber yields the best charcoal.[3]

The Greeks derived the almond from Asia Minor, and from Greece it was apparently introduced into Italy.[4] In the northern part of Media, the people subsisted upon the produce of trees, making cakes of apples, sliced and dried, and bread of roasted almonds.[5] A certain quantity of dried sweet almonds was to be furnished daily for the table of the Persian kings.[6] The fruit is mentioned in Pahlavi literature (above, p. 193).

The *Yiṅ yai šeṅ lan* mentions almonds among the fruit grown in Aden.[7] The Arabic name is *lewze* or *lauz*. Under this name the medicinal properties of the fruit are discussed in the Persian pharmacopœia of Abu Mansur, who knew both the sweet almond (*bādām-i šīrīn*) and the bitter one (*bādām-i tälx*).[8] It is curious that bitter almonds were used as currency in the empire of the Moguls. They were brought into the

---

[1] G. WATT, Commercial Products of India, p. 905; and Dictionary, Vol. VI, p. 343. JORET, Plantes dans l'antiquité, Vol. II, p. 279. W. ROXBURGH (Flora Indica, p. 403) concluded that the almond is a native of Persia and Arabia, whereas it does not succeed in India, requiring much nursing to keep it alive.

[2] Terminologie, p. 33.

[3] A really wild almond is said to be very common in Palestine and Syria (A. AARONSOHN, Agric. and Bot. Explorations in Palestine, p. 14).

[4] HEHN, Kulturpflanzen, pp. 393, 402; FLÜCKIGER and HANBURY, Pharmacographia, pp. 244, 245.

[5] STRABO, XI. XIII, 11.

[6] Polyaenus, Strategica, IV, 32.

[7] ROCKHILL, T'oung Pao, 1915, p. 609.

[8] ACHUNDOW, Abu Mansur, p. 128.

405

province of Gujarat from Persia, where they grow in dry and arid places between rocks; they are as bitter as colocynth, and there is no fear that children will amuse themselves by eating them.[1]

What WATTERS[2] has stated about the almond is for the greater part inexact or erroneous. "For the almond which does not grow in China the native authors and others have apparently only the Persian name which is Bádán. This the Chinese transcribe *pa-tan* 八擔 or 巴旦 and perhaps also, as suggested by Bretschneider, *pa-lan* 杷欖." First, the Persian name for the almond is *bādām*; second, the Chinese characters given by Watters are not apt to transcribe this word, as the former series answers to ancient *pat-dam, the latter to *pa-dan. Both 八 and 巴 only had an initial labial surd, but never a labial sonant, and for this reason could not have been chosen for the transcription of a foreign *ba* in the T'ang period, when the name of the almond made its début in China. Further, the character 旦, which was not possessed of a final labial nasal, would make a rather bad reproduction of the required element *dam*. In fact, the characters given by Watters are derived from the *Pen ts'ao kan mu*,[3] and represent merely a comparatively modern readjustment of the original form made at a time when the transposition of sonants into surds had taken effect. The first form given by Watters, as stated in the *Pen ts'ao* itself, is taken from the *Yin šan čeň yao* (see p. 236), written by Ho Se-hwi during the Yüan period; while the second form is the work of Li Ši-čen, as admitted by himself, and accordingly has no phonetic value whatever.[4] Indeed, we have a phonetically exact transcription of the Iranian term, handed down from the T'ang period, when the Chinese still enjoyed the possession of a well-trained ear, and, in view of the greater wealth of sounds then prevailing in their speech, also had the faculty of reproducing them with a fair degree of precision. This transcription is presented by 婆淡 *p'o-tan*, *bwa-dam, almond (*Amygdalus communis* or *Prunus amygdalus*), which actually reproduces Middle Persian *vadam*, New Persian *bādām* (Kurd *badem, beïv* and *baíf*, "almond-tree").[5] This term,

---

[1] TAVERNIER, Travels in India, Vol. I, p. 27.

[2] Essays on the Chinese Language, p. 348.

[3] Ch. 29, p. 4. Hence adopted also by the Japanese botanists (MATSUMURA, No. 2567), but read *amendo* (imitation of our word).

[4] He further gives as name for the almond *hu-lu-ma* 忽鹿麻 = Persian *xurmā* (*khurmā*), but this word properly refers to the date (p. 385). From the *Ta Miň i t'uň či* (Ch. 89, p. 24), where the almonds of Herat are mentioned, it appears that *hu-lu-ma* (*xurmā*) was the designation of a special variety of almond, "resembling a jujube and being sweet."

[5] The assertion of STUART (Chinese Materia Medica, p. 40), that *pa-tan* may refer to some country in Asia Minor or possibly be another name for Persia, is erroneous.

as far as I know, is first mentioned in the *Yu yan tsa tsu*,[1] where it is said, "The flat peach 偏桃 grows in the country Po-se (Persia), where it is styled *p'o-tan*. The tree reaches a height of from fifty to sixty feet, and has a circumference of four or five feet. Its leaves resemble those of the peach, but are broader and larger. The blossoms, which are white in color, appear in the third month. When the blossoms drop, the formation of the fruit has the appearance of a peach, but the shape is flat. Hence they are called 'flat peaches.' The meat is bitter and acrid, and cannot be chewed; the interior of the kernel, however, is sweet, and is highly prized in the Western Regions and all other countries." Although the fact of the introduction of the plant into China is not insisted upon by the author, Twan Č'en-ši, his description, which is apparently based on actual observation, may testify to a cultivation in the soil of his country. This impression is corroborated by the testimony of the Arabic merchant Soleiman, who wrote in A.D. 851, and enumerates almonds among the fruit growing in China.[2] The correctness of the Chinese reproduction of the Iranian name is confirmed by the Tibetan form *ba-dam*, Uigur and Osmanli *badam*, and Sanskrit *vātāma* or *bādāma*, derived from the Middle Persian.[3]

The fundamental text of the *Yu yan tsa tsu* has unfortunately escaped Li Ši-čen, author of the *Pen ts'ao kan mu*, and he is accordingly led to the vague definition that the almond comes from the old territory of the Mohammedans; in his time, he continues, the tree occurred in all places West of the Pass (Kwan si; that is, Kan-su and Šen-si). The latter statement is suppressed in BRETSCHNEIDER's translation of the text,[4] probably because it did not suit his peremptory opinion that the almond-tree does not occur in China. He did not know, either, of the text of the *Yu yan tsa tsu*, and his vague data were adopted by A. DE CANDOLLE.[5]

LOUREIRO[6] states that the almond is both wild and cultivated in

---

[1] Ch. 18, p. 10 b.

[2] M. REINAUD, Relation des voyages, Vol. I, p. 22.

[3] Cf. the writer's Loan-Words in Tibetan, No. 111. It should be repeated also in this place that the Tibetan term *p'a-lin*, which only means "dried apricots," bears no relation to the Persian designation of the almond, as wrongly asserted by Watters.—The almond is also known to the Lo-lo (Nyi Lo-lo *ñi-ma*, Ahi Lo-lo *i-ni-zo, i-sa*).

[4] *Chinese Recorder*, 1870, p. 176.

[5] Origin of Cultivated Plants, p. 219. He speaks erroneously of the *Pen ts'ao* published in the tenth or eleventh century. Bretschneider, of course, meant the *Pen ts'ao* of the sixteenth century.

[6] Flora cochinchinensis, p. 316. PERROT and HURRIER (Matière médicale et pharm. sino-annamites, p. 153) have an *Amygdalus cochinchinensis* for Annam.

China. Bunge says that it is commonly cultivated in North China; but
that recent botanists have not seen it in South China, and the one
cultivated near Peking is *Prunus davidiana*, a variety of *P. persica*.[1]
These data, however, are not in harmony with Chinese accounts which
attribute the cultivation of the almond to China; and it hardly sounds
plausible that the Chinese should confound with this tree the apricot,
which has been a native of their country from time immemorial.
WATTERS asserts that "the Chinese have mixed up the foreign almond
with their native apricot. The name of the latter is *hiṅ* 杏, and the
kernels of its fruit, when dried for food, are called *hiṅ-žen* 杏仁. This
name is given also to the kernels of almonds as imported into China
from their resemblance in appearance and to some extent in taste to
the seeds of apricots." The fact that almond-meat is styled "apricot-
kernel" does not prove that there is a confusion between *hiṅ* and *hiṅ-*
žen, or between almond and apricot. The confusion may be on the
part of foreigners who take apricot-kernels for almonds.[2]

It has been stated by BRETSCHNEIDER[3] that the word *pa-lan* 杷欖
(*pa-lam), used by the travellers Ye-lu Č'u-ts'ai and Č'aṅ Č'un, might
transcribe the Persian word *bādām*. This form first appears in the *Suṅ*
ši (Ch. 490) in the account of Fu-lin, where the first element is written
phonetically 巴,[4] so that the conclusion is almost warranted that this
word was transmitted from a language spoken in Fu-lin. In all probability, the question is of a Fu-lin word of the type *palam* or *param* (perhaps *faram, fram, or even *spram).

The fruit *pa-lan* must have been known in China during the Sung,
for it is mentioned by Fan Č'eṅ-ta 范成大 (1126–93), in his *Kwei hai*
yü heṅ či,[5] in the description of the *ši li* 石栗 (*Aleurites triloba*), which

---

[1] BRETSCHNEIDER, Early Researches into the Flora of China, p. 149; FORBES
and HEMSLEY, *Journal Linnean Soc.*, Vol. XXIII, p. 217. W. C. BLASDALE (Description of Some Chinese Vegetable Food Materials, p. 48, Washington, 1899) mentions a peculiar variety of the almond imported from China into San Francisco.
The almond is cultivated in China according to K. v. SCHERZER (Berichte österr.
Exped. nach Siam, China und Japan, p. 96). L. DE REINACH (Le Laos, p. 280)
states that almond-trees grow in the northern part of Laos.

[2] F. N. MEYER (Agricultural Explorations in the Orchards of China, p. 53)
supposes erroneously that the consumption of apricot-kernels has given rise to the
statement that almonds grow in China. Cf. SCHLEGEL's Nederlandsch-Chineesch
Woordenboek, Vol. I, p. 226.

[3] Mediæval Researches, Vol. I, p. 20.

[4] Cf. HIRTH, China and the Roman Orient, p. 63. His identification with
Greek βάλανος, which refers only to the acorn, a wild fruit, is hardly satisfactory,
for phonetic and historical reasons. For Hirth's translation of 杏 by "almonds"
in the same clause read "apricots."

[5] Ed. of *Či pu tsu čai ts'uṅ šu*, p. 24.

is said to be like *pa-lan-tse*. In the Gazetteer of Č'eṅ-te fu, *pa-lan žen* 仁 is given as a variety of apricot.[1]

Ho Yi-hiṅ, in his *Čeṅ su wen*, published in 1884,[2] observes that "at present the people of the capital style the almond *pa-ta* 巴達, which is identical with *pa-tan* 巴旦. The people of Eastern Ts'i 東齊 (Šan-tuṅ) call the almond, if it is sweet and fine, *žen hiṅ* 榛杏 (hazel-nut apricot), because it has the taste of hazel-nuts.[3] According to the *Hiaṅ tsu pi ki* 香祖筆記, a certain kind of almond, styled 'almond of the *I wu hui Park*' 異物彙苑, is exported from Herat 哈烈. At present it occurs in the northern part of China. The fruit offered in the capital is large and sweet, that of Šan-tuṅ is small with thin and scant meat."

The old tradition concerning the origin of the almond in Persia is still alive in modern Chinese authors. The Gazetteer of Šaṅ-se čou in the prefecture of T'ai-p'iṅ, Kwaṅ-si Province, states that the flat peach is a cultivation of the country Po-se (Persia).[4] The tree is (or was) cultivated in that region. Also the *Hwa mu siao či* 花木小志 (p. 29 b)[5] testifies to indigenous cultivation by saying that almond-trees grow near the east side of mountains. It may be, of course, that the almond has shared the fate of the date-palm, and that its cultivation is now extinct in China.[6]

---

[1] O. FRANKE, Beschreibung des Jehol-Gebietes, p. 75.

[2] Ch. 12, p. 5 b (see above, p. 399).

[3] This observation is also made by Li Ši-čen.

[4] *Šaṅ-se čou či* 上思州志, Ch. 14, p. 7 b (published in 1835).

[5] Published in the *Č'un ts'ao t'aṅ tsi* 春草堂集 during the period Tao-kwaṅ (1820–50).

[6] HAUER (Erzeugnisse der Provinz Chili, *Mitt. Sem. or. Spr.*, 1908, p. 14) mentions almonds, large and of sweet flavor, as a product of the district of Mi-yün in Či-li, and both sweet and bitter almonds as cultivated in the district of Lwan-p'iṅ in the prefecture of Č'eṅ-te (Jehol), the annual output of the latter locality being given as a hundred thousand catties,—a hardly credible figure should almonds really be involved. Hauer's article is based on the official reports submitted by the districts to the Governor-General of the Province in 1904; and the term rendered by him "almond" in the original is *ta pien fen* 大扁分, apparently a local or colloquial expression which I am unable to trace in any dictionary. It is at any rate questionable whether it has the meaning "almond." O. FRANKE, in his description of the Jehol territory, carefully deals with the flora and products of that region without mentioning almonds, nor are they referred to in the Chinese Gazetteer of Č'eṅ-te fu.

THE FIG

42. The fig (*Ficus carica*) is at present cultivated in the Yang-tse valley as a small, irregular shrub, bearing a fruit much smaller and inferior in quality to the Persian species.[1] According to the *Pen ts'ao kań mu*, its habitat is Yań-čou (the lower Yang-tse region) and Yün-nan. In his time, Li Ši-čen continues, it was cultivated also in Če-kiań, Kiań-su, Hu-pei, Hu-nan, Fu-kien, and Kwań-tuń (吳楚閩越) by means of twigs planted in the ground. The latter point is of particular interest in showing that the process of caprification has remained unknown to the Chinese, and, in fact, is not mentioned in their works. The fig is not indigenous to China; but, while there is no information in Chinese records as to the when and how of the introduction, it is perfectly clear that the plant was introduced from Persia and India, not earlier than the T'ang period.

The following names for the fig are handed down to us:—

(1) Po-se (Persian) 阿駔 *a-ži*, **a-žit(žir)* (or 阿驛 *a-yi*, **a-yik*),[2] corresponds to an Iranian form without *n*, as still occurs in Kurd *hežir* or *ezir*. There is another reading, 駔 *tsań*, which is not at the outset to be rejected, as has been done by WATTERS[3] and HIRTH.[4] The *Pen ts'ao kań mu*[5] comments that the pronunciation of this character (and this is apparently an ancient gloss) should be 楚 *č'u*, **dzu*, **tsu*, **ts'u*, so that we obtain **adzu*, **atsu*, **ats'u*. This would correspond to an ancient Iranian form **aju*. At any rate, the Chinese transcriptions, in whatever form we may adopt them, have nothing to do with New Persian *anjir*, as asserted by Hirth, but belong to an older stage of Iranian speech, the Middle Persian.

(2) 映日 *yiń-ži*,[6] **ań-žit(r)*. This is not "apparently a tran-

---

[1] STUART, Chinese Materia Medica, p. 174. The *Či wu miń ši t'u k'ao* (Ch. 36, p. 2), however, speaks of the fig of Yün-nan as a large tree. According to F. N. MEYER (Agricultural Explorations in the Orchards of China, p. 47), the fig is grown in northern China only as an exotic, mostly in pots and tubs. In the milder parts of the country large specimens are found here and there in the open. He noticed black and white varieties. They are cultivated in Šan-hwa 善化 in the prefecture of Č'ań-ša, Hu-nan (*Šan hwa hien či*, Ch. 16, p. 15 b, ed. 1877), also in the prefecture of Šun-t'ien, Či-li (*Kwań-sü Šun t'ien fu či*, Ch. 50, p. 10).

[2] *Yu yań tsa tsu*, Ch. 18, p. 13.

[3] Essays on the Chinese Language, p. 349.

[4] *Journal Am. Or. Soc.*, Vol. XXX, p. 20.

[5] Ch. 31, p. 9.

[6] *Pen ts'ao kań mu*, Ch. 31, p. 26.

410

scription of Hindustani añjir," as affirmed by Hirth, but of New Persian *anjīr* or *enjir*, the Hindustānī (as well as Sanskrit *añjīra*) being simply borrowed from the Persian; Bukhārā *injir*, Afghan *intsir;* Russian *indžaru.*

(3) Fu-lin 底 楠 *ti-ni* or *ti-čen* 珍 or 瑠 (*ti-tsen, *ti-ten); the latter variant is not necessarily to be rejected, as is done by Hirth. Cf. Assyrian *tittu* (from *tintu); Phœnician *tīn;* Hebrew *ti'nu, te'ēnāh;*[1] Arabic *tīn, tine, tima;* Aramaic *ts'īntā, tēnta, tena;* Pahlavi *tin* (Semitic loan-word). The Semitic name is said to have taken its starting-point from south-eastern Arabia, where also, in the view of the botanists, the origin of fig-culture should be sought; but in view of the Assyrian word and the antiquity of the fig in Assyria,[2] this theory is not probable. There is no doubt that the Chinese transcription answers to a Semitic name; but that this is the Aramaic name, as insisted on by Hirth in favor of his theory that the language of Fu-lin should have been Aramaic, is not cogent. The transcription *ti-ni*, on the contrary, is much nearer to the Arabic, Phœnician, and Hebrew forms.[3]

(4) 優曇鉢 (or better 跋) *yu-t'an-po*, *u-dan-pat(par), *u-dan-bar = Sanskrit *udambara (Ficus glomerata).*[4] According to Li Ši-čen, this name is current in Kwaṅ-tuṅ.

(5) 無 花 果 *wu hwa kwo* ("flowerless fruit"),[5] Japanese *ičijiku.* The erroneous notion that the fig-tree does not bloom is not peculiar to Albertus Magnus, as Hirth is inclined to think, but goes back to times of antiquity, and occurs in Aristotle and Pliny.[6] This wrong observation arose from the fact that the flowers, unlike those of most fruit-trees, make no outward appearance, but are concealed within the

---

[1] In the so-called histories of the fig concocted by botanists for popular consumption, one can still read the absurdity that Latin *ficus* is to be derived from Hebrew *feg.* Such a Hebrew word does not exist. What does exist in Hebrew, is the word *pag*, occurring only in Canticle (II, 13), which, however, is not a general term for the fig, but denotes only a green fig that did not mature and that remained on the tree during the winter. Phonetically it is impossible to connect this Hebrew word with the Latin one. In regard to the fig among the Semites, see, above all, the excellent article of E. Levesque in the Dictionnaire de la Bible (Vol. II, col. 2237).

[2] E. Bonavia, Flora of the Assyrian Monuments, p. 14.

[3] It is surprising to read Hirth's conclusion that "*ti-ni* is certainly much nearer the Aramean word than the Greek συκῆ [better σῦκον] for fig, or ἐρινεός for caprificus." No one has ever asserted, or could assert, that these Greek words are derived from Semitic; their origin is still doubtful (see Schrader in Hehn, Kulturpflanzen, p. 100).

[4] *Fan yi miṅ yi tsi,* Ch. 8, p. 5.

[5] Also other fruits are described under this name (see *Či wu miṅ ši t'u k'ao,* Ch. 16, pp. 58–60). The terms under 4 and 5 are identified by Kao Ši-ki 高 士 奇 in his *T'ien lu ši yü* 天 祿 識 餘 (Ch. A, p. 60, published in 1690, ed. of Šwo liṅ).

[6] XVI, 39.

fruit on its internal surface. On cutting open a fig when it has attained little more than one-third its size, the flowers will be seen in full development.[1]

The common fig-tree (*Ficus carica*) is no less diffused over the Iranian plateau than the pomegranate. The variety *rupestris* is found in the mountains Kuh-Kiluyeh; and another species, *Ficus johannis*, occurs in Afghanistan between Tebbes and Herat, as well as in Baluchistan.[2] In the mountain districts of the Taurus, Armenia, and in the Iranian table-lands, fig-culture long ago reached a high development. Toward the east it has spread to Khorasan, Herat, Afghanistan, as well as to Merw and Khiwa.[3] There can be no doubt, either, that the fig was cultivated in Sasanian Persia; for it is mentioned in Pahlavi literature (above, p. 192), and we have a formal testimony to this effect in the Annals of the Liang dynasty, which ascribe *udambara* to Po-se (Persia) and describe the blossoms as charming.[4] In India, as stated, this term refers to *Ficus glomerata;* in China, however, it appears to be also used for *Ficus carica*. Hüan Tsaṅ[5] enumerates *udambara* among the fruits of India.

Strabo[6] states that in Hyrcania (in Bactria) each fig-tree annually produced sixty medimni (one bushel and a half) of fruit. According to Herodotus,[7] Croesus was dissuaded from his expedition against Cyrus on the plea that the Persians did not even drink wine, but merely water, nor did they have figs for sustenance. This, of course, is an anecdote without historical value, for we know surely enough that the ancient Persians possessed both grapes and wine. Another political anecdote of the Greeks is that of Xerxes, who, by having Attic figs served at his meals, was daily reminded of the fact that the land where they grow was not yet his own. The new discovery of the presence of figs in ancient Babylonia warrants the conclusion that they were likewise known and consumed in ancient Persia.

We have no means of ascertaining as to when and how the fig spread from Iran to China. The *Yu yaṅ tsa tsu* is reticent as to the transmission, and merely describes the tree as existing in Fu-lin and

---

[1] LINDLEY and MOORE, Treasury of Botany, pt. I, p. 492.

[2] C. JORET, Plantes dans l'antiquité, Vol. II, p. 45.

[3] G. EISEN, The Fig: Its History, Culture, and Curing, p. 20 (U. S. Department of Agriculture, Washington, 1901).

[4] *Liaṅ šu*, Ch. 54, p. 14 b. Read *yu-t'an-po* instead of *yu-po-t'an*, as there printed through an oversight.

[5] *Ta T'aṅ si yü ki*, Ch. 2, p. 8.

[6] II. 1, 14.

[7] I, 71.

Persia.[1] We have, however, the testimony of the Arabic merchant Soleiman, who wrote in A.D. 851, to the effect that the fig then belonged to the fruits of China.[2]

Bretschneider has never written on the subject, but did communicate some notes to the botanist Solms-Laubach, from whom they were taken over by G. EISEN.[3] Here we are treated to the monstrous statement, "The fig is supposed to have reached China during the reign of the Emperor Tschang-Kien [sic!], who fitted out an expedition to Turan in the year 127 A.D." [sic!]. It is safe to say that Bretschneider could not have perpetrated all this nonsense; but, discounting the obvious errors, there remains the sad fact that again he credited Čan K'ien with an introduction which is not even ascribed to him by any Chinese text. It is not necessary to be more Chinese than the Chinese, and this Changkienomania is surely disconcerting. What a Hercules this Čan K'ien must have been! It has never happened in the history of the world that any individual ever introduced into any country such a stupendous number of plants as is palmed off on him by his epigone admirers.

Li Ši-čen, in his notice of the "flowerless fruit," does not fall back on any previous *Pen ts'ao;* of older works he invokes only the *Yu yan tsa tsu* and the *Fan yü či* 方輿志, which mention the *udambara* of Kwan-si.

The fig of Yün-nan deserves special mention. Wu K'i-tsün, author of the excellent botanical work *Či wu min ši t'u k'ao,* has devoted a special chapter (Ch. 36) to the plants of Yün-nan, the first of these being the *yu-t'an* (udambara) flower, accompanied by two illustrations. From the texts assembled by him it becomes clear that this tree was introduced into Yün-nan from India by Buddhist monks. Among other stories, he repeats that regarding the monk P'u-t'i(Bodhi)-pa-po, which has been translated by C. SAINSON;[4] but whereas Yan Šen, in his *Nan čao ye ši,* written in 1550, said that one of these trees planted by the monk was still preserved in the Temple of the Guardian Spirit 土主廟 of Yün-nan fu, Wu K'i-tsün states after the *Yün-nan t'un či* that for a long time none remained in existence, owing to the ravages and burnings of troops. Judging from the illustration, the fig-tree of Yün-nan is a species different from *Ficus carica.* The genus *Ficus*

---

[1] Contrary to what is stated by A. DE CANDOLLE (Origin of Cultivated Plants, p. 296) after Bretschneider. But the description of the fig in that Chinese work leaves no doubt that the author speaks from observation, and that the fig, accordingly, was cultivated in the China of his time.

[2] M. REINAUD, Relation des voyages, Vol. I, p. 22.

[3] *Op. cit.,* p. 20.

[4] Histoire du Nan-Tchao, p. 196.

comprises nearly a hundred and sixty species, and of the cultivated fig there is a vast number of varieties.

According to the *Yamato-honzō*[1] of 1709, figs (*ičijiku*) were first introduced into Nagasaki in the period Kwan-ei 寛永 (1624–44) from the islands in the South-Western Ocean. This agrees with E. KAEMPFER'S[2] statement that figs were brought into Japan and planted by Portuguese.

---

[1] Ch. 10, p. 26 b.
[2] History of Japan, Vol. I, p. 180 (ed. reprinted Glasgow, 1906).

THE OLIVE

43. The *Yu yan tsa tsu*[1] has the following notice of an exotic plant: "The *ts'i-t'un* 齊墩 (*dzi-tun, *zi-tun) tree has its habitat in the country Po-se (Persia), likewise in the country Fu-lin (Syria). In Fu-lin it is termed 齊厤 *ts'i-t'i*[2] (*dzi, zi-ti). The tree grows to a height of twenty or thirty feet. The bark is green, the flowers are white, resembling those of the shaddock (*yu* 柚, *Citrus grandis*), and very fragrant. The fruit is similar to that of the *yan-t'ao* 楊桃 (*Averrhoa carambola*) and ripens in the fifth month. The people of the Western countries press an oil out of it for frying cakes and fruit, in the same manner as sesame seeds (*kü-šen* 巨勝)[3] are utilized in China."

The transcription *ts'i-t'un* has been successfully identified by HIRTH[4] with Persian *zeitun*, save that we have to define this form as Middle Persian; and Fu-lin *ts'i-t'i* with Aramaic *zaitā* (Hebrew *zayiθ*). This is the olive-tree (*Olea Europaea*).[5] The Persian word is a loan from the Semitic, the common Semitic form being *zeitu (Arabic *zeitun*). It is noteworthy that the Fu-lin form agrees more closely with Grusinian and Ossetic *zet'i*, Armenian *jēt*, *dzēt* ("olive-oil"), *zeit* ("olive"), Arabic *zait*,[6] than with the Aramaic word. The olive-tree, mentioned in Pahlavi literature (above, p. 193), grows spontaneously in Persia and Baluchistan, but the cultivated species was in all likelihood received by the Iranians (as well as by the Armenians) from the Semites. The olive-tree was known in Mesopotamia at an early date: objects in clay in the form of an olive belonging to the time of Urukagina, one of the pre-Sargonic rulers of Lagash, are still extant.[7]

---

[1] Ch. 18, p. 11.

[2] A gloss thus indicates the reading of this character by the *fan ts'ie* 湯分.

[3] See above, p. 292.

[4] *Journal Am. Or. Soc.*, Vol. XXX, 1910, p. 19.

[5] See, for instance, the illustrated article "olivier" in DUJARDIN-BEAUMETZ and EGASSE, Plantes médicinales indigènes et exotiques (p. 492, Paris, 1889), which is a very convenient and commendable reference-book, particularly valuable for its excellent illustrations. Cf. also S. KRAUSS, Talmudische Archäologie, Vol. II, p. 214; S. FRAENKEL, Die aramäischen Fremdwörter im Arabischen, p. 147.

[6] W. MILLER, Sprache der Osseten, p. 10; HÜBSCHMANN, Arm. Gram., p. 309.

[7] HANDCOCK, Mesopotamian Archæology, p. 13. The contributions which A. ENGLER has made to the olive in Hehn's Kulturpflanzen (p. 118) are just as singular as his notions of the walnut. Leaves of the olive-tree have been found in Pliocene deposits near Mongardino north-west of Bologna, and this is sufficient for Engler to "prove" the autochthonous character of the tree in Italy. All it proves, if the

415

SCHLIMMER[1] says that *Olea europaea* is largely cultivated by the inhabitants of Mendjil between Besht and Ghezwin in Persia, and that the olives are excellent; nevertheless the oil extracted is very bad and unfit to eat. The geographical distribution of the tree in Iran has well been traced by F. SPIEGEL.[2]

The word *ts'i-t'un* has been perpetuated by the lexicographers of the Emperor K'ien-luṅ (1736–95). It makes its appearance in the Dictionary of Four Languages, in the section "foreign fruit."[3] For the Tibetan and Mongol forms, one has chosen the transcriptions *č'i-tun siu* (transcribing *tse* 子) and *čitun jimin* respectively; while it is surprising to find a Manchu equivalent *ulusun*, which has been correctly explained by H. C. v. d. Gabelentz and Sakharov. In the Manchu-Chinese Dictionary *Ts'iṅ wen pu hui*, published in 1771, we find the

---

fact be correct, is that a wild olive once occurred in the Pliocene of Italy, which certainly does not exclude the idea and the well-established historical fact that the cultivated olive was introduced into Italy from Greece in historical times. The notice of Pliny (xv, 1) weighs considerably more in this case than any alleged palæontological wisdom, and the Pliocene has nothing to do with historical times of human history. The following is truly characteristic of Engler's uncritical stand-point and his inability to think historically: "Since the fruits of the olive-tree are propagated by birds, and in many localities throughout the Mediterranean the con-ditions for the existence of the tree were prepared, it was quite natural also that the tree settled in the localities suitable for it, before the Oriental civilized nations made one of the most important useful plants of it." If the birds were the sole propagators of the tree, why did they not carry it to India, the Archipelago, and China, where it never occurred? The distribution of the olive shows most clearly that it was brought about by human activity, and that we are confronted with a well-defined geographical zone as the product of human civilization,—Western Asia and the Mediterranean area. There is nothing in Engler like the vision and breadth of thought of a de Candolle, in whose Origin of Cultivated Plants we read (p. 280), "The question is not clearly stated when we ask if such and such olive-trees of a given locality are really wild. In a woody species which lives so long and shoots again from the same stock when cut off by accident, it is impossible to know the origin of the individuals observed. They may have been sown by man or birds at a very early epoch, for olive-trees of more than a thousand years old are known. The effect of such sowing is a naturalization, which is equivalent to an extension of area. The point in question is, therefore, to discover what was the home of the species in very early prehistoric times, and how this area has grown larger by dif-ferent modes of transport. It is not by the study of living olive-trees that this can be answered. We must seek in what countries the cultivation began, and how it was propagated. The more ancient it is in any region, the more probable it is that the species has existed wild there from the time of those geological events which took place before the coming of prehistoric man." Here we meet a thinker of critical acumen, possessed of a fine historical spirit, and striving for truth nobly and honestly; and there, a dry pedant, who thinks merely in terms of species and genera, and is unwilling to learn and to understand history.

[1] Terminologie, p. 406.

[2] Eranische Altertumskunde, Vol. I, pp. 257–258.

[3] Appendix, Ch. 3, p. 10.

following definition of *ulusun* in Chinese: "*Ts'i-t'un* is a foreign fruit, which is produced in the country Po-se (Persia). The bark of the tree is green, the flowers are white and aromatic. Its fruit ripens in the fifth month and yields an oil good for frying cakes." This is apparently based on the notice of the *Yu yan tsa tsu*. The Manchu word *ulusun* (-*sun* being a Manchu ending) seems to be an artificial formation based on Latin *oleum* (from Greek *elaion*), which was probably conveyed through the Jesuit missionaries.

The olive remained unknown to the Japanese; their modern botanical science calls it *oreifu* 阿列布, which reproduces our "olive."[1] The Japanese botanists, without being aware of the meaning of *ts'i-tun*, avail themselves of the characters for this word (reading them *ego-no-ki*) for the designation of *Styrax japonica*.[2]

The so-called Chinese olive, *kan-lan* 橄欖, has no affinity with the true olive of the West-Asiatic and Mediterranean zone, although its appearance comes very near to this fruit.[3] The name *kan-lan* applies to *Canarium album* and *C. pimela*, belonging to the order *Burseraceae*, while the olive ranks in that of the *Oleaceae*.[4] Ma Či, who, in his *K'ai*

---

[1] MATSUMURA, No. 2136.

[2] *Ibid.*, No. 3051.

[3] The *kan-lan* tree itself is suspected to be of foreign origin; it was most probably introduced from Indo-China into southern China. Following are briefly the reasons which prompt me to this opinion. 1. According to Li Ši-čen, the meaning of the name *kan-lan* remains unexplained, and this comment usually hints at a foreign word. The ancient pronunciation was *kam-lam or *kam-ram, which we still find in Annamese as *kam-lan*. The tree abounds in Annam, the fruit being eatable and preserved in the same manner as olives (PERROT and HURRIER, Mat. méd. et pharmacopée sino-annamites, p. 141). Moreover, we meet in Pa-yi, a T'ai language spoken in Yün-nan, a word (*mak*)-*k'am*, which in a Pa-yi-Chinese glossary is rendered by Chinese *kan-lan* (the element *mak* means "fruit"; see F. W. K. MÜLLER, T'oung Pao, Vol. III, p. 27). The relationship of Annamese to the T'ai languages has been clearly demonstrated by H. MASPERO, and it seems to me that Chinese *kam-lam is borrowed from Annam-T'ai. There are many more such Chinese botanical names, as I hope to show in the near future. 2. The plant appears in Chinese records at a comparatively recent date. It is first described in the *Nan čou i wu či* of the third century as a plant of Kwaň-tuň and Fu-kien and in the *Nan faň ts'ao mu čwaň* (Ch. c, p. 3 b). It is mentioned as a tree of the south in the *Kin lou tse* of the Emperor Yüan of the Liang in the sixth century (see above, p. 222). A description of it is due to Liu Sün in his *Liň piao lu i* (Ch. B, p. 5 b). In the materia medica it first appears in the *K'ai pao pen ts'ao* of the end of the tenth century. 3. The tree remained always restricted to the south-eastern parts of China bordering on Indo-China. According to the *San fu hwaň t'u*, it belonged to the southern plants brought to the Fu-li Palace of the Han Emperor Wu after the conquest of Nan Yüe (cf. above, p. 262).

[4] The fruit of *Canarium* is a fleshy drupe from three to six cm in length, which contains a hard, triangular, sharp-pointed seed. Within this are found one or more oily kernels. The flesh of the fresh, yellowish-green fruit, like that of the true olive, is somewhat acrid and disagreeable, and requires special treatment before it can

pao pen ts'ao (written between A.D. 968 and 976), describes the *kan-lan*, goes on to say that "there is also another kind, known as *Po-se kan-lan* ('Persian *kan-lan*'), growing in Yuṅ čou 邕 州,[1] similar to *kan-lan* in color and form, but different in that the kernel is divided into two sections; it contains a substance like honey, which is soaked in water and eaten." The *Šaṅ se čou či*[2] mentions the plant as a product of Šaṅ-se čou in Kwaṅ-si. It would be rather tempting to regard this tree as the true olive, as tentatively proposed by STUART;[3] but I am not ready to subscribe to this theory until it is proved by botanists that the olive-tree really occurs in Kwaṅ-si. Meanwhile it should be pointed out that weighty arguments militate against this supposition. First of all, the *Po-se kan-lan* is a wild tree: not a word is said to the effect that it is cultivated, still less that it was introduced from Po-se. If it had been introduced from Persia, we should most assuredly find it as a cultivation; and if such an introduction had taken place, why should it be confined to a few localities of Kwaṅ-si? Li Ši-čen does not express an opinion on the question; he merely says that the *faṅ* 方 *lan*, another variety of *Canarium* to be found in Kwaṅ-si (unidentified), is a kind of *Po-se kan-lan*, which proves distinctly that he regards the latter as a wild plant. The T'ang authors are silent as to the introduction of the olive; nevertheless, judging from the description in the *Yu yaṅ tsa tsu*, it may be that the fruit was imported from Persia under the T'ang. Maybe the *Po-se kan-lan* was so christened on account of a certain resemblance of its fruit to the olive; we do not know. There is one specific instance on record that the Po-se of Ma Či applies to the Malayan Po-se (below, p. 483); this may even be the case here, but the connection escapes our knowledge.

S. JULIEN[4] asserts that the Chinese author from whom he derives his information describes the olive-tree and its fruit, but adds that the use of it is much restricted. The Chinese name for the tree is not given. Finally, it should be pointed out that Ibn Baṭūṭa of the four-

---

be made palatable. Its most important constituent is fat, which forms nearly one-fourth of the total nutritive material. Cf. W. C. BLASDALE, Description of Some Chinese Vegetable Food Materials, p. 43, with illustration (U. S. Department of Agriculture, Bull. No. 68, 1899). The genus *Canarium* comprises about eighty species in the tropical regions of the Old World, mostly in Asia (ENGLER, Pflanzenfamilien, Vol. III, pt. 4, p. 240).

[1] Name under the T'ang dynasty of the present prefecture Nan-niṅ in Kwaṅ-si Province.

[2] Ch. 14, p. 7 b (see above, p. 409).

[3] Chinese Materia Medica, p. 89.

[4] Industries de l'empire chinois, p. 120.

teenth century positively denies the occurrence of olives in China.[1] Of course, this Arabic traveller is not an authority on Chinese affairs: many of his data concerning China are out and out absurd. He may even not have visited China, as suggested by G. Ferrand; notwithstanding, he may be right in this particular point. Likewise the Archbishop of Soltania, who wrote about 1330, states, "There groweth not any oil olive in that country."[2]

---

[1] YULE, Cathay, Vol. IV, p. 118.
[2] Ibid., Vol. III, p. 96.

CASSIA PODS AND CAROB

44. In his *Pen ts'ao ši i*, written during the first half of the eighth century, Č'en Ts'aṅ-k'i has this notice regarding an exotic plant: "*A-lo-p'o* 阿勒勃 (*a-lak-bwut*) grows in the country Fu-lin (Syria), its fruit resembling in shape that of the *tsao kia* 皂莢 (*Gleditschia* or *Gymnocladus sinensis*), save that it is more rounded and elongated. It is sweet of taste and savory."[1]

In the *Čeṅ lei pen ts'ao*[2] we read that "*a-lo-p'o* grows in the country Fu-ši 佛近"; that is, Bhoja, Sumatra. Then follows the same description as given above, after Č'en Ts'aṅ-k'i. The name *p'o-lo-men tsao kia* 婆羅門皂莢 is added as a synonyme. Li Ši-čen[3] comments that P'o-lo-men is here the name of a Si-yü 西域 ("Western Regions") country, and that Po-se is the name of a country of the south-western barbarians; that is, the Malayan Po-se. The term *p'o-lo-men tsao kia*, which accordingly would mean "*Gleditschia* of the P'o-lo-men country," he ascribes to Č'en Ts'aṅ-k'i, but in his quotation from this author it does not occur. The country P'o-lo-men here in question is the one mentioned in the *Man šu*.[4]

A somewhat fuller description of this foreign tree is contained in the *Yu yaṅ tsa tsu*,[5] as follows: "The Persian *tsao kia* (*Gleditschia*) has its habitat in the country Po-se (Persia), where it is termed *hu-ye-yen-mo* 忽野簷默, while in Fu-lin it is styled *a-li-k'ü-fa* 阿梨去伐.[6] The tree has a height of from thirty to forty feet, and measures from four to five feet in circumference. The leaves resemble those of *Citrus medica* (*kou yüan* 枸緣), but are shorter and smaller. During the cold season it does not wither.[7] It does not flower, and yet bears fruit.[8] Its pods are two feet long. In their interior are shells (*ko ko* 隔隔). Each of these encloses a single seed of the size of a finger, red of color,

---

[1] *Pen ts'ao kaṅ mu*, Ch. 31, p. 9 b, where the name of the plant is wrongly written *a-p'o-lo*. The correct form *a-lo-p'o* is given in the *Čeṅ lei pen ts'ao*.

[2] Ch. 12, p. 56 (ed. of 1587).

[3] *Pen ts'ao kaṅ mu*, Ch. 31, p. 9 b.

[4] See below, p. 468.

[5] Ch. 18, p. 12. Also Li Ši-čen has combined this text with the preceding one under the heading *a-p'o-lo* (instead of *a-lo-p'o*).

[6] The *Pen ts'ao kaṅ mu* (Ch. 31, p. 9 b), in quoting this text, gives the Po-se name as *hu-ye-yen* and the Fu-lin name only as *a-li*.

[7] This means, it is an evergreen.

[8] This is due to erroneous observation.

420

and extremely hard. The interior [the pulp] is as black as [Chinese] ink and as sweet as sugar-plums. It is eatable, and is also employed in the pharmacopœia."

The tree under consideration has not yet been identified, at least not from the sinological point of view.[1] The name a-lo-p'o is Sanskrit; and the ancient form *a-lak(rak, rag)-bwut(bud) is a correct and logical transcription of Sanskrit aragbadha, aragvadha, āragvadha, or ārgvadha, the Cassia or Cathartocarpus fistula (Leguminosae), already mentioned by the physician Caraka, also styled suvarṇaka ("gold-colored") and rājataru ("king's tree").[2] This tree, called the Indian laburnum, purging cassia, or pudding pipe tree from its peculiar pods (French canéficier), is a native of India, Ceylon, and the Archipelago[3] (hence Sumatra and Malayan Po-se of the Chinese), "uncommonly beautiful when in flower, few surpassing it in the elegance of its numerous long, pendulous racemes of large, bright-yellow flowers, intermixed with the young, lively green foliage."[4] The fruit, which is common in most bazars of India, is a brownish pod, about sixty cm long and two cm thick. It is divided into numerous cells, upwards of forty, each containing one smooth, oval, shining seed. Hence the Chinese comparison with the pod of the Gleditschia, which is quite to the point. These pods are known as cassia pods. They are thus described in the "Treasury of Botany": "Cylindrical, black, woody, one to two feet long, not splitting, but marked by three long furrows, divided in the interior into a number of compartments by means of transverse partitions, which project from the placentæ. Each compartment of the fruit contains a single seed, imbedded in pulp, which is used as a mild laxative." Whether the tree is cultivated in Asia I do not know; GARCIA DA ORTA affirms that he saw it only in a wild state.[5] The description of the tree and fruit in the Yu yaṅ tsa tsu is fairly correct. Cassia fistula is indeed from twenty to thirty feet high (in Jamaica even fifty feet). The seed, as stated there, is of a reddish-brown color, and the pulp is of a dark viscid substance.

---

[1] STUART (Chinese Materia Medica, p. 496) lists the name a-p'o-lo (instead of a-lo-p'o) among "unidentified drugs." Bretschneider has never noted it.

[2] A large number of Sanskrit synonymes for the tree are enumerated by RÖDIGER and POTT (Zeitschrift f. d. K. d. Morg., Vol. VII, p. 154); several more may be added to this list from the Bower Manuscript.

[3] GARCIA DA ORTA (Markham, Colloquies, p. 114) adds Malacca and Sofala. In Javanese it is teṅguli or treṅguli.

[4] W. ROXBURGH, Flora Indica, p. 349.

[5] Likewise F. PYRARD (Vol. II, p. 361, ed. of Hakluyt Society), who states that "it grows of itself without being sown or tended."

When I had established the above identification of the Sanskrit name, it was quite natural for me to lay my hands on MATSUMURA'S "Shokubutsu mei-i" and to look up *Cassia fistula* under No. 754: it was as surprising as gratifying to find there, "*Cassia fistula* 阿勃勒 *namban-saikachi*." This Japanese name means literally the "*Gleditschia japonica* (*saikači*=Chinese *tsao-kia-tse*) of the Southern Barbarians" (Chinese Nan Fan). The Japanese botanists, accordingly, had succeeded in arriving at the same identification through the description of the plant; while the philological equation with the Sanskrit term escaped them, as evidenced by their adherence to the wrong form *a-p'o-lo*, sanctioned by the *Pen ts'ao kan mu*. The case is of methodological interest in showing how botanical and linguistic research may supplement and corroborate each other: the result of the identification is thus beyond doubt; the rejection of *a-p'o-lo* becomes complete, and the restitution of *a-lo-p'o*, as handed down in the *Čen lei pen ts'ao*, ceases to be a mere philological conjecture or emendation, but is raised into the certainty of a fact.

The Arabs know the fruit of this tree under the names *xarnub hindi* ("Indian carob")[1] and *xiyār šanbār* ("cucumber of necklaces," from its long strings of golden flowers).[2] Abu'l Abbās, styled en-Nebāti ("the Botanist"), who died at Sevilla in 1239, the teacher of Ibn al-Baiṭār, who preserved extracts from his lost work Rihla ("The Voyage"), describes *Cassia fistula* as very common in Egypt, particularly in Alexandria and vicinity, whence the fruit is exported to Syria;[3] it commonly occurs in Bassora also, whence it is exported to the Levant and Irak. He compares the form of the tree to the walnut and the fruit to the carob. The same comparison is made by Išak Ibn Amrān, who states in Leclerc's translation, "Dans chacun de ces tubes est renfermée une pulpe noire, sucrée et laxative. Dans chaque compartiment est un noyau qui a le volume et la forme de la graine de caroubier. La partie employée est la pulpe, à l'exclusion du noyau et du tube."

The Persians received the fruit from the Arabs on the one hand, and from north-western India on the other. They adopted the Arabic word *xiyār-šanbār*[4] in the form *xiyār-čambar* (compare also Armenian *xiar-*

---

[1] LECLERC, Traité des simples, Vol. II, p. 17.

[2] *Ibid.*, p. 64. Also *qitta hindi* ("Indian cucumber"), *ibid.*, Vol. III, p. 62.

[3] GARCIA DA ORTA says that it grows in Cairo, where it was also found by Pierre Belon. In ancient times, however, the tree did not occur in Egypt: LORET, in his Flore pharaonique, is silent about it. It was no doubt brought there by the Arabs from India.

[4] GARCIA DA ORTA spells it *hiar-xamber*.

šamb, Byzantine Greek χιαρσάμβερ, χεασαμπάρ); and it is a Middle-Persian variation of this type that is hidden in the "Persian" transcription of the *Yu yan tsa tsu*, *hu-ye-yen-mo* 忽野簷獣, anciently *xut(xur)-ya-džem(dzem)-m'wăk(băk, băx). The prototype to be restored may have been *xaryadžambax. There is a New-Persian word for the same tree and fruit, *bakbar*. It is also called *kābuli* ("coming from Kabul").

The Fu-lin name of the plant is 阿梨去伐 *a-li-k'ü-fa*, *a-li(ri)-go-vaδ. I. LOEW[1] does not give an Aramaic name for *Cassia fistula*, nor does he indicate this tree, neither am I able to find a name for it in the relevant dictionaries. We have to take into consideration that the tree is not indigenous to western Asia and Egypt, and that the Arabs transplanted it there from India (cf. the Arabic terms given above, "Indian carob," and "Indian cucumber"). The Fu-lin term is evidently an Indian loan-word, for the transcription *a-ri-go-vaδ corresponds exactly to Sanskrit *ārgvadha*, answering to an hypothetical Aramaic form *arigbada or *arigfada. In some editions of the *Yu yan tsa tsu*, the Fu-lin word is written *a-li* or *a-li-fa*, *a-ri-vaδ. These would likewise be possible forms, for there is also a Sanskrit variant *ārevata* and an Indian vernacular form *ali* (in Panjābī).

The above texts of Č'en Ts'an-k'i and Twan Č'en-ši, author of the *Yu yan tsa tsu*, give occasion for some further comments. PELLIOT[2] maintained that the latter author, who lived toward the end of the ninth century, frequently derived his information from the former, who wrote in the first part of the eighth century;[3] from the fact that Č'en in many cases indicates the foreign names of exotic plants, Pelliot is inclined to infer that Twan has derived from him also his nomenclature of plants in the Fu-lin language. This is by no means correct. I have carefully read almost all texts preserved under the name of Č'en (or his work, the *Pen ts'ao ši i*) in the *Čen lei pen ts'ao* and *Pen ts'ao kan mu*, and likewise studied all notices of plants by Twan; with the result that Twan, with a few exceptions, is independent of Č'en. As to Fu-lin names, none whatever is recorded by the latter, and the above text is the only one in which the country Fu-lin figures, while he gives the plant-name solely in its Sanskrit form. In fact, all the foreign names noted by Č'en come from the Indo-Malayan area. The above case shows plainly that Twan's information does not at all depend on Č'en's

---

[1] Aramaeische Pflanzennamen.

[2] *T'oung Pao*, 1912, p. 454.

[3] The example cited to this effect (*Bull. de l'Ecole française*, Vol. IV, p. 1130) is not very lucky, for in fact the two texts are clearly independent.

passage: the two texts differ both as to descriptive matter and nomenclature. In regard to the Fu-lin information of Twan, HIRTH's opinion[1] is perfectly correct: it was conveyed by the monk Wan, who had hailed directly from Fu-lin.[2] The time when he lived is unknown, but most probably he was a contemporary of Twan. The Fu-lin names, accordingly, do not go back to the beginning of the eighth century, but belong to the latter half of the ninth.

An interesting point in connection with this subject is that both the Iranian and the Malayan Po-se play their rôle with reference to the plant and fruit in question. This, as far as I know, is the only instance of this kind. Fortunately, the situation is perfectly manifest on either side. The fact that Twan Č'eṅ-ši hints at the Iranian Po-se (Persia) is well evidenced by his addition of the Iranian name; while the tree itself is not found in Persia, and merely its fruit was imported from Syria or India. The Po-se, alluded to in the *Čeṅ lei pen ts'ao* and presumably traceable to Č'en Ts'aṅ-k'i, unequivocally represents the Malayan Po-se: it is joined to the names of Sumatra and P'o-lo-men; and *Cassia fistula* is said to occur there, and indeed occurs in the Malayan zone. Moreover, Li Ši-čen has added such an unambiguous definition of the location of this Po-se, that there is no room for doubt of its identity.

45. Reference has been made to the similarity of cassia pods to carob pods, and it would not be impossible that the latter were included in the "Persian Gleditschia" of the Chinese.

Ceratonia siliqua, the carob-tree, about thirty feet in height, is likewise a genus of the family *Leguminosae*, a typical Mediterranean cultivation. The pods, called carob pods, carob beans, or sometimes sugar pods, contain a large quantity of mucilaginous and saccharine matter, and are commonly employed in the south of Europe for feeding live-stock, and occasionally, in times of scarcity, as human food. The popular names "locust-pods" or "St. John's Bread" rest on the supposition that the pods formed the food of St. John in the wilderness (LUKE, xv, 16); but there is better reason to believe that the locusts of St. John were the animals so called, and these are still eaten in the Orient. The common Semitic name for the tree and fruit is Assyrian *xarūbu*, Aramaic *xārūbā*, Arabic *xarrūb* and *xarnub*.[3] New Persian *xurnūb* (*khurnūb*) or *xarnūb*, also *xarrūb* (hence Osmanli *xarúp*,[4] Neo-

---

[1] *Journal Am. Or. Soc.*, Vol. XXX, 1910, p. 18.

[2] Cf. above, p. 359.

[3] Egyptian *džarudž, garuta, darruga;* Coptic *garate*, are Greek loan-words (the tree never existed in Egypt, as already stated by Pliny, XIII, 16), from κεράτια.

[4] Also *ketšibujnuzu* ("goat's horn").

Greek χαρούπιον, Italian *carrobo* or *carrubo*, Spanish *algarrobo*, French *caroube* or *carouge*), is based on the Semitic name. *Lelekī* is another Persian word for the tree, according to SCHLIMMER,[1] peculiar to Gilan.

The Arabs distinguish three varieties of carob, two of which are named *saidalāni* and *šābuni*.[2] There is no doubt that the Arabs who were active in transplanting the tree to the west conveyed it also to Persia. A. de Candolle does not mention the occurrence of the carob in that country. It is pointed out, however, by the Mohammedan writers on Persia. It is mentioned as a cultivation of the province Sābūr by Muqaddasī[3] and Yāqūt.[4] Abu Mansur discusses the medicinal properties of the fruit in his pharmacopœia; he speaks of a Syrian and a Nabathæan *xarnūb*.[5] SCHLIMMER[6] remarks that the tree is very common in the forest of Gilan; the pods serve the cows as food, and are made into a sweet and agreeable syrup. No Sanskrit name for the tree exists, and the tree itself did not anciently occur in India.[7]

A botanical problem remains to be solved in connection with *Cassia fistula*. DuHALDE[8] mentions cassia-trees (*Cassia fistula*) in the province of Yün-nan toward the kingdom of Ava. "They are pretty tall, and bear long pods; whence 'tis called by the Chinese, Chang-ko-tse-shu, the tree with long fruit (長 菓 子 樹); its pods are longer than those we see in Europe, and not composed of two convex shells, like those of ordinary pulse, but are so many hollow pipes, divided by partitions into cells, which contain a pithy substance, in every respect like the cassia in use with us." S. W. WILLIAMS[9] has the following: "*Cassia fistula*, 槐 花 青 *hwai hwa ts'in*, is the name for the long cylindrical pods of the senna tree (*Cathartocarpus*), known to the Chinese as *č'an kwo-tse šu*, or tree with long fruit. They are collected in Kwan-si for their pulp and seeds, which are medicinal. The pulp is reddish and sweet, and not so drastic as the American sort; if gathered before the seeds are ripe, its taste is somewhat sharp. It is not exported, to any great

---

[1] Terminologie, p. 120. The pods are also styled *tarmiš*.

[2] L. LECLERC, Traité des simples, Vol. II, p. 16.

[3] P. SCHWARZ, Iran, p. 32.

[4] BARBIER DE MEYNARD, Dictionnaire géographique de la Perse, p. 294.

[5] ACHUNDOW, Abu Mansur, p. 59.

[6] Terminologie, p. 119.

[7] The alleged word for the carob, *çimbibheda*, given in the English-Sanskrit Dictionary of A. BOROOAH, is a modern artificial formation from *çimbi* or *çimba* ("pod"). According to WATT, the tree is now almost naturalized in the Salt Range and other parts of the Panjab.

[8] Description of the Empire of China, Vol. I, p. 14 (or French ed., Vol. I, p. 26).

[9] Chinese Commercial Guide, p. 114 (5th ed., 1863).

extent, west of the Cape." F. P. SMITH,[1] with reference to this statement of Williams, asserts that the drug is unknown in Central China, and has not been met with in the pages of the *Pen ts'ao*. Likewise STUART,[2] on referring to DuHalde and Williams, says, "No other authorities are found for this plant occurring in China, and it is not mentioned in the *Pen ts'ao*. The Customs Lists do not mention it; so, if exported as Williams claims, it must be by land routes. The subject is worthy of investigation." *Cassia fistula* is not listed in the work of Forbes and Hemsley.

There is no doubt that the trees described by DuHalde and Williams exist, but the question remains whether they are correctly identified. The name *hwai* used by Williams would rather point to a *Sophora*, which likewise yields a long pod containing one or five seeds, and his description of the pulp as reddish does not fit *Cassia fistula*. Contrary to the opinions of Smith and Stuart, the species of Williams is referred to in the *Pen ts'ao kan mu*.[3] As an appendix to his *a-p'o-lo* (instead of *a-lo-p'o*), Li Ši-čen treats of the seeds of a plant styled *lo-wan-tse* 羅望子, quoting the *Kwei hai yü hen či* by Fan Č'en-ta (1126–93) as follows: "Its habitat is in Kwan-si. The pods are several inches long, and are like those of the *fei tsao* 肥皂 (*Gleditschia* or *Gymnocladus sinensis*) and the *tao tou* 刀豆 (*Canavallia ensiformis*). The color [of the pulp] is standard red 正丹. Inside there are two or three seeds, which when baked are eatable and of sweet and agreeable flavor."[4] This *lo-wan* is identified with *Tamarindus indica*;[5] and this, I believe, is also the above plant of Williams, which must be dissociated from *Cassia fistula*; for, while Li Ši-čen notes the latter as a purely exotic plant, he does not state that it occurs in China; as to *lo-wan*, he merely regards it as a kindred affair on account of the peculiar pods: this does not mean, of course, that the trees yielding these pods are related species. The fruit of *Tamarindus indica* is a large swollen pod from four to six inches long, filled with an acid pulp. In India it is largely used as food, being a favorite ingredient in curries and chutnies, and for pickling fish. It is also employed in making a cooling drink or sherbet.[6]

---

[1] Contributions towards the Materia Medica of China, p. 53.

[2] Chinese Materia Medica, p. 96.

[3] Ch. 31, p. 9 b.

[4] The text is exactly reproduced (see the edition in the *Či pu tsu čai ts'un šu*, p. 24).

[5] MATSUMURA, No. 3076 (in Japanese *čōsen-modama-rabōši*).

[6] WATT, Commercial Products of India, p. 1067.

NARCISSUS

46. The *Yu yaṅ tsa tsu*[1] contains the following notice: "The habitat of the *nai-k'i* 㮈 祇 is in the country Fu-lin (Syria). Its sprouts grow to a height of three or four feet. Its root is the size of a duck's egg. Its leaves resemble those of the garlic (*Allium sativum*). From the centre of the leaves rises a very long stem surmounted by a six-petaled flower of reddish-white color.[2] The heart of this flower is yellow-red, and does not form fruit. This plant grows in the winter and withers during the summer. It is somewhat similar to shepherd's-purse (*tsi* 薺, *Capsella bursa-pastoris*) and wheat.[3] An oil is pressed from the flowers, with which they anoint the body as a preventive of colds, and is employed by the king of Fu-lin and the nobles in his country."

Li Ši-čen, in his *Pen ts'ao kaṅ mu*,[4] has placed this extract in his notice of *šwi sien* 水 仙 (*Narcissus tazetta*),[5] and after quoting it, adds this comment: "Judging from this description of the plant, it is similar to Narcissus; it cannot be expected, of course, that the foreign name should be identical with our own."[6] He is perfectly correct, for the description answers this flower very well, save the comparison with *Capsella*. Dioscorides also compares the leaves of *Narcissus* to those of *Allium*, and says that the root is rounded like a bulb.[7]

The philological evidence agrees with this explanation; for *nai-k'i*, *nai-gi, apparently answers to Middle Persian *nargi, New Persian *nargis (Arabic *narjis*),[8] Aramaic *narkim*, Armenian *nargēs* (Persian

---

[1] Ch. 18, p. 12 b.

[2] Cf. the description of Theophrastus (Hist. plant., VII, 13): "In the case of narcissus it is only the flower-stem which comes up, and it immediately pushes up the flower." Also Dioscorides (IV, 158) and Pliny (XXI, 25) have given descriptions of the flower.

[3] This sentence is omitted (and justly so) in the text, as reprinted in the *Pen ts'ao kaṅ mu;* for these comparisons are lame.

[4] Ch. 13, p. 16.

[5] Also this species is said to have been introduced from abroad (*Hwa mu siao či* 花 木 小 志, p. 19 b, in *Č'un ts'ao t'aṅ tsi*, Ch. 25).

[6] In another passage of his work (Ch. 14, p. 10) he has the same text under *šan nai* 山 柰 (*Kæmpferia galanga*), but here he merely adds that the description of the *Yu yaṅ tsa tsu* is "a little like *šan nai.*"

[7] LECLERC, Traité des simples, Vol. III, p. 368.

[8] According to HÜBSCHMANN (Armen. Gram., p. 201), the New-Persian form would presuppose a Pahlavi *narkis. In my opinion, Greek νάρκισσος is derived from an Iranian language through the medium of an idiom of Asia Minor, not *vice versâ*, as believed by NOELDEKE (Persische Studien, II, p. 43).

427

loan-word), denoting *Narcissus tazetta*, which is still cultivated in Persia and employed in the pharmacopœia.[1] Oil was obtained from the narcissus, which is called ναρκίσσιον in the Greek Papyri.[2]

HIRTH[3] has erroneously identified the Chinese name with the nard. Aside from the fact that the description of the *Yu yan tsa tsu* does not at all fit this plant, his restoration, from a phonetic viewpoint, remains faulty. K'an-hi does not indicate the reading *not* for the first character, as asserted by Hirth, but gives the readings *nai, ni,* and *yin*. The second character reads *k'i,* which is evolved from *gi, but does not represent *ti,* as Hirth is inclined to make out.[4]

For other reasons it is out of the question to see the nard in the term *nai-k'i;* for the nard, a product of India, is well known to the Chinese under the term *kan sun hian* 甘松香.[5] The Chinese did not have to go to Fu-lin to become acquainted with a product which reached them from India, and which the Syrians themselves received from India by way of Persia.[6] Hebrew *nērd* (Canticle), Greek νάρδος,[7] Persian *nard* and *nārd,* are all derived from Sanskrit *nalada,* which already appears in the Atharvaveda.[8] Hirth's case would also run counter to his theory that the language of Fu-lin was Aramaic, for the word *nard* does not occur there.

---

[1] SCHLIMMER, Terminologie, p. 390. Narcissus is mentioned among the aromatic flowers growing in great abundance in Bišavûr, province of Fars, Persia (G. LE STRANGE, Description of the Province of Fars, p. 51). It is a flower much praised by the poets Hafiz and Jami.

[2] T. REIL, Beiträge zur Kenntnis des Gewerbes im hellenistischen Aegypten, p. 146. Regarding narcissus-oil, see Dioscorides, I, 50; and LECLERC, Traité des simples, Vol. II, p. 103.

[3] *Journal Am. Or. Soc.,* Vol. XXX, 1910, p. 22.

[4] See particularly PELLIOT, *Bull. de l'Ecole française,* Vol. IV, p. 291.

[5] STUART, Chinese Materia Medica, p. 278.

[6] I. LOEW, Aram. Pflanzennamen, pp. 368–369.

[7] First in Theophrastus, Hist. plant., IX. VII, 2.

[8] See p. 455.

47. The *Yu yan tsa tsu*[1] has the following notice of an exotic plant referred exclusively to Syria: "The plant 阿勃参 *a-p'o-ts'an* (*a-bwut-sam) has its habitat in the country Fu-lin (Syria). The tree is over ten feet high. Its bark is green and white in color. The blossoms are fine 細, two being opposite each other (biflorate). The flowers resemble those of the rape-turnip, *man-tsin* 蔓菁 (*Brassica rapa-depressa*), being uniformly yellow. The seeds resemble those of the pepper-plant, *hu-tsiao* 胡椒 (*Piper nigrum*). By chopping the branches, one obtains a juice like oil, that is employed as an ointment, serving as a remedy for ringworm, and is useful for any disease. This oil is held in very high esteem, and its price equals its weight in gold."

As indicated in the *Pen ts'ao kan mu si i*,[2] the notice of the plant *a-p'o-san* has been adopted by two works,— the *C'en fu t'un hwi* 程賦統會, which simply notes that it grows in Fu-lin; and the *Hwa i hwa mu k'ao* 華夷花木考 ("Investigations into the Botany of China and Foreign Countries"), which has copied the account of the *Yu yan tsa tsu* without acknowledgment. Neither of these books gives any additional information, and the account of the *Yu yan tsa tsu* remains the only one that we possess.

The transcription *a-bwut(bwur)-sam, which is very exact, leads to Aramaic and Talmudic *afursama* אפורסמא[3] (Greek βάλσαμον, Arabic *balessān*), the balm of Gilead (*Amyris gileadensis, Balsamodendron giliadense*, or *Commiphora opobalsamum*, family *Burseraceae*) of ancient fame. This case splendidly corroborates Hirth's opinion that the language of Fu-lin (or rather one of the languages of Fu-lin) was Aramaic. The last two characters *p'o-ts'an* (*bwut-sam) could very well transcribe Greek *balsam;* but the element 阿 excludes Greek and any other language in which this word is found, and admits no other than Aramaic. In Syriac we have *apursāmā* and *pursāmā* (*pursmā*), hence Armenian *aprsam* or *aprasam*.[4] In Neo-Hebrew, *afobalsmōn* or

---

[1] Ch. 18, p. 12.

[2] Ch. 4, p. 15.

[3] I. Loew, Aramaeische Pflanzennamen, p. 73. Also *afarsma* and *afarsmōn* (J. Buxtorf, Lexicon chaldaicum, p. 109; J. Levy, Neuhebr. Wörterbuch, Vol. I, p. 151). Cf. S. Krauss, Talmudische Archäologie, Vol. I, pp. 234-236.

[4] Hübschmann, Armenische Grammatik, p. 107. I do not believe in the Persian origin of this word, as tentatively proposed by this author.

afofalsmōn is derived from the Greek ὁποβάλσαμον.[1] It is supposed also
that Old-Testament Hebrew *bāsām* refers to the balsam, and might
represent the prototype of Greek *balsamon*, while others deny that the
Hebrew word had this specific meaning.[2] In my opinion, the Greek
l cannot be explained from the Hebrew word.

Twan Č'eṅ-ši's description of the tree, made from a long-distance
report, is tolerably exact. The *Amyris gileadensis* or balsam-tree is an
evergreen shrub or tree of the order *Amyridaceae*, belonging to the
tropical region, chiefly growing in southern Arabia, especially in the
neighborhood of Mecca and Medina, and in Abyssinia. As will be seen,
it was transplanted to Palestine in historical times, and Twan was
therefore justified in attributing it to Fu-lin. The height of the tree is
about fourteen feet, with a trunk eight or ten inches in diameter. It
has a double bark,—an exterior one, thin and red, and an interior one,
thick and green; when chewed, it has an unctuous taste, and leaves an
aromatic odor. The blossoms are biflorate, and the fruit is of a gray
reddish, of the size of a small pea, oblong, and pointed at both ends.
The tree is very rare and difficult to cultivate. Twan's oil, of course,
is the light green, fragrant gum exuded from the branches, always highly
valued as a remedy, especially efficacious in the cure of wounds.[3] It
was always a very costly remedy, and Twan's valuation (equaling its
weight in gold) meets its counterpart in the statement of Theophrastus
that it sells for twice its weight in silver.

Flavius Josephus (first century A.D.)[4] holds that the introduction
of the balsam-tree into Palestine, which still flourished there in his
time, is due to the queen of Saba. In another passage[5] he states that
the opobalsamum (sap of the tree) grows at Engedi, a city near the lake
Asphaltitis, three hundred furlongs from Jerusalem; and again,[6] that it
grows at Jericho: the balsam, he adds in the latter passage, is of all
ointments the most precious, which, upon any incision made in the wood
with a sharp stone, exudes out like juice.

From the time of Solomon it was cultivated in two royal gardens.

---

[1] J. LEVY, *op. cit.*, Vol. I, p. 137.

[2] E. LEVESQUE in Dictionnaire de la Bible, Vol. I, col. 1517. The *rapprochement* of *bāsām* and *balsamon* has already been made by D'HERBELOT (Bibliothèque
orientale, Vol. I, p. 377), though he gives *basam* only as Persian. The Arabic form
is derived from the Greek.

[3] Jeremiah, VIII, 22. Regarding its employment in the pharmacology of the
Arabs, see LECLERC, Traité des simples, Vol. I, pp. 255–257.

[4] Antiquitates judaicae, VIII. VI, 6.

[5] *Ibid.*, IX. I, 2.

[6] *Ibid.*, XIV. IV, I.

This fact was already known to Theophrastus,[1] who gives this account: "Balsam grows in the valley of Syria. They say that there are only two parks in which it grows, one of about four acres, the other much smaller. The tree is as tall as a good-sized pomegranate, and is much branched; it has a leaf like that of rue, but it is pale; and it is evergreen. The fruit is like that of the terebinth in size, shape, and color, and this too is very fragrant, indeed more so than the gum. The gum, they say, is collected by making incisions, which is done with bent pieces of iron at the time of the Dog-star, when there is scorching heat; and the incisions are made both in the trunks and in the upper parts of the tree. The collecting goes on throughout the summer; but the quantity which flows is not very large: in a day a single man can collect a shell-full. The fragrance is exceedingly great and rich, so that even a small portion is perceived over a wide distance. However, it does not reach us in a pure state: what is collected is mixed with other substances; for it mixes freely with such, and what is known in Hellas is generally mixed with something else.[2] The boughs are also very fragrant. In fact, it is on account of these boughs, they say, that the tree is pruned (as well as for a different reason), since the boughs cut off can be sold for a good price. In fact, the culture of the trees has the same motive as the irrigation (for they are constantly irrigated). And the cutting of the boughs seems likewise to be partly the reason why the trees do not grow tall; for, since they are often cut about, they send out branches instead of putting out all their energy in one direction. Balsam is said not to grow wild anywhere. From the larger park are obtained twelve vessels containing each about three pints, from the other only two such vessels. The pure gum sells for twice its weight in silver, the mixed sort at a price proportionate to its purity. Balsam then appears to be of exceptional value."

As the tree did not occur wild in Palestine, but only in the state of cultivation, and as its home is in southern Arabia, the tradition of Josephus appears to be well founded, though it is not necessary to connect the introduction with the name of the Queen of Saba.

Strabo,[3] describing the plain of Jericho, speaks of a palace and the garden of the balsamum. "The latter," he says, "is a shrub with an aromatic odor, resembling the cytisus (*Medicago arborea*) and the terminthus (terebinth-tree). Incisions are made in the bark, and vessels

---

[1] Hist. plant., IX, 6 (cf. the edition and translation of A. HORT, Vol. II, p. 245).

[2] E. WIEDEMANN (*Sitzber. phys.-med. Soz. Erl.*, 1914, pp. 178, 191) has dealt with the adulteration of balsam from Arabic sources.

[3] XVI. II, 41.

are placed beneath to receive the sap, which is like oily milk. When collected in vessels, it becomes solid. It is an excellent remedy for headache, incipient suffusion of the eyes, and dimness of sight. It bears therefore a high price, especially as it is produced in no other place."

Dioscorides[1] asserts erroneously that balsam grows only in a certain valley of India and in Egypt; while Ibn al-Baiṭār,[2] in his Arabic translation of Dioscorides, has him correctly say that it grows only in Judæa, in the district called Rūr (the valley of the Jordan). It is easily seen how Judæa in Greek writing could be misread for India.

To Pliny,[3] balsamum was only known as a product of Judæa (uni terrarum Iudaeae concessum). He speaks of the two gardens after Theophrastus, and gives a lengthy description of three different kinds of balsamum.

In describing Palestine, Tacitus[4] says that in all its productions it equals Italy, besides possessing the palm and the balsam; and the far-famed tree excited the cupidity of successive invaders. Pompey exhibited it in the streets of Rome in 65 B.C., and one of the wonderful trees accompanied the triumph of Vespasian in A.D. 79. During the invasion of Titus, two battles took place at the balsam-groves of Jericho, the last being intended to prevent the Jews from destroying the trees. They were then made public property, and were placed under the protection of an imperial guard; but it is not recorded how long the two plantations survived. In this respect, the Chinese report of the *Yu yan tsa tsu* is of some importance, for it is apt to teach that the balm of Gilead must still have been in existence in the latter part of the ninth century. It further presents clear-cut evidence of the fact that Judæa was included in the Chinese notion of the country Fu-lin.

Abd al-Laṭīf (1161–1231)[5] relates how in his time balsam was collected in Egypt. The operation was preferably conducted in the summer. The tree was shorn of its leaves, and incisions were made in the trunk, precaution being taken against injuring the wood. The sap was collected in jars dug in the ground during the heat, then they were taken out to be exposed to the sun. The oil floated on the surface and was cleaned of foreign particles. This was the true and purest balsam, forming only the tenth part of the total quantity produced by a tree. At present, in Arabia leaves and branches of the tree are boiled. The first

---

[1] I, 18.

[2] LECLERC, Traité des simples, Vol. I, 255.

[3] XII, 25, § 111.

[4] Hist., v, 6.

[5] SILVESTRE DE SACY, Relation de l'Egypte, p. 20 (Paris, 1810).

floating oil is the best, and reserved for the harem; the second is for commerce.

The tree has existed in Egypt from the eleventh to the beginning of the seventeenth century. It was presumably introduced there by the Arabs. D'HERBELOT[1] cites an Arabic author as saying that the balm of Mathara near Cairo was much sought by the Christians, owing to the faith they put in it. It served them as the chrism in Confirmation.

The Irish pilgrim Symon Semeonis, who started on his journey to the Holy Land in 1323, has the following interesting account of the balsam-tree of Egypt:[2] "To the north of the city is a place called Matarieh, where is that famous vine said to have been formerly in Engaddi (cf. Cant., 1, 13), which distils the balsam. It is diligently guarded by thirty men, for it is the source of the greater portion of the Sultan's wealth. It is not like other vines, but is a small, low, smooth tree, and odoriferous, resembling in smoothness and bark the hazel tree, and in leaves a certain plant called *nasturcium aquaticum*. The stalk is thin and short, usually not more than a foot in length; every year fresh branches grow out from it, having from two to three feet in length and producing no fruit. The keepers of the vineyard hire Christians, who with knives or sharp stones break or cut the tops of these branches in several places and always in the sign of a cross. The balsam soon distils through these fractures into glass bottles. The keepers assert that the flow of balsam is more abundant when the incision is made by a Christian than by a Saracen." [3]

In 1550 PIERRE BELON[4] still noted the tree in Cairo. Two specimens were still alive in 1612. In 1615, however, the last tree died.

The Semitic word introduced into China by the *Yu yan tsa tsu* seems to have fallen into oblivion. It is not even mentioned in the *Pen ts'ao kan mu*. The word "balsam," however, was brought back to China by the early Jesuits. In the famous work on the geography of the world, the *Či fan wai ki* 職方外紀,[5] first draughted by Pantoja, and after his death enlarged and edited in 1623 by Giulio Aleni (1582–1649), the Peru balsam is described under the name *pa'r-sa-mo* 拔爾撒摩. The same word with reference to the same substance is employed by

---

[1] Bibliothèque orientale, Vol. I, p. 392.

[2] M. ESPOSITO, The Pilgrimage of Symon Semeonis: A Contribution to the History of Mediæval Travel (*Geographical Journal*, Vol. LI, 1918, p. 85).

[3] Cf. the similar account of K. v. MEGENBERG (Buch der Natur, p. 358, written in 1349–50).

[4] Observations de plusieurs singularitez et choses memorables, trouvées en Grece, Asie, Iudée, Egypte, Arabie, p. 246.

[5] Ch. 4, p. 3 (ed. of *Šou šan ko ts'un šu*).

Ferdinand Verbiest (1623–88) in his *K'un yü t'u šwo* 坤輿圖說, and was hence adopted in the pharmacopœia of the Chinese, for it figures in the *Pen ts'ao kan mu ši i*.[1] The Chinese Gazetteer of Macao[2] mentions *pa'r-su-ma* aromatic 巴爾酥麻香 as a kind of benjoin. In this case we have a transcription of Portuguese *bálsamo*.

---

[1] Ch. 6, p. 19. See, further, WATTERS, Essays on the Chinese Language, p. 339.
[2] *Ao-men či lio*, Ch. B, p. 41 (cf. WYLIE, Notes on Chinese Literature, p. 60).

NOTE ON THE LANGUAGE OF FU-LIN

48. The preceding notes on Fu-lin plants have signally confirmed Hirth's opinion in regard to the language of Fu-lin, which was Aramaic. There now remains but one Fu-lin plant-name to be identified. This is likewise contained in the *Yu yan tsa tsu*.[1] The text runs as follows:—

"The *p'an-nu-se* 槃砮稽 tree has its habitat in Po-se (Persia), likewise in Fu-lin. In Fu-lin it is styled *k'ün-han* 羣漢. The tree is thirty feet high, and measures from three to four feet in circumference. Its leaves resemble those of the *si žun* 細榕 (the Banyan tree, *Ficus retusa*). It is an evergreen. The flowers resemble those of the citrus, *kü* 橘, and are white in color. The seeds are green and as large as a sour jujube, *swan tsao* 酸棗 (*Diospyros lotus*). They are sweet of taste and glossy (fat, greasy). They are eatable. The people of the western regions press oil out of them, to oint their bodies with to ward off ulcers."

The transcription *p'an-nu-se* answers to ancient *bwan-du-sek; and *k'ün-han*, to ancient g'win-xan. Despite a long-continued and intensive search, I cannot discover any Iranian plant-name of the type *bandusek* or *wandusek*, nor any Aramaic word like *ginxan*. The botanical characteristics are too vague to allow of a safe identification. Nevertheless I hope that this puzzle also will be solved in the future.[2]

In the Fu-lin name *a-li-k'ü-fa* we recognized an Indian loan-word in Aramaic (p. 423). It would be tempting to regard as such also the Fu-lin word for "pepper" *a-li-xa-da 阿梨訶陀 (*a-li-ho-t'o*), which may be restored to *alixada, arixada, arxad; but no such word is known from Indian or in Aramaic. The common word for "pepper" in Aramaic is *filfol* (from Sanskrit *pippala*). In certain Kurd dialects J. DE MORGAN[3] has traced a word *alat* for "pepper," but I am not certain that this is

---

[1] Ch. 18, p. 10 b.

[2] My colleague, Professor M. Sprengling at the University of Chicago, kindly sent me the following information: "Olive-oil was used to ward off ulcers (see WINER, Bibl. Realwörtb., Vol. II, p. 170; and KRAUSS, Archaeologie des Talmud, Vol. I, pp. 229, 233, 683). Neither in Krauss nor elsewhere was I able to find the name of an oil-producing tree even remotely resembling *ginxan*. There is a root *qnx* ('to wipe, to rub, to anoint'). It is theoretically possible that *q* is pronounced voiced and thus becomes a guttural *g*, and that from this root, by means of the suffix *-an*, may be derived a noun *qinxan, *ginxan to which almost any significance derived from 'rubbing, anointing' might be attached. But for the existence of such a noun or adjective I have not the slightest evidence."

[3] Mission scientifique en Perse, Vol. V, p. 132.

435

connected with our Fu-lin word, which at any rate represents a loan-word.

There is another Fu-lin word which has not yet been treated correctly. The T'ang Annals, in the account of Fu-lin (Ch. 221), mention a mammal, styled ts'uṅ 貚, of the size of a dog, fierce, vicious, and strong.[1] BRETSCHNEIDER,[2] giving an incorrect form of the name, has correctly identified this beast with the hyena, which, not being found in eastern Asia, is unknown to the Chinese. Ma Twan-lin adds that some of these animals are reared,[3] and the hyena can indeed be tamed. The character for the designation of this animal is not listed in K'aṅ-hi's Dictionary; but K'aṅ-hi gives it in the form 貚[4] with the pronunciation hien (fan-ts'ie 黃棟, sound equivalent 縣), quoting a commentary to the dictionary Er ya, which is identical with the text of Ma Twan-lin relative to the animal ts'uṅ. This word hien (or possibly hüan) can be nothing but a transcription of Greek ύαινα, hyaena, or ύαίνη. On the other hand, it should be noted that this Greek word has also passed as a loan into Syriac;[5] and it would therefore not be impossible that it was Syrians who transmitted the Greek name to the Chinese. This question is altogether irrelevant; for we know, and again thanks to Hirth's researches, that the Chinese distinguished two Fu-lin,— the Lesser Fu-lin, which is identical with Syria, and the Greater Fu-lin, the Byzantine Empire with Constantinople as capital.[6] Byzantine Greek, accordingly, must be included among the languages spoken in Fu-lin.

As to the origin of the name Fu-lin, I had occasion to refer to Pelliot's new theory, according to which it would be based on Rōm, Rūm.[7] I am of the same opinion, and perfectly in accord with the fundamental principles by which this theory is inspired. In fact, this is the method followed throughout this investigation: by falling back on the ancient phonology of Chinese, we may hope to restore correctly the prototypes of the Chinese transcriptions. Pelliot starts from the Old-Armenian form Hrom or Hrōm,[8] in which h represents

---

[1] HIRTH, China and the Roman Orient, pp. 60, 107, 220.

[2] Knowledge possessed by the Ancient Chinese of the Arabs, p. 24.

[3] HIRTH (op. cit., p. 79) translates, "Some are domesticated like dogs." But the phrase 似狗 following 有養者 forms a separate clause. In the text printed by Hirth (p. 115, Q 22) the character 方 is to be eliminated.

[4] Thus reproduced by PALLADIUS in his Chinese-Russian Dictionary (Vol. I, p. 569) with the reading süan.

[5] R. P. SMITH, Thesaurus syriacus, Vol. I, col. 338.

[6] Cf. HIRTH, Journal Am. Or. Soc., Vol. XXXIII, 1913, pp. 202–208.

[7] The Diamond (this volume, p. 8). PELLIOT's notice is in Journal asiatique, 1914, I, pp. 498–500.

[8] Cf. HÜBSCHMANN, Armen. Gram., p. 362.

the spiritus asper of the initial Greek *r*. In some Iranian dialects the spiritus asper is marked by an initial vowel: thus in Pahlavi Arūm, in Kurd Urum. The ancient Armenian words with initial *hr*, as explained by A. Meillet, were borrowed from Parthian dialects which transformed initial Iranian *f* into *h*: for instance, Old Iranian *framana* (now *ferman*, "order") resulted in Armenian *hraman*, hence from Parthian **hraman*. Thus **Frōm*, probably conveyed by the Sogdians, was the prototype from which Chinese Fu-lin, **Fu-lim*, was fashioned. In my opinion, the Chinese form is not based on **Frōm*, but on **Frim* or **Frīm*. Rīm must have been an ancient variant of Rūm; Rim is still the Russian designation of Rome.[1] What is of still greater importance is that, as has been shown by J. J. Modi,[2] there is a Pahlavi name Sairima, which occurs in the Farvardin Yašt, and is identified with Rum in the Būndahišn; again, in the Šāhnāmeh the corresponding name is Rum. This country is said to have derived its name from Prince Selam, to whom it was given; but this traditional opinion is not convincing. A form Rima or Rim has accordingly existed in Middle Persian; and, on the basis of the Chinese transcription **Fu-lim* or **Fu-rim*, it is justifiable to presuppose the Iranian (perhaps Parthian) prototype **Frim*, from which the Chinese transcription was made.

---

[1] What Pelliot remarks on the Tibetan names Ge-sar and P'rom is purely hypothetical, and should rather be held in abeyance for the present. We know so little about the Ge-sar epic, that no historical conclusions can be derived from it. For the rest, the real Tibetan designation for Byzance or Turkey, in the same manner as in New Persian, is Rum (*T'oung Pao*, 1916, p. 491). In regard to the occurrence of this name in Chinese transcriptions of more recent date, see BRETSCHNEIDER, Mediæval Researches, Vol. II, p. 306; and HIRTH, Chau Ju-kua, p. 141.

[2] Asiatic Papers, p. 244 (Bombay, 1905).

THE WATER–MELON

49. This Cucurbitacea (*Citrullus vulgaris* or *Cucurbita citrullus*) is known to the Chinese under the name *si kwa* 西瓜 ("melon of the west"). The plant now covers a zone from anterior Asia, the Caucasus region, Persia to Turkistan and China, also southern Russia and the regions of the lower Danube. There is no evidence to lead one to suppose that the cultivation was very ancient in Iran, India, Central Asia, or China; and this harmonizes with the botanical observation that the species has not been found wild in Asia.[1]

A. ENGLER[2] traces the home of the water-melon to South Africa, whence he holds it spread to Egypt and the Orient in most ancient times, and was diffused over southern Europe and Asia in the pre-Christian era. This theory is based on the observation that the water-melon grows spontaneously in South Africa, but it is not explained by what agencies it was disseminated from there to ancient Egypt. Nevertheless the available historical evidence in Asia seems to me to speak in favor of the theory that the fruit is not an Asiatic cultivation; and, since there is no reason to credit it to Europe, it may well be traceable to an African origin.

The water-melon is not mentioned by any work of the T'ang dynasty; notably it is absent from the *T'ai p'iñ hwan yü ki*. The earliest allusion to it is found in the diary of Hu Kiao 胡嶠, entitled *Hien lu ki* 陷虜記, which is inserted in chapter 73 of the History of the Five Dynasties (*Wu tai ši*), written by Ñou-yañ Siu 歐陽修 (A.D. 1017–72) and translated by E. CHAVANNES.[3] Hu Kiao travelled in the country of the Kitan from A.D. 947 to 953, and narrates that there for the first time he ate water-melons (*si kwa*).[4] He goes on to say, "It is told that the Kitan, after the annihilation of the Uigur, obtained this cultivation. They cultivated the plant by covering the seeds with cattle-manure and placing mats over the beds. The fruit is as large as that of the

---

[1] A. DE CANDOLLE, Origin of Cultivated Plants, p. 263.

[2] In Hehn, Kulturpflanzen, p. 323.

[3] Voyageurs chinois chez les Khitan (*Journal asiatique*, 1897, I, pp. 390–442).

[4] Chavannes' translation "melons" (p. 400) is inadequate; the water-melon is styled in French *pastèque* or *melon d'eau*. Hu Kiao, of course, was acquainted with melons in general, but what he did not previously know is this particular species. During Napoleon's expedition to Egypt, "on mangeait des lentilles, des pigeons, et un melon d'eau exquis, connu dans les pays méridionaux sous le nom de *pastèque*. Les soldats l'appelaient *sainte pastèque*" (THIERS, Histoire de la révolution française).

438

tuṅ kwa 冬 瓜 (*Benincasa cerifera*)[1] and of sweet taste."[2] The water-melon is here pointed out as a novelty discovered by a Chinese among the Kitan, who then occupied northern China, and who professed to have received it from the Turkish tribe of the Uigur. It is not stated in this text that Hu Kiao took seeds of the fruit along or introduced it into China proper. This should be emphasized, in view of the conclusion of the *Pen ts'ao kaṅ mu* (see below), and upheld by Bretschneider and A. de Candolle, that the water-melon was in China from the tenth century. At that time it was only in the portion of China held by the Kitan, but still unknown in the China of the Chinese.[3]

---

[1] "Cultivated in China, Japan, India and Africa, and often met with in a wild state: but it is uncertain whether it is indigenous" (FORBES and HEMSLEY, *Journal Linnean Society*, Vol. XXIII, p. 315).

[2] Hu Kiao was a good observer of the flora of the northern regions, and his notes have a certain interest for botanical geography. Following his above reference to the water-melon, he continues, "Going still farther east, we arrived at Niao-t'an, where for the first time willows [Jurči *suxei*] are encountered, also water-grass, luxuriant and fine; the finest of this kind is the grass *si-ki* 息 雞 with large blades. Ten of these are sufficient to satisfy the appetite of a horse. From Niao-t'an we advanced into high mountains which it took us ten days' journey to cross. Then we passed a large forest, two or three *li* long, composed entirely of elms, *wu-i* 蕪 荑 (*Ulmus macrocarpa*), the branches and leaves of which are set with thorns like arrow-feathers. The soil is devoid of grass." *Si-ki* apparently represents the transcription of a Kitan word. Three species of elm occur in the Amur region,—*Ulmus montana*, *U. campestris*, and *U. suberosa* (GRUM-GRŽIMAILO, Opisanie Amurskoi Oblasti, p. 316). In regard to the locality T'aṅ-č'eṅ-tien, Hu Kiao reports, "The climat there is very mild, so that the Kitan, when they suffer from great cold, go there to warm up. The wells are pure and cool; the grass is soft like down, and makes a good sleeping-couch. There are many peculiar flowers to be found, of which two species may be mentioned,—one styled *han-kin* 旱 金, the size of the palm of a hand, of gold color so brilliant that it dazzles man; the other, termed *ts'iṅ žaṅ* 青 囊, like the *kin t'eṅ* 金 燈 (*Orithia edulis*) of China, resembling in color an *Indigofera* (*lan* 藍) and very pleasing." The term *han-kin* appears to be the transcription of a Kitan word; so is perhaps also *ts'iṅ žaṅ*, although, according to STUART (Chinese Materia Medica, p. 404), the leaves of *Sesamum* are so called; this plant, however, cannot come here into question.

[3] The *Pien tse lei pien* cites the *Wu tai ši* to the effect that Siao Han 蕭 翰, after the subjugation of the Uigur, obtained the seeds of water-melons and brought them back, and that the fruit as a product of the Western Countries (*Si yü*, that is, Central Asia) was called "western melon" (*si kwa*). I regret not having been able to trace this text in the *Wu tai ši*. The biography of Siao Han inserted in the *Kiu Wu tai ši* (Ch. 98, pp. 6 b-7 a) contains nothing of the kind. The statement itself is suspicious for two reasons. Siao Han, married to A-pu-li, sister of the Emperor Wu-yü, in A.D. 948 was involved in a high-treason plot, and condemned to death in the ensuing year (cf. H. C. V. D. GABELENTZ, Geschichte der grossen Liao, p. 65; and CHAVANNES, *op. cit.*, p. 392). Hu Kiao was secretary to Siao Han, and in this capacity accompanied him to the Kitan. After his master's death, Hu Kiao was without support, and remained among the Kitan for seven years (up to the year 953). It was in the course of these peregrinations that, as related above, he was first introduced to water-melons. Now, if Siao Han had really introduced this fruit into

The man who introduced the fruit into China proper was Hun Hao
洪皓 (A.D. 1090–1155), ambassador to the Kin or Jurči, among whom he
remained for fifteen years (1129–43). In his memoirs, entitled *Sun mo
ki wen* 松漠紀聞, he has the following report:[1] "The water-melon
(*si kwa*) is in shape like a flat *Acorus* (*p'u* 蒲), but rounded. It is very
green in color, almost blue-green. In the course of time it will change
into yellow. This Cucurbitacea (*t'ie* 瓞) resembles the sweet melon (*tien
kwa* 甜瓜, *Cucumis melo*), and is sweet and crisp.[2] Its interior is filled

China during his lifetime (that is, prior to the year 949), we might justly assume
that his secretary Hu Kiao must have possessed knowledge of this fact, and would
hardly speak of the fruit as a novelty. Further, the alleged introduction of the
fruit by Siao Han conflicts with the tradition that this importation is due to Hun
Hao in the twelfth century (see above). It would be nothing striking, of course, if, as
the fruit was cultivated by the Kitan, several Chinese ambassadors to this people
should have carried the seeds to their country; but, as a rule, such new acquisitions
take effect without delay, and if Siao Han had imported the seeds, there was no
necessity for Hun Hao to do so again. Therefore it seems preferable to think either
that the text of the above quotation is corrupted, or that the tradition, if it existed,
is a subsequent makeshift or altogether erroneous.

[1] Not having access to an edition of this work, I avail myself of the extract, as
printed in the *Kwan k'ün fan p'u* (Ch. 14, p. 17 b), the texts of which are generally
given in a reliable form.

[2] In regard to the melon (*Cucumis melo*), A. DE CANDOLLE (Origin of Cultivated
Plants, p. 261) says with reference to a letter received from Bretschneider in 1881,
"Its introduction into China appears to date only from the eighth century of our
era, judging from the epoch of the first work which mentions it. As the relations
of the Chinese with Bactriana, and the north-west of India by the embassy of
Chang-Kien, date from the second century, it is possible that the culture of the
species was not then widely diffused in Asia." Nothing to the effect is to be found in
Bretschneider's published works. In his Bot. Sin. (pt. II, p. 197) he states that all
the cucurbitaceous plants now cultivated for food in China are probably indigenous
to the country, with the exception of the cucumber and water-melon, which, as their
Chinese names indicate, were introduced from the West. In the texts assembled
in the *Pen ts'ao kan mu* regarding *tien kwa*, no allusion is made to foreign origin.
Concerning the gourd or calabash (*Lagenaria vulgaris*), A. DE CANDOLLE (*l. c.*,
p. 246) states after a letter of Bretschneider that "the earliest work which mentions
the gourd is that of Tchong-tchi-chou, of the first century before Christ, quoted in
a work of the fifth or sixth century." This seems to be a confusion with the *Čun
šu šu* of the T'ang period (BRETSCHNEIDER, Bot. Sin., pt. I, p. 79). The gourd, of
course, occurs in ancient canonical literature (Bot. Sin., pt. II, p. 198). The history
of this and other cucurbitaceous plants requires new and critical investigation, the
difficulty of which is unfortunately enhanced by a constant confusion of terms in
all languages, the name of one species being shifted to another. It means very little,
of course, that at present, as recently emphasized again by H. J. SPINDEN (Pro-
ceedings Nineteenth Congress of Americanists, p. 271, Washington, 1917), *Lagenaria*
is distributed over the New and Old Worlds alike; the point is, where the centre of the
cultivation was (according to A. de Candolle it was in India; see, further, ASA GRAY,
Scientific Papers, Vol. I, p. 330), and how it spread, or whether the wild form had a
wide geographical range right from the beginning, and was cultivated independently
in various countries. In view of the great antiquity of the cultivation both in India
and China, the latter assumption would seem more probable; but all this requires
renewed and profound investigation.

with a juice which is very cold. Huṅ Hao, when he went out as envoy, brought the fruit back to China. At present it is found both in the imperial orchards and in village gardens. It can be kept for several months, aside from the fact that there is nothing to prevent it from assuming a yellow hue in course of time. In P'o-yaṅ 鄱 陽[1] there lived a man who for a long time was afflicted with a disease of the eyes. Dried pieces of water-melon were applied to them and caused him relief, for the reason that cold is a property of this fruit." Accordingly the water-melon was transplanted into China proper only in the latter part of the twelfth century. Also the Ši wu ki yüan 事 物 紀 原,[2] which says that in the beginning there were no water-melons in China, attributes their introduction to Huṅ Hao. The Kin or Jurči, a nation of Tungusian origin, appear to have learned the cultivation from the Kitan. From a Jurči-Chinese glossary we know also the Jurči designation of the water-melon, which is xeko, corresponding to Manchu xengke, a general term for cucurbitaceous plants. In Golde, xinke (in other Tungusian dialects kemke, kenke) denotes the cucumber, and seho or sego the water-melon. The proper Manchu word for the water-melon is dungga or dunggan. The Tungusian tribes, accordingly, did not adopt the Persian-Turkish word karpuz (see below) from the Uigur, but applied to the water-melon an indigenous word, that originally denoted another cucurbitaceous species.

Following is the information given on the subject in the Pen ts'ao kaṅ mu.

Wu Žui 吳瑞, a physician from the province of Če-kiaṅ in the thirteenth century, author of the Ži yuṅ pen ts'ao 日 用 本 草, is cited in this work as follows: "When the Kitan had destroyed the Uigur, they obtained this cultivation. They planted this melon by covering the seeds with cattle-manure. The formation of this fruit is like the peck tou 斗; it is large and round like a gourd, and in color like green jade. The seeds have a color like gold, but some like black hemp. In the northern part of our country the fruit is plentiful." Li Ši-čen observes, "According to the Hien lu ki by Hu Kiao (see p. 438), this cultivation was obtained after the subjugation of the Uigur. It is styled 'western melon' (si kwa). Accordingly it is from the time of the Wu-tai (A.D. 907–960) that it was first introduced into China.[3] At present it occurs both in the south and north of the country, though the southern

---

[1] In the prefecture of Žao-čou, Kiaṅ-si.

[2] The work of Kao Č'eṅ 高 承 of the Sung dynasty.

[3] The same opinion is expressed by Yaṅ Šen (1488–1559) in his Tan k'ien tsuṅ lu (above, p. 331).

fruit is inferior in taste to that of the north." He distinguishes sweet, insipid, and sour varieties.

In the *T'ao huṅ kiṅ ču* 陶宏景注[1] it is stated that in Yuṅ-kia 永嘉 (in the prefecture of Wen-čou, Či-li) there were *han kwa* 寒瓜 ("cold melons") of very large size, which could be preserved till the coming spring, and which are regarded as identical with the water-melon. Li Ši-čen justly objects to this interpretation, commenting that, if the water-melon was first introduced in the Wu-tai period, the name *si kwa* could not have been known at that time. This objection must be upheld, chiefly for the reason that we have no other records from the fourth century or even the T'ang period which mention the water-melon: it is evidently a post-T'ang introduction.[2]

Ye Tse-k'i, in his *Ts'ao mu tse* 草木子 written in 1378, remarked that water-melons were first introduced under the Yüan, when the Emperor Ši-tsu 世祖 (Kubilai) subjugated Central Asia. This view was already rejected under the Ming in the *Čen ču č'wan* 珍珠船 by Č'en Ki-žu 陳繼儒, who aptly referred to the discovery of the fruit by Hu Kiao, and added that it is not mentioned in the *Er ya*, the various older *Pen ts'ao*, the *Ts'i min yao šu*, and other books of a like character, it being well known that the fruit did not anciently exist in China. As to this point, all Chinese writers on the subject appear to be agreed; and its history is so well determined, that it has not given rise to attempts of antedating or "changkienizing" the introduction.

The Chinese travellers during the Mongol period frequently allude to the large water-melons of Persia and Central Asia.[3] On the other hand, Ibn Baṭūṭa mentions the excellent water-melons of China, which are like those of Khwarezm and Ispahan.[4]

According to the Manchu officers Fusambô and Surde, who published an account of Turkistan about 1772,[5] the water-melon of this region, though identical with that of China, does not equal the latter in taste; on the contrary, it is much inferior to it. Other species of melon belong to the principal products of Turkistan; some are called by the Chinese "Mohammedan caps" and "Mohammedan eyes." The so-called "Hami melon," which is not a water-melon, and ten varieties of which are distinguished, enjoys a great reputation. Probably it is

---

[1] Apparently a commentary to the works of T'ao Huṅ-kiṅ (A.D. 451–536).

[2] The alleged synonyme *han kwa* for the water-melon, adopted also by BRETSCHNEIDER (*Chinese Recorder*, 1871, p. 223) and others, must therefore be weeded out.

[3] Cf. BRETSCHNEIDER, Mediæval Researches, Vol. I, pp. 20, 31, 67, 89.

[4] YULE, Cathay, new ed., Vol. IV, p. 109.

[5] *Hui k'iaṅ či*, see above, p. 230; and below, p. 562.

a variety of sweet melon (*Cucumis melo*), called in Uigur and Djagatai *kogun, kavyn,* or *kaun,* in Turkī *qāwa* and *qawāq.*

It is said to have been introduced into China as late as the K'ań-hi era (1662–1721), and was still expensive at that time, but became ubiquitous after the subjugation of Turkistan.[1] Of other foreign countries that possess the water-melon, the *Yiń yai šen lan* mentions Su-men-ta-la (Sumatra), where the fruit has a green shell and red seeds, and is two or three feet in length,[2] and Ku-li 古里 (Calicut) in India, where it may be had throughout the year.[3] In the country of the Mo-ho the fruits are so heavy that it takes two men to lift them. They are said to occur also in Camboja.[4] If it is correct that the first report of the water-melon reached the Chinese not earlier than the tenth century (and there is no reason to question the authenticity of this account), this late appearance of the fruit would rather go to indicate that its arrival in Central Asia was almost as late or certainly not much earlier; otherwise the Chinese, during their domineering position in Central Asia under the T'ang, would surely not have hesitated to appropriate it. This state of affairs is confirmed by conditions in Iran and India, where only a mediæval origin of the fruit can be safely supposed.

The point that the water-melon may have been indigenous in Persia from ancient times is debatable. Such Persian terms as *hindewāne* ("Indian fruit") [Afghan *hindwānā*] or *battīx indi* ("Indian melon")[5] raise the suspicion that it might have been introduced from India.[6] GARCIA DA ORTA states, "According to the Arabs and Persians, this fruit was brought to their countries from India, and for that reason they

---

[1] *Hui k'iań ĕi,* Ch. 2; and *Či wu miń ši t'u k'ao,* Ch. 16, p. 85.

[2] Malayan *mandelīkei, tambīkei,* or *semańka* (Javanese *semoñka,* Čam *samkai*). Regarding other Malayan names of cucurbitaceous plants, see R. BRANDSTETTER, Mata-Hari, p. 27; cf. also J. CRAWFURD, History of the Indian Archipelago, Vol. I, p. 435.

[3] Regarding other cucurbitaceous plants of Calicut, see ROCKHILL, T'oung Pao, 1915, pp. 459, 460; but *tuń kwa* is not, as there stated, the cucumber, it is *Benincasa cerifera.*

[4] *Kwań k'ün fań p'u,* Ch. 14, p. 18. Cf. PELLIOT, Bull. de l'Ecole française, Vol. II, p. 169. Water-melons are cultivated in Siam (PALLEGOIX, Description du royaume Thai, Vol. I, p. 126).

[5] From the Arabic; Egyptian *bettu-ka,* Coptic *betuke;* hence Portuguese and Spanish *pasteca,* French *pastèque.* The *battīx hindi* has already been discussed by Ibn al-Baiṭār (L. LECLERC, Traité des simples, Vol. I, p. 240) and by Abu Mansur (ACHUNDOW, p. 23). Armenian *ttum* bears no relation to the *dudaim* of the Bible, as tentatively suggested by E. SEIDEL (Mechithar, p. 121). The latter refers to the mandragora.

[6] Thus also SPIEGEL, Eranische Altertumskunde, Vol. I, p. 259.

call it Batiec Indi, which means 'melon of India,' and Avicenna so calls it in many places."[1] Nor does Persian *herbuz*,[2] Middle Persian *harbōjīnā* or *xarbūzak* (literally, "donkey-cucumber") favor the assumption of an indigenous origin. VÁMBÉRY[3] argues that Turkish *karpuz* or *harbuz* is derived from the Persian, and that accordingly the fruit hails from Persia, though the opposite standpoint would seem to be equally justifiable, and the above interpretation may be no more than the outcome of a popular etymology. But Vámbéry, after all, may be right; at least, by accepting his theory it would be comparatively easy to account for the migration of the water-melon. In this case, Persia would be the starting-point from which it spread to the Turks of Central Asia and finally to China.[4] A philological argument may support the opinion that the Turkish word was derived from Persia: besides the forms with initial guttural, we meet an alternation with initial dental, due to phonetic dissimilation. The Uigur, as we know from the Uigur-Chinese vocabulary, had the word as *karpuz;* but the Mongols term the water-melon *tarbus*. Likewise in Turkī we have *tarbuz*, but also *qarpuz*. This alternation is not Mongol-Turkish, but must have pre-existed in Persian, as we have *tarambuja* in Neo-Sanskrit, and in Hindustānī there is *xarbūza* and *tarbūza* (also *tarbūz* and *tarmus*), and correspondingly *tarbuz* in West-Tibetan. In Puštu, the language of the Afghans, we have *tarbuja* in the sense of "water-melon," and *xarbuja* designating various kinds of musk-melon.[5] Through Turkish mediation the same word reached the Slavs (Russian *arbŭz*,[6] Bulgarian *karpŭz*, Polish *arbuz, garbuz, harbuz*) and Byzantines (Greek καρπούσια), and Turkish tribes appear to have been active in disseminating the fruit east and west.

It would therefore be plausible also that, as stated by JORET,[7] the fruit may have been propagated from Iran to India, although the date of this importation is unknown. From Indian sources, on the other hand, nothing is to be found that would indicate any great antiquity of the cultivation of this species. Of the alleged Sanskrit word *chayapula*,

---

[1] C. MARKHAM, Colloquies by Garcia da Orta, p. 304.

[2] From which Armenian *xarpzag* is derived.

[3] Primitive Cultur des turko-tatarischen Volkes, pp. 217–218.

[4] Vámbéry, of course, is wrong in designating Persia and India as the mother-country of this cultivation. The mother-country was ancient Egypt or Africa in a wider sense.

[5] H. W. BELLEW, Report on the Yusufzais, p. 255 (Lahore, 1864).

[6] In the dialects of northern Persia we also find such forms as *arhuz* and *arhos* (J. DE MORGAN, Mission en Perse, Vol. V, p. 212).

[7] Plantes dans l'antiquité, Vol. II, p. 252.

which A. DE CANDOLLE introduces as evidence for the early diffusion
of the cultivation into Asia, I cannot find any trace. The Sanskrit
designations of the water-melon, *nāṭāmra* ("mango of the Naṭa"?),
goḍumba, tarambuja, sedu, are of recent origin and solely to be found in
the lexicographers; while others, like *kāliṅga* (*Benincasa cerifera*), originally refer to other cucurbitaceous plants. WATT gives only modern
vernacular names.

Chinese *si kwa* has been equated with Greek σικύα by HIRTH,[1] who
arbitrarily assigns to the latter the meaning "water-melon." This
philological achievement has been adopted by GILES in his Chinese
Dictionary (No. 6281). The Greek word, however, refers only to the
cucumber, and the water-melon remained unknown to the Greeks of
ancient times.[2] A late Greek designation for the fruit possibly is πέπων,
which appears only in Hippocrates.[3] A. DE CANDOLLE[4] justly remarked
that the absence of an ancient Greek name which may with certainty
be attributed to this species seems to show that it was introduced into
the Graeco-Roman world about the beginning of the Christian era.
The Middle and Modern Greek word χαρπουζά or καρπούσια, derived
from Persian or Turkish, plainly indicates the way in which the Byzantine world became acquainted with the water-melon. There is,
further, no evidence that the Greek word σικύα ever penetrated into
Asia and reached those peoples (Uigur, Kitan, Jurči) whom the Chinese
make responsible for the transmission of the water-melon. The Chinese
term is not a transcription, but has the literal meaning "western melon";
and the "west" implied by this term does not stretch as far as Greece, but,
as is plainly stated in the *Wu tai ši*, merely alludes to the fact that the
fruit was produced in Turkistan. *Si kwa* is simply an abbreviation
for *Si yü kwa* 西或瓜; that is, "melon of Turkistan."[5]

According to the *Yamato-honzō*[6] of 1709, water-melons were first
introduced into Japan in the period Kwan-ei (1624–44).

---

[1] Fremde Einflüsse in der chinesischen Kunst, p. 17.

[2] A. DE CANDOLLE, Géographie botanique, p. 909.

[3] Even this problematic interpretation is rejected by L. LECLERC (Traité des
simples, Vol. I, p. 239), who identifies the Greek word with the common gourd.
Leclerc's controversy with A. de Candolle should be carefully perused by those
who are interested in the history of the melon family.

[4] Origin of Cultivated Plants, p. 264.

[5] Illustrations of Chinese water-melon fields may be seen in F. H. KING, Farmers of Forty Centuries, pp. 282, 283.

[6] Ch. 8, p. 3.

FENUGREEK

50. In regard to the fenugreek (*Trigonella foenum-graecum*, French *fenugrec*), Chinese *hu-lu-pa* (Japanese *koroha*) 胡蘆巴, STUART[1] states without further comment that the seeds of this leguminous plant were introduced into the southern provinces of China from some foreign country. But BRETSCHNEIDER[2] had correctly identified the Chinese name with Arabic *hulba* (*xulba*). The plant is first mentioned in the *Pen ts'ao* of the Kia-yu period (A.D. 1056–64) of the Sung dynasty, where the author, Čaṅ Yü-si 掌禹錫, says that it grows in the provinces of Kwaṅ-tuṅ and Kwei-čou, and that, according to some, the species of Liṅ-nan represents the seeds of the foreign *lo-po* (*Raphanus sativus*), but that this point has not yet been investigated. Su Suṅ, in his *T'u kiṅ pen ts'ao*, states that "the habitat of the plant is at present in Kwaṅ-tuṅ, and that in the opinion of some the seeds came from Hai-nan and other barbarians; passengers arriving on ships planted the seeds in Kwaṅ-tuṅ (Liṅ-wai), where the plant actually grows, but its seeds do not equal the foreign article; the seeds imported into China are really good." Then their employment in the pharmacopœia is discussed.[3] The drug is also mentioned in the *Pen ts'ao yen i*.[4]

The transcription *hu-lu-pa* is of especial interest, because the element *hu* forms part of the transcription, but may simultaneously imply an allusion to the ethnic name Hu. The form of the transcription shows that it is post-T'ang; for under the T'ang the phonetic equivalent of the character 胡 was still possessed of an initial guttural, and a foreign element *xu* would then have been reproduced by a quite different character.

The medical properties of the plant are set forth by Abu Mansur in his Persian pharmacopœia under the name *hulbat*.[5] The Persian name

---

[1] Chinese Materia Medica, p. 442.

[2] Bot. Sin., pt. I, p. 65.

[3] STUART (*l. c.*) says wrongly that the seeds have been in use as a medicine since the T'ang dynasty; this, however, has been the case only since the Sung. I do not know of any mention of the plant under the T'ang. This negative documentary evidence is signally confirmed by the transcription of the name, which cannot have been made under the T'ang.

[4] Ch. 12, p. 4 b (ed. of Lu Sin-yüan).

[5] ACHUNDOW, Abu Mansur, p. 47. Another Persian form is *hulya*. In Armenian it is *hulbā* or *hulbe* (E. SEIDEL, Mechithar, p. 183). See also LECLERC, Traité

446

is *šanbalīd*, *šanbalīle* in Ispahan, and *šamlīz* in Shiraz, which appears in India as *šamli*. As is well known, the plant occurs wild in Kashmir, the Panjāb, and in the upper Gangetic plain, and is cultivated in many parts of India, particularly in the higher inland provinces. The Sanskrit term is *methī, methikā,* or *methini*.[1] In Greek it is βουκέρας ("ox-horn"),[2] Middle Greek χούλπεν (from the Arabic), Neo-Greek τῆλυ; Latin *foenum graecum*.[3] According to A. DE CANDOLLE,[4] the species is wild (besides the Panjāb and Kashmir) in the deserts of Mesopotamia and of Persia, and in Asia Minor. JOHN FRYER[5] enumerates it among the products of Persia.[6]

Another West-Asiatic plant introduced by the Arabs into China under the Sung is 押不蘆 *ya-pu-lu*, first mentioned by Čou Mi 周密 (1230-1320) as a poisonous plant growing several thousand *li* west from the countries of the Mohammedans (*Kwei sin tsa ši, sü tsi* A, p. 38, ed. of *Pai hai;* and *Či ya t'an tsa č'ao*, Ch. A, p. 40 b, ed. of *Yüe ya t'an ts'un šu*). This name is based on Arabic *yabruh* or *abruh* (Persian *jabrūh*), the mandragora or mandrake. This subject has been discussed by me in detail in a monograph "La Mandragore" (in French), *T'oung Pao*, 1917, pp. 1-30.

---

des simples, Vol. I, p. 443. SCHLIMMER (Terminologie, p. 547) remarks, "L'infusion de la semence est un remède favori des médecins indigènes dans les blennorhagies urethriques chroniques."

[1] It occurs, for instance, as a condiment in an Indian tale of King Vikramāditya (A. WEBER, *Abh. Berl. Akad.*, 1877, p. 67).

[2] Hippocrates; Theophrastus, Hist. plant., IV. IV, 10; or τῆλις: *ibid.*, III. XVI, 2; Dioscorides, II, 124.

[3] Pliny, XXIV, 120.

[4] Origin of Cultivated Plants, p. 112.

[5] New Account of East India and Persia, Vol. II, p. 311.

[6] For further information see FLÜCKIGER and HANBURY, Pharmacographia, p. 172.

NUX–VOMICA

51. The nux-vomica or strychnine tree (*Strychnos nux-vomica*) is mentioned in the *Pen ts'ao kaṅ mu* under the name 番木鼈 *fan mu-pie* ("foreign *mu-pie*," *Momordica cochinchinensis*, a cucurbitaceous plant), with the synonymes 馬錢子 *ma ts'ien-tse* ("horse-coins," referring to the coins on a horse's bridle, hence Japanese *mačin*), 苦實 把豆 *k'u ši pa tou* ("*pa-tou* [*Croton tiglium*] with bitter fruits"),[1] and 火失刻把都 *hwo-ši-k'o pa-tu*. The latter term, apparently of foreign origin, has not yet been identified; and such an attempt would also have been futile, as there is an error in the transcription. The correct mode of writing the word which is given in the *Čo keṅ lu*,[2] written in A.D. 1366, is 火失剌 *hwo-ši-la*, and this is obviously a transcription of Persian *kučla* or *kučula* ("nux-vomica"), a name which is also current in India (thus in Hindustānī; Bengālī *kučila*). The second element *pa-tu* is neither Persian nor Arabic, and, in my opinion, must be explained from Chinese *pa-tou* (*Croton tiglium*).

The text of the *Čo keṅ lu* is as follows: "As regards *hwo-ši-la pa-tu*, it is a drug growing in the soil of Mohammedan countries. In appearance it is like *mu-pie-tse* (*Momordica cochinchinensis*), but smaller. It can cure a hundred and twenty cases; for each case there are special ingredients and guides." This is the earliest Chinese mention of this drug that I am able to trace; and as it is not yet listed in the *Čeṅ lei pen ts'ao* of 1108, the standard work on materia medica of the Sung period, it is justifiable to conclude that it was introduced into China only in the age of the Mongols, during the fourteenth century. This is further evidenced by the very form of the transcription, which is in harmony with the rules then in vogue for writing foreign words. The *Kwaṅ k'ün faṅ p'u*[3] cites no other source relative to the subject than the *Pen ts'ao kaṅ mu*, which indeed appears to be the first and only

---

[1] This name does not mean, as asserted by STUART (Chinese Materia Medica, p. 425), "bitter-seeded Persian bean." STUART (*ibid.*, p. 132) says that the Arabic name for *Croton tiglium* is "batoo, which was probably derived from the Chinese name *pa tou* 巴豆." True it is that the Arabs are acquainted with this plant as an importation from China (L. LECLERC, Traité des simples, Vol. II, p. 95), but only under the name *dend*. I fail to trace a word *batu* in any Arabic dictionary or in Ibn al-Baiṭār.

[2] Ch. 7, p. 5 b. See above, p. 386.

[3] Ch. 6, p. 7.

Pen ts'ao to notice it. The point is emphasized that the drug serves for the poisoning of dogs. The plant now grows in Se-č'wan.

The Sanskrit term for nux-vomica is *kupīlu*, from which is derived Tibetan *go-byi-la* or *go-bye-la*.[1] The latter is pronounced *go-ji-la*, hence the Mongols adopted it as *gojila*. It is uncertain whether the Sanskrit name is related to Persian *kučla* or not.

According to FLÜCKIGER and HANBURY,[2] the tree is indigenous to most parts of India, especially the coast districts, and is found in Burma, Siam, Cochin-China, and northern Australia. The use of the drug in India, however, does not seem to be of ancient date, and possibly was taught there by the Mohammedans. It is mentioned in the Persian pharmacopœia of Abu Mansur (No. 113) under the Arabic name *jauz ul-qei*.[3] SCHLIMMER[4] gives also the terms *azaragi* and *gatel el-kelbe*, and observes, "Son emploi dans la paralysie est d'ancienne date, car l'auteur du *Mexzen el-Edviyeh* en parle déjà, ajoutant en outre que la noix vomique est un remède qui change le tempérament froid en tempérament chaud; le même auteur recommande les cataplasmes avec sa poudre dans la coxalgie et dans les maladies articulaires."

The Arabs, who say that the tree occurs only in the interior of Yemen, were well acquainted with the medicinal properties of the fruit.[5] Nux-vomica is likewise known in Indo-China (Čam *salaiñ* and *phun akam*, Khmer *slêñ*, Annamese *ku-či;* the latter probably a transcription of *kučila*).[6]

The *Kew Bulletin* for 1917 (p. 341) contains the following notice on *Strychnos nux-vomica* in Cochin-China: "In *K. B.* 1917 (pp. 184, 185), some evidence is given as to the occurrence of this species in Cochin-China in the wild state. Since the account was written a letter and a packet of undoubted nux-vomica seeds have been received from the Director, Agricultural and Commercial Services, Cochin-China, with the information that the seeds were obtained from trees growing wild in the country. H. B. M.'s Consul, Saigon, also sends the following information about *S. nux-vomica* in Cochin-China which he has received from Monsieur Morange, Director of the Agricultural and Commercial

---

[1] Cf. Loan-Words in Tibetan, No. 50 (*T'oung Pao*, 1916, p. 457).

[2] Pharmacographia, p. 428.

[3] ACHUNDOW, Abu Mansur, p. 43.

[4] Terminologie, p. 402.

[5] L. LECLERC, Traité des simples, Vol. I, p. 380.

[6] Cf. E. PERROT and P. HURRIER, Matière médicale et pharmacopée sino-annamites, p. 171; the Chinese and Annamese certainly did not avail themselves of this drug "from time immemorial," as stated by these authors. See, further, C. FORD, *China Review*, Vol. XV, 1887, p. 220.

Services of Cochin-China, and also a sample of the seeds obtained from a Chinese exporter. The tree exists in the Eastern provinces of Cochin-China, principally in the forests of Baria. The seeds are bought by Chinese from the savage tribes known as Mois, who collect them in the forest; the Chinese then export them to China or sell them again to firms exporting to Europe. The time of fruiting is in November and December. M. Morange considers that the tree is certainly indigenous in Cochin-China, and was not introduced by early traders." If the tree is indigenous there, it was certainly discovered there, as far as the Chinese are concerned, only after the Mongol period. H. MAITRE[1] deals with the poisons used by the Moi for their arrows, and arrives at the conclusion that they are derived from the upas tree (*Antiaris*). He does not mention *Strychnos*.

---

[1] Les régions Moï du sud indo-chinois, pp. 119–121 (Paris, 1909).

THE CARROT

52. The carrot[1] (*Daucus carota*), *hu lo-po* (Japanese *ninjin*) 胡 蘿 蔔 ("Iranian turnip"), a native of northern Europe, was first introduced into China at the time of the Yüan dynasty (A.D. 1260–1367). This is the opinion of Li Ši-čen, who states that the vegetable first appeared at the time of the Yüan from the land of the Hu; and it is likewise maintained in the *Kwaṅ k'üṅ faṅ p'u*[2] that the carrot first came from the countries beyond the frontier 邊 塞. I know of no text that would give a more detailed account of its introduction or allude to the country of its origin. Nevertheless it is very likely that this was some Iranian region. Li Ši-čen states that in his time it was abundantly cultivated in the northern part of the country and in Šan-tuṅ, likewise in middle China.[3]

The history of the carrot given by WATT[4] after G. Birdwood suffers from many defects. A fundamental error underlies the statement, "In fact, the evidence of cultivation would lead to the inference that the carrot spread from Central Asia to Europe, and if so it might be possible to trace the European names from the Indian and Persian." On the contrary, the carrot is a very ancient, indigenous European cultivation, which is by no means due to the Orient. Carrots have been found in the pile-dwellings of Robenhausen.[5] It is not to the point, either, that, as stated by Watt and Birdwood, "indeed the carrot seems to have been grown and eaten in India, while in Europe it was scarcely known as more than a wild plant." The Anglo-Saxons cultivated the carrot in their original habitat of Schleswig-Holstein at a time when, in my opinion, the carrot was not yet cultivated in India; and they con-

---

[1] From French *carote*, now *carotte*, Italian *carota*, Latin *caróta;* Greek καρωτόν (in Diphilus). This word has supplanted Anglo-Saxon *moru*, from *morhu (Old High German *moraha, morha;* Russian *morkov'*, Slovenian *mrkva*). Regarding the origin of the word *lo-po*, cf. *T'oung Pao*, 1916, pp. 83–86.

[2] Ch. 4, p. 24.

[3] A designation for the carrot not yet indicated is *fu* 伏 *lo-po*, derived from the three *fu* 三 伏, the three decades of the summer, extending from about the middle of July to the middle of August: during the first *fu* the seeds of the carrot are planted, in the second *fu* the carrots are pale red, in the third they are yellow (*Šan hwa hien či* 善 化 縣 志, Ch. 16, p. 14 b, ed. 1877).

[4] Commercial Products of India, p. 489, or Dictionary, Vol. III, p. 45.

[5] J. Hoops, Waldbäume und Kulturpflanzen, p. 297; G. Buschan, Vorgeschichtliche Botanik, p. 148.

451

tinued to cultivate it in England.[1] Moreover, the carrot grows wild in Britain and generally in the north temperate zone of Europe and Asia, and no doubt represents the stock of the cultivated carrot, which can be developed from it in a few generations.[2] It is impossible to connect Anglo-Saxon *moru* (not *mora*, as in Watt) with Sanskrit *mūla* or *mūlaka*. No evidence is given for the bold assertion that "the carrot appears to have been regularly used in India from fairly ancient times." The only sources quoted are Baber's Memoirs[3] and the Ain-i Akbari, both works of the sixteenth century. I fail to see any proof for the alleged antiquity of carrot cultivation in India. There is no genuine Sanskrit word for this vegetable. It is incorrect that "the Sanskrit *garjaru* originated the Persian *zardak* and the Arabic *jegar*" (*sic*, for *jezer*). Boehtlingk gives for *garjara* only the meaning "kind of grass." As indicated below, it was the Arabs who carried the carrot to Persia in the tenth century, and I do not believe that it was known in India prior to that time. According to Watt, *Daucus carota* is a native of Kashmir and the western Himalaya at altitudes of from 5000 to 9000 feet; and throughout India it is cultivated by Europeans, mostly from annually imported seed, and by the natives from an acclimatised if not indigenous stock. Also N. G. MUKERJI[4] observes, "The English root-crop which has a special value as a nourishing famine-food and fodder is the carrot. Up-country carrot or *gajra* is not such a nourishing and palatable food as European carrot, and of all the carrots experimented with in this country, the red Mediterranean variety grown at the Cawnpore Experimental Farm seems to be the best."

W. ROXBURGH[5] states that *Daucus carota* "is said to be a native of Persia; in India it is only found in a cultivated state." He gives two Sanskrit names,— *grinjana* and *gargara*, but his editor remarks that he finds no authority for these. In fact, these and Watt's alleged Sanskrit names are not at all Sanskrit, but merely Hindi (Hindī *gājara*); and this word is derived from Persian (not the Persian derived from Sanskrit, as alleged by Watt). The only Sanskrit terms for the carrot known to me are *yavana* ("Greek or foreign vegetable") and *pītakanda* (literally, "yellow root"), which appears only in the Rājanighaṇṭu, a work from the beginning of the fifteenth century. This

---

[1] HOOPS, *op. cit.*, p. 600.

[2] A. DE CANDOLLE, Géographie botanique, p. 827.

[3] Baber ate plenty of carrots on the night (December 21, 1526) when an attempt was made to poison him. Cf. H. BEVERIDGE, The Attempt to Poison Babur Padshah (*Asiatic Review*, Vol. XII, 1917, pp. 301–304).

[4] Handbook of Indian Agriculture, 2d ed., p. 304.

[5] Flora Indica, p. 270.

descriptive formation is sufficient to show that the cultivated carrot was foreign to the Hindu. Also W. AINSLIE[1] justly concludes, "Carrots appear to have been first introduced into India from Persia."

According to SCHWEINFURTH,[2] *Daucus carota* should display a very peculiar form in Egypt,—a sign of ancient cultivation. This requires confirmation. At all events, it does not prove that the carrot was cultivated by the ancient Egyptians. Neither Loret nor Woenig mentions it for ancient Egypt.

In Greek the carrot is σταφυλῖνος (hence Syriac *istaflīn*). It is mentioned by Theophrastus[3] and Pliny;[4] δαῦκος or δαῦκον was a kind of carrot or parsnip growing in Crete and used in medicine; hence Neo-Greek τὸ δαφκί ("carrot"), Spanish *dauco*. A. DE CANDOLLE[5] is right in saying that the vegetable was little cultivated by the Greeks and Romans, but, as agriculture was perfected, took a more important place.

The Arabs knew a wild and a cultivated carrot, the former under the name *nehšel* or *nehsel*,[6] the knowledge of which was transmitted to them by Dioscorides,[7] the latter under the names *jezer*, *sefanariya* (in the dialect of Magreb *zorudiya*), and *sabāhīa*.[8] The Arabic word *dauku* or *dūqū*, derived from Greek δαῦκος, denotes particularly the seed of the wild carrot.[9]

JORET[10] presumes that the carrot was known to the ancient Iranians. The evidence presented, however, is hardly admissible: *Daucus maximus* which grows in Western Persia is only a wild species. This botanical fact does not prove that the Iranians were acquainted with the cultivated *Daucus carota*. An Iranian name for this species is not known. Only in the Mohammedan period does knowledge of it spring up in Persia; and the Persians then became acquainted with the carrot under the Arabic name *jazar* or *jezer*, which, however, may have been derived from Persian *gazar* (*gezer*). It is mentioned under the Arabic name in the Persian pharmacopœia of Abu Mansur,[11] who apparently copied from Arabic sources. He further points out a wild species under the

---

[1] Materia Indica, Vol. I, p. 57.

[2] Z. f. Ethnologie, Vol. XXIII, 1891, p. 662.

[3] Hist. plant., IX. xv, 5.

[4] xx, 15.

[5] Géographie botanique, p. 827.

[6] L. LECLERC, Traité des simples, Vol. III, p. 380.

[7] LECLERC, op. cit., Vol. I, p. 353.

[8] LECLERC, ibid., and p. 367.

[9] LECLERC, ibid., p. 138.

[10] Plantes dans l'antiquité, Vol. II, p. 66.

[11] ACHUNDOW, Abu Mansur, p. 42.

name *šašqāqul*, which, according to ACHUNDOW, is *Eryngium campestre*. It is therefore very probable that it was the Arabs who introduced the carrot into Persia during the tenth century. Besides *gazar* (*gezer*), Persian names are *zardak*[1] and *šawandar;* the latter means "beet-root" and "carrot."

JOHN FRYER, who travelled in India and Persia from 1672 to 1681, enumerates carrots among the roots of Persia.[2] The late arrival of the vegetable in Persia is signally confirmed by the Chinese tradition regarding its introduction under the Mongols. This is the logical sequence of events.[3]

SCHLIMMER[4] has the following note on the subject: "Ce légume, formé en compôte, est considéré par les Persans comme un excellent aphrodisiaque, augmentant la quantité et améliorant la qualité du sperme. L'alimentation journalière avec des carottes est fortement prônée dans les hydropisies; les carottes cuites, conservées au vin aigre, dissiperaient l'engorgement de la rate." Only the yellow variety of carrot, with short, spindle-shaped roots, occurs in Fergana.[5]

---

[1] Possibly derived from *zard* ("yellow"). Persian *mūrāmūn* is said to denote a kind of wild carrot. In Osmanli the carrot is called *hawuj*.

[2] New Account of East India and Persia, Vol. II, p. 310 (Hakluyt Soc., 1912).

[3] Regarding the Tibetan names of the carrot, see my notes in *T'oung Pao*, 1916, pp. 503–505.

[4] Terminologie, p. 176.

[5] S. KORŽINSKI, Vegetation of Turkistan (in Russian), p. 51.

AROMATICS

53. The *Sui šu*[1] mentions two aromatics or perfumes peculiar to K'an (Sogdiana),— *kan hian* 肌[2]香 and *a-sa-na hian* 阿薩郁香. Fortunately we have a parallel text in the *T'ai p'in hwan yü ki*,[3] where the two aromatics of K'an are given as 甘松香阿薩那香. Hence it follows that the *kan* of the Sui Annals is no more than an abbreviation of *kan sun*, which is well known as an aromatic, and identical with the true spikenard furnished by *Nardostachys jatamansi*. It is Sanskrit *nalada*, Tibetan *span spos*, Persian *nard* or *sunbul*, Armenian *sumbul*, *smbul*, *snbul*, etc.[4] It is believed that the nard found by Alexander's soldiers in Gedrosia[5] represents the same species, while others hold that it was an *Andropogon*.[6]

The Sanskrit term *nalada* is found in the *Fan yi min yi tsi*[7] in the form 那羅陀 *na-lo-t'o*, *na-la-da. It is accompanied by the fanciful analysis *nara-dhara* ("held or carried by man"), because, it is said, people carry the fragrant flower with them in their girdles. The word *nalada* is of ancient date, for it appears in the Atharvaveda.[8] Hebrew *nērd*, Greek *nardos*,[9] Persian *nard* and *nārd*, are derived therefrom.[10] Being used in the Bible, the word was carried to all European languages.

---

[1] Ch. 83, p. 4 b.

[2] This character is not listed in K'an-hi, but the phonetic element 甘 leaves no doubt that its phonetic value is *kan*, *kam.

[3] Ch. 183, p. 4.

[4] ABU MANSUR (Achundow's translation, pp. 82, 241) mentions *sunbul-i-hindī*, the nard of India. SCHLIMMER (Terminologie, p. 36) identifies this name as *Andropogon nardoides* or *Nardus indica*. On the other hand, he says (p. 555) that *Nardostachys* or *Valeriana jatamansi* has not yet been found in Persia, but that it could be replaced in therapeutics by *Valeriana sisymbrifolia*, found abundantly in the mountains north of Teheran.

[5] Arrian, Anabasis, VI. XXII, 5.

[6] JORET, Plantes dans l'antiquité, Vol. II, p. 648. See, further, Periplus, 48; and Pliny, XII, 28; WATT, Commercial Products of India, p. 792. MARCO POLO (ed. of YULE, Vol. I, pp. 115, 272, 284) mentions spikenard as a product of Bengal, Java, and Sumatra. The Malayan word *nārāwastu*, mentioned by YULE (*ibid.*, p. 287), must be connected with Sanskrit *nalada*.

[7] Ch. 8, p. 4 b.

[8] MACDONELL and KEITH, Vedic Index, Vol. I, p. 437; H. ZIMMER, Altindisches Leben, p. 68.

[9] First mentioned by Theophrastus, IX. VIII, 2, 3.

[10] See above, p. 428.

455

According to Stuart,[1] this plant is found in the province of Yün-nan and on the western borders of Se-č'wan, but whether indigenous or transplanted is uncertain. If it should not occur in other parts of China, it is more likely that it came from India, especially as Yün-nan has of old been in contact with India and abounds in plants introduced from there.

54. 阿薩郍[2] *a-sar(sat)-na (Sui šu), 阿薛那 a-sie-na (Wei šu, Ch. 102, p. 9), is not explained. There is no doubt that this word represents the transcription of an Iranian, more specifically Sogdian, name; but the Sogdian terms for aromatics are still unknown to us. Hypothetical restorations of the name are *asarna, axšarna, asna.

55. Storax, an aromatic substance (now obtained from Liquidambar orientalis; in ancient times, however, from Styrax officinalis), is first mentioned by Herodotus[3] as imported into Hellas by the Phœnicians. It is styled by the Chinese 蘇合 su-ho, *su-gap (giep), su-gab (Japanese sugō), being mentioned both in the Wei lio and in the Han Annals as a product of the Hellenistic Orient (Ta Ts'in).[4] It is said there, "They mix a number of aromatic substances and extract from them the sap by boiling, which is made into su-ho" (合會諸香煎其汁以爲蘇合).[5] It is notable that this clause opens and ends with the same word ho 合; and it would thus not be impossible that the explanation is merely the result of punning on the term su-ho, which is doubtless the transcription of a foreign word. Aside from this semasiological interpretation, we have a geographical theory expressed in the Kwaṅ či, written prior to A.D. 527, as follows: "Su-ho is produced in the country Ta Ts'in; according to others, in the country Su-ho. The natives of this country gather it and press the juice out of it to make it into an aromatic, fatty substance. What is sold are the sediments

---

[1] Chinese Materia Medica, p. 278.

[2] This character is not in K'aṅ-hi. It appears again on the same page of the Sui šu (4 b) in the name of the river *Na-mit 郍密 (Zarafšan) in the kingdom Ṅan 安, and on p. 4 a in 郍色波國, the country Na-se-po (*Na-sek-pwa; according to Chavannes, Documents sur les Tou-kiue, p. 146, Nakhšab or Nasaf). On pp. 6 b and 7 a the river Na-mit is written 那. Cf. also Chavannes and Pelliot, Traité manichéen, pp. 58, 191.

[3] III, 107.

[4] Hou Han šu, Ch. 118, pp. 4 b—5 a. E. H. Parker (China Review, Vol. XV, p. 372) indicates in an anecdote relative to Čwaṅ-tse that he preferred the dung-beetle's dung-roll to a piece of storax, and infers that indirect intercourse with western Asia must have begun as early as the fourth century B.C., when Čwaṅ-tse flourished. The source for this story is not stated, and it may very well be a product of later times.

[5] The Sü Han šu gives the same text with the variant, "call it su-ho."

of this product."[1] Nothing is known, however, in Chinese records about this alleged country Su-ho (*Su-gab); hence it is probable that this explanation is fictitious, and merely inspired by the desire to account in a seemingly plausible way for the mysterious foreign word.

In the Annals of the Liang Dynasty,[2] storax is enumerated among the products of western India which are imported from Ta Ts'in and An-si (Parthia). It is explained as "the blending of various aromatic substances obtained by boiling their saps; it is not a product of nature."[3] Then follows the same passage relating to the manufacture in Ta Ts'in as in the *Kwaṅ či;* and the *Liaṅ šu* winds up by saying that the product passes through the hands of many middlemen before reaching China, and loses much of its fragrancy during this process.[4] It is likewise on record in the same Annals that in A.D. 519 King Jayavarman of Fu-nan (Camboja) sent among other gifts storax to the Chinese Court.[5]

Finally, *su-ho* is enumerated among the products of Sasanian Persia.[6] Judging from the commercial relations of Iran with the Hellenistic Orient and from the nature of the product involved, we shall not err in assuming that it was traded to Persia in the same manner as to India.

The Chinese-Sanskrit dictionaries contain two identifications of the name *su-ho*. In the third chapter of the *Yü k'ie ši ti lun* 瑜伽師地論 (Yogācāryabhūmiçāstra),[7] translated in A.D. 646–647 by Hüan Tsaṅ, we find the name of an aromatic in the form 窣堵魯迦 *su-tu-lu-kia*, *sut-tu-lu-kyie*; that is, Sanskrit *sturuka* = storax.[8] It is identified by Yüan Yiṅ with what was formerly styled 兜樓婆 *tou-lou-p'o*, *du-lyu-bwa*.[9] It is evident that the transcription *su-tu-lu-kia* is based on a form corresponding to Greek *styrak-s, storak-s, styrákion* of the Papyri (Syriac *stiraca, astorac*). This equation presents the

---

[1] *Fan yi miṅ yi tsi*, Ch. 8, p. 9; *T'ai p'iṅ yü lan*, Ch. 982, p. 1 b.

[2] *Liaṅ šu*, Ch. 54, p. 7 b.

[3] The *Fan yi miṅ yi tsi*, which reproduces this passage, has, "It is not a single (or homogeneous) substance."

[4] Cf. HIRTH, China and the Roman Orient, p. 47.

[5] Cf. PELLIOT, *Bull. de l'Ecole française*, Vol. III, p. 270.

[6] *Sui šu*, Ch. 83, p. 7 b; or *Čou šu*, Ch. 50, p. 6. It does not follow from these texts, that, as assumed by HIRTH (Chao Ju-kua, pp. 16, 262), *su-ho* or any other product of Persia was imported thence to China. The texts are merely descriptive in saying that these are products to be found in Persia.

[7] BUNYIU NANJIO, Catalogue of the Chinese Tripitaka, No. 1170.

[8] *Yi ts'ie kiṅ yin i*, Ch. 22, p. 3 b (cf. PELLIOT, *T'oung Pao*, 1912, pp. 478–479). This text has been traced by me independently. I do not believe that this name is connected with *turuṣka*.

[9] Probably Sanskrit *dūrvā* (cf. *Journal asiatique*, 1918, II, pp. 21–22).

strongest evidence for the fact that the *su-ho* of the Chinese designates the storax of the ancients.[1]

The *Fan yi miñ yi tsi* (*l.c.*) identifies Sanskrit 咄魯瑟劍 *tu-lu-se-kien*, *tu-lu-sŏt-kiam, answering to Sanskrit *turuṣkam*, with *su-ho*. In some works this identification is even ascribed to the *Kwaṅ či* of the sixth century (or probably earlier). In the *Pien tse lei pien*,[2] where the latter work is credited with this Sanskrit word, we find the character 竭 *kie*, *g'iaδ, in lieu of the second character *lu*. The term *turuṣka* refers to real incense (olibanum).[3] It is very unlikely that this aromatic was ever understood by the word *su-ho*, and it rather seems that some ill-advised adjustment has taken place here.

T'ao Huṅ-kiṅ (A.D. 451–536) relates a popular tradition that *su-ho* should be lion's ordure, adding that this is merely talk coming from abroad, and untrue.[4] Č'en Ts'aṅ-k'i of the eighth century states,[5] "Lion-ordure is red or black in color; when burnt, it will dissipate the breath of devils; when administered, it will break stagnant blood and kill worms. The perfume *su-ho*, however, is yellow or white in color: thus, while the two substances are similar, they are not identical. People say that lion-ordure is the sap from the bark of a plant in the western countries brought over by the Hu. In order to make people prize this article, this name has been invented." This tradition as yet unexplained is capable of explanation. In Sanskrit, *rasamala* means "excrement," and this word has been adopted by the Javanese and Malayans for the designation of storax.[6] Thus this significance of the word may have given the incentive for the formation of that trade-trick,— examples of which are not lacking in our own times.

Under the T'ang, *su-ho* was imported into·China also from Malayan regions, especially from K'un-lun (in the Malayan area), described as

---

[1] The most important pharmacological and historical investigation of the subject still remains the study of D. HANBURY (Science Papers, pp. 127–150), which no one interested in this matter should fail to read.

[2] Ch. 195, p. 8 b.

[3] Cf. Language of the Yüe-chi, p. 7.

[4] He certainly does not say, as BRETSCHNEIDER (Bot. Sin., pt. III, p. 463) wrongly translates, "but the foreigners assert that this is not true." Only the foreigners could have brought this fiction to China, as is amply confirmed by Č'en Ts'aṅ-k'i. Moreover, the *T'aṅ pen ču* 唐本注 says straight, "This is a falsehood of the Hu."

[5] *Čeṅ lei pen ts'ao*, Ch. 12, p. 52 (ed. of 1587).

[6] BRETSCHNEIDER (*l. c.*) erroneously attributes to Garcia da Orta the statement that Rocamalha should be the Chinese name for the storax, and STUART (Chinese Materia Medica, p. 243) naturally searched in vain for a confirmation of this name in Chinese books. GARCIA says in fact that liquid storax is here (that is, in India) called Rocamalha (MARKHAM, Colloquies, p. 63), and does not even mention China in this connection.

purple-red of color, resembling the *tse t'an* 紫檀 (*Pterocarpus santalinus*, likewise ascribed to K'un-lun), strong, solid, and very fragrant.[1] This is *Liquidambar altingiana* or *Altingia excelsa*, a lofty deciduous tree growing in Java, Burma, and Assam, with a fragrant wood yielding a scented resin which hardens upon exposure to the air. The Arabs imported liquid storax during the thirteenth century to Palembang on Sumatra;[2] and the *T'ai p'in hwan yü ki* states that *su-ho* oil is produced in Annam, Palembang (*San-fu-ts'i*), and in all barbarous countries, from a tree-resin that is employed in medicine. The *Mon ki pi t'an* discriminates between the solid storax of red color like a hard wood, and the liquid storax of glue-like consistency which is in general use.[3]

The Chinese transcription *su-ho*, *su-gap, has not yet been explained. HIRTH's[4] suggestion that the Greek στύραξ should have been "mutilated" into *su-ho* is hardly satisfactory, for we have to start from the ancient form *su-gab, which bears no resemblance to the Greek word save the first element. In the Papyri no name of a resin has as yet been discovered that could be compared to *su-gab.[5] Nor is there any such Semitic name (cf. Arabic *lubnā*). In view of this situation, the question may be raised whether *su-gab would not rather represent an ancient Iranian word. This supposition, however, cannot be proved, either, in the present state of science. Storax appears in the Persian materia medica of Abu Mansur under the Arabic name *mī'a*.[6] The storax called rose-maloes is likewise known to the Persians, and is said to be derived

---

[1] *Čen lei pen ts'ao, l. c.* This tree is mentioned in the *Ku kin ču* (Ch. c, p. 1 b, as a product of Fu-nan, and by Čao Žu-kwa as a variety of sandal-wood (HIRTH) Chao Ju-kua, p. 208). Li Ši-čen (*Pen ts'ao kan mu*, Ch. 34, p. 12) says that the people of Yün-nan call *tse t'an* by a peculiar word, 勝 *šen*; this is pronounced *sen* in Yün-nan, and accordingly traceable to a dialectic variation of *čandan*, *sandan*, *sandal*. The Japanese term is *šitan* (MATSUMURA, No. 2605).

[2] HIRTH, Chao Ju-kua, p. 61.

[3] Cf. *Pien tse lei pien*, Ch. 195, p. 8 b; BRETSCHNEIDER, Bot. Sin., pt. III, p. 464. The *Hian p'u* quoted in the *Pen ts'ao* is the work of Ye T'in-kwei 葉廷珪, not the well-known work by Hun Č'u, in which the passage in question does not occur (see p. 2, ed. of *T'an Sun ts'un šu*, where it is said that it is difficult to recognize the genuine article). For further information on liquid storax, see HIRTH, Chao Ju-kua, p. 200.

[4] Chao Ju-kua, p. 200.

[5] MUSS-ARNOLT (*Transactions Am. Phil. Assoc.*, Vol. XXIII, p. 117) derives the Greek word from Hebrew *z'ri;* the Greek should have assimilated the Semitic loan-word to στύραξ ("spike"). This is pure fantasy. The Hebrew word, moreover, does not relate to storax, but, according to GESENIUS, denotes a balsam or resin like mastic (above, p. 252). The Hebrew word for *Styrax officinalis* is said to be *nātāf* (EXODUS, xxx, 34), Septuaginta στακή, Vulgata *stacte* (E. LEVESQUE in Dictionnaire de la Bible, Vol. V, col. 1869-70).

[6] ACHUNDOW, Abu Mansur, p. 138.

from a tree growing on the Island of Cabros in the Red Sea (near Kadez, three days' journey from Suez), the product being obtained by boiling the bark in salt water until it obtains the consistency of glue.[1]

56–57. The earliest notice of myrrh is contained in the *Nan čou ki* 南州記 of Sü Piao 徐表 (written before the fifth century A.D., but only preserved in extracts of later works), if we may depend on the *Hai yao pen ts'ao*, in which this extract is contained.[2] Sü Piao is made to say there that "the myrrh grows in the country Po-se, and is the pine-tree resin of that locality. In appearance it is like 神香 *šen hiaṅ* ('divine incense') and red-black in color. As to its taste, it is bitter and warm." Li Ši-čen annotates that he is ignorant of what the product *šen hiaṅ* is. In the *Pei ši*, myrrh is ascribed to the country Ts'ao (Jāguḍa) north of the Ts'uṅ-liṅ (identical with the Ki-pin of the Han), while this product is omitted in the corresponding text of the *Sui šu*. Myrrh, further, is ascribed to Ki-pin.[3] The *Čeṅ lei pen ts'ao* gives a crude illustration of the tree under the title *mu yao* of Kwaṅ-čou (Kwaṅ-tuṅ), saying that the plant grows in Po-se and resembles benjoin (*ṅan-si hiaṅ*, p. 464), being traded in pieces of indefinite size and of black color.

In regard to the subject, Li Ši-čen[4] cites solely sources of the Sung period. He quotes K'ou Tsuṅ-ši, author of the *Pen ts'ao yen i* (A.D. 1116), to the effect that myrrh grows in Po-se, and comes in pieces of indefinite size, black in color, resembling benjoin. In the text of this work, as edited by Lu Sin-yüan,[5] this passage is not contained, but merely the medicinal properties of the drug are set forth.[6] Su Suṅ observes that "myrrh now occurs in the countries of the Southern Sea (Nan-hai) and in Kwaṅ-čou. Root and trunk of the tree are like those of *Canarium* (*kan-lan*). The leaves are green and dense. Only in the course of years does the tree yield a resin, which flows down into the soil, and hardens into larger or smaller pieces resembling benjoin. They may be gathered at any time."

A strange confusion occurs in the *Yu yaṅ tsa tsu*,[7] where the myrtle (*Myrtus communis*) is described under its Aramaic name *asa* (Arabic

---

[1] SCHLIMMER, Terminologie, p. 495.

[2] *Čeṅ lei pen ts'ao*, Ch. 13, p. 39; *Pen ts'ao kaṅ mu*, Ch. 34, p. 17.

[3] *T'ai p'iṅ hwan yü ki*, Ch. 182, p. 12 b.

[4] *Pen ts'ao kaṅ mu, l. c.*

[5] Ch. 14, p. 4 b.

[6] In all probability, there is an editorial error in the edition of the *Pen ts'ao* quoted; in other editions the same text is ascribed to Ma Či, one of the collaborators in the *K'ai pao pen ts'ao*.

[7] Ch. 18, p. 12.

as), while this section opens with the remark, "The habitat of the myrrh tree 没 is in Po-se."[1] It may be, however, that, as argued by HIRTH, *mu* may be intended in this case to transcribe Middle and New Persian *mūrd*, which means "myrtle" (not only in the Bŭndahišn, but generally).[2] Myrrh and myrtle have nothing to do with each other, belonging not only to different families, but even to different orders; nor does the myrtle yield a resin like myrrh. It therefore remains doubtful whether myrrh was known to the Chinese during the T'ang period; in this case, the passage cited above from the *Nan čou ki* (like many another text from this work) must be regarded as an anachronism. Čao Žu-kwa gives the correct information that myrrh is produced on the Berbera coast of East Africa and on the Hadramaut littoral of Arabia; he has also left a fairly correct description of how the resin is obtained.[3]

Li Ši-čen[4] thinks that the transcription 没 or 末 represents a Sanskrit word. This, of course, is erroneous: myrrh is not an Indian product, and is only imported into India from the Somali coast of Africa and from Arabia. The former Chinese character answers to ancient *mut or *mur; the latter, to *mwat, mwar, or mar. The former no doubt represents attempts at reproducing the Semito-Persian name,— Hebrew *mōr*, Aramaic *murā*, Arabic *murr*, Persian *mor* (Greek σμύρα, σμύρον, μύρον, Latin *myrrha*).[5]

Whether the Chinese transcribed the Arabic or Persian form, remains uncertain: if the transcription should really appear as late as the age of the Sung, it is more probable that the Arabic yielded the prototype; but if it can be carried back to the T'ang or earlier, the assumption is in favor of Iranian speech.

---

[1] Cf. HIRTH, *Journal Am. Or. Soc.*, Vol. XXX, p. 20. Owing to a curious misconception, the article of the *Yu yan tsa tsu* has been placed under *mi hian* 蜜香 ("gharu-wood") in the *Pen ts'ao kan mu* (Ch. 34, p. 10 b), for *mu* 没 *hian* is wrongly supposed to be a synonyme of *mi hian*.

[2] Another New-Persian word for this plant is *anībā* or *anītā*. In late Avestan it is *muštemeša* (BARTHOLOMAE, Altiran. Wört., col. 1189). I do not believe that the Persian word and Armenian *murt* are derived from Greek μυρσίνη (SCHRADER in Hehn, Kulturpflanzen, p. 238) or from Greek μύρτος (NÖLDEKE, Persische Studien, II, p. 43).

[3] HIRTH, Chau Ju-kua, p. 197.

[4] *Pen ts'ao kan mu*, Ch. 34, p. 17.

[5] Pliny, XII, 34–35; LECLERC, Traité des simples, Vol. III, p. 300; V. LORET, Flore pharaonique, p. 95. The transcription *mwat appears to transcribe Javanese and Bali *madu* ("myrrh"; Malayan *manisan lebah*). In an Uigur text translated from Sogdian or Syriac appears the word *zmurna* or *zmuran* ("myrrh"), connected with the Greek word (F. W. K. MÜLLER, Uigurica, pp. 5–7).

Theophrastus[1] mentions in the country Aria a "thorn" on which is found a gum resembling myrrh in appearance and odor, and this drops when the sun shines on it. STRABO[2] affirms that Gedrosia produced aromatics, particularly nard and myrrh, in such quantity that Alexander's army used them, on the march, for tent-coverings and beds, and thus breathed an air full of odors and more salubrious. Modern botanists, however, have failed to find these plants in Gedrosia or any other region of Iran;[3] and the Iranian myrrh of the ancients, in all probability, represents a different species of *Balsamodendron* (perhaps *B. pubescens* or *B. mukul*). According to W. GEIGER,[4] *Balsamodendron mukul* is called in Baluči *bōd, bōδ,* or *bōz,* a word which simply means "odor, aroma." It is a descendant of Avestan *baoiδi,* which we find in Pahlavi as *bōd, bōi,* Sogdian *fraβōδan, βōδa,* New Persian *bōi, bō* (Ossetic *bud,* "incense").[5]

It is noteworthy also that the ancient Chinese accounts of Sasanian Persia do not make mention of myrrh. The botanical evidence being taken into due consideration, it appears more than doubtful that the statement of the *Nan čou ki, Yu yan tsa tsu, K'ai pao pen ts'ao,* and *Čen lei pen ts'ao,* that the myrrh-tree grows in Po-se, can be referred to the Iranian Po-se. True it is, the tree does not occur, either, in the Malayan area; but, since the product was evidently traded to China by way of Malaysia, the opinion might gain ground among the Chinese that the home of the article was the Malayan Po-se.

The Japanese style the myrrh *mirura,* which is merely a modern transcription of "myrrha."[6]

58. *Ts'in mu hian* 青木香 ("dark-wood aromatic") is attributed to Sasanian Persia.[7] What this substance was, is not explained; and merely from the fact that the name in question, as well as *mu hian* 木香 ("tree aromatic") and *mi hian* 蜜香, usually refer to costus root or putchuck (also pachak), we may infer that the Persian aromatic was of a similar character. Thus it is assumed by HIRTH;[8] but the matter remains somewhat hypothetical. The Chinese term, indeed, has

---

[1] Hist. plant., IV. IV, 13.

[2] XV. II, 3.

[3] C. JORET, Plantes dans l'antiquité, Vol. I, p. 48.

[4] Etymologie des Baluči, p. 46.

[5] In regard to the use of incense on the part of the Manichæans, see CHAVANNES and PELLIOT, Traité manichéen, pp. 302-303, 311.

[6] J. MATSUMURA, Shokubutsu mei-i, No. 458.

[7] *Wei šu,* Ch. 102, p. 5 b; *Sui šu,* Ch. 83, p. 7 b.

[8] Chau Ju-kua, p. 221. Putchuck is not the root of *Aucklandia costus,* but of *Saussurea lappa* (see WATT, Commercial Products of India, p. 980).

no botanical value, being merely a commercial label covering different roots from most diverse regions. If Čao Žu-kwa compares the putchuck-yielding plant with *Luffa cylindrica*, a *Cucurbitacea* of southern China, with which he compares also the cardamom, it is perfectly clear that he does not visualize the genuine costus-root of *Saussurea lappa*, a tall, stout herb, indigenous to the moist, open slopes surrounding the valley of Kashmir, at an elevation of eight or nine thousand feet. If he further states that the product is found in Hadramaut and on the Somali coast, it is, in my opinion, not logical to reject this as "wrong," for a product of the name *mu hian* was certainly known in the China of his time from that region. And why not? Also Dioscorides mentions an Arabian costus, which is white and odoriferous and of the best quality; besides, he has an Indian costus, black and smooth, and a Syrian variety of wax color, dusky, and of strong odor. It is obvious that these three articles correspond to the roots of three distinct species, which have certain properties in common; and it has justly been doubted that the modern costus is the same thing as that of the ancients. The Arabs have adopted the nomenclature of Dioscorides.[1] The Sheikh Daūd distinguishes an Indian species, white; a black one from China; and a red, heavy one, adding that it is said to be a tree of the kind of *Agallochum*. Nearly everywhere in Asia have been found aromatic roots which in one way or another correspond to the properties of the Indian *kuṣṭha*. Thus in Tibet and Mongolia the latter is adjusted with the genus *Inula;* and the Tibetan word *ru-rta*, originally referring to an *Inula,* was adopted by the Buddhist translators as a rendering of Sanskrit *kuṣṭha*.[2] In the same manner, the Chinese term *mu hian* formerly denoted an indigenous plant of Yün-nan, which, according to the ancient work *Pie lu*, grew in the mountain-valleys of Yuṅ-č'aṅ.[3] The correctness of this tradition is confirmed by the *Man šu*, which mentions a mountain-range, three days' journey south of Yuṅ-č'aṅ, by name Ts'iṅ-mu-hiaṅ ("Dark-Wood Aromatic"), and owing its name to the great abundance of this root.[4] The *Man šu*, further, extends its occurrence to the country

---

[1] LECLERC, Traité des simples, Vol. III, pp. 85–86.

[2] H. LAUFER, Beiträge zur Kenntnis der tibetischen Medicin, p. 61.

[3] Also Wu K'i-tsün (*Či wu miṅ ši t'u k'ao*, Ch. 25, p. 11) observes correctly that this species is not the putchuck coming from the foreign barbarians. His three illustrations, putchuck from Hai-čou in Kiaṅ-su, from Kwaṅ-tuṅ, and from Č'u-čou in Ṅan-hwi, are reproduced from the *T'u šu tsi č'eṅ* (XX, Ch. 117), and represent three distinct plants.

[4] The *Tien hai yü heṅ či* (Ch. 3, p. 1; see above, p. 228) states that *mu hiaṅ* is produced in the native district Č'ö-li 車 里 土 司, formerly called Č'an-li 產 里, of Yün-nan.

K'un-lun of the Southern Sea;[1] and Su Kuṅ of the T'ang says that, of
the two kinds of *mu-hiaṅ* (known to him), that of K'un-lun is the best,
while that from the West Lake near Haṅ-čou is not good.[2] In the time
of T'ao Huṅ-kiṅ (A.D. 451–536) the root was no longer brought from
Yuṅ-č'aṅ; but the bulk of it was imported on foreign ships, with the
report that it came from Ta Ts'in (the Hellenistic Orient),[3] — hence
presumably the same article as the Arabian or Syrian costus of Dioscorides. The *Nan faṅ ts'ao mu čwaṅ* is cited by Čen Kwan of the seventh
century as saying that the root is produced in India, being the product
of an herbaceous plant and of the appearance of licorice. The same
text is ascribed to the *Nan čou i wu či* of the third century in the *T'ai
p'iṅ yü lan*,[4] while the *Kwaṅ či* attributes the product to Kiao-čou
(Tonking) and India. A different description of the plant is again given
by Su Suṅ. Thus it is no wonder that the specimens from China
submitted for identification have proved to be from different plants,
as *Aplotaxis auriculata, Aristolochia kaempferi, Rosa banksia*, etc.[5] If,
accordingly, costus (to use this general term) was found not only in
India and Kashmir, but also in Arabia, Syria, Tibet, Mongolia, China,
and Malacca, it is equally possible also that Persia had a costus of her
own or imported it from Syria as well as from India.[6] This is a question
which cannot be decided with certainty. The linguistic evidence is
inconclusive, for the New-Persian *kust* is an Arabic loan-word, the
latter, of course, being traceable to Sanskrit *kuṣṭha*, which has obtained
a world-wide propagation.[7] Like so many other examples in the history of commerce, this case illustrates the unwillingness of the world
to tolerate monopolies for any length of time. The real costus was
peculiar (and still is) to Kashmir, but everywhere attempts were constantly made to trace equivalents or substitutes. The trade-mark
remained the same, while the article was subjected to changes.

59. Under the term *ṅan* (or *an*)-*si hiaṅ* 安西香 the Chinese have

---

[1] PELLIOT, *Bull. de l'Ecole française*, Vol. IV, p. 226.

[2] The attribution of the root to K'un-lun is not fiction, for this tradition is
confirmed by Garcia da Orta, who localizes pucho on Malacca, whence it is exported
to China.

[3] This text is doubtless authentic; it is already recorded in the *T'ai p'iṅ yü lan*
(Ch. 991, p. 11).

[4] Ch. 982, p. 3.

[5] HANBURY, Science Papers, p. 257; STUART, Chinese Materia Medica, p. 43.

[6] In the sixteenth century, as we learn from GARCIA (Markham, Colloquies,
p. 150), costus was shipped from India to Ormuz, and thence carried to Persia and
Khorasan; it was also brought into Persia and Arabia by way of Aden.

[7] In Tokharian it is found in the form *kaṣṣu* (S. LÉVI, *Journal asiatique*, 1911,
II, p. 138).

combined two different aromatics,— an ancient product of Iranian regions, as yet unidentified; and the benjoin yielded by the *Styrax benjoin*, a small tree of the Malay Archipelago.[1] It is necessary to discriminate sharply between the two, and to understand that the ancient term originally relating to an Iranian aromatic, when the Iranian importation had ceased, was subsequently transferred to the Malayan article, possibly on account of some outward resemblance of the two, but that the two substances have no botanical and historical interrelation. The attempt of Čao Žu-kwa to establish a connection between the two, and to conjecture that the name is derived from An-si (Parthia), but that the article was imported by way of San-fo-ts'i (Palembang on Sumatra),[2] must be regarded as unfounded; for the question is not of an importation from Parthia or Persia to Sumatra, but it is the native product of a plant actually growing in Sumatra, in Borneo, and other Malayan islands.[3] The product is called in Malayan *kamiñan* (GARCIA: *cominham*), Javanese *meñan*, Sunda *miñan*. The duplicity of the article and the sameness of the term have naturally caused a great deal of confusion among Chinese authors, and perhaps no less among European writers. At least, the subject has not yet been presented clearly, and least of all by BRETSCHNEIDER.[4]

According to Su Kuṅ, *ñan-si hiaṅ* is produced among the Western Žuṅ 西戎 (Si-žuṅ),—a vague term, which may allude to Iranians (p. 203). Li Sün, in his *Hai yao pen ts'ao*, written in the second half of the eighth century, states that the plant grows in Nan-hai ("Southern Sea"; that is, the Archipelago) and in the country Po-se. The coordination with Nan-hai renders it probable that he hints at the Malayan Po-se rather than at Persia, the more so, as Li Ši-čen himself states that the plant now occurs in Annam, Sumatra, and all foreign countries.[5] The reason why the term *ñan-si* was applied to the Malayan

---

[1] The word "benjoin" is a corruption of Arabic *lubān jāwī* ("incense of Java"; that is, Sumatra of the Arabs). The Portuguese made of this *benzawi*, and further *beijoim, benjoim* (in Vasco da Gama and Duarte Barbosa); Spanish *benjui, menjui;* Italian *belzuino, belguino;* French *benjoin.* Cf. R. DOZY and W. H. ENGELMANN, Glossaire des mots espagnols et portugais dérivés de l'arabe, p. 239; S. R. DALGADO, Influência do vocabulário português, p. 27.

[2] HIRTH, Chao Ju-kua, p. 201.

[3] According to GARCIA (C. Markham, Colloquies, p. 49), benjoin is only known in Sumatra and Siam. According to F. PYRARD (Vol. II, p. 360, ed. of Hakluyt Society), who travelled from 1601 to 1610, it is chiefly produced in Malacca and Sumatra.

[4] Bot. Sin., pt. III, No. 313.

[5] As the Malayan product does not fall within the scope of the present investigation, this subject is not pursued further here (see HIRTH, Chau Ju-kua, pp. 201–202). In Bretschneider's translation of this matter, based on the unreliable

product may be explained from the fact that to the south-west of China, west of the Irawaddy, there was a city Ṅan-si 安西, mentioned in the Itinerary of Kia Tan and in the *Man šu* of the T'ang period.[1] The exact location of this place is not ascertained. Perhaps this or another locality of an identical name lent its name to the product; but this remains for the present a mere hypothesis. The *Tien hai yü heñ či*[2] states that *ñan-si* is produced in the native district Pa-po ta-tien 八百大旬土司, formerly called 八百媳婦地, of Yün-nan.

The *Yu yañ tsa tsu*[3] contains the following account: "The tree furnishing the *ñan-si* aromatic is produced in the country Po-se.[4] In Po-se it is termed *p'i-sie* 辟邪 tree ('tree warding off evil influences').[5] The tree grows to a height of thirty feet, and has a bark of a yellow-black color. The leaves are oblong,[6] and remain green throughout the winter. It flowers in the second month. The blossoms are yellow. The heart of the flower is somewhat greenish (or bluish). It does not form fruit. On scraping the tree-bark, the gum appears like syrup, which is called *ñan-si* aromatic. In the sixth or seventh month, when this substance hardens, it is fit for use as incense, which penetrates into the abode of the spirits and dispels all evil." Although I am not a botanist, I hardly believe that this description could be referred to *Styrax benjoin*. This genus consists only of small trees, which never reach a height of thirty feet; and its flowers are white, not yellow. Moreover, I am not convinced that we face here any Persian plant, but I think that the Po-se of the *Yu yañ tsa tsu*, as in some other cases, hints at the Malayan Po-se.[7]

---

text of the *Pen ts'ao*, occurs a curious misunderstanding. The sentence 燒之能集鼠者爲眞 is rendered by him, "By burning the true *an-si hiang* incense rats can be allured (?)." The interrogation-mark is his. In my opinion, this means, "In burning it, that kind which attracts rodents is genuine."

[1] Cf. Pelliot, *Bull. de l'Ecole française*, Vol. IV, pp. 178, 371.

[2] Ch. 3, p. 1 (see above, p. 228).

[3] Ch. 18, p. 8 b.

[4] Both Bretschneider (Bot. Sin., pt. III, p. 466) and Hirth (Chao Ju-kua, p. 202) identify this Po-se with Persia, without endeavoring, however, to ascertain what tree is meant; and *Styrax benzoin* does not occur in Persia. Garcia already stated that benjuy (as he writes) is not found in Armenia, Syria, Africa, or Cyrene, but only in Sumatra and Siam.

[5] *P'i-sie* is not the transcription of a foreign word; the ancient form *bik-dza would lead to neither a Persian nor a Malayan word.

[6] Bretschneider, who was a botanist, translates this clause (葉有四角), "The leaves spread out into four corners (!)." Literally it means "the leaves have four corners"; that is, they are rectangular or simply oblong. The phrase *se leñ* 四稜 with reference to leaves signifies "four-pointed," the points being understood as acute.

[7] See the following chapter on this subject.

An identification of *ṅan-si* to which PELLIOT[1] first called attention is given in the Chinese-Sanskrit dictionary *Fan yi miṅ yi tsi*,[2] where it is equated with Sanskrit *guggula*. This term refers to the gum-resin obtained from *Boswellia serrata* and the produce of *Balsamodendron mukul*, or *Commiphora roxburghii*, the *bdellion* of the Greeks.[3] Perhaps also other Balsamodendrons are involved; and it should be borne in mind that *Balsamodendron* and *Boswellia* are two genera belonging to the same family, *Burseraceae* or *Amyrideae*. Pelliot is quite right in assuming that in this manner it is easier to comprehend the name *ṅan-si hiaṅ*, which seems to be attached to the ancient Chinese name of the Persia of the Arsacides. In fact, we meet on the rocks of Baluchistan two incense-furnishing species, *Balsamodendron pubescens* and *B. mukul*,[4] observed by the army of Alexander in the deserts of Gedrosia, and collected in great quantity by the Phœnician merchants who accompanied him.[5]

While it is thus possible that the term *ṅan-si hiaṅ* was originally intended to convey the significance "Parthian aromatic," we must not lose sight of the fact that it is not mentioned in the ancient historical documents relative to Parthia (An-si) and Persia (Po-se),— a singular situation, which must furnish food for reflection. The article is pointed out only as a product of Kuča in Turkistan and the Kingdom of Ts'ao 漕 (Jāguḍa) north of the Ts'uṅ-liṅ.[6]

Aside from the geographical explanation, the Chinese have attempted also a literal etymology of the term. According to Li Ši-čen, this aromatic "wards off evil and sets at rest 安息 all demoniacal influences 諸邪; hence its name. Others, however, say that *ṅan-si* is the name of a country." This word-for-word interpretation is decidedly forced and fantastic.

---

[1] *T'oung Pao*, 1912, p. 480.

[2] Ch. 8, p. 10 b.

[3] Cf. *T'oung Pao*, 1914, p. 6.

[4] JORET, Plantes dans l'antiquité, Vol. II, p. 48. The former species is called in Baluči *bayi* or *bai*.

[5] *Ibid.*, p. 649.

[6] *Sui šu*, Ch. 83, pp. 5 b, 7 b.

THE MALAYAN PO-SE AND ITS PRODUCTS

On the preceding pages reference has repeatedly been made to the fact that besides the Iranian Po-se 波斯, transcribing the ancient name Parsa, the Chinese were also acquainted with another country and people of the same name, and always written in like manner, the location of which is referred to the Southern Ocean, and which, as will be seen, must have belonged to the Malayan group. We have noted several cases in which the two Po-se are confounded by Chinese writers; and so it is no wonder that the confusion has been on a still larger scale among European sinologues, most of whom, if the Malayan Po-se is involved in Chinese records, have invariably mistaken it for Persia. It is therefore a timely task to scrutinize more closely what is really known about this mysterious Po-se of the Southern Sea. Unfortunately the Chinese have never co-ordinated the scattered notices of the southern Po-se; and none of their cyclopædias, as far as I know, contains a coherent account of the subject. Even the mere fact of the duplicity of the name Po-se never seems to have dawned upon the minds of Chinese writers; at least, I have as yet failed to trace any text insisting on the existence of or contrasting the two Po-se. Groping my way along through this matter, I can hardly hope that my study of source-material is complete, and I feel sure that there are many other texts relative to the subject which have either escaped me or are not accessible.

The Malayan Po-se is mentioned in the *Man šu* 蠻書 (p. 43 b),[1] written about A.D. 860 by Fan Čo 樊綽, who says, "As regards the country P'iao 驃 (Burma), it is situated seventy-five days' journey (or two thousand *li*) south of the city of Yuṅ-č'aṅ.[2] . . . It borders on Po-se 波斯 and P'o-lo-men 婆羅門 (Brāhmaṇa);[3] in the west, however, on the city Še-li 舍利." It is clearly expressed in this document that Po-se, as known under the T'ang, was a locality somewhere conterminous with Burma, and on the mainland of Asia.

---

[1] Regarding this work, see WYLIE, Notes on Chinese Literature, p. 40; and PELLIOT, *Bull. de l'Ecole française*, Vol. II, p. 156; Vol. IV, p. 132.

[2] In Yün-nan. The *T'ai p'iṅ hwan yü ki* gives the distance of P'iao from that locality as 3000 *li* (cf. PELLIOT, *Bull. de l'Ecole française*, Vol. IV, p. 172). The text of the *Man šu* is reproduced in the same manner in the *Su kien* of Kwo Yün-t'ao (Ch. 10, p. 10 b), written in 1236.

[3] I do not believe that this term relates to India in general, but take it as denoting a specific country near the boundary of Burma.

468

In another passage of the *Man šu* (p. 29), the question is of a place Ta-yin-k'un 大銀孔 (evidently a silver-mine), not well determined, probably situated on the Gulf of Siam, to the south of which the people of the country P'o-lo-men (Brāhmaṇa), Po-se, Še-p'o (Java), P'o-ni (Borneo), and K'un-lun, flock together for barter. There are many precious stones there, and gold and musk form their valuable goods.[1] There is no doubt that the Malayan Po-se is understood here, and not Persia, as has been proposed by PELLIOT.[2] A similar text is found in the *Nan i či* 南夷志 ("Records of Southern Barbarians"), as quoted in the *T'ai p'iṅ yü lan,*[3] "In Nan-čao there are people from P'o-lo-men, Po-se, Še-p'o (Java), P'o-ni (Borneo), K'un-lun, and of many other heretic tribes, meeting at one trading-mart, where pearls and precious stones in great number are exchanged for gold[4] and musk." This text is identical with that of the *Man šu*, save that the trading centre of this group of five tribes is located in the kingdom of Nan-čao (in the present province of Yün-nan). E. H. PARKER[5] has called attention to a mention of Po-se in the T'ang Annals, without expressing, however, an opinion as to what Po-se means in this connection. In the chapter on P'iao (Burma) it is there stated that near the capital of that country there are hills of sand and a barren waste which borders on Po-se and P'o-lo-men, — identical with the above passage of the *Man šu*.[6]

In A.D. 742, a Buddhist priest from Yaṅ-čou on the Yangtse, Kien-čen 鑑眞 by name, undertook a voyage to Japan, in the course of which he also touched Canton in 748. In the brief abstract of his diary given by the Japanese scholar J. TAKAKUSU,[7] we read, "Dans la rivière de Canton, il y avait d'innombrables vaissaux appartenant aux brahmanes, aux Persans, aux gens de Koun-loun (tribu malaise)." The text of the work in question is not at my disposal, but there can be no doubt that it contains the triad P'o-lo-men, Po-se, K'un-lun, as mentioned in the *Man šu*, and that the question is not of Brahmans, but of the country

---

[1] In another passage (p. 34 b) Fan Čo states that musk is obtained in all mountains of Yuṅ-č'aṅ and Nan-čao, and that the natives use it as a means of exchange.

[2] *Bull. de l'Ecole française*, Vol. IV, p. 287, note 2.

[3] Ch. 981, p. 5 b.

[4] The text has 養金. I do not know what *šu* ("to boil") could mean in this connection. It is probably a wrong reading for 黃, as we have it in the text of the *Man šu*.

[5] Burma with Special Reference to Her Relations with China, p. 14 (Rangoon, 1893).

[6] This passage is not contained in the notice of P'iao in the *Kiu T'aṅ šu* (Ch. 197, p. 7 b).

[7] Premier Congrès International des Etudes d'Extrême-Orient, p. 58 (Hanoi, 1903); cf. G. FERRAND, Textes relatifs à l'Extrême-Orient, Vol. II, p. 638.

and people P'o-lo-men on the border of Burma, the Po-se likewise on the border of Burma, and the Malayan K'un-lun. In the first half of the eighth century, accordingly, we find the Malayan Po-se as a seafaring people trading with the Chinese at Canton. Consequently also the alleged "Persian" settlement on the south coast of Hainan, struck by the traveller, was a Malayan-Po-se colony. In view of this situation, the further question may be raised whether the pilgrim Yi Tsiṅ in A.D. 671 sought passage at Canton on a Persian ship.[1] This vessel was bound for Palembang on Sumatra, and sailed the Malayan waters; again, in my opinion, the Malayan Po-se, not the Persians, are here in question.

The Malayan Po-se were probably known far earlier than the T'ang period, for they appear to have been mentioned in the *Kwaṅ ǰi* written before A.D. 527. In the *Hiaṅ p'u* 香 譜 of Huṅ Č'u 洪 芻 of the Sung,[2] this work is quoted as saying that *žu hiaṅ* 乳 香 (a kind of incense)[3] is the sap of a pine-tree in the country Po-se in the Southern Sea. This Po-se is well enough defined to exclude the Iranian Po-se, where, moreover, no incense is produced.[4]

The same text is also preserved in the *Hai yao pen ts'ao* of Li Sün of the eighth century,[5] in a slightly different but substantially identical wording: "*Žu hiaṅ* grows in Nan-hai [the countries of the Southern Sea]: it is the sap of a pine-tree in Po-se. That kind which is red like cherries and transparent ranks first." K'ou Tsuṅ-ši, who wrote the *Pen ts'ao yen i* in A.D. 1116, says that the incense of the Southern Barbarians (Nan Fan) is still better than that of southern India. The Malayan Po-se belonged to the Southern Barbarians. The fact that these, and not the Persians, are to be understood in the accounts relating to incense, is brought out with perfect lucidity by Č'en Č'eṅ 陳 承, who wrote the *Pen ts'ao pie šwo* 本 草 別 説 in A.D. 1090, and who says, "As regards the west, incense is produced in India (T'ien-ču); as re-

---

[1] CHAVANNES, Religieux éminents, p. 116; J. TAKAKUSU, I-Tsing, p. XXVIII.

[2] Ed. of *T'aṅ Suṅ ts'uṅ šu*, p. 5.

[3] Not necessarily from *Boswellia*, nor identical with frankincense. The above text says that *žu hiaṅ* is a kind of *hün-lu*. The latter is simply a generic term for incense, without referring to any particular species. I strictly concur with PELLIOT (*T'oung Pao*, 1912, p. 477) in regarding *hün-lu* as a Chinese word, not as the transcription of a foreign word, as has been proposed.

[4] If *hün lu* is enumerated in the *Sui šu* among the products of Persia, this means that incense was used there as an import-article, but it does not follow from this that "it was brought to China on Persian ships" (HIRTH, Chau Ju-kua, p. 196). The "Persian ships," it seems, must be relegated to the realm of imagination. Only from the Mohammedan period did really Persian ships appear in the far east. The best instance to this effect is contained in the notes of Hwi Čao of the eighth century (HIRTH, *Journal Am. Or. Soc.*, 1913, p. 205).

[5] *Pen ts'ao kaṅ mu*, Ch. 34, p. 16.

gards the south, it is produced in Po-se and other countries. That of the west is yellow and white in color, that of the south is purple' or red." It follows from this text that the southern Po-se produced a kind of incense of their own; and it may very well be, that, as stated in the *Kwan či*, a species of pine was the source of this product.

The *Kwan či* contains another interesting reference to Po-se. It states that the tree 柯 *ko*, *ka (*Quercus cuspidata*), grows in the mountains and valleys of Kwan-tun and Kwan-si, and that Po-se people use its timber for building boats.[1] These again are Malayan Po-se. The *Kwan či* was possibly written under the Tsin dynasty (A.D. 265–420),[2] and the Iranian Po-se was then unknown to China. Its name first reached the Chinese in A.D. 461, when an embassy from Persia arrived at the Court of the Wei.[3] It should be borne in mind also that Persia's communications with China always took place overland by way of Central Asia; while the Malayan Po-se had a double route for reaching China, either by land to Yün-nan or by sea to Canton. It would not be impossible that the word *ka for this species of oak, and also its synonyme 木 奴 *mu-nu*, *muk-nu, are of Malayan-Po-se origin.

The *Kiu yü či* 九 域 志, published by Wan Ts'un 王 存 in A.D. 1080, mentions that the inhabitants of Po-se wear a sort of cotton kerchief, and make their sarong (*tu-man* 都 縵) of yellow silk.[4]

In A.D. 1103, three countries, Burma, Po-se, and K'un-lun, presented white elephants and perfumes to the King of Ta-li in Yün-nan. Again, this is not Persia, as translated by C. SAINSON.[5] Persia never had any relations with Yün-nan, and how the transportation of elephants from Persia to Yün-nan could have been accomplished is difficult to realize. We note that the commercial relations of these Po-se with Yün-nan, firmly established toward the end of the ninth century under the T'ang, were continued in the twelfth century under the Sung.

In the History of the Sung Dynasty occurs an incidental mention of Po-se.[6] In A.D. 992 an embassy arrived in China from Java, and it is said that the envoys were dressed in a way similar to those of Po-se, who

---

[1] This passage is transmitted by Li Sūn of the eighth century in his *Hai yao pen ts'ao* (*Pen ts'ao kan mu*, Ch. 35 B, p. 14), who, as will be seen, mentions several plants and products of the Malayan Po-se.

[2] PELLIOT, *Bull. de l'Ecole française*, Vol. IV, p. 412.

[3] Cf. DEVÉRIA in Centenaire de l'Ecole des Langues Orientales, p. 306.

[4] E. H. PARKER, who made this text known (*China Review*, Vol. XIX, 1890, p. 191), remarked, "It seems probable that not Persia, but one of the Borneo or Malacca states, such as P'o-li or P'o-lo, is meant."

[5] Histoire du Nan-tchao, p. 101 (translation of the *Nan čao ye ši*, written by Yan Šen in 1550).

[6] *Sun ši*, Ch. 489.

had brought tribute before. The Javanese could hardly be expected to have been dressed like Persians, as rashly assumed by GROENEVELDT;[1] but they were certainly dressed like their congeners, the Malayan Po-se.

Čou K'ü-fei, in his *Lin wai tai ta*,[2] written in 1178, gives the following description of the country Po-se: "In the South-Western Ocean there is the country Po-se. The inhabitants have black skin and curly hair. Both their arms are adorned with metal bracelets, and they wrap around their bodies a piece of cotton-cloth with blue patterns. There are no walled towns. Early in the morning, the king holds his court, being seated cross-legged on a bench covered with a tiger-skin, while his subjects standing beneath pay him homage. In going out he is carried in a litter (軟兜 *žwan tou*), or is astride an elephant. His retinue consists of over a hundred men, who, carrying swords and shouting (to clear the way), form his body-guard. They subsist on flour products, meat, and rice, served in porcelain dishes, and eat with their fingers." The same text has been reproduced by Čao Žu-kwa with a few slight changes. His reading that Po-se is situated "above the countries of the south-west" is hardly correct.[3] At all events, the geographical definition of the Sung authors is too vague to allow of a safe conclusion. The expression of the *Lin wai tai ta* does not necessarily mean that Po-se was located on an island, and Hirth infers that we might expect to find it in or near the Malay Peninsula. However vague the above description may be, it leaves no doubt of the fact that the tribe in question is one of Malayan or Negrito stock.

As far as I know, no mention is made of the Malayan Po-se in the historical and geographical texts of the Ming, but the tradition regarding that country was kept alive. In discussing the *a-lo-p'o* (*Cassia fistula*) of Č'en Ts'an-k'i, as noted above (p. 420), Li Ši-čen annotates that Po-se is the name of a country of the barbarians of the south-west 波斯西南夷國名也.

There is some evidence extant that the language of Po-se belongs to the Malayan family. TSUBOI KUMAZO[4] has called attention to the numerals of this language, as handed down in the *Kōdanšō* (Memoirs of Oye), a Japanese work from the beginning of the twelfth century. These are given in Japanese transcription as follows:—

1	*sasaa, sasaka*	6	*namu*	20	*toaro*
2	*toa*	7	*toku, tomu*	30	*akaro, akafuro*
3	*naka, maka*	8	*jembira,* or *gemmira*	40	*hiha-furo*
4	*namuha (nampa)*	9	*sa-i-bira,* or *sa-i-mi-ra*	100	*sasarato, sasaratu*
5	*rima (lima)*	10	*sararo,* or *šararo*	1000	*sasaho, sasahu*

---

[1] Notes on the Malay Archipelago, p. 144.

[2] Ch. 3, p. 6 b.

[3] Ch. A, p. 33 b; HIRTH's translation, p. 152.

[4] Actes du Douzième Congrès des Orientalistes, Rome 1899, Vol. II, p. 121.

Florenz has correctly recognized in this series the numerals of a Malayan language, though they cannot throughout be identified (and this could hardly be expected) with the numerals of any known dialect. Various Malayan languages must be recruited for identification, and some forms even then remain obscure. The numeral 1 corresponds to Malayan *sa, satu;* 2 to *dua;* 4 to *ampat;* 5 to *lima;* 6 to *namu;* 7 to *tujoh;* 9 to *sembilan;* 10 to *sa-puloh.* The numeral 20 is composed of *toa* 2 and *ro* 10 (Malayan *puloh*); 30 *aka* (=*naka,* 3) and *ro* or *furo* 10. The numeral 100 is formed of *sasa* 1 and *rato*=Malayan -*ratus.*

Two Po-se words are cited in the *Yu yaṅ tsa tsu,*[1] which, as formerly pointed out by me, cannot be Persian, but betray a Malayan origin.[2] There it is said that the Po-se designate ivory as 白暗 *pai-ṅan,* and rhinoceros-horn as 黑暗 *hei-ṅan.* The former corresponds to ancient *bak-am; the latter, to *hak-am or *het-am. The latter answers exactly to Jarai *hötam,* Bisaya *itom,* Tagalog *ītim,* Javanese *item,* Makasar *etaṅ,* Čam *hutam* (*hatam* or *hutum*), Malayan *hītam,* all meaning "black."[3] The former word is not related to the series *putih, pūteh,* as I was previously inclined to assume, but to the group: Čam *bauṅ, boṅ,* or *bhuṅ;* Senoi *biūg,* other forms in the Sakei and Semang languages of Malakka *biok, biäk, biēg, begiäk, bekuṅ, bekog;*[4] Alfur, Boloven, Kon tu, Kaseng, Lave, and Niah *bok,* Sedeng *röboṅ,* Stieng *bōk* ("white"); Bahnar *bak* (Mon *bu*).[5] It almost seems, therefore, as if the speech of Po-se bears some relationship to the languages of the tribes of Malacca. The Po-se distinguished rhinoceros-horn and ivory as "black" and "white." However meagre the linguistic material may be, it reveals, at any rate, Malayan affinities, and explodes BRETSCHNEIDER'S theory[6] that the Po-se of the Archipelago, alleged to have been on Sumatra, owes its origin to the fact that "the Persians carried on a great trade with Sumatra, and probably had colonies there." This is an unfounded speculation, justly rejected also by G. E. GERINI:[7] these Po-se were not Persians, but Malayans.

The Po-se question has been studied to some extent by G. E. GERINI,[8] who suggests its probable identity with the Vasu state located by the Bhāgavata Purāṇa in Kuçadvīpa, and who thinks it may be

---

[1] Ch. 16, p. 14.

[2] Chinese Clay Figures, p. 145.

[3] Cf. CABATON and AYMONIER, Dictionnaire čam-français, p. 503.

[4] P. SCHMIDT, *Bijdragen tot de Taal-, Land- en Volkenkunde,* Vol. VIII, 1901, p. 420.

[5] *Ibid.,* p. 344.

[6] Knowledge possessed by the Chinese of the Arabs, p. 16.

[7] Researches on Ptolemy's Geography of Eastern Asia, p. 471.

[8] *Ibid.,* p. 682.

Lambesi; i.e., Besi or Basi (*lam* meaning "village"), a petty state on the west coast of Sumatra immediately below Acheh, upon which it borders. This identification is impossible, first of all, for phonetic reasons: Chinese *po* 波 was never possessed of an ancient labial sonant, but solely of a labial surd (*pwa).[1]

TSUBOI KUMAZO[2] regards Po-se as a transcription of Pasi, Pasei, Pasay, Pazze, or Pacem, a port situated on northern Sumatra near the Diamond Cape, which subsequently vied in wealth with Majapahit and Malacca, and called Basma by Marco Polo.[3]

C. O. BLAGDEN[4] remarks with reference to this Po-se, "One is very much tempted to suppose that this stands for Pose (or Pasai) in north-eastern Sumatra, but I have no evidence that the place existed as early as 1178." If this be the case, the proposed identification is rendered still more difficult; for, as we have seen, Po-se appears on the horizon of the Chinese as early as from the seventh to the ninth century under the T'ang, and probably even at an earlier date. The only text that gives us an approximate clew to the geographical location of Po-se is the *Man šu;* and I should think that all we can do under the circumstances, or until new sources come to light, is to adhere to this definition; that is, as far as the T'ang period is concerned. Judging from the movements of Malayan tribes, it would not be impossible that, in the age of the Sung, the Po-se had extended their seats from the mainland to the islands of the Archipelago, but I am not prepared for the present either to accept or to reject the theory of their settlement on Sumatra under the Sung.

Aside from the references in historical texts, we have another class of documents in which the Malayan Po-se is prominent, the *Pen-ts'ao* literature and other works dealing with plants and products. I propose to review these notices in detail.

60. In regard to alum, F. P. SMITH[5] stated that apart from native localities it is also mentioned as reaching China from Persia, K'un-lun,

---

[1] On p. 471 Gerini identifies Po-se with the Basīsi tribe in the more southern parts of the Malay Peninsula. On the other hand, it is difficult to see why Gerini searched for Po-se on Sumatra, as he quotes after Parker a Chinese source under the date A.D. 802, to the effect that near the capital of Burma there were hills of sand, and a barren waste which borders on Po-se and P'o-lo-men (see above, p. 469).

[2] Actes du Douzième Congrès des Orientalistes, Rome 1899, Vol. II, p. 92.

[3] Cf. YULE, Marco Polo, Vol. II, pp. 284–288. Regarding the kings of Pase, see G. FERRAND, Textes relatifs à l'Extrême-Orient, Vol. II, pp. 666–669.

[4] *Journal Royal As. Soc.*, 1913, p. 168.

[5] Contributions towards the Materia Medica of China, p. 10.

and Ta Ts'in. J. L. SOUBEIRAN[1] says, "L'alun, qui était tiré primitivement de la Perse, est aujourd'hui importé de l'Occident." F. DE MÉLY[2] translates the term *Po-se ts'e fan* by "*fan* violet de Perse." All this is wrong. HIRTH[3] noted the difficulty in the case, as alum is not produced in Persia, but principally in Asia Minor. Pliny[4] mentions Spain, Egypt, Armenia, Macedonia, Pontus, and Africa as alum-producing countries. Hirth found in the *P'ei wen yün fu* a passage from the *Hai yao pen ts'ao*, according to which *Po-se fan* 波斯礬 ("Persian alum," as he translates) comes from Ta Ts'in. In his opinion, "Persian alum" is a misnomer, Persia denoting in this case merely the emporium from which the product was shipped to China. The text in question is not peculiar to the *Hai yao pen ts'ao* of the eighth century, but occurs at a much earlier date in the *Kwan čou ki* 廣州記, an account of Kwan-tun, written under the Tsin dynasty (A.D. 265–419), when the name of Persia was hardly known in China. This work, as quoted in the *Čen lei pen ts'ao*,[5] states that *kin sien* 金線 *fan* ("alum with gold threads") is produced 生 in the country Po-se, and in another paragraph that the white alum of Po-se (*Po-se pai fan*) comes from Ta Ts'in.[6] The former statement clearly alludes to the alum discolored by impurities, as still found in several localities of India and Upper Burma.[7] Accordingly the Malayan Po-se (for this one only can come into question here) produced an impure kind of alum, and simultaneously was the transit mart for the pure white alum brought from western Asia by way of India to China. It is clear that, because the native alum of Po-se was previously known, also the West-Asiatic variety was named for Po-se. A parallel to the *Po-se fan* is the *K'un-lun fan*, which looks like black mud.[8]

61. The *Wu lu* 吳錄, written by Čan Po 張勃 in the beginning of the fourth century, contains the following text on the subject of "ant-lac" (*yi tsi* 蟻漆):[9] "In the district of Kü-fun 居風 (in Kiu-čen, Ton-

---

[1] Etudes sur la matière médicale chinoise (Minéraux), p. 2 (reprint from *Journal de pharmacie et de chimie*, 1866).

[2] Lapidaire chinois, p. 260.

[3] Chinesische Studien, p. 257.

[4] XXXV, 52.

[5] Ch. 3, p. 40 b.

[6] Also in the text of the *Hai yao pen ts'ao*, as reproduced in the *Pen ts'ao kan mu* (Ch. 11, p. 15 b), two Po-se alums are distinguished.

[7] WATT, Commercial Products of India, p. 61.

[8] *Pen ts'ao kan mu, l. c.*

[9] *T'ai p'in hwan yü ki*, Ch. 171, p. 5.

king)[1] there are ants living on coarse creepers. The people, on examining the interior of the earth, can tell the presence of ants from the soil being freshly broken up; and they drive tree-branches into these spots, on which the ants will crawl up, and produce a lac that hardens into a solid mass." Aside from the absurd and fantastic notes of Aelian,[2] this is the earliest allusion to the lac-insect which is called in Annamese *con môi*, in Khmer *kandter*, in Čam *mū, mur*, or *muor*.[3] The Chinese half-legendary account[4] agrees strikingly with what Garcia reports as the Oriental lore of this wonder of nature: "I was deceived for a long time. For they said that in Pegu the channels of the rivers deposit mud into which small sticks are driven. On them are engendered very large ants with wings, and it is said that they deposit much lacre[5] on the sticks. I asked my informants whether they had seen this with their own eyes. As they gained money by buying rubies and selling the cloths of Paleam and Bengal, they replied that they had not been so idle as that, but that they had heard it, and it was the common fame. Afterwards I conversed with a respectable man with an enquiring mind, who told me that it was a large tree with leaves like those of a plum tree, and that the large ants deposit the lacre on the small branches. The ants are engendered in mud or elsewhere. They deposit the gum on the tree, as a material thing, washing the branch as the bee makes honey; and that is the truth. The branches are pulled off the tree and put in the shade to dry. The gum is then taken off and put into bamboo joints, sometimes with the branch."[6]

In the *Yu yan tsa tsu*[7] we read as follows: "The *tse-kun* tree 紫鉚[8] 樹 has its habitat in Camboja (Čen-la), where it is called 勒佉 *lo-k'ia*, *lak-ka (that is, *lakka, lac*).[9] Further, it is produced in the country

---

[1] Regarding this locality, cf. H. MASPERO, Etudes d'histoire d'Annam, V, p. 19 (*Bull. de l'Ecole française*, 1918, No. 3).

[2] Nat. Anim., IV, 46. There is no other Greek or Latin notice of the matter.

[3] Cf. AYMONIER and CABATON (Dictionnaire čam-français, p. 393), who translate the term "termite, pou de bois, fourmi blanche."

[4] Much more sensible, however, than that of Aelian.

[5] The Portuguese word for "lac, lacquer," the latter being traceable to *lacre*. The ending -re is unexplained.

[6] C. MARKHAM, Colloquies, p. 241.

[7] Ch. 18, p. 9.

[8] The *Pai-hai* edition has erroneously the character 鉥.

[9] From Pāli *lākhā* (Sanskrit *lākṣā, laktaka*); Čam *lak*, Khmer *lāk*; Siamese *rak* (cf. PALLEGOIX, Description du royaume Thai, Vol. I, p. 144). We are thus entitled to trace the presence of this Indian word in the languages of Indo-China to the age of the T'ang. The earliest and only classical occurrence of the word is in the Periplus (Ch. 6: λάκκος). Cf. also Prākrit *lakkā;* Kawi and Javanese *lākā;* Tagalog *lakha*.

Po-se 波斯. The tree grows to a height of ten feet, with branches dense and luxuriant. Its leaves resemble those of the Citrus and wither during the winter. In the third month it flowers, the blossoms being white in color. It does not form fruit. When heavy fogs, dew, and rain moisten the branches of this tree, they produce *tse-kuṅ*. The envoys of the country Po-se, Wu-hai 烏海 and Ša-li-šen 沙利深 by name, agreed in their statement with the envoys from Camboja, who were a *če č'uṅ tu wei* 折衝都尉[1] and the çramaṇa 施沙尼拔陁 Ši-ša-ni-pa-t'o (Çiçanibhadra?). These said, 'Ants transport earth into the ends of this tree, digging nests in it; the ant-hills moistened by rain and dew will harden and form *tse-kuṅ*.[2] That of the country K'un-lun is the most excellent, while that of the country Po-se ranks next.' "[3]

---

[1] Title of a military officer.

[2] "The gum-lac which comes from Pegu is the cheapest, though it is as good as that of other countries; what causes it to be sold cheaper is that the ants, making it there on the ground in heaps, which are sometimes of the size of a cask, mix with it a quantity of dirt" (TAVERNIER, Travels in India, Vol. II, p. 22).

[3] The story of lacca and the ants producing it was made known in England at the end of the sixteenth century. JOHN GERARDE (The Herball or Generall Historie of Plantes, p. 1349, London, 1597, 1st ed; or, enlarged and amended by Thomas Johnson, p. 1533, London, 1633) tells it as follows: "The tree that bringeth forth that excrementall substance, called *Lacca*, both in the shops of Europe and elsewhere, is called of the Arabians, Persians and Turkes *Loc Sumutri*, as who should say *Lacca* of Sumutra: some which have so termed it, have thought that the first plentie thereof came from Sumutra, but herein they have erred; for the abundant store thereof came from Pegu, where the inhabitants thereof do call it *Lac*, and others of the same province *Trec*. The history of which tree, according to that famous Herbarist Clusius is as followeth. There is in the countrey of Pegu and Malabar, a great tree, whose leaves are like them of the Plum tree, having many small twiggie branches; when the trunke or body of the tree waxeth olde, it rotteth in sundrie places, wherein do breed certaine great ants or Pismires, which continually worke and labour in the time of harvest and sommer, against the penurie of winter: such is the diligence of these Ants, or such is the nature of the tree wherein they harbour, or both, that they provide for their winter foode, a lumpe or masse of substance, which is of a crimson colour, so beautifull and so faire, as in the whole world the like cannot be seene, which serveth not onely to phisicall uses, but is a perfect and costly colour for Painters, called by us, Indian Lack. The Pismires (as I said) worke out this colour, by sucking the substance or matter of Lacca from the tree, as Bees do make honie and waxe, by sucking the matter thereof from all herbes, trees, and flowers, and the inhabitants of that countrie, do as diligently search for this Lacca, as we in England and other countries, seeke in the woods for honie; which Lacca after they have found, they take from the tree, and drie it into a lumpe; among which sometimes there come over some sticks and peeces of the tree with the wings of the Ants, which have fallen amongst it, as we daily see. The tree which beareth Lacca groweth in Zeilan and Malavar, and in other partes of the East Indies." The second edition of 1633 has the following addition, "The Indian Lacke or Lake which is the rich colour used by Painters, is none of that which is used in shops, nor here figured or described by Clusius, wherefore our Author was much mistaken in that he here confounds together things so different; for this is of a resinous substance, and a faint red colour, and wholly unfit for Painters, but used alone and in composition to make the best hard

The question here is of gum-lac or stick-lac (Gummi lacca; French *laqne en bâtons*), also known as kino, produced by an insect, *Coccus* or *Tachardia lacca*, which lives on a large number of widely different trees,[1] called 紫鉚 or 梗 *tse-kuṅ* or *tse-keṅ*. Under the latter name it is mentioned in the "Customs of Camboja" by Čou Ta-kwan;[2] under the former, in the *Pen ts'ao yen i*.[3] At an earlier date it occurs as 紫鑛 in the *T'aṅ hui yao*,[4] where it is said in the notice of P'iao (Burma), that there the temple-halls are coated with it. In all probability, this word represents a transcription: Li Ši-čen assigns it to the Southern Barbarians.

The Po-se in the text of the *Yu yaṅ tsa tsu* cannot be Persia, as is sufficiently evidenced by the joint arrival of the Po-se and Camboja envoys, and the opposition of Po-se to the Malayan K'un-lun. Without any doubt we have reference here to the Malayan Po-se. The product itself is not one of Persia, where the lac-insect is unknown.[5] It should be added that the *Yu yaṅ tsa tsu* treats of this Po-se product along with the plants of the Iranian Po-se discussed on the preceding pages; and there is nothing to indicate that Twan Č'eṅ-ši, its author, made a distinction between the two homophonous names.[6]

62. The Malayan Po-se, further, produced camphor (*Dryobalanops aromatica*), as we likewise see from the *Yu yaṅ tsa tsu*,[7] where the tree

---

sealing wax. The other seemes to be an artificiall thing, and is of an exquisite crimson colour, but of what it is, or how made, I have not as yet found any thing that carries any probabilitie of truth." Gerarde's information goes back to Garcia, whose fundamental work then was the only source for the plants and drugs of India.

[1] WATT, Commercial Products of India, p. 1053; not necessarily *Erythrina*, as stated by STUART (Chinese Materia Medica, p. 489). Sir C. MARKHAM (Colloquies, p. 241) says picturesquely that the resinous exudation is produced by the puncture of the females of the lac-insect as their common nuptial and accouchement bed, the seraglio of their multi-polygamous bacchabunding lord, the male *Coccus lacca;* both the males and their colonies of females live only for the time they are ceaselessly reproducing themselves, and as if only to dower the world with one of its most useful resins, and most glorious dyes, the color "lake."

[2] PELLIOT, *Bull. de l'Ecole française*, Vol. II, p. 166.

[3] Ch. 14, p. 4 b (ed. of Lu Sin-yüan).

[4] Ch. 100, p. 18 b. Also Su Kuṅ and Li Sün of the T'ang describe the product.

[5] The word *lak* (Arabic) or *rānglāk* (Persian) is derived from Indian, and denotes either the Indian product or the gum of *Zizyphus lotus* and other plants (ACHUNDOW, Abu Mansur, p. 265). In the seventeenth century the Dutch bought gum-lac in India for exportation to Persia (TAVERNIER, *l. c.*). Cf. also LECLERC, Traité des simples, Vol. III, p. 241; and G. FERRAND, Textes relatifs à l'Extrême-Orient, p. 340.

[6] In regard to stick-lac in Tibet, see H. LAUFER, Beiträge zur Kenntnis der tibetischen Medicin, pp. 63–64.

[7] Ch. 18, p. 8 b.

is ascribed to Bali 婆利 (P'o-li, *Bwa-li)[1] and to Po-se. Camphor is not produced in Persia;[2] and HIRTH[3] is not justified in here rendering Po-se by Persia and commenting that camphor was brought to China by Persian ships.

63. The confusion as to the two Po-se has led Twan Č'eṅ-ši[4] to ascribe the jack-fruit tree (*Artocarpus integrifolia*) to Persia, as would follow from the immediate mention of Fu-lin; but this tree grows neither in Persia nor in western Asia. It is a native of India, Burma, and the Archipelago. The mystery, however, remains as to how the author obtained the alleged Fu-lin name.[5]

Pepper (*Piper longum*), according to Su Kuṅ of the T'ang, is a product of Po-se. This cannot be Persia, which does not produce pepper.[6]

In the chapter on the walnut we have noticed that the *Pei hu lu*, written about A.D. 875 by Twan Kuṅ-lu, mentions a wild walnut as growing in the country Čan-pei (*Čambi, Jambi), and gathered and eaten by the Po-se. The *Liṅ piao lu i*, written somewhat later (between 889 and 904), describes the same fruit as growing in Čan-pi (*Cambir, Jambir), and gathered by the Hu. This text is obviously based on the older one of the *Pei hu lu;* and Liu Sün, author of the *Liṅ piao lu i*, being under the impression that the Iranian Po-se is involved, appears to have substituted the term Hu for Po-se. The Iranian Po-se, however, is out of the question: the Persians did not consume wild walnuts; and, for all we know about Čan-pi, it must have been some Malayan region.[7] I have tentatively identified the plant in question with *Juglans cathayensis* or, which is more probable, *Canarium commune;* possibly another genus is intended. As regards the situation of Čan-pi (or -pei) and Po-se of the T'ang, much would depend on the botanical evidence. I doubt that any wild walnut occurs on Sumatra.

The *Hai yao pen ts'ao*, written by Li Sün in the second half of the eighth century, and as implied by the title, describing the drugs from

---

[1] Its Bali name is given as 固不婆律 *ku-pu-p'o-lü*, *ku-put-bwa-lwut, which appears to be based on a form related to the Malayan type *kāpor-bārus*. Cf. also the comments of PELLIOT (*T'oung Pao*, 1912, pp. 474–475).

[2] SCHLIMMER (Terminologie, p. 98) observes, "Les auteurs indigènes persans recommendent le camphre de Borneo comme le meilleur. Camphre de menthe, provenant de la Chine, se trouve depuis peu dans le commerce en Perse." Camphor was imported into Sīrāf (W. OUSELEY, Oriental Geography of Ebn Haukal, p. 133; G. LE STRANGE, Description of the Province of Fars, p. 42).

[3] Chau Ju-kua, p. 194.

[4] *Yu yaṅ tsa tsu*, Ch. 18, p. 10.

[5] Cf. HIRTH, Chau Ju-kua, p. 213.

[6] See above, pp. 374, 375.

[7] See the references given above on p. 268.

the countries beyond the sea and south of China, has recorded several products of Po-se, which, as we have seen, must be interpreted as the Malayan region of this name. Such is the case with benjoin (p. 464) and cummin (p. 383).

We noticed (p. 460) that the *Nan čou ki* and three subsequent works attribute myrrh to Po-se, but that this can hardly be intended for the Iranian Po-se, since myrrh does not occur in Persia. Here the Malayan Po-se is visualized, inasmuch as the trade in myrrh took its route from East Africa and the Hadramaut coast of Arabia by way of the Malay Archipelago into China, and thus led the Chinese (erroneously) to the belief that the tree itself grew in Malaysia.

64. The case of aloes (*Aloe vulgaris* and other species) presents a striking analogy to that of myrrh, inasmuch as this African plant is also ascribed to Po-se, and a substitute for it was subsequently found in the Archipelago. Again it is Li Sün of the T'ang period who for the first time mentions its product under the name *lu-wei* 蘆薈, stating that it grows in the country Po-se, has the appearance of black confectionery, and is the sap of a tree.[1] Su Suṅ of the Sung dynasty observes, "At present it is only shipped to Canton. This tree grows in the mountain-wilderness, its sap running down like tears and coagulating. This substance is gathered regardless of the season or month." Li Ši-čen feels doubtful as to whether the product is that of a tree or of an herb 草: he points out that, according to the *Ta Miṅ i t'uṅ či*, aloes, which belongs to the class of herbs, is a product of Java, Sumatra (San-fu-ts'i), and other countries, and that this is contradictory to the data of the T'ang and Sung Pen-ts'ao. It was unknown to him, however, that the first author thus describing the product is Čao Žu-kwa,[2] who indeed classifies *Aloe* among herbs, and derives it from the country Nu-fa 奴發, a dependency of the Arabs, and in another passage from an island off the Somali coast, evidently hinting at Socotra. This island is the home of the *Aloe perryi*, still imported into Bombay.[3]

The name *lu-wei* is traced by Hirth to Persian *alwā*. This theory is difficult to accept for many reasons. Nowhere is it stated that *lu-wei* is a Persian word. Li Ši-čen, who had good sense in diagnosing foreign words, remarks that *lu-wei* remains unexplained. The Chinese historical texts relative to the Iranian Po-se do not attribute to it this product, which, moreover, did not reach China by land, but exclusively

---

[1] *Pen ts'ao kaṅ mu*, Ch. 34, p. 21 b. The juice of *Aloe abyssinica* is sold in the form of flat circular cakes, almost black in color.

[2] *Ču fan či*, Ch. B, p. 11 (cf. HIRTH's translation, p. 225).

[3] Regarding the history of aloes, see especially FLÜCKIGER and HANBURY, Pharmacographia, p. 680.

over the maritime route to Canton. Aloes was only imported to Persia,[1] but it is not mentioned by Abu Mansur. The two names *sebr zerd* and *sebr sugutri* (=Sokotra), given by SCHLIMMER,[2] are of Arabic and comparatively modern origin; thus is likewise the alleged Persian word *alwā*. The Persians adopted it from the Arabs; and the Arabs, on their part, admit that their *alua* is a transcription of the Greek word ἀλόη.[3] We must not imagine, of course, that the Chinese, when they first received this product during the T'ang period, imported it themselves directly from the African coast or Arabia. It was traded to India, and from there to the Malayan Archipelago; and, as intimated by Li Sün, it was shipped by the Malayan Po-se to Canton. Another point overlooked by Hirth is that *Aloe vera* has been completely naturalized in India for a long time, although not originally a native of the country.[4] GARCIA DA ORTA even mentions the preparation of aloes in Cambay and Bengal.[5] Thus we find in India, as colloquial names for the drug, such forms as *alia, ilva, eilya, elio, yalva,* and *aliva* in Malayan, which are all traceable to the Arabic-Greek *alua, alwā*. This name was picked up by the Malayan Po-se and transmitted by them with the product to the Chinese, who simply eliminated the initial *a* of the form *aluwa* or *aluwe* and retained *luwe*.[6] Besides *lu-wei*, occur also the transcriptions 奴 or 訥會 *nu* or *no hwi*, the former in the *K'ai-pao pen ts'ao* of the Sung, perhaps suggested by the Nu-fa country or to be explained by the phonetic interchange of *l* and *n*. It is not intelligible to me why Hirth says that in the Ming dynasty *lu-wei* "was, as it is now, catechu, a product of the *Acacia catechu* (Sanskrit *khadira*)." No authority for this theory is cited; but this is quite impossible, as catechu or cutch was well known to the Chinese under the names *er-č'a* or *hai'r-č'a*.[7]

65. A plant, 縮砂蔤 *so-ša-mi*, *suk-ša-m'it(m'ir), Japanese *šukušamitsu* (*Amomum villosum* or *xanthioides*), is first mentioned by Li Sün as "growing in the countries of the Western Sea (Si-hai) as well as in Si-žun 西戎 and Po-se, much of it coming from the Nan-tuṅ circuit

---

[1] W. OUSELEY, Oriental Geography of Ebn Haukal, p. 133.

[2] Terminologie, p. 22.

[3] LECLERC, Traité des simples, Vol. II, p. 367.

[4] G. WATT, Commercial Products of India, p. 59.

[5] C. MARKHAM, Colloquies, p. 6.

[6] WATTERS (Essays on the Chinese Language, p. 332), erroneously transcribing *lu-hui*, was inclined to trace the Chinese transcription directly to the Greek *aloe;* this of course, for historical reasons, is out of the question.

[7] See STUART, Chinese Materia Medica, p. 2; and my Loan-Words in Tibetan, No. 107, where the history of these words is traced.

安東道."¹ According to Ma Či, it grows in southern China, and, according to Su Suṅ, in the marshes of Liṅ-nan; thus it must have been introduced between the T'ang and Sung dynasties. In regard to the name, which is no doubt of foreign origin, Li Ši-čen observes that its significance is as yet unexplained. Certainly it is not Iranian, nor is it known to me that *Amomum* occurs in Persia. On the contrary, the plant has been discovered in Burma, Siam, Camboja, and Laos.² Therefore Li Sün's Po-se obviously relates again to the Malayan Po-se; yet his addition of Si-hai and Si-žuṅ is apt to raise a strong suspicion that he himself confounded the two Po-se and in this case thought of Persia. I have not yet succeeded in tracing the foreign word on which the Chinese transcription is based, but feel sure that it is not Iranian. The present colloquial name is *ts'ao ša žen* 草砂仁.³

66. There is a plant styled 婆羅得 *p'o-lo-te*, *bwa-ra-tïk, or 婆羅勒 *p'o-lo-lo*, *bwa-ra-lak(lok, lek), not yet identified. Again our earliest source of information is due to Li Sün, who states, "P'o-lo-te grows in the countries of the Western Sea (Si-hai) and in Po-se. The tree resembles the Chinese willow; and its seeds, those of the castor-oil plant (*pei-ma tse, Ricinus communis*, above, p. 403); they are much used by druggists."⁴ Li Ši-čen regards the word as Sanskrit, and the elements of the transcription hint indeed at a Sanskrit name. It is evidently Sanskrit *bhallātaka*, from which are derived Newārī *pālāla*, Hindustānī *belatak* or *bhelā*, Persian *balādur*, and Arabic *belādur* (GARCIA: *balador*). Other Sanskrit synonymes of this plant are *aruṣka, bījapādapa, vīravṛkṣa, viṣāsyā*, and *dahana*. It is mentioned in several passages of the Bower Manuscript.

This is the marking-nut tree (*Semecarpus anacardium*, family *Anacardiaceae*), a genus of Indian trees found throughout the hotter parts of India as far east as Assam, also distributed over the Archipelago as far as the Philippines⁵ and North Australia. It does not occur in Burma or Ceylon, nor in Persia or western Asia. The fleshy receptacle bearing the fruit contains a bitter and astringent substance, which is universally used in India as a substitute for marking-ink. The Chinese

---

¹ *Pen ts'ao kaṅ mu*, Ch. 14, p. 13 b.

² STUART, Chinese Materia Medica, p. 38. LOUREIRO (*so-xa-mi*) mentions it for Cochin-China (PERROT and HURRIER, Mat. méd. et pharmacopée sino-annamites, p. 97).

³ *Či wu miṅ ši t'u k'ao*, Ch. 25, p. 72.

⁴ *Pen ts'ao kaṅ mu*, Ch. 35, p. 7; *Čeṅ lei pen ts'ao*, Ch. 5, p. 14 b. In the latter work Li Sün attributes the definition "Western Sea and Po-se" to Sü Piao, author of the *Nan čou ki*.

⁵ M. BLANCO, Flora de Filipinas, p. 216.

say expressly that it dyes hair and mustache black.[1] It gives to cotton fabrics a black color, which is said to be insoluble in water, but soluble in alcohol. The juice of the pericarp is mixed with lime water as a mordant before it is used to mark cloth. In some parts of Bengal the fruits are regularly used as a dye for cotton cloths.[2] The fleshy cups on which the fruit rests, roasted in ashes, and the kernels of the nuts, are eaten as food. They are supposed to stimulate the mental powers, especially the memory. The acrid juice of the pericarp is a powerful vesicant, and the fruit is employed medicinally.

In regard to the Persian-Arabic *balādur*, Ibn al-Baiṭār states expressly that this is an Indian word,[3] and there is no doubt that it is derived from Sanskrit *bhallātaka*. The term is also given by Abu Mansur, who discusses the application of the remedy.[4] The main point in this connection is that *p'o-lo-te* is a typical Indian plant, and that the Po-se of the above Chinese text cannot refer to Persia. Since the tree occurs in the Malayan area, however, it is reasonable to conclude that again the Malayan Po-se is intended. The case is analogous to the preceding one, and the Malayan Po-se were the mediators. At any rate, the transmission to China of an Indian product with a Sanskrit name by way of the Malayan Po-se is far more probable than by way of Persia. I am also led to the general conclusion that almost all Po-se products mentioned in the *Hai yao pen ts'ao* of Li Sün have reference to the Malayan Po-se exclusively.

67. A drug, by the name 補骨脂 *pu-ku-či* (*bu-kut-tši), identified with *Psoralea corylifolia*, is first distinctly mentioned by Ma Či 馬志, collaborator in the *K'ai pao pen ts'ao* (A.D. 968–976) of the Sung period, as growing in all districts of Liṅ-nan (Kwaṅ-tuṅ) and Kwaṅ-si, and in the country Po-se. According to Ta Miṅ 大明, author of the *Ži hwa ču kia pen ts'ao* 日華諸家本草, published about A.D. 970, the drug would have been mentioned in the work *Nan čou ki* by Sü Piao (prior to the fifth century),[5] who determined it as 胡韭子 *hu kiu-tse*, the "*Allium odorum* of the Hu." This, however, is plainly an anachronism, as neither the plant, nor the drug yielded by it, is mentioned by any T'ang writers, and for the first time looms up in the pharmacopœia of the Sung. Su Suṅ, in his *T'u kiṅ pen ts'ao*, observes that the plant now occurs abundantly on the mountain-slopes of southern China,

---

[1] *Čeṅ lei pen ts'ao*, Ch. 5, p. 14 b.

[2] Cf. WATT, Dictionary, Vol. VI, pt. 2, p. 498.

[3] LECLERC, Traité des simples, Vol. I, pp. 162, 265.

[4] ACHUNDOW, Abu Mansur, p. 30.

[5] See above, p. 247.

also in Ho-čou 合州 in Se-č'wan, but that the native product does not come up to the article imported on foreign ships.[1] Ta Min defines the difference between the two by saying that the drug of the Southern Barbarians is red in color, while that of Kwan-tun is green. Li Ši-čen annotates that the Hu name for the plant is 婆固脂 *p'o-ku-či* (*bwa-ku-či, bakuči), popularly but erroneously written 破故紙 *p'o-ku-či* (*pa-ku-či), that it is the "*Allium odorum* of the Hu," because the seeds of the two plants are similar in appearance, but that in fact it is not identical with the *Allium* growing in the land of the Hu. These are all the historical documents available. STUART[2] concludes that the drug comes from Persia; but there is neither a Persian word *bakuči*, nor is it known that the plant (*Psoralea corylifolia*) exists in Persia. The evidence presented by the Chinese sources is not favorable, either, to this conclusion, for those data point to the countries south of China, associated in commerce with Kwan-tun. The isolated occurrence of the plant in a single locality of Se-č'wan is easily explained from the fact that a large number of immigrants from Kwan-tun have settled there. In fact, the word *bakuči yielded by the Chinese transcription is of Indian origin: it answers to Sanskrit *vākucī*, which indeed designates the same plant, *Psoralea corylifolia*.[3] In Bengālī and Hindustānī it is *hakūč*[4] and *bāvacī*, Uriyā *bākucī*, Panjāb *bābcī*, Bombay *bawacī*, Marathī *bavacya* or *bavacī*, etc. According to WATT, it is a common herbaceous weed found in the plains from the Himalaya through India to Ceylon. According to AINSLIE, this is a dark brown-colored seed, about the size of a large pin-head, and somewhat oval-shaped; it has an aromatic, yet unctuous taste, and a certain degree of bitterness. The species in question is an annual plant, seldom rising higher than three feet; and is common in southern India. It has at each joint one leaf about two inches long, and one and a half broad; the flowers are of a pale flesh color, being produced on long, slender, axillary peduncles. In Annam it is known as *hot-bo-kot-či* and *p'a-ko-či*.[5] It is therefore perfectly obvious

---

[1] According to the Gazetteer of Šen-si Province (*Šen-si t'un či*, Ch. 43, p. 31), the plant occurs in the district Ši-ts'üan 石泉 in the prefecture Hin-nan.

[2] Chinese Materia Medica, p. 359; likewise F. P. SMITH (Contributions, p. 179) and PERROT and HURRIER (Matière médicale et pharmacopée sino-annamites, p. 150).

[3] W. AINSLIE, Materia Indica, Vol. II, p. 141.

[4] This name is also given by W. ROXBURGH (Flora Indica, p. 588). See, further, WATT, Dictionary of the Economic Products of India, Vol. VI, p. 354.

[5] PERROT and HURRIER, Mat. méd. et pharmacopée sino-annamites, p. 150. According to these authors, the plant is found in the south and west of China as well as in Siam. Wu K'i-tsün says that physicians now utilize it to a large extent in lieu of cinnamon (*Či wu min ši t'u k'ao*, Ch. 25, p. 65).

that the designation "Allium of the Hu" is a misnomer, and that the plant in question has nothing to do with the Hu in the sense of Iranians, nor with Persia. The Po-se of Ma Či, referred to above, in fact represents the Malayan Po-se.

68. In the *Pen ts'ao kaṅ mu*, a quotation is given from the *Ku kin ču*, which is not to be found in the accessible modern editions of this work. The assertion is made there with reference to that work that ebony 烏文木 is brought over on Po-se ships. It is out of the question that Po-se in this case could denote Persia, as erroneously assumed by STUART,[1] as Persia was hardly known under that name in the fourth century, when the *Ku kin ču* was written, or is supposed to have been written, by Ts'ui, Pao;[2] and, further, ebony is not at all a product of Persia.[3] Since the same work refers ebony to Kiao-čou (Tonking), it may be assumed that this Po-se is intended for the Malayan Po-se; but, even in this case, the passage may be regarded as one of the many interpolations from which the *Ku kin ču* has suffered.

Chinese *wu-men* 烏楠 (*u-mon), "ebony" (timber of *Diospyros ebenum* and *D. melanoxylon*) is not a transcription of Persian *ābnūs*, as proposed by HIRTH.[4] There is no phonetic coincidence whatever. Nowhere is it stated that the Chinese word is Persian or a foreign word at all. There is, further, no evidence to the effect that ebony was ever traded from Persia to China; on the contrary, according to Chinese testimony, it came from Indo-China, the Archipelago, and India; according to Li Ši-čen, from Hai-nan, Yün-nan, and the Southern Barbarians.[5] The speculation that the word had travelled east and west with the article from "one of the Indo-Chinese districts," is untenable; for the ebony of western Asia and Greece did not come from Indo-China, but from Africa and India. The above Chinese term is not a transcription at all: the second character *men* is simply a late substitution of the Sung period for the older 文, as used in the *Ku kin ču*, *wu wen* meaning "black-streaked wood." In the *Pen ts'ao kaṅ mu*[6] it is said

---

[1] Chinese Materia Medica, p. 253.

[2] Persia under the name Po-se is first mentioned in A.D. 461, on the occasion of an embassy sent from there to the Court of the Wei (compare above, p. 471).

[3] It was solely imported into Persia (W. OUSELEY, Oriental Geography of Ebn Haukal, p. 133).

[4] Chau Ju-kua, p. 216.

[5] The *Ko ku yao lun* (Ch. 8, p. 5 b; ed. of *Si yin hüan ts'uṅ šu*) gives Hai-nan, Nan-fan ("Southern Barbarians"), and Yün-nan as places of provenience, and adds that there is much counterfeit material, dyed artificially. The poles of the tent of the king of Camboja were made of ebony (*Sui šu*, Ch. 82, p. 3).

[6] Ch. 35 B, p. 13.

that the character *men* should be pronounced in this case 漫 *man*, that the name of the tree is 文木 (thus written in the *Nan fan ts'ao mu čwan*), and that the southerners, because they articulate 文 like 槿, have substituted the latter. This is a perfectly satisfactory explanation. The *Ku kin ču*,[1] however, has preserved a transcription in the form 欝木欝 *i-muk-i or 欝 *bu (wu), which must have belonged to the language of Kiao-čou 交州 (Tonking), as the product hailed from there. Compare Khmer *mak pen* and Čam *mökiā* ("ebony," *Diospyros ebenaster*).[2]

Ebony was known in ancient Babylonia, combs being wrought from this material.[3] It is mentioned in early Egyptian inscriptions as being brought from the land of the Negroes on the upper Nile. Indeed, Africa was the chief centre that supplied the ancients with this precious wood.[4] From Ethiopia a hundred billets of ebony were sent every third year as tribute to Darius, king of Persia. Ezekiel[5] alludes to the ebony of Tyre. The Periplus (36) mentions the shipping of ebony from Barygaza in India to Ommana in the Persian Gulf. Theophrastus,[6] who is the first to mention the ebony-tree of India, makes a distinction between two kinds of Indian ebony, a rare and nobler one, and a common variety of inferior wood. According to Pliny,[7] it was Pompey who displayed ebony in Rome at his triumph over Mithridates; and Solinus, who copies this passage, adds that it came from India, and was then shown for the first time. According to the same writer, ebony was solely sent from India, and the images of Indian gods were sometimes carved from this wood entirely, likewise drinking-cups.[8] Thus the ancients were acquainted with ebony as a product of Africa and India at a time when Indo-China was still veiled to them, nor is any reference made to the far east in any ancient western account of the subject. The word itself is of Egyptian origin: under the name *heben*, ebony formed an important article with the country Punt. Hebrew *hobnīm* is related to this word or directly borrowed from it, and Greek ἔβενος is derived from Semitic. Arabic-Persian *'abnūs* is taken as a loan from the Greek, and Hindī *ābanūsa* is the descendant of *abnūs*.

---

[1] Ch. c, p. 1 b. The product is described as coming from Kiao-čou, being of black color and veined, and also called "wood with black veins" (*wu wen mu*).

[2] AYMONIER and CABATON, Dictionnaire čam-français, p. 366.

[3] HANDCOCK, Mesopotamian Archæology, p. 349.

[4] Herodotus, III, 97.

[5] XXVII, 15.

[6] Hist. plant., IV. IV, 6.

[7] XII, 4, § 20.

[8] Solinus, ed. MOMMSEN, pp. 193, 221.

It is thus obvious that the term Po-se in Chinese records demands great caution, and must not be blindly translated "Persia." Whenever it is used with reference to the Archipelago, the chances are that Persia is not in question. The Malayan Po-se has become a fact of historical significance. He who is intent on identifying this locality and people must not lose sight of the plants and products attributed to it. I disagree entirely with the conclusion of HIRTH and ROCKHILL[1] that from the end of the fourth to the beginning of the seventh centuries all the products of Indo-China, Ceylon, India, and the east coast of Africa were classed by the Chinese as "products of Persia (Po-se)," the country of the majority of the traders who brought these goods to China. This is a rather grotesque generalization, inspired by a misconception of the term Po-se and the Po-se texts of the *Wei šu* and *Sui šu*. The latter, as already emphasized, do not speak at all of any importation of Persian goods to China, but merely give a descriptive list of the articles to be found in Persia. Whenever the term Po-se is prefixed to the name of a plant or a product, it means only one of two things,—Persia or the Malayan Po-se,— but this attribute is never fictitious. Not a single case is known to me where a specific product of Ceylon or India is ever characterized by the addition Po-se.

---

[1] Chau Ju-kua, p. 7.

PERSIAN TEXTILES

69. Brocades, that is, textiles interwoven with gold or silver threads, were manufactured in Irān at an early date. Gold rugs are mentioned in the Avesta (*zaranaēne upasterene*, Yašt xv, 2). Xerxes is said to have presented to citizens of Abdera a tiara interwoven with gold.[1] The historians of Alexander give frequent examples of such cloth in Persia.[2] Pliny,[3] speaking of gold textiles of the Romans, traces this art to the Attalic textures, and stamps it as an invention of the kings of Asia (Attalicis vero iam pridem intexitur, invento regum Asiae).[4] The accounts of the ancients are signally confirmed by the Chinese.

Persian brocades 波斯錦 are mentioned in the Annals of the Liang as having been sent as tribute in A.D. 520 to the Emperor Wu from the country Hwa 滑.[5] The king of Persia wore a cloak of brocade, and brocades were manufactured in the country.[6] Textiles woven with gold threads 金縷織成 are expressly mentioned;[7] this term almost reads like a translation of Persian *zar-bāf* (literally, "gold weaving").[8] Persian brocades, together with cotton stuffs from An-si (Parthia) 安西 白氍, are further mentioned at the time of the Emperor Ši Tsuṅ 世宗 (A.D. 954-958) of the Hou Čou dynasty, among tribute-gifts sent from Kwa čou 瓜州 in Kan-su.[9] The Kirgiz received precious materials for the dress of their women from An-si (Parthia), Pei-t'iṅ 北廷 (Bišbalik, in Turkistan), and the Ta-ši 大食 (Tadžik, the Arabs). The Arabs made pieces of brocade of such size that the weight of each equalled that of twenty camel-loads. Accordingly these large pieces were cut up into

---

[1] Herodotus, VIII, 120.

[2] YATES, Textrinum Antiquorum, pp. 366-368.

[3] XXXIII, 19, § 63.

[4] At the Court of the Persian kings there was a special atelier for the weaving of silken, gold, and silver fabrics,—styled *siār bāf xāne* (E. KAEMPFER, Amoenitatum exoticarum fasciculi V, p. 128, Lemgoviae, 1712).

[5] *Liaṅ šu*, Ch. 54, p. 13 b. Hwa is the name under which the Ephthalites first appear in Chinese history (CHAVANNES, Documents sur les Tou-kiue occidentaux, p. 222).

[6] *Kiu T'aṅ šu*, Ch. 198, p. 10 b (see also *Liaṅ šu*, Ch. 54, p. 14 b; and *Sui šu* Ch. 83, p. 7 b). Hūan Tsaṅ refers to brocade in his account of Persia (*Ta T'aṅ si yü ki*, Ch. 11, p. 17 b, ed. of *Šou šan ko ts'uṅ šu*).

[7] *Sui šu, l. c.;* 金線錦袍師子錦袴 in *Liaṅ šu, l. c.*

[8] Cf. Loan-Words in Tibetan, No. 118.

[9] *Wu tai ši*, Ch. 74, p. 3 b; *Kiu Wu Tai ši*, Ch. 138, p. 1 b.

488

twenty smaller ones, so that they could be accommodated on twenty camels, and were presented once in three years by the Arabs to the Kirgiz. The two nations had a treaty of mutual alliance, shared also by the Tibetans, and guaranteeing protection of their trade against the brigandage of the Uigur.[1] The term *hu kin* 胡錦 ("brocades of the Hu," that is, Iranians) is used in the *Kwan yü ki* 廣輿記[2] with reference to Khotan.[3] The Iranian word for these textiles, though not recognized heretofore, is also recorded by the Chinese. This is 疊 *tie*, anciently *džiep, dziep, diep, dib*,[4] being the equivalent of a Middle-Persian form *dīb* or *dēp*,[5] corresponding to the New-Persian word *dībā* ("silk brocade," a colored stuff in which warp and woof are both made of silk), *dībāh* ("gold tissue"), Arabicised *dībādž* ("vest of brocade, cloth of gold"). The fabric as well as the name come from Sasanian Persia, and were known to the Arabs at Mohammed's time.[6] The Chinese term occurs as a textile product of Persia in the *Sui šu* (Ch. 83, p. 7ᵇ). At a much earlier date it is cited in the Han Annals (*Hou Han šu*, Ch. 116, p. 8) as a product of the country of the Ai-lao in Yün-nan. This is not surprising in view of the fact that at that period Yün-nan, by way of India, was in communication with Ta Ts'in: in A.D. 120 Yün Yu Tiao 雍由調, King of the country T'an 撣, presented to the Chinese emperor musicians and jugglers, who stated that "they had come from the Mediterranean 海西, which is the same as Ta Ts'in, and that south-west from the Kingdom of T'an there is communication with Ta Ts'in." The commentator of the Han Annals refers to the *Wai kwo čwan* 外國傳[7] as saying that the women of Ču-po 諸薄 (Java) make white *tie* and ornamented cloth 花布. The character 帛 *po* ("silk"), preceding the term *tie* in the Han Annals, represents a separate item, and

---

[1] *T'an šu*, Ch. 217 B, p. 18; *T'ai p'in hwan yü ki*, Ch. 199, p. 14. Cf. DEVÉRIA, in Centenaire de l'Ecole des Langues Orientales, p. 308.

[2] Ch. 24, p. 7 b. Regarding the various editions of this work, see p. 251.

[3] Likewise in the Sung Annals with reference to a tribute sent from Khotan in 961 (CHAVANNES and PELLIOT, Traité manichéen, p. 274). Regarding Persian brocades mentioned by mediæval writers, see FRANCISQUE-MICHEL, Recherches sur le commerce, la fabrication et l'usage des étoffes de soie d'or et d'argent, Vol. I, pp. 315-317, Vol. II, pp. 57-58 (Paris, 1852, 1854).

[4] According to the *Yi ts'ie kin i* (Ch. 19, p. 9 b), the pronunciation of the character *tie* was anciently identical with that of 氎 (see No. 70), and has the *fan ts'ie* 徒頰; that is, *t'iap, *diab, d'ab*. The *T'an šu ši yin* (Ch. 23, p. 1 b) indicates the same *fan ts'ie* by means of 徒協. The phonetic element 疊 serves for the transcription of Sanskrit *dvīpa* (PELLIOT, Bull. de l'Ecole française, Vol. IV, p. 357).

[5] A Pahlavi form *dēpāk* is indicated by WEST (Pahlavi Texts, Vol. I, p. 286); hence Armenian *dipak*.

[6] C. H. BECKER, Encyclopædia of Islam, Vol. I, p. 967.

[7] Cf. *Journal asiatique*, 1918, II, p. 24.

is not part of the transcription, any more than the word 錦 *kin*, which precedes it in the Sui Annals; but the combination of both *po* and *kin* with *tie* indicates and confirms very well that the latter was a brocaded silk. Hirth[1] joins *po* with *tie* into a compound in order to save the term for his pets the Turks. "The name *po-tie* is certainly borrowed from one of the Turki languages. The nearest equivalent seems to be the Jagatai Turki word for cotton, *pakhta*." There are two fundamental errors involved here. First, the Cantonese dialect, on which Hirth habitually falls back in attempting to restore the ancient phonetic condition of Chinese, does not in fact represent the ancient Chinese language, but is merely a modern dialect in a far-advanced stage of phonetic decadence. The sounds of ancient Chinese can be restored solely on the indications of the Chinese phonetic dictionaries and on the data of comparative Indo-Chinese philology. Even in Cantonese, *po-tie* is pronounced *pak-tip*, and it is a prerequisite that the foreign prototype of this word terminates in a final labial. The ancient phonetics of 帛 疊 is not *pak-ta*, but **bak-dzip* or **dip*, and this bears no relation to *pakhta*. Further, it is impossible to correlate a foreign word that appears in China in the Han period with that of a comparatively recent Turkish dialect, especially as the Chinese data relative to the term do not lead anywhere to the Turks; and, for the rest, the word *pakhta* is not Turkish, but Persian, in origin.[2] Whether the term *tie* has anything to do with cotton, as already stated by Chavannes,[3] is uncertain; but, in view of the description of the plant as given in the *Nan ši*[4] or *Liaň šu*,[5] it may be granted that the term *po-tie* was subsequently transferred to cotton.

The ancient pronunciation of *po-tie* being **bak-dib*, it would not be impossible that the element *bak* represents a reminiscence of Middle Persian *pambak* ("cotton"), New Persian *panpa* (Ossetic *bambag*, Armenian *bambak*). This assumption being granted, the Chinese term *po-tie*(=Middle Persian **bak-dib*=*pambak dīp*) would mean "cotton brocade" or "cotton stuff." Again, *po-tie* was a product of Iranian regions: *kin siu po tie* 金 繡 白 疊 is named as a product of K'aṅ (Sogdiana) in the Sasanian era;[6] and, as has been shown, *po-tie* from Parthia

---

[1] Chao Ju-kua, p. 218.

[2] Steingass, Persian–English Dictionary, p. 237.

[3] Documents sur les Tou-kiue occidentaux, p. 352.

[4] Ch. 79, p. 6 b.

[5] Ch. 54, p. 13 b. Cf. Chavannes, *ibid.*, p. 102; see also F. W. K. Müller, Uigurica, II, pp. 70, 105.

[6] *Sui šu*, Ch. 83, p. 4. Hence **bak-dīb* may also have been a Sogdian word.

is specially named. *Po-tie*, further, appears in India;[1] and as early as A.D. 430 Indian *po-tie* was sent to China from Ho-lo-tan 阿 羅 單 on Java.[2] According to a passage of the *Kiu T'an šu*,[3] the difference between *ku-pei* (Sanskrit *karpāsa*)[4] and *po-tie* was this, that the former was a coarse,

---

[1] *Nan Ši*, Ch. 78, p. 7 a.

[2] *Suň šu*, Ch. 97, p. 2 b.

[3] Ch. 197, p. 1 b, indicated by PELLIOT (*Bull. de l'Ecole française*, Vol. III, p. 269).

[4] It is evident that the transcription *ku-pei* is not based directly on Sanskrit *karpāsa;* but I do not believe with WATTERS (Essays on the Chinese Language, p. 440) and HIRTH (Chau Ju-kua, p. 218) that Malayan *kāpas* is at the root of the Chinese form, which, aside from the lack of the final *s*, shows a peculiar vocalism that cannot be explained from Malayan. Of living languages, it is Bahnar *kŏpaiḥ* ("cotton") which presents the nearest approach to Chinese *ku-pei* or *ku-pai*. It is therefore my opinion that the Chinese received the word from a language of Indo-China.

The history of cotton in China is much in need of a revision. The following case is apt to show what misunderstandings have occurred in treating this subject. *Ku-čuň* (*ku-džuň, *ku-duň) 古 終 is the designation of a cotton-like plant grown in the province of Kwei-čou 桂 州; the yarn is dyed and made into *pan pu* 斑 布. This is contained in the *Nan Yüe či* 南 越 志 by Šen Hwai-yüan 沈 懷 遠 of the fifth century (*Pen ts'ao kan mu*, Ch. 36, p. 24). SCHOTT (Altaische Studien, III, *Abh. Berl. Akad.*, 1867, pp. 137, 138; he merely refers to the source as "a description of southern China," without citing its title and date), although recognizing that the question is of a local term, proposed, if it were permitted to read *kutuň* instead of *kučuň*, to regard the word as an indubitable reproduction of Arabic *quṭun*, which resulted in the *coton, cotton, kattun*, etc., of Europe. MAYERS then gave a similar opinion; and HIRTH (Chau Ju-kua, p. 219), clinging to a Fu-čou pronunciation *ku-tüň* (also WATTERS, Essays, p. 440, transcribes *ku-tuň*), accepted the alleged derivation from the Arabic. This, of course, is erroneous, as in the fifth century there was no Arabic influence on China, nor did the Arabs themselves then know cotton. It would also be difficult to realize how a plant of Kwei-čou could have been baptized with an Arabic name at that or any later time. Moreover, *ku-čuň* is not a general term for "cotton" in Chinese; the above work remains the only one in which it has thus far been indicated. *Ku-čuň*, as Li Ši-čen points out, is a tree-cotton 木 綿 (*Bombax malabaricum*), which originated among the Southern Barbarians (Nan Fan 南 番), and which at the end of the Sung period was transplanted into Kiaň-nan. It is very likely that, as stated by STUART (Chinese Materia Medica, p. 197), the cotton-tree was known in China from very ancient times, and that its product was used in the manufacture of cloth before the introduction of the cotton-plant (*Gossypium herbaceum*). In fact, the same work *Nan yüe či* reports, "None of the Man tribes in the kingdom Nan-čao rear silkworms, but they merely obtain the seeds of the *so-lo* (*sa-la) 娑 羅 tree, the interior of which is white and contains a floss that can be wrought like silk and spun into cloth; it bears the name *so-lo luň twan* 娑 羅 籠 段." The *Faň yü či* 方 輿 志 of Ču Mu 覩 穆 of the Sung period alludes to the same tree, which is said to be from thirty to fifty feet in height. The *Ko ku yao lun* (Ch. 8, p. 4 b; ed. of *Si yin hüan ts'uň šu*) speaks of cotton stuffs 兜 羅 錦 (= 綿; *tou-lo* = Sanskrit *tūla*) which come from the Southern Barbarians, Tibet (Si-fan), and Yün-nan, being woven from the cotton in the seeds of the *so-lo* tree, resembling velvet, five to six feet wide, good for making bedding and also clothes. The *Tien hi* writes the word 梭 羅 (G. SOULIÉ, *Bull. de l'Ecole française*, Vol. VIII, p. 343). *Sa-la* is the indigenous name of the tree; *sa-la* is still the Lo-lo designation

and the latter a fine textile. In the Glossary of the T'ang Annals the word *tie* is explained as "fine hair" 絪毛 and "hair cloth" 毛布; these terms indeed refer to cotton stuffs, but simultaneously hint at the fact that the real nature of cotton was not yet generally known to the Chinese of the T'ang period. In the *Kwan yü ki*, *po-tie* is named as a product of Turfan; the threads, it is said, are derived from wild silkworms, and resemble fine hemp.

Russian *altabás* ("gold or silver brocade," "Persian brocade": DAL'), Polish *altembas*, and French *altobas*, in my opinion, are nothing but reproductions of Arabic-Persian *al-dībādž*, discussed above. The explanation from Italian *alto-basso* is a jocular popular etymology; and the derivation from Turkish *altun* ("gold") and *b'az* ("textile")[1] is likewise a failure. The fact that textiles of this description were subsequently manufactured in Europe has nothing to do, nor does it conflict, with the derivation of the name which Inostrantsev wrongly seeks in Europe.[2] In the seventeenth century the Russians received *altabás* from the Greeks; and Ibn Rosteh, who wrote about A.D. 903, speaks then of Greek *dībādž*.[3] According to Makkari, *dībādž* were manufactured by the Arabs in Almeria, Spain,[4] the centre of the Arabic silk industry.[5]

70. 毾氈 *t'a-ten*, *dap (=毾)[6]-dan (=氈), tap-tan, woollen rugs. The name of this textile occurs in the *Wei lio* of the third century A.D. as a product of the anterior Orient (Ta Ts'in),[7] and in the Han Annals

---

for cotton (VIAL, Dictionnaire français lo-lo, p. 97). Likewise it is *sa-la* in P'u-p'a, *sö-lö* in Čö-ko (*Bull. de l'Ecole française*, Vol. IX, p. 554). In the same manner I believe that *ku-džun was the name of the same or a similar tree in the language of the aborigines of Kwei-čou. Compare Lepcha *ka-čuk ki kun* ("cotton-tree"), Sin-p'o *ga-dun* ("cotton-tree"), given by J. F. NEEDHAM (Outline Grammar of the Singpho Language, p. 90, Shillong, 1889), and Meo *čoa* ("cotton"), indicated by M. L. PIERLOT (Vocabulaire méo, Actes du XIV⁰ Congrès int. des Orientalistes Alger 1905, pt. I, p. 150).

[1] Proposed by SAVEL'EV in *Erman's Archiv*, Vol. VII, 1848, p. 228.

[2] K. INOSTRANTSEV, Iz istorii starinnyx tkanei (*Zapiski Oriental Section Russian Archaeol. Soc.*, Vol. XIII, 1901, pp. 081–084).

[3] G. JACOB, Handelsartikel, p. 7; Waren beim arabisch-nordischen Verkehr, p. 16.

[4] G. MIGEON, Manuel d'art musulman, Vol. II, p. 420.

[5] DEFREMERY, *Journal asiatique*, 1854, p. 168; FRANCISQUE–MICHEL, Recherches sur le commerce, la fabrication et l'usage des étoffes de soie, d'or et d'argent, Vol. I, pp. 232, 284–290 (Paris, 1852).

[6] The *fan ts'ie* is 徒頰; that is, *du-kiap=d'iap (*Yi ts'ie kin yin i*, Ch. 19, p. 9 b), or 它闔 *du-hap=dap (*Hou Han šu*, Ch. 118, p. 5 b).

[7] F. HIRTH, China and the Roman Orient, pp. 71, 112, 113, 255. *T'a-ten* of five and nine colors are specified.

as a product of India.[1] In the Sui Annals it appears as a product of Persia.[2] CHAVANNES has justly rejected the fantastic explanation given in the dictionary *Ši min*, which merely rests on an attempt at punning. The term, in fact, represents a transcription that corresponds to a Middle-Persian word connected with the root √tāb ("to spin"): cf. Persian *tāftan* ("to twist, to spin"), *tābaδ* ("he spins"), *tāfta* or *tāfte* ("garment woven of linen, kind of silken cloth, taffeta"). Greek τάπης and ταπήτιον (frequent in the Papyri; ταπιδυφοι, "rug-weavers") are derived from Iranian.[3] There is a later Attic form δάπις. The Middle-Persian form on which the Chinese transcription is based was perhaps *tāptan, tāpetan, -an being the termination of the plural. The Persian word resulted in our *taffeta* (med. Latin *taffata*, Italian *taffetà*, Spanish *tafetan*).

71. To the same type as the preceding one belongs another Chinese transcription, 拓壁 *čo(t'o)-pi*, 柘胖 *tso-p'i*, or 柘必 *tso-pi*, dance-rugs sent to China in A.D. 718 and 719 from Maimargh and Bukhārā respectively.[4] These forms correspond to an ancient *ta-bik (壁 or 胖) or *ta-biδ (必), and apparently go back to two Middle-Persian forms *tābiχ and *tābeδ or *tābiδ (or possibly with medial *p*).[5]

72. More particularly we hear in the relations of China with Persia about a class of textiles styled *yüe no pu* 越諾布.[6] As far as I know, this term occurs for the first time in the Annals of the Sui Dynasty (A.D. 590–617), in the notice on Po-se (Persia).[7] This indicates that the object in question, and the term denoting it, hailed from Sasanian Persia.

---

[1] E. CHAVANNES, Les Pays d'occident d'après le Heou Han Chou (*T'oung Pao*, 1907, p. 193). Likewise in the *Nan ši* (Ch. 78, p. 5 b) and in Čao Žu-kwa (translation of HIRTH and ROCKHILL, p. 111).

[2] *Sui šu*, Ch. 83, p. 7 b.

[3] P. HORN, Grundriss iran. Phil., Vol. I, pt. 2, p. 137. NÖLDEKE's notion (Persische Studien, II, p. 40) that Persian *tanbasa* ("rug, carpet") should be derived from the Greek word, in my opinion, is erroneous.

[4] CHAVANNES, *T'oung Pao*, 1904, p. 34.

[5] These two parallels possibly are apt to shed light on the Old High-German duplicates *teppih* and *teppid*. The latter has been traced directly to Italian *tappeto* (Latin *tapēte, tapētum*), but the origin of the spirant χ in *teppih* has not yet been explained, and can hardly be derived from the final *t*. Would derivation from an Iranian source, direct or indirect, be possible?

[6] According to HIRTH (Chau Ju-kua, p. 220), "a light cotton gauze or muslin, of two kinds, pure white, and spangled with gold"; but this is a doubtful explanation.

[7] *Sui šu*, Ch. 83, p. 7 b. This first citation of the term has escaped all previous writers on the subject,—Hirth, Chavannes, and Pelliot. From the *Sui šu* the text passed into the *T'ai p'in hwan yü ki* (Ch. 185, p. 18 b).

In the T'ang Annals we read that in the beginning of the period
K'ai-yüan (A.D. 713–741) the country of K'an (Sogdiana), an Iranian
region, sent as tribute to the Chinese Court coats-of-mail, cups of rock-
crystal, bottles of agate, ostrich-eggs, textiles styled *yüe no*, dwarfs,
and dancing-girls of Hu-sūan 胡旋 (Xwārism).[1] In the *Ts'e fu yüan kwei*
the date of this event is more accurately fixed in the year 718.[2] The
Man šu, written by Fan Čo of the T'ang period, about A.D. 860,[3] mentions *yüe no* as a product of the Small P'o-lo-men 小婆羅門 (Brāhmaṇa) country, which was conterminous with P'iao 驃 (Burma) and
Mi-č'en (*Midžen) 彌臣.[4] This case offers a parallel to the presence
of *tie* in the Ai-lao country in Yün-nan.

The Annals of the Sung mention *yüe no* as exported by the Arabs
into China.[5] The *Lin wai tai ta*,[6] written by Čou K'ü-fei in 1178, mentions white *yüe-no* stuffs in the countries of the Arabs, in Bagdād, and
yüe-no stuffs in the country Mi 憹.

HIRTH[7] was the first to reveal the term *yüe no* in Čao Žu-kwa, who
attributes white stuffs of this name to Bagdād. His transcription *yüt-
nok*, made on the basis of Cantonese, has no value for the phonetic
restoration of the name, and his hypothetical identification with *cut-
tanee* must be rejected; but as to his collocation of the second element
with Marco Polo's *nac*, he was on the right trail. He was embarrassed,
however, by the first element *yüe*, "which can in no way be explained
from Chinese and yet forms part of the foreign term." Hence in his
complete translation of the work[8] he admits that the term cannot as
yet be identified. His further statement, that in the passage of the
T'an šu, quoted above, the question is possibly of a country *yüe-no*
(Bukhārā), rests on a misunderstanding of the text, which speaks only
of a textile or textiles. The previous failures in explaining the term
simply result from the fact that no serious attempt was made to restore

---

[1] Cf. CHAVANNES, Documents sur les Tou-kiue occidentaux, pp. 136, 378,
with the rectification of PELLIOT (*Bull. de l'Ecole française*, Vol. IV, 1904, p. 483).
Regarding the dances of Hu-sūan, see *Kin ši hwi yüan kiao k'an ki* 近事會元校
勘記 (p. 3), Critical Annotations on the *Kin ši hwi yüan* by Li Šan-kiao 李上交
of the Sung (in *Ki fu ts'un šu*, *t'ao* 10).

[2] CHAVANNES, *T'oung Pao*, 1904, p. 35.

[3] See above, p. 468.

[4] *Man šu*, p. 44 b (ed. of *Yün-nan pei čen či*). Regarding Mi-č'en, see PELLIOT,
Bull. de l'Ecole française, Vol. IV, p. 171.

[5] *Sun ši*, Ch. 490; and BRETSCHNEIDER, Knowledge possessed by the Chinese
of the Arabs, p. 12. Bretschneider admitted that this product was unknown to him.

[6] Ch. 3, pp. 2–3.

[7] Länder des Islam, p. 42 (Leiden, 1894).

[8] Chau Ju-kua, p. 220.

it to its ancient phonetic condition.[1] Moreover, it was not recognized that *yüe no* represents a combination of two Iranian words, and that each of these elements denotes a particular Iranian textile.

(1) The ancient articulation of what is now sounded *yüe* 越 was *vat, vaδ, wiāδ, or, with liquid final, *var or *val.[2] Thus it may well be inferred that the Chinese transcription answers to a Middle-Persian form of a type *vār or *vāl. There is a Persian word *barnū* or *barnūn* ("brocade"), *vālā*, which means "a kind of silken stuff,"[3] and *bālās*, "a kind of fine, soft, thin armosin silk, an old piece of cloth, a kind of coarse woollen stuff."[4]

(2) 諾 *no* corresponds to an ancient *nak,[5] and is easily identified with Persian *nax* (*nakh*), "a carpet beautiful on both sides, having a long pile; a small carpet with a short pile; a raw thread of yarn of any sort,"[6] but also "brocade." The early mention of the Chinese term, especially in the Sui Annals, renders it quite certain that the word *nak* or *nax* was even an element of the Middle-Persian language. Hitherto it had been revealed only in mediæval authors, the *Yüan č'ao pi ši*,

---

[1] DE GOEJE's identification of *yüe-no pu* with *djannābī* (in HIRTH, Länder des Islam, p. 61) is a complete failure: *pu* ("cloth") does not form part of the transcription, which can only be read *vaδ-nak, var-nak,* or *val-nak*. TSUBOI KUMAZO (Actes XII° Congrès international des Orientalistes Rome 1899, Vol. II, p. 112) has already opposed this unfortunate suggestion.

[2] For examples, see CHAVANNES, Mémoires historiques de Se-ma Ts'ien, Vol. IV, p. 559; and particularly cf. PELLIOT, *Journal asiatique*, 1914, II, p. 392.

[3] STEINGASS, Persian-English Dictionary, p. 1453. HORN (Grundriss iran. Phil., Vol. I, pt. 2, p. 29) translates the word "a fine stuff," and regards it as a loan-word from Greek βῆλον ("veil"), first proposed, I believe, by NÖLDEKE (Persische Studien, II, p. 39). This etymology is not convincing to me. On the contrary, *vāla* is a genuine Persian word, meaning "eminent, exalted, high, respectable, sublime, noble"; and it is quite plausible that this attribute was transferred to a fine textile. It was, further, the Persians who taught the Greeks lessons in textile art, but not the reverse. F. JUSTI (Iranisches Namenbuch, p. 516) attributes to *vāla* also the meaning "banner of silk."

[4] STEINGASS, *op. cit.*, p. 150. The Iranian character of this word is indicated by WaxI *palás*, SariqolI *palūs* ("coarse woollen cloth") of the Pamir languages. Perhaps also Persian *bat* ("stuff of fine wool"), WaxI *böt*, SariqolI *bél* (cf. W. TOMA-SCHEK, Pamirdialekte, Sitzber. Wiener Akad., 1880, p. 807) may be enlisted as possible prototypes of Chinese *vat, val; but I do not believe with Tomaschek that this series bears any relation to Sanskrit *paṭṭa* and *lāṭa* or Armenian *lōtik* ("mantle"). The latter, in my opinion, is a loan-word from Greek λῶδιξ ("cover, rug"), that appears in the Periplus (§ 24) and in the Greek Papyri of the first century A.D. (T. REIL, Beiträge zur Kenntnis des Gewerbes im hellenistischen Ægypten, p. 118).

[5] See, for instance, *T'oung Pao*, 1914, p. 77, and 1915, p. 8, where the character in question serves for transcribing Tibetan *nag*. It further corresponds to *nak* in Annamese, Korean, and Japanese, as well as in the transcriptions of Sanskrit words.

[6] STEINGASS, Persian-English Dictionary, p. 1391.

Yüan ši, Ibn Baṭūṭa, Rubruk, Marco Polo, Pegoletti, etc.[1] W. BANG has shown in a very interesting essay[2] that also the Codex Cumanicus contains the term *nac* (Cumanian), parallel with Persian *nagh* and Latin *nachus*, in the sense of "gold brocades," and that the introitus natorum et nascitorum of the books of tax-rates of Genoa about 1420 refers to these textiles, and has nothing to do with the endowment of the newborn, as had been translated. Bang points out also "náchi, a kinde of slight silke wouen stuffe" in Florio, "Queen Anna's New World of Words" (London, 1611). In mediæval literature the term *nac, nak, naque*, or *nachiz* occurs as early as the eleventh century, and figures in an inventory of the Cathedral of Canterbury of the year 1315.

73. 護那 *hu-na*, **ɣu-na*, a textile product of Persia[3] (or 護冄).[4] An ancient Iranian equivalent is not known to me, but must be supposed to have been **ɣuna* or **guna*. This word may be related to Šighnan (Pāmir language) *ghâun* ("coarse sack"), Kashmir *gun*, Sanskrit *goṇi*;[5] Anglo-Indian *gunny, gunny-bag*, trading-name of the coarse sacking and sacks made from the fibre of the jute.[6]

74. 檀 *t'an*, **dan*, **tan*, a textile product of Persia, likewise mentioned in the Sui Annals. This is doubtless the Middle-Persian designation of a textile connected with the root √tan ("to spin"), of which several Middle-Persian forms are preserved.[7] Compare Avestan *tanva*, Middle Persian *tanand*, Persian *tanīðan, tanandō* ("spider"), and, further, Persian *tan-basa, tan-bīsa* ("small carpet, rug"); *tanīd* ("a web"); *tānīdan* ("to twist, weave, spin").

75. 措哈剌 *sa-ha-la* or 瑣哈嘓[8] *so-ha-la*, of green color, is men-

---

[1] See E. BRETSCHNEIDER, Notices of the Mediæval Geography, p. 288, or Mediæval Researches, Vol. II, p. 124; YULE, Cathay, new ed. by CORDIER, Vol. III, pp. 155-156, 169; YULE, Marco Polo, Vol. I, pp. 63, 65, 285; W. HEYD, Histoire du commerce du levant au moyen âge, p. 698; and, above all, F.-MICHEL, Recherches sur le commerce etc., des étoffes de soie, Vol. I, pp. 261-264. A. HOUTUM-SCHINDLER (*Journal As. Soc. Bengal*, Vol. VI, 1910, p. 265) states that *nax* occurs in a letter of Rašid-eddin.

[2] Ueber den angeblichen "Introitus natorum et nascitorum" in den Genueser Steuerbüchern, in *Bull. de la Classe des Lettres de l'Académie royale de Belgique*, No. 1, 1912, pp. 27-32.

[3] *Sui šu*, Ch. 83, p. 7 b.

[4] *T'ai p'iň hwan yü ki*, Ch. 185, p. 18 b.

[5] W. TOMASCHEK, Pamirdialekte (*Sitzber. Wiener Akad.*, 1880, p. 808).

[6] YULE, Hobson-Jobson, p. 403.

[7] SALEMANN, Grundriss iran. Phil., Vol. I, pt. 1, p. 303.

[8] This transcription is given in the *Č'aň wu či* 長物志 by Wen Čen-heň 文 震亨 of the Ming (Ch. 8, p. 1 b; ed. of *Yüe ya t'aň ts'uň šu*). He describes the material as resembling sheep-wool, as thick as felt, coming from the Western Regions, and very expensive.

tioned in the Ming history as having been sent as a present in 1392 from Samarkand. The Ming Geography, as stated by BRETSCHNEIDER,[1] mentions this stuff as a manufacture of Bengal and So-li, saying that it is woven from wool and is downy. There is a red and a green kind. Bretschneider's view, that by sa-ha-la the Persian *šāl* is intended, must be rejected.[2] In the *Yiṅ yai šeṅ lan* of 1416, sa-ha-la is enumerated among the goods shipped from Malacca, being identified by GROENE-VELDT with Malayan *saklat* or *sahalat*.[3] *Sa-ha-la* is further mentioned for Ormuz and Aden.[4]

In the *Ko ku yao lun* 格古要論, written by Ts'ao Čao 曹昭 in 1387, revised and enlarged in 1459 by Waṅ Tso 王佐,[5] we meet this word in the transcription 灑 (＝洒) 海剌 sa-hai-la,[6] which is said to come from Tibet 西番 in pieces three feet in width, woven from wool, strong and thick like felt, and highly esteemed by Tibetans. Under the heading *p'u-lo* 普羅 (＝Tibetan *p'rug*)[7] it is said in the same work that this Tibetan woollen stuff resembles *sa-hai-la*.

Persian *sakirlāt*, *sagirlāt*, has been placed on a par with Chinese *sa-ha-la* by T. WATTERS[8] and A. HOUTUM-SCHINDLER;[9] it is not this Persian word, however, that is at the root of Chinese *sa-ha-la*, but *saqalāt* or *saqallāt*, also *saqalāṭ*, *saqallāṭ* ("scarlet cloth"). Dr. E. D. Ross[10] has been so fortunate as to discover in a Chinese-Persian vocabulary of 1549 the equation: Chinese *sa-ha-la* = Persian *saqalat*. This settles the problem definitely. There is, further, Persian *saqlāṭūn* or *saqlāṭin*, said to mean "a city in Rūm where scarlet cloth is made, scarlet cloth or dress made from it." The latter name is mentioned as early as A.D. 1040 and 1150 by Baihaki and Edrīsī respectively.[11] According to Edrīsī, it was a silk product of Almeria in Spain, which is doubtless meant by the city of Rūm. Yāqūt tells of its manufacture in Tabrīz,

[1] Mediæval Researches, Vol. II, p. 258.

[2] Regarding the Chinese transcription of this Persian word, see ROCKHILL, *T'oung Pao*, 1915, p. 459.

[3] Notes on the Malay Archipelago, p. 253.

[4] ROCKHILL, *T'oung Pao*, 1915, pp. 444, 606, 608. It does not follow from the text, however, that sa-ha-la was a kind of thin veiling or gauze, as the following term (or terms) 摸紗 is apparently a matter in itself.

[5] Ch. 8, p. 4 b (ed. of *Si yin hüan ts'uṅ šu*).

[6] This mode of writing is also given in the *Č'aṅ wu či*, cited above.

[7] *T'oung Pao*, 1914, p. 91.

[8] Essays on the Chinese Language, p. 342.

[9] *Journal As. Soc. Bengal*, Vol. VI, 1910, p. 265.

[10] *Journal As. Soc. Bengal*, Vol. IV, 1908, p. 403.

[11] YULE, Hobson-Jobson, p. 861.

so that the Chinese reference to Samarkand becomes intelligible. The Chinese reports of *sa-ha-la* in India, Ormuz, and Aden, however, evidently refer to European broadcloth, as does also Tibetan *sag-lad*.[1]

The Ain-i Akbari speaks of *suklāt* (*saqalāt*) of Rūm (Turkey), Farangī (Europe), and Purtagālī (Portugal); and the Persian word is now applied to certain woollen stuffs, and particularly to European broadcloth.

The Persian words *sakirlāt* and *saqalāt* are not interrelated, as is shown by two sets of European terms which are traced to the two Persian types: *sakirlāt* is regarded as the ancestor of "scarlet" (med. Latin *scarlatum, scarlata;* Old French *escarlate,* New French *écarlate,* Middle English *scarlat,* etc.); *saqlāṭūn* or *siqlāṭūn* is made responsible for Old French *siglaton,* Provençal *sisclaton* (twelfth century), English obs. *ciclatoun* (as early as 1225), Middle High German *ciclāt* or *siglāt.* Whether the alleged derivations from the Persian are correct is a debatable point, which cannot be discussed here; the derivation of *siglaton* from Greek κυκλάς (*cyclas*), due to Du Cange, is still less plausible.[2] Dr. Ross (*l.c.*) holds that "the origin of the word scarlet seems to be wrapped in mystery, and there seems to be little in favor of the argument that the word can be traced to Arabic or Persian sources."

76. Toward the close of the reign of Kao Tsuṅ 高宗, better known as Wen Č'eṅ 文成 (A.D. 452–465) of the Hou Wei dynasty (386–532), the king of Su-le (Kashgar) sent an emissary to present a garment (*kāṣāya*) of Çākyamuni Buddha, over twenty feet in length. On examination, Kao Tsuṅ satisfied himself that it was a Buddha robe. It proved a miracle, for, in order to get at the real facts, the Emperor had the cloth put to a test and exposed to a violent fire for a full day, but it was not consumed by the flames. All spectators were startled and spell-bound.[3] This test has repeatedly been made everywhere with asbestine cloth, of which many examples are given in my article "Asbestos and Salamander."[4] The Chinese themselves have recognized without difficulty that this Buddha relic of Kashgar was made of an asbestine material. In the *Lu č'aṅ kuṅ ši k'i*,[5] a modern work,

---

[1] See Loan-Words in Tibetan, No. 119.

[2] Cf. also F.-MICHEL, Recherches sur le commerce etc., des étoffes de soie, Vol. I, pp. 233–235. The Greek word in question does not refer to a stuff, but to a robe (κυκλάς, "round, circular," scil., εσθής, "a woman's garment with a border all round it"). *Cycladatus* in Suetonius (Caligula, LII) denotes a tunic with a rich border.

[3] *Wei šu,* Ch. 102, p. 4 b.

[4] *T'oung Pao,* 1915, pp. 299–373.

[5] Ed. of *Ts'iṅ čao t'aṅ ts'uṅ šu,* p. 40 (see above, p. 346). On p. 41 b there is a notice of fire-proof cloth, consisting of quotations from earlier works, which are all contained in my article.

which contains a great number of valuable annotations on subject-matters mentioned in the Annals, the *kāṣāya* of Kashgar is identified with the fire-proof cloth of the Western Regions and Fu-nan (Camboja); that is, asbestos.

During the Kʻai-yüan and Tʻien-pao periods (A.D. 713–755), Persia sent ten embassies to China, offering among other things "embroideries of fire-hair" (*hwo mao siu* 火 毛 繡).[1] CHAVANNES[2] translates this term "des broderies en laine couleur de feu." In my opinion, asbestos is here in question. Thus the term was already conceived by ABEL-RÉMUSAT.[3] I have shown that asbestos was well known to the Persians and Arabs, and that the mineral came from Badaxšān.[4] An additional

---

[1] *Tʻaṅ šu*, Ch. 221 B, p. 7. In the *Tʻaṅ hui yao* (Ch. 100, p. 4) this event is fixed in the year 750.

[2] Documents sur les Tou-kiue, p. 173.

[3] Nouveaux mélanges asiatiques, Vol. I, p. 253. The term *hwo pu* 火 布 ("fire-cloth") for asbestos appears in the *Suṅ šu* (Ch. 97, p. 10). The Chinese notions of textiles made from an "ice silkworm," possibly connected with Persia (cf. H. MASPERO, *Bull. de l'Ecole française*, Vol. XV, No. 4, 1915, p. 46), in my opinion, must be dissociated from asbestos; the Chinese sources (chiefly *Wei lio*, Ch. 10, p. 2 b) say nothing to the effect that this textile was of the nature of asbestos. Maspero's argumentation (*ibid.*, pp. 43–45) in regard to the alleged asbestos from tree-bark, which according to him should be a real asbestine stuff, appears to me erroneous. He thinks that I have been misled by an inexact translation of S. W. WILLIAMS. First, this translation is not by Williams, but, as expressly stated by me (*l. c.*, p. 372), the question is of a French article of d'Hervey-St.-Denys, translated into English by Williams. If an error there is (the case is trivial enough), it is not due to Williams or myself, but solely to the French translator, who merits Maspero's criticism. Second, Maspero is entirely mistaken in arguing that this translation should have influenced my interpretation of the text on p. 338. This is out of the question, as all this was written without knowledge of the article of St.-Denys and Williams, which became accessible to me only after the completion and printing of the manuscript, and was therefore relegated to the Addenda inserted in the proofs. Maspero's interpretation leads to no tangible result, in fact, to nothing, as is plainly manifest from his conclusion that one sort of asbestos should have been a textile, the other a kind of felt. There is indeed no asbestos felt. How Maspero can deny that Malayan bark-cloth underlies the Chinese traditions under notice, which refer to Malayan regions, is not intelligible to me. Nothing can be plainer than the text of the Liang Annals: "On Volcano Island there are trees which grow in the fire. The people in the vicinity of the island peel off the bark, and spin and weave it into cloth hardly a few feet in length. This they work into kerchiefs, which do not differ in appearance from textiles made of palm and hemp fibres," etc. (pp. 346, 347). What else is this but bark-cloth? And how could we assume a Malayan asbestine cloth if asbestos has never been found and wrought anywhere in the Archipelago? I trust that M. Maspero, for whose scholarship I have profound respect, will pardon me for not accepting his opinion in this case, and for adhering to my own interpretation. I may add here a curious notice from J. A. DE MANDELSLO's Voyages into the East Indies (p. 133, London, 1669): "In the Moluccaes there is a certain wood, which, laid in the fire, burns, sparkles, and flames, yet consumes not, and yet a man may rub it to powder betwixt his fingers."

[4] *Tʻoung Pao*, 1915, pp. 327–328.

---

text to this effect may be noted here. Ibn al-Faqīh, who wrote in
A.D. 902, has this account: "In Kirmān there is wood that is not burnt
by fire, but comes out undamaged.[1] A Christian[2] wanted to commit
frauds with such wood by asserting that it was derived from the cross of
the Messiah. Christian folks were thus almost led into temptation. A
theologian, noting this man, brought them a piece of wood from Kirmān, which was still more impervious to fire than his cross-wood."
According to P. Schwarz,[3] to whom we owe the translation of this
passage, the question here is of fossilized forests. Most assuredly, however, asbestos is understood. The above text of the *Wei šu* is thus by
far the earliest allusion to asbestos from an Iranian region.

The following notes may serve as additional information to my
former contribution. Čou Mi 周密 (1230–1320), in his *Či ya t'aṅ tsa
t'ao* 志雅堂雜鈔, mentions asbestine stuffs twice.[4] In one passage
he relates that in his house there was a piece of fire-proof cloth (*hwo
hwan pu*) over a foot long, which his maternal grandfather had once
obtained in Ts'üan čou 泉州 (Fu-kien Province).[5] Visitors to his house
were entertained by the experiment of placing it on the fire of a brazier.
Subsequently Čao Moṅ-i 趙孟議 borrowed it from him, but never
returned it. In the other text he quotes a certain Ho Ts'iṅ-fu 霍清夫
to the effect that fire-proof cloth is said to represent the fibres of the
mineral coal of northern China, burnt and woven, but not the hair of
the fire-rodent (salamander). This is accompanied by the comment
that coal cannot be wrought into fibres, but that now *pu-hwei-mu*
不灰木 (a kind of asbestos) is found in Pao-tiṅ (Či-li).[6] A brief notice
of asbestos is inserted in the *Ko ku yao lun*,[7] where merely the old fables
are reiterated. Information on the asbestos of Či-li Province will be

---

[1] Qazwīnī adds to this passage, "even if left in fire for several days."

[2] Qazwīnī speaks in general of charlatans.

[3] *Iran im Mittelalter*, p. 214.

[4] Ch. A, p. 20 b; and Ch. B, p. 25 b (ed. of *Yüe ya t'aṅ ts'uṅ šu*).

[5] This locality renders it almost certain that this specimen belonged to those
imported by the Arabs into China during the middle ages (p. 331 of my article).
The asbestos of Mosul is already mentioned in the *Liṅ wai tai ta* (Ch. 3, p. 4).

[6] The term *pu-hwei-mu* ("wood burning without ashes, incombustible wood")
appears as early as the Sung period in the *Čeṅ lei pen ts'ao* (Ch. 5, p. 35): it comes
from Šaṅ-taṅ (south-east portion of Šan-si and part of Ho-nan), and is now found
in the Tse-lu mountains 澤潞山. It is a kind of stone, of green and white color,
looking like rotten wood, and cannot be consumed by fire. Some call it the root of
soapstone.

[7] Ch. 8, p. 4 (ed. of *Si yin hüan ts'uṅ šu*). In Ch. 7, p. 17, there is a notice on
pu-hwei-mu stone, stated to be a product of Tse-čou and Lu-ṅan in Šan-si, and employed for lamps.

found in the *Ki fu t'un či*,[1] on asbestos of Se-č'wan in the *Se č'wan t'un či*.[2] In the eighteenth century the Chinese noticed asbestos among the Portuguese of Macao, but the article was rarely to be found in the market.[3] Hanzō Murakami discusses asbestos (石 綿, "stone cotton") as occurring in the proximity of Kin-čou 金 洲 in Šeň-kiň, Manchuria.[4]

In regard to the salamander, FRANCISQUE-MICHEL[5] refers to "Traditions tératologiques de Berger de Xivrey" (Paris, Imprimerie royale, 1836, pp. 457, 458, 460, 463) and to an article of Duchalais entitled "L'Apollon sauroctone" (*Revue archéologique*, Vol. VI, 1850, pp. 87–90); further to Mahudel in *Mémoires de littérature tirés des registres de l'Académie royale des inscriptions et belles-lettres*, Vol. IV, pp. 634–647. Quoting several examples of salamander stuff from mediæval romances, Francisque-Michel remarks, "Ces étoffes en poil de salamandre, qui vraisemblablement étaient passées des fables des marchands dans celles des poètes, venaient de loin, comme ceux qui avaient par là beau jeu pour mentir. On en faisait aussi des manteaux; du moins celui de dame Jafite, du *Roman de Gui le Gallois*, en était."

No one interested in this subject should fail to read chapter LII of book III of Rabelais' *Le Gargantua et Le Pantagruel*, entitled "Comment doibt estre preparé et mis en œuvre le celebre Pantagruelion."

77. The word "drugget," spelled also droggitt, drogatt, druggit (Old French *droguet*, Spanish *droguete*, Italian *droghetto*) is thus defined in the new Oxford English Dictionary: "Ulterior origin unknown. Littré suggests derivation from *drogue* drug as 'a stuff of little value'; some English writers have assumed a derivation from Drogheda in Ireland, but this is mere wanton conjecture, without any historical basis. Formerly kind of stuff, all of wool, or mixed of wool and silk or wool and linen, used for wearing apparel. Now, a coarse woollen stuff for floor-coverings, table-cloths, etc." The Century Dictionary says, "There is nothing to show a connection with drug."

Our lexicographers have overlooked the fact that the same word occurs also in Slavic. F. MIKLOSICH[6] has indicated a Serbian *doroc* ("pallii genus") and Magyar *darócz* ("a kind of coarse cloth"), but neglected to refer to the well-known Russian word *dorógi* or *dórogi*, which apparently represents the source of the West-European term. The latter has been dealt with by K. INOSTRANTSEV[7] in a very interesting

---

[1] Ch. 74, pp. 10 b, 13.

[2] Ch. 74, p. 25.

[3] *Ao-men či lio*, Ch. B, p. 41.

[4] *Journal Geol. Soc. Tokyo*, Vol. XXIII, No. 276, 1916, pp. 333–336. The same journal, Vol. XXV, No. 294, March, 1918, contains an article on asbestos in Japan and Korea by K. OKADA.

[5] Recherches sur le commerce, la fabrication et l'usage des étoffes de soie, d'or et d'argent, Vol. II, pp. 90, 462 (Paris, 1854).

[6] Fremdwörter in den slavischen Sprachen, *Denk. Wiener Akad.*, Vol. XV, 1867, p. 84.

[7] Iz istorii starinnix tkanei, *Zapiski of the Russian Arch. Soc.*, Vol. XIII, 1902, p. 084.

study on the history of some ancient textiles. According to this author, the *dorógi* of the Russians were striped silken fabrics, which came from Gilan, Kašan, Kizylbaš, Tur, and Yas in Persia. DAL′ says in his Russian Dictionary that this silk was sometimes interwoven with gold and silver. In 1844 VELTMAN proposed the identity of Russian *dorógi* with the Anglo-French term. BEREZIN derived it from Persian *darādža* ("kaftan"), which is rejected, and justly so, by Inostrantsev. On his part, he connects the word with Persian *dārāi* ("a red silken stuff"),[1] and invokes a passage in VESELOVSKI'S "Monuments of Diplomatic and Commercial Relations of Moscovite Rus with Persia," in which the Persian word *dārāi* is translated by Russian *dorógi*. This work is unfortunately not accessible to me, so I cannot judge the merits of the translation; but the mere fact of rendering *dorógi* by *dārāi* would not yet prove the actual derivation of the former from the latter. For philological reasons this theory seems to me improbable: it is difficult to realize that the Russians should have made *dorógi* out of a Persian *dārāi*. All European languages have consistently preserved the medial *g*, and this cannot be explained from *dārāi*. Another prototype therefore, it seems to me, comes into question; and this probably is Uigur *torgu*, Jagatai *torka*, Koibal *torga*, Mongol *torga(n)*, all with the meaning "silk."[2] It remains to search for the Turkish dialect which actually transmitted the word to Slavic.

---

[1] Mentioned, for instance, in the list of silks in the Ain-i Akbari (BLOCHMANN'S translation, Vol. I, p. 94).

[2] Cf. *T'oung Pao*, 1916, p. 489.

IRANIAN MINERALS, METALS, AND PRECIOUS STONES

78. 呼洛 hu-lo, *χu-lak, perhaps also *fu-lak, *fu-rak, a product of Persia,[1] which is unexplained. In my opinion, this word may correspond to a Middle Persian *furak = New Persian būrak, būra, Armenian porag ("borax"). Although I am not positive about this identification, I hope that the following notes on borax will be welcome. It is well known that Persia and Tibet are the two great centres supplying the world-market with borax. The ancient Chinese were familiar with this fact, for in the article on Po-se (Persia) the T'ai p'in hwan yü ki[2] states that "the soil has salty lakes, which serve the people as a substitute for salt" (地有鹹池人代鹽味). Our own word "borax" (the x is due to Spanish, now written borraj) comes from Persian, having been introduced into the Romanic languages about the ninth century by the Arabs. Russian burá was directly transmitted from Persia. Likewise our "tincal, tincar" (a crude borax found in lake-deposits of Persia and Tibet) is derived from Persian tinkār, tankāl,[3] or tangār, Sanskritized ṭankaṇa, ṭanka, ṭanga, ṭagara;[4] Malayan tingkal; Kirgiz dänäkär, Osmanli tängar.[5] Another Persian word that belongs to this category, šora ("nitre, saltpetre"), has been adopted by the Tibetans in the same form šo-ra, although they possess also designations of their own, ze-ts'wa, ba-ts'wa ("cow's salt"), and ts'a-la. The Persian word is Sanskritized into sorāka, used in India for nitre, saltpetre, or potassium nitrate.[6]

79. The relation of Chinese nao-ša ("sal ammoniac, chloride of sodium")[7] to Persian nušādir or naušādir is rather perspicuous; nevertheless it has been asserted also that the Persian word is derived from

---

[1] Sui šu, Ch. 83, p. 7 b.

[2] Ch. 185, p. 19.

[3] It is not a Tibetan name, as supposed by ROEDIGER and POTT (Z. f. K. Morg., Vol. IV, p. 268).

[4] These various attempts at spelling show plainly that the term has the status of a loan-word, and that the Sanskrit term has nothing to do with the name of the people who may have supplied the product, the Τάγγανοι in the Himalaya of Ptolemy (YULE, Hobson-Jobson, p. 923). How should borax be found in the Himalaya!

[5] KLAPROTH, Mémoires relatifs à l'Asie, Vol. III, p. 347.

[6] See, further, T'oung Pao, 1914, pp. 88–89.

[7] D. HANBURY, Science Papers, pp. 217, 276.

503

the Chinese. F. DE MÉLY[1] argues that nao-ša is written ideographically, and that the text of the *Pen ts'ao kaṅ mu* adds, "Il vient de la province de Chen-si; on le tire d'une montagne d'où il sort continuellement des vapeurs rouges et dangereuses et très difficile à aborder par rapport à ces mêmes vapeurs. Il en vient aussi de la Tartarie, on le tire des plaines où il y a beaucoup de troupeaux, de la même façon que le salpêtre de houssage; les Tartares et gens d'au delà de la Chine salent les viandes avec ce sel." Hence F. de Mély infers that the Persians, on their part, borrowed from the Chinese their nao-ša, to which they added the ending *dzer*, as in the case of the bezoar styled in Persian *badzeher*.[2]

The case, however, is entirely different. The term nao-ša is written phonetically, not ideographically, as shown by the ancient transcription 鐃沙 in the Sui Annals (see below) and the variant 磠砂 (properly *nuṅ-ša*, but indicated with the pronunciation nao-ša);[3] also the synonymes *ti yen* 狄鹽 ("salt of the barbarians") and *Pei-t'iṅ ša* 北庭砂 ("ore of Pei-t'iṅ," in Turkistan), which appear as early as the Sung period in the *T'u kiṅ pen ts'ao* of Su Suṅ, allude to the foreign origin of the product. The term is thus plainly characterized as a foreign loan in the *Pen ts'ao kaṅ mu*. This, further, is brought out by the history of the subject. The word is not found in any ancient Chinese records. The Chinese learned about nao-ša in Sogdiana and Kuča for the first time during the sixth century A.D. The *Pen ts'ao* of the T'ang period is the earliest pharmacopœia that mentions it. Su Kuṅ 蘇恭, the reviser of this work, and the author of the *Čeṅ lei pen ts'ao*, know of but one place of provenience, the country of the Western Žuṅ 西戎 (F. de Mély's "Tartary"). It is only Su Suṅ 蘇頌 of the Sung period, who in his *T'u kiṅ pen ts'ao* remarks, "At present it occurs also in Si-liaṅ and in the country Hia [Kan-su] as well as in Ho-tuṅ [Šan-si], Šen-si, and in the districts of the adjoining regions" 今西涼夏國及河東陝西近邊州郡亦有之 [note the additions of 今 "at present" and 亦 "also"]. And he hastens to add, "However (然), the pieces coming from the Western Žuṅ are clear and bright, the largest having the size of a fist and being from three to five ounces in weight, the smallest

---

[1] L'Alchimie chez les Chinois (*Journal asiatique*, 1895, II, p. 338) and Lapidaire chinois, p. LI.

[2] All this is rather lack of criticism or poor philology. The Persian word in question is *pāzahr*, literally meaning "antidote" (see below, p. 525). Neither this word nor *nušadir* has an ending like *dzer*, and there is no analogy between the two.

[3] According to the *Pie pen ču* 別本注, cited in the *Čeṅ lei pen ts'ao* (Ch. 5, p. 10, ed. of 1587), the transcription *nuṅ-ša* should represent the pronunciation of the Hu people; that is, Iranians. Apparently it was an Iranian dialectic variation with a nasalized vowel *u*. It is indicated as a synonyme of nao-ša in the *Ši yao er ya* of the T'ang period (see Beginnings of Porcelain, p. 115).

reaching the size of a finger and being used for medical purposes."[1]
It is accordingly the old experience that the Chinese, as soon as they
became acquainted with a foreign product, searched for it on their own
soil, and either discovered it there, or found a convenient substitute.
In this case, Su Sun plainly indicates that the domestic substitute was
of inferior quality; and there can be no doubt that this was not sal
ammoniac, which is in fact not found in China, but, as has been demonstrated by D. HANBURY,[2] chloride of sodium. As early as the eighteenth
century it was stated by M. COLLAS[3] that no product labelled *nao-ša*
in Peking had any resemblance to our sal ammoniac.

H. E. STAPLETON,[4] author of a very interesting study on the employment of sal ammoniac in ancient chemistry, has hazarded an etymo-
logical speculation as to the term *nao-ša*. Persian *nūšādur* appears to
him to be the Chinese word *nau-ša*, suffixed by the Persian word *dārū*
("medicine"),[5] and the Sanskrit *navasāra* would also seem to be simply
the Chinese name in a slightly altered form. H. E. Stapleton is a
chemist, not a philologist; it therefore suffices to say that these speculations, as well as his opinion "that the syllables *nau-ša* appear to be
capable of complete analysis into Chinese roots,"[6] are impossible.

The Hindustānī name can by no means come into question as the
prototype of the Chinese term, as proposed by F. P. SMITH[7] and T.
WATTERS;[8] for the Chinese transcription was framed as early as the
sixth century A.D., when Hindustānī was not yet in existence. The
Hindustānī is simply a Persian loan-word of recent date, as is
likewise Neo-Sanskrit *naiçadala;* while Sanskrit *navasāra, navasādara,*
or *narasāra,* the vacillating spelling of which betrays the character
of a loan-word, is traceable to a more ancient Iranian form (see
below).

In the *Sui šu*[9] we meet the term in the form 磠沙 *nao-ša,* stated to

---

[1] See also *Pen ts'ao yen i,* Ch. 6, p. 4 b (ed. of Lu Sin-yūan).

[2] Science Papers, pp. 217, 276.

[3] Mémoires concernant les Chinois, Vol. XI, 1786, p. 330.

[4] Sal Ammoniac: a Study in Primitive Chemistry (*Memoirs As. Soc. Bengal,*
Vol. I, 1905, pp. 40–41).

[5] He starts from the popular etymology *nūš dārū* ("life-giving medicine"),
which, of course, is not to be taken seriously.

[6] Even if this were the case, it would not tend to prove that the word is of
Chinese origin. As is now known to every one, there is nothing easier to the Chinese
than to transcribe a foreign word and to choose such characters as will convey a
certain meaning.

[7] Contributions toward the Materia Medica of China, p. 190.

[8] Essays on the Chinese Language, p. 350.

[9] Ch. 83, pp. 4 b and 5 b.

be a product of K'an (Sogdiana) and Kuča.[1] The fact that this transcription is identical with 碙 we recognize from the parallel passage in the *Pei ši*,[2] where it is thus written. The text of the Sui Annals with reference to Iranian regions offers several such unusual modes of writing, where the *Pei ši* has the simple types subsequently adopted as the standard. The variation of the Sui Annals, at all events, demonstrates that the question is of reproducing a foreign word; and, since it hails from Sogdiana, there can be no doubt that it was a word of the Sogdian language of the type *navša or *nafša (cf. Sanskrit *navasāra*, Armenian *navt'*, Greek νάφθα); Persian *našādir*, *nušādir*, *naušādir*, *naušādur*, *nōšādur*, being a later development. It resulted also in Russian *nušatyr*. In my opinion, the Sogdian word is related to Persia *neft* ("naphta"), which may belong to Avestan *napta* ("moist").[3]

Tribute-gifts of *nao-ša* are not infrequently mentioned in the Chinese Annals. In A.D. 932, Wan Žen-mei 王仁美, Khan of the Uigur, presented to the Court among other objects *ta-p'en ša* ("borax")[4] and sal ammoniac (*kan ša*).[5] In A.D. 938 Li Šen-wen 李聖文, king of Khotan, offered *nao-ša* and *ta-p'en ša* ("borax") to the Court; and in A.D. 959 jade and *nao-ša* were sent by the Uigur.[6] The latter event is recorded also in the *Kiu Wu Tai ši*,[7] where the word is written 硇砂, phonetically *kan-ša*, but apparently intended only as a graphic variant for *nao-ša*.[8] The same work ascribes sal ammoniac (written in the same manner) to the T'u-fan (Tibetans) and the Tan-hian (a Tibetan tribe in the Kukunōr region).[9] In the T'ang period the substance was well

---

[1] According to Masūdi (BARBIER DE MEYNARD, Les Prairies d'or, Vol. I, p. 347), sal-ammoniac mines were situated in Soghd, and were passed by the Mohammedan merchants travelling from Khorasan into China. Kuča still yields sal ammoniac (A. N. KUROPATKIN, Kashgaria, pp. 27, 35, 76). This fact is also noted in the *Hui k'ian či* (Ch. 2), written about 1772 by two Manchu officials, Fusambô and Surde, who locate the mine 45 *li* west of Kuča in the Šartatsi Mountains, and mention a red and white variety of sal ammoniac. Cf. also M. REINAUD, Relation des voyages faits par les Arabes et les Persans dans l'Inde et à la Chine, Vol. I, p. CLXIII.

[2] Ch. 97, p. 12.

[3] Cf. P. HORN, Neupersische Etymologie, No. 1035; H. HÜBSCHMANN, Persische Studien, p. 101, and Armen. Gram., p. 100.

[4] As I have shown on a former occasion (*T'oung Pao*, 1914, p. 88), Chinese *p'en* (*bun) is a transcription of Tibetan *bul*.

[5] *Ts'e fu yüan kwei*, Ch. 972, p. 19.

[6] *Wu Tai hui yao*, Chs. 28, p. 10 b; and Ch. 29, p. 13 b (ed. of *Wu yin tien*).

[7] Ch. 138, p. 3.

[8] The character *kan* is not listed in K'an-hi's Dictionary.

[9] Ch. 138, pp. 1 b, 3 a.

known. The *Ši yao er ya*[1] gives a number of synonymes of Chinese origin, as *kin tsei* 金賊, *č'i ša* 赤砂 ("red gravel"), *pai hai tsin* 白海精 ("essence of the white sea").

Sal ammoniac is found in Dimindān in the province of Kirmān. Yāqūt (1179–1229) gives after Ibn al-Faqīh (tenth century) a description of how *nušādir* is obtained there, which in the translation of C. BARBIER DE MEYNARD[2] runs as follows:—

"Cette substance se trouve principalement dans une montagne nommée Donbawend, dont la hauteur est évaluée à 3 farsakhs. Cette montagne est à 7 farsakhs de la ville de Guwašir. On y voit une caverne profonde d'où s'échappent des mugissements semblables à ceux des vagues et une fumée épaisse. Lorsque cette vapeur, qui est le principe du sel ammoniac, s'est attachée aux parois de l'orifice, et qu'une certaine quantité s'est solidifiée, les habitants de la ville et des environs viennent la recueillir, une fois par mois ou tous les deux mois. Le sulthan y envoie des agents qui, la récolte faite, en prélèvent le cinquième pour le trésor; les habitants se partagent le reste par la voie du sort. Ce sel est celui qu'on expédie dans tous les pays."

Ibn Haukal describes the mines of Setrušteh thus:[3] "The mines of sal ammoniac are in the mountains, where there is a certain cavern, from which a vapor issues, appearing by day like smoke, and by night like fire. Over the spot whence the vapor issues, they have erected a house, the doors and windows of which are kept so closely shut and plastered over with clay that none of the vapor can escape. On the upper part of this house the copperas rests. When the doors are to be opened, a swiftly-running man is chosen, who, having his body covered over with clay, opens the door; takes as much as he can of the copperas, and runs off; if he should delay, he would be burnt. This vapor comes forth in different places, from time to time; when it ceases to issue from one place, they dig in another until it appears, and then they erect that kind of house over it; if they did not erect this house, the vapor would burn, or evaporate away."

Taxes are still paid in this district with sal ammoniac. Abu Mansur sets forth its medicinal properties.[4]

---

[1] See Beginnings of Porcelain (this volume, p. 115).

[2] Dictionnaire géographique de la Perse, p. 235 (Paris, 1861). Ibn al-Faqīh's text is translated by P. SCHWARZ (Iran im Mittelalter, p. 252). According to Ibn Haukal (W. OUSELEY, Oriental Geography of Ebn Haukal, p. 233), sal-ammoniac mines were located in Maweralnahr (Transoxania).

[3] W. OUSELEY, *op. cit.*, p. 264.

[4] ACHUNDOW, Abu Mansur, p. 144.—ABEL-RÉMUSAT (Mélanges asiatiques, Vol. I, p. 209, 1825), translating from the Japanese edition of the cyclopædia *San ts'ai t'u hui*, gave the following interesting account: "Le sel nommé (en chinois)

The Tibetans appear to have received sal ammoniac from India, as shown at least by their term *rgya ts'wa* ("Indian salt"), literally translated into Mongol *Änätkäk dabusu*. Mongol *Änätkäk* is a reproduction of Chinese *In-duk-kwok ("country of India"). The informants of M. COLLAS[1] stated that the *nao-ša* of the Peking shops came from Tibet or adjacent places. Lockhart received in Peking the information that it is brought from certain volcanic springs in Se-č'wan and in Tibet.[2]

80. 密陀僧 *mi-t'o-seṅ*, *m'it(m'ir)-da-saṅ, and 沒多僧 *mu-to-seṅ*, *mut(mur)-ta-saṅ, litharge, dross of lead, is an exact reproduction of Persian *mirdāsang* or *murdāsang* of the same meaning.[3] Both transcriptions are found in the *Pen ts'ao* of the T'ang dynasty, written about the middle of the seventh century.[4] Therefore we are entitled to extend the Persian word into the period of Middle Persian. Su Kuṅ, the reviser of the *T'aṅ pen ts'ao*, states expressly that both *mi-t'o* and *mu-to* are words from the language of the Hu or Iranians (胡言也), and that the substance comes from or is produced in Persia, being in shape like the teeth of the yellow dragon, but stronger and heavier; there is also some of white color with veins as in Yün-nan marble. Su Suṅ of the Sung period says that then ("at present") it was also found

---

nao-cha (en persan *nouchader*) et aussi sel de Tartarie, sel volatil, se tire de deux montagnes volcaniques de la Tartarie centrale; l'une est le volcan de Tourfan, qui a donné à cette ville (ou pour mieux dire à une ville qui est située à trois lieues de Tourfan, du côté de l'est) le nom de Ho-tcheou, ville de feu; l'autre est la montagne Blanche, dans le pays de Bisch-balikh; ces deux montagnes jettent continuellement des flammes et de la fumée. Il y a des cavités dans lesquelles se ramasse un liquide verdâtre. Exposé à l'air, ce liquide se change en un sel, qui est le *nao-cha*. Les gens du pays le recueillent pour s'en servir dans la préparation des cuirs. Quant à la montagne de Tourfan, on en voit continuellement sortir une colonne de fumée; cette fumée est remplacée le soir par une flamme semblable à celle d'un flambeau. Les oiseaux et les autres animaux, qui en sont éclairés, paraissent de couleur rouge. On appelle cette montagne le Mont-de-Feu. Pour aller chercher le *nao-cha*, on met des sabots, car des semelles de cuir seraient trop vite brûlées. Les gens du pays recueillent aussi les eaux-mères qu'ils font bouillir dans des chaudières, et ils en retirent le sel ammoniac, sous la forme de pains semblables à ceux du sel commun. Le *nao-cha* le plus blanc est réputé le meilleur; la nature de ce sel est très-pénétrante. On le tient suspendu dans une poêle au-dessus du feu pour le rendre bien sec; on y ajoute du gingembre pour le conserver. Exposé au froid ou à l'humidité, il tombe en déliquescence, et se perd." Waṅ Yen-te, who in A.D. 981 was sent by the Chinese emperor to the ruler of Kao-č'aṅ, was the first to give an account of the sal-ammoniac mountain of Turkistan (BRETSCHNEIDER, Mediæval Researches, Vol. II, p. 190). See also F. DE MÉLY, Lapidaire chinois, p. 140; W. SCHOTT, Zur Uigurenfrage, II, p. 45 (*Abh. Berl. Akad.*, 1875) and Ueber ein chinesisches Mengwerk (*ibid.*, 1880, p. 6); GEERTS, Produits, p. 322.

[1] Mémoires concernant les Chinois, Vol. XI, p. 331.

[2] D. HANBURY, Science Papers, p. 277.

[3] Cf. HÜBSCHMANN, Armen. Gram., p. 270.

[4] *Čeṅ lei pen ts'ao*, Ch. 4, p. 31; and *Pen ts'ao kaṅ mu*, Ch. 8, p. 8 b.

in the silver and copper foundries of Kwaṅ-tuṅ and Fu-kien. It is further mentioned briefly in the *Pen ts'ao yen i* of 1116,[1] which maintains that the kind with a color like gold is the best.

According to Yāqūt, mines of antimony, known under the name *razi*, litharge, lead, and vitriol, were in the environs of Donbawend or Demawend in the province of Kirmān.[2] In the Persian pharmacopœia of Abu Mansur, the medicinal properties of litharge are described under the Arabicized name *murdāsanj*, to which he adds the synonymous term *murtak*.[3] Pegoletti, in the fourteenth century, gives the word with a popular etymology as *morda sangue*.[4] The Dictionary of Four Languages[5] correlates Chinese *mi-t'o-seṅ* with Tibetan *gser-zil* (literally, "gold brightness"),[6] Manchu *čirčan*, and Mongol *jildunur*.[7]

81. PALLADIUS[8] offers a term 紫磨金 *tse-mo kin* with the meaning "gold from Persia," no source for it being cited. In the *Pen ts'ao kaṅ mu*,[9] the *tse-mo kin* of Po-se (Persia) is given as the first in a series of five kinds of gold of foreign countries,[10] without further explanation. The term occurs also in Buddhist literature: CHAVANNES[11] has found it in the text of a Jātaka, where he proposes as hypothetical translation, "un amas d'or raffiné rouge." It therefore seems to be unknown what the term signifies, although a special kind of gold or an alloy of gold is apparently intended. The *Šwi kiṅ ču* 水經注[12] says that the first quality of gold, according to Chinese custom, is styled *tse-mo kin* (written as above); according to the custom of the barbarians, however, *yaṅ-mai* 陽邁. From this it would appear that *tse-mo* is a Chinese term, not a foreign one.

---

[1] Ch. 5, p. 6 b (ed. of Lu Sin-yūan).

[2] BARBIER DE MEYNARD, *op. cit.*, p. 237.

[3] ACHUNDOW, Abu Mansur, p. 139. This form goes back to Middle Persian *murtak* or *martak*.

[4] YULE, Cathay, new ed., Vol. III, p. 167.

[5] Ch. 22, p. 71.

[6] JAESCHKE, in his Tibetan Dictionary, was unable to explain this term.

[7] KOVALEVSKI, in his Mongol Dictionary, explains this word wrongly by "mica."

[8] Chinese-Russian Dictionary, Vol. II, p. 203.

[9] Ch. 8, p. 1 b.

[10] The four others are, the dark gold of the eastern regions, the red gold of Lin-yi, the gold of the Si-žuṅ, and the gold of Čan-č'eṅ (Camboja). The five kinds of foreign gold are mentioned as early as the tenth century in the *Pao ts'aṅ lun* 寶藏論.

[11] Fables et contes de l'Inde, in Actes du XIV⁰ Congrès des Orientalistes, Vol. I, 1905, p. 103.

[12] Ch. 36, p. 18 b (ed. Wu-č'aṅ, 1877). See p. 622.

The *Ko ku yao lun*[1] has a notice of *tse kin* 紫金 ("purple gold")
as follows: "The ancients say that the *pan-liaṅ* 半 兩 money[2] is *tse
kin*. The people of the present time make it by mixing copper with
gold, but our contemporaries have not yet seen genuine *tse kin*."
The same alloy is mentioned as a product of Ma-k'o-se-li in the
Tao i či lio, written in 1349 by Waṅ Ta-yüan.[3] I am not sure, of
course, that this *tse kin* is identical with *tse-mo kin*.

In the same manner as the Chinese speak of foreign gold, they also
offer a series of foreign silver. There are four kinds; namely, silver of
Sin-ra (in Korea), silver of Po-se (Persia), silver of Lin-yi, and silver
of Yün-nan. Both gold and silver are enumerated among the products
of Sasanian Persia. The *Hai yao pen ts'ao* cites the *Nan yüe či* of the
fifth century to the effect that the country Po-se possesses a natural
silver-dust 銀 屑, employed as a remedy, and that remedies are tested
by means of finger-rings.[4] Whether Persia is to be understood here
seems doubtful to me. Gold-dust is especially credited to the country
of the Arabs.[5]

82. 鹽綠 *yen-lü* ("the green of salt," various compositions with
copper-oxide) is mentioned as a product of Sasanian Persia[6] and of
Kuča.[7] Su Kuṅ of the T'ang (seventh century) points it out as a product
of Karašar (Yen-či 焉 耆), found in the water on the lower surface of
stones. Li Sün, who wrote in the second half of the eighth century,
states that "it is produced in the country Po-se (Persia) adhering to
stones, and that the kind imported on ships is called *ši-lü* 石 綠 ('the
green of the stone'); its color is resistant for a long time without changing; the imitation made in China from copper and vinegar must not
be employed in the pharmacopœia, nor does it retain its color long."
Li Ši-čen employs the term "green salt of Po-se."[8] The substance was
employed as a remedy in eye-diseases.

This is Persian *zingār* (Arabic *zinjar*), described in the stone-book
of Pseudo-Aristotle as a stone extracted from copper or brass by means

---

[1] Ch. 6, p. 12 b.

[2] See Beginnings of Porcelain, p. 83.

[3] ROCKHILL, *T'oung Pao*, 1915, p. 622.

[4] *Čeṅ lei pen ts'ao*, Ch. 4, p. 23.

[5] *Ibid.*, Ch. 4, p. 21 b.

[6] *Sui šu*, Ch. 83, p. 7 b.

[7] *Čou šu*, Ch. 50, p. 5; *Sui šu*, Ch. 83, p. 5 b.

[8] Cf. also GEERTS, Produits, p. 634; F. DE MÉLY, Lapidaire chinois, pp. 134,
243. According to Geerts, the term is applied in Japan to acetate of copper, formerly
imported, but now prepared in the country.

of vinegar, and employed as an ingredient in many remedies for eye-diseases.[1]

83. The Emperor Yan (A.D. 605–616) of the Sui dynasty, after his succession to the throne, despatched Tu Han-man 杜 行 滿 to the Western Countries. He reached the kingdom of Nan 安 (Bukhārā), obtained manicolored salt (*wu se yen*), and returned.[2] Istaxrī relates that in the district of Dārābejird there are mountains of white, yellow, green, black, and red salts; the salt in other regions originates from the interior of the earth or from water which forms crystals; this, however, is salt from mountains which are above the ground. Ibn Haukal adds that this salt occurs in all possible colors.[3]

The *Pei hu lu*[4] distinguishes red, purple, black, blue, and yellow salts. *Č'i yen* 赤 鹽 ("red salt") like vermilion, and white salt like jade, are attributed to Kao-č'an (Turfan).[5] Black salt (*hei yen*) was a product of the country Ts'ao (Jāguḍa) north of the Ts'un-lin.[6] It is likewise attributed to southern India.[7] These colored salts may have been impure salt or minerals of a different origin.

84. 鍮石 *t'ou-ši* is mentioned as a metallic product of Sasanian Persia (enumerated with gold, silver, copper, *pin*, iron, and tin) in the *Sui šu*.[8] It is further cited as a product of Nü kwo, the Women's Realm south of the Ts'un-lin;[9] of A-lo-yi-lo 阿 羅 伊 羅 in the north of Uḍḍi-yāna,[10] and of the Arabs (Ta-ši).[11] Hüan Tsan's Memoirs contain the term three times, once as a product found in the soil of northern India (together with gold, silver, copper, and iron), and twice as a material from which Buddhist statues were made.[12] According to the *Kin č'u*

---

[1] J. RUSKA, Steinbuch des Aristoteles, p. 182; and Steinbuch des Qazwīnī, p. 25.

[2] *Sui šu*, Ch. 83, p. 4 b.

[3] P. SCHWARZ, Iran, p. 95.

[4] Ch. 2, p. 11 (ed. of Lu Sin-yüan).

[5] *Sui šu*, Ch. 83, p. 3 b. In the *T'ai p'in hwan yü ki* (Ch. 180, p. 11 b) the same products are assigned to Kü-ši 車 師 (Turfan).

[6] *Sui šu*, Ch. 83, p. 8.

[7] *T'an šu*, Ch. 221 A, p. 10 b.

[8] Ch. 83, p. 7 b.

[9] *T'ai p'in hwan yü ki*, Ch. 186, p. 9.

[10] *Ibid.*, p. 12 b.

[11] *Ibid.*, p. 15 b.

[12] Cf. S. JULIEN, Mémoires sur les contrées occidentales, Vol. I, pp. 37, 189, 354. Julien is quite right in translating the term by *laiton* ("brass"). PALLADIUS (Chinese-Russian Dictionary, Vol. II, p. 16) explains it as "brass with admixture of lead, possessing attractive power." The definition of Giles ("rich ore brought from Persia supposed to be an ore of gold and copper, or bronze") is inexact. *T'ou-*

swi ši ki 荆楚歲時記, written in the sixth century, the needles used by women on the festival of the seventh day of the seventh month[1] were made of gold, silver, or *t'ou-ši*.[2] Under the T'ang, *t'ou-ši* was an officially adopted alloy, being employed, for instance, for the girdles of the officials of the eighth and ninth grades.[3] It was sent as tribute from Iranian regions; for instance, in A.D. 718, from Māimargh (northwest of Samarkand).[4]

The *Ko ku yao lun* states, "*T'ou-ši* is the essence of natural copper. At present zinc-bloom is smelted to make counterfeit *t'ou*. According to Ts'ui Fan 崔昉, one catty of copper and one catty of zinc-bloom will yield *t'ou-ši*. The genuine *t'ou* is produced in Persia. It looks like gold, and, when fired, assumes a red color which will never turn black." This is clearly a description of brass which is mainly composed of copper and zinc. Li Ši-čen[5] identifies *t'ou-ši* with the modern term *hwaṅ t'uṅ* ("yellow copper"); that is, brass. According to T'an Ts'ui,[6] *t'ou-ši* is found in the Č'ŏ-li 車里 t'u-se of Yün-nan.

The Chinese accounts of *t'ou* or *t'ou-ši* agree with what the Persians and Arabs report about *tūtiya*. It was in Persia that zinc was first mined, and utilized for a new copper alloy, brass. Ibn al-Faqīh, who wrote about A.D. 902, has left a description of the zinc-mines situated in a mountain Dunbāwand in the province of Kirmān. The ore was (and still is) a government monopoly.[7] Jawbarī, who wrote about 1225, has described the process of smelting.[8] The earliest mention of the term occurs in the Arabic stone-book of Pseudo-Aristotle (ninth century),[9] where the stone *tūtiyā* is explained as belonging to the stones found in mines, with numerous varieties which are white, yellow, and green;

ši is only said to resemble gold, and the notion that brass resembles gold turns up in all Oriental writers. See also BEAL, Records of the Western World, Vol. I, p. 51; and CHAVANNES (*T'oung Pao*, 1904, p. 34), who likewise accepts the only admissible interpretation, "brass."

[1] Cf. W. GRUBE, Zur Pekinger Volkskunde, p. 76; J. PRZYLUSKI, *T'oung Pao*, 1914, p. 215.

[2] *P'ei wen yün fu*, Ch. 100 A, p. 25.

[3] Jade, p. 286; cf. also *Ta T'aṅ leu tien*, Ch. 8, p. 22.

[4] CHAVANNES, *T'oung Pao*, 1904, p. 34.

[5] *Pen ts'ao kaṅ mu*, Ch. 8, pp. 3 and 4. Cf. also GEERTS, Produits, p. 575.

[6] *Tien hai yü heṅ či*, Ch. 2, p. 3 b.

[7] P. SCHWARZ, Iran im Mittelalter, p. 252.

[8] G. FERRAND, Textes relatifs à l'Extrême-Orient, p. 610 (cf. also pp. 225, 228; and LECLERC, Traité des simples, Vol. I, p. 322).

[9] J. RUSKA, Steinbuch des Aristoteles, p. 175. J. BECKMANN (Beyträge zur Geschichte der Erfindungen, Vol. III, p. 388) states that the word first occurs in Avicenna of the eleventh century.

the quarries are located on the shores of Hind and Sind. This is probably intended for vitriol or sulphate of copper.[1]

In Chinese *t'ou-ši*, the second element *ši* ("stone") does not form part of the transcription; the term means simply "*t'ou* stone," and *t'ou* (**tu*) reproduces the first syllable of Persian *tūtiya*, which, on the basis of the Sui Annals, we are obliged to assign also to the Middle-Persian language. To derive the Chinese word from Turkish *tūj*, as proposed by WATTERS,[2] and accepted without criticism by HIRTH,[3] is utterly impossible. The alleged Turkish word occurs only in Osmanli and other modern dialects, where it is plainly a Persian loan-word, but not in Uigur, as wrongly asserted by Hirth. This theory seems to imply that the element *ši* should form part of the transcription; this certainly is out of the question, as 石 represents ancient **šek or **sak, **zak, and could not reproduce a palatal. For the rest, the Chinese records point to Iran, not to the Turks, who had no concern whatever with the whole business.[4] Two variations of the Persian word have penetrated into the languages of Europe. The Arabs carried their *tūtiyā* into Spain, where it appears as *atutia* with the Arabic article; in Portuguese we have *tutia*, in French *tutie*, in Italian *tuzia*, in English *tutty*. A final palatal occurs in the series Osmanli *tuj* or *tunč*, Neo-Greek τούντζι, Albanian *tuč*, Serbian and Bulgarian *tuč*, Rumanian *tuciŭ*. Whether Sanskrit *tuttha*, as has been assumed, is to be connected with the Persian word, remains doubtful to me: the Sanskrit word refers only to green or blue vitriol.[5] It is noteworthy that Persian *birinj* ("brass"), a more recent variant of *piriṅ* (Kurd *pirinjok*, Armenian *plinj*),[6] has not migrated into any foreign language, for I am far from being convinced that our word "bronze" should be traceable to this type.[7]

The Japanese pronunciation of 鍮 石 is *čūseki*. The Japanese used

---

[1] A curious error occurs in FELDHAUS' Technik (col. 1367), where it is asserted, "Qazwīnī says about 600 that zinc is known in China, and could also be made flexible there." Qazwīnī wrote his cyclopædia in 1134, and says nothing about zinc in China (cf. RUSKA, Steinbuch des Qazwīnī, p. 11); but he mentions a *tūtiyā* mine in Spain (G. JACOB, Studien in arabischen Geographen, p. 13).

[2] Essays on the Chinese Language, p. 359.

[3] Chau Ju-kua, p. 81. *T'ou-ši* does not mean "white copper" in the passage under notice, but means "brass." "White copper" is a Chinese and quite different alloy (see below, p. 555).

[4] It is likewise odd to connect Italian *tausia* (properly *taunia*) and German *tauschieren* with this word. This is just as well as to derive German *tusche* from an alleged Chinese *t'u-se* (HIRTH, Chines. Studien, p. 226).

[5] P. C. RAY, History of Hindu Chemistry, 2d ed., Vol. II, p. 25.

[6] HÜBSCHMANN, Persische Studien, p. 27.

[7] O. SCHRADER, Sprachvergleichung und Urgeschichte, Vol. II, p. 73.

to import the alloy from China, and their Honzŏ (*Pen ts'ao*) give formulas for its preparation.[1] The Koreans read the same word *not* or *not-si*. The French missionaries explain it as "composition de différents métaux qui sert à faire les cuillères, etc. Airain, cuivre jaune (première qualité). Cuivre rouge et plomb."[2]

The history of zinc in the East is still somewhat obscure; at least, it so appears from what the historians of the metal have written about the subject. I quote from W. R. INGALLS:[3] "It is unknown to whom is due the honor of the isolation of zinc as a metal, but it is probable that the discovery was first made in the East. In the sixteenth century zinc was brought to Europe from China and the East Indies under the name of tutanego (whence the English term tutenegue), and it is likely that knowledge of it was obtained from that source at an earlier date. . . . The production of zinc on an industrial scale was first begun in England; it is said that the method applied was Chinese, having been introduced by Dr. Isaac Lawson, who went to China expressly to study it. In 1740 John Champion erected works at Bristol and actually began the manufacture of spelter, but the production was small, and the greater part used continued to come from India and China." The fact that in the eighteenth century the bulk of zinc which came to Europe was shipped from India is also emphasized by J. BECKMANN,[4] who, writing in 1792, regretted that it was then unknown where, how, and when this metal was obtained in India, and in what year it had first been brought over to Europe. According to the few notices of the subject, he continues, it originates from China, from Bengal, from Malakka, and from Malabar, whence also copper and brass are obtained. On the other hand, W. AINSLIE[5] states that by far the greater part of zinc which is met with in India is brought from Cochin-China or China, where both the calamine and blende are common. Again, S. JULIEN[6] informs us that zinc is not mentioned in ancient books, and appears to have been known in China only from the beginning of the seventeenth century.

W. HOMMEL[7] pleaded for the origin of zinc-production in India, whence it was obtained by the Chinese. He does not know, of course, that there is no evidence for such a theory in Chinese sources. The

---

[1] GEERTS, Produits, p. 641; F. DE MÉLY, Lapidaire chinois, p. 42.

[2] Dictionnaire coréen-français, p. 291.

[3] Production and Properties of Zinc, pp. 2–3 (New York and London, 1902).

[4] *Op. cit.*, Vol. III, p. 408.

[5] Materia Indica, Vol. I, p. 573.

[6] Industries de l'empire chinois, p. 46.

[7] *Chemiker-Zeitung*, 1912, p. 905.

Indian hypothesis, I believe, has been accepted by others. In my opinion, the art of zinc-smelting originated neither in India nor in China, but in Persia. We noted from Ibn al-Faqīh that the zinc-mines of Kirmān were wrought in the tenth century; and the early Chinese references to *t'ou-ši* would warrant the conclusion that this industry was prominent under the Sasanians, and goes back at least to the sixth century.

Li Ši-čen[1] states that the green copper of Persia can be wrought into mirrors. I have no other information on this metal.

85. 鑌 or 鑌鐵 *pin t'ie, pin* iron, is mentioned as a product of Sasanian Persia,[2] also ascribed to Ki-pin (Kashmir).[3] Mediæval authors like Č'aṅ Te mention it also for India and Hami.[4] The *Ko ku yao lun*[5] says that *pin t'ie* is produced by the Western Barbarians (Si Fan), and that its surface exhibits patterns like the winding lines of a conch or like sesame-seeds and snow. Swords and other implements made from this metal are polished by means of gold threads, and then these patterns become visible; the price of this metal exceeds that of silver. This clearly refers to a steel like that of Damascus, on which fine dark lines are produced by means of etching acids.[6]

Li Ši-čen[7] states that *pin t'ie* is produced by the Western Barbarians (Si Fan), and cites the *Pao ts'aṅ lun* 寶藏論, by Hien Yüan-šu 軒轅述 of the tenth century, to the effect that there are five kinds of iron, one of these being *pin t'ie*, which is so hard and sharp that it can cut metal and hard stone. K'aṅ-hi's Dictionary states that *pin* is wrought into sharp swords. Previous investigators have overlooked the fact that this metal is first mentioned for Sasanian Persia, and have merely pointed to the late mediæval mention in the Sung Annals.[8]

The word *pin* has not yet been explained. Even the Pan-Turks have not yet discovered it in Turkish. It is connected with Iranian *spaina, Pamir languages *spin*, Afghan *ōspīna* or *ōspana*, Ossetic *äfsän*.[9] The

---

[1] *Pen ts'ao kaṅ mu,* Ch. 8, p. 3 b.

[2] *Čou šu,* Ch. 50, p. 6; *Sui šu,* Ch. 83, p. 7 b.

[3] *T'ai p'iṅ hwan yü ki,* Ch. 182, p. 12 b.

[4] BRETSCHNEIDER, Mediæval Researches, Vol. I, p. 146; *Kwaṅ yü ki,* Ch. 24, p. 5 b.

[5] Ch. 6, p. 14 b (ed. of *Si yin hūan ts'uṅ šu*).

[6] A reference to *pin t'ie* occurs also in the *Šan kü sin hwa* 山居新話, written by Yaṅ Yü 楊瑀 in 1360 (p. 19, ed. of *Či pu tsu čai ts'uṅ šu*).

[7] *Pen ts'ao kaṅ mu,* Ch. 8, p. 11 b.

[8] BRETSCHNEIDER, On the Knowledge possessed by the Chinese of the Arabs, p. 12, and *China Review,* Vol. V, p. 21; W. F. MAYERS, *China Review,* Vol. IV, p. 175.

[9] HÜBSCHMANN, Persische Studien, p. 10.

character *pin* has been formed *ad hoc*, and, as already remarked by Mayers, is written also without the classifier; that is, in a purely phonetic way.

86. 瑟瑟 *se-se*, *sit-sit (Japanese *šitsu-šitsu*), hypothetical restoration *sirsir, a precious stone of Sasanian Persia, which I have discussed at some length in my "Notes on Turquois in the East" (pp. 25–35, 45–55, 67–68). For this reason only a brief summary is here given, with some additional information and corrections. I no longer believe that *se-se* might be connected with Shignan (p. 47) or Arabic *jaza* (p. 52), but am now convinced that *se-se* represents the transcription of an Iranian (most probably Sogdian) word, the original of which, however, has not yet been traced. Chinese records leave us in the dark as to the character of the Iranian *se-se*. It is simply enumerated in a list of precious stones of Persia and Sogdiana (K'añ).[1] The T'ang Annals locate the *se-se* mines to the south-east of the Yaxartes in Sogdiana;[2] and the stones were traded to China by way of Khotan.[3] Possibly the Nestorians were active in bringing to China these stones which were utilized for the decoration of their churches. The same history ascribes columns of *se-se* to the palaces of Fu-lin (Syria);[4] in this case the question is of a building-stone. In ancient Tibet, *se-se* formed part of the official costume, being worn by officials of the highest rank in strings suspended from the shoulder. The materials ranking next to this stone were gold, plated silver, silver, and copper,[5] — a clear index of the fact that *se-se* was regarded in Tibet as a precious stone of great value, and surpassing gold. The Tibetan women used to wear beads of this stone in their tresses, and a single bead is said to have represented the equivalent of a noble horse.[6] Hence arose the term *ma kia ču* 馬價珠 ("pearl or bead equalling a horse in price"). These beads are treated in the *Ko ku yao lun*[7] as a separate item, and distinct from turquois.[8]

In the T'ang period, *se-se* stones were also used as ornaments by the

---

[1] *Pei ši*, Ch. 97, pp. 7 b, 12; *Čou šu*, Ch. 50, p. 6; *Sui šu*, Ch. 83, p. 7 b; *Wei šu*, Ch. 102, pp. 5 a, 9 b.

[2] *T'añ šu*, Ch. 221 B, p. 2 b.

[3] *T'añ šu*, Ch. 221 A, p. 10 b.

[4] *Kiu T'añ šu*, Ch. 198, p. 11 b; *T'añ šu*, Ch. 221 B, p. 7 b.

[5] *T'añ šu*, Ch. 216 A, p. 1 b (not in *Kiu T'añ šu*).

[6] *Sin Wu Tai ši*, Ch. 74, p. 4 b.

[7] Ch. 6, p. 5 b.

[8] As justly said by GEERTS (Produits de la nature japonaise et chinoise, p. 481), it is possible that *ma kia ču* (Japanese *bakašu*) is merely a synonyme of the emerald. Also in the *Pen ts'ao kañ mu* (Ch. 8, p. 17 b) a distinction is made between the two articles, *tien-tse* 靛子 being characterized as *pi* 碧, *ma kia ču* as *ts'ui* 翠.

women of the Nan Man (the aboriginal tribes of southern China), being fastened in their hair;[1] and were known in the kingdom of Nan-čao.[2] Likewise the women of Wei-čou 維州 in Se-č'wan wore strung se-se in their hair.[3] Further, we hear at the same time of se-se utilized by the Chinese and even mined in Chinese soil. In some cases it seems that a building-stone is involved; in others it appears as a transparent precious stone, strung and used for curtains and screens, highly valued, and on a par with genuine pearls and precious metals.[4] Under the year 786, the T'ang Annals state, "The Kwan-č'a-ši 離察使[5] of San-čou 陜州 (in Ho-nan), Li Pi 李泌 by name, reported to the throne that the foundries of Mount Lu-ši 盧氏 produce se-se, and requested that it should be prohibited to accept these stones in the place of taxes; whereupon the Emperor (Te Tsuṅ) replied, that, if there are se-se not produced by the soil, they should be turned over to the people, who are permitted to gather them for themselves." The question seems to be in this text of a by-product of metallic origin; and this agrees with what Kao Se-sun remarks in his Wei lio, that the se-se of his time (Sung period) were made of molten stone.

I have given two examples of the employment of se-se in objects of art from the K'ao ku t'u and Ku yü t'u p'u (p. 31). Meanwhile I have found two instances of the use of the word se-se in the Po ku t'u lu, published by Waṅ Fu in 1107–11. In one passage of this work,[6] the patina of a tiṅ 鼎, attributed to the Čou period, is compared with the color of se-se: since patinas occur in green, blue, and many other hues, this does not afford conclusive evidence as to the color of se-se. In another case[7] a small tiṅ dated in the Han period is described as being decorated with inlaid gold and silver, and decorated with the seven jewels (saptaratna) and se-se of very brilliant appearance. This is striking, as se-se are not known to be on record under the Han, but first appear in the accounts of Sasanian Persia: either the bronze vessel in question was not of the Han, but of the T'ang; or, if it was of the Han, the stone thus diagnosed by the Sung author cannot have been identical with what was known by this name under the T'ang. I already had occasion to state (p. 33) that the Sung writers knew no longer what the

---

[1] T'aṅ šu, Ch. 222 A, p. 2.

[2] Man šu, p. 48.

[3] T'ai p'iṅ hwan yü ki, Ch. 78, p. 9 b.

[4] Miṅ hwaṅ tsa lu, Ch. B, p. 4; Wei lio, Ch. 5, p. 3; Tu yaṅ tsa pien, Ch. A, pp. 3, 8; Ch. C, pp. 5, 9 b, 14 b.

[5] Official designation of a Tao-t'ai.

[6] Ch. 3, p. 15 b.

[7] Ch. 5, p. 46 b.

se-se of the T'ang really were, that the T'ang *se-se* were apparently lost in the age of the Sung, and that substitutes merely designated by that name were then in vogue.

Under the Yüan or Mongol dynasty the word *se-se* was revived. Č'aṅ Te, the envoy who visited Bagdad in 1259, reported *se-se* among the precious stones of the Caliph, together with pearls, lapis lazuli, and diamonds. A stone of small or no value, found in Kin-čou (in Šeṅ-kiṅ, Manchuria), was styled *se-se;*[1] and under the reign of the Emperor Č'eṅ-tsuṅ (1295–1307) we hear that two thousand five hundred catties of *se-se* were palmed off on officials in lieu of cash payments, a practice which was soon stopped by imperial command.[2] Under the Ming, *se-se* was merely a word vaguely conveying the notion of a precious stone of the past, and transferred to artifacts like beads of colored glass or clay.[3]

The Chinese notices of *se-se* form a striking analogy to the accounts of the ancients regarding the emerald (*smaragdos*), which on the one hand is described as a precious stone, chiefly used for rings, on the other hand as a building-stone. Theophrastus[4] states, "The emerald is good for the eyes, and is worn as a ring-stone to be looked at. It is rare, however, and not large. Yet it is said in the histories of the Egyptian kings that a Babylonian king once sent as a gift an emerald of four cubits in length and three cubits in width; there is in the temple of Jupiter an obelisk composed of four emeralds, forty cubits high, four cubits wide, and two cubits thick. The false emerald occurs in well-known places, particularly in the copper-mines of Cyprus, where it fills lodes crossing one another in many ways, but only seldom is it large enough for rings." H. O. Lenz[5] is inclined to understand by the latter kind malachite. Perhaps the *se-se* of Iran and Tibet was the emerald; the *se-se* used for pillars in Fu-lin, malachite. No Chinese definition of what *se-se* was has as yet come to light, and we have to await further information before venturing exact and positive identifications.

In Buddhist literature the emerald appears in the transcription *mo-lo-k'ie-t'o* 摩羅伽陀,[6] corresponding to Sanskrit *marakata*. In the transcription 助木剌 *ču-mu-la*, in the seventeenth century written 祖母綠 *tsu-mu-lü*, the emerald appears to be first mentioned in the

---

[1] *Yüan ši*, Ch. 24, p. 2 b.

[2] *Ibid.*, Ch. 21, p. 7 b.

[3] Cf. Notes on Turquois, p. 34.

[4] *De lapidibus*, 42.

[5] Mineralogie der Griechen und Römer, p. 20.

[6] *Fan yi miṅ yi tsi*, Ch. 8, p. 14 b.

Čo ken lu, written in 1366.[1] The Dictionary in Four Languages[2] writes this word *tsie-mu-lu* 租 姆 磲. This is a transcription of Persian *zumurrud*.

The word itself is of Semitic origin. In Assyrian it has been traced in the form *barraktu* in a Babylonian text dated in the thirty-fifth year of Artaxerxes I (464–424 B.C.).[3] In Hebrew it is *bāreket* or *bārkat*, in Syriac *borko*, in Arabic *zummurud*, in Armenian *zemruxt;* in Russian *izumrud*. The Greek *maragdos* or *smaragdos* is borrowed from Semitic; and Sanskrit *marakata* is derived from Greek, Tibetan *mar-gad* from Sanskrit.[4] The Arabic-Persian *zummurud* appears to be based directly on the Greek form with initial sibilant.

87. In regard to turquois I shall be brief. The Persian turquois, both that of Nīšāpūr and Kirmān, is first mentioned under the name *tien-tse* 甸 子 in the *Čo ken lu* of 1366. This does not mean that the Chinese were not acquainted with the Persian turquois at a somewhat earlier date. It is even possible that the Kitan were already acquainted with turquois.[5] I do not believe that *pi-lu* 碧 磲 represents a transcription of Persian *firūza* ("turquois"), as proposed by WATTERS[6] without indicating any source for the alleged Chinese word, which, if it exists, may be restricted to the modern colloquial language. I have not yet traced it in literature.[7] As early as 1290 turquoises were mined in Hui-č'wan, Yün-nan.[8] The Geography of the Ming dynasty indicates a turquois-mine in Nan-nin čou 安 寧 州 in the prefecture of Yün-nan,

---

[1] Ch. 7, p. 5 b; *Wu li siao ši*, Ch. 7, p. 14. The author of this work cites the writing of the Yüan work as the correct one, adding *tsu-mu-lü*, which he says is at present in vogue, as an erroneous form. It is due to an adjustment suggested by popular etymology, the character *lü* ("green") referring to the green color of the stone, whose common designation is *lü pao ši* 綠 寶 石 ("green precious stone"); see GEERTS, Produits, p. 481.

[2] Ch. 22, p. 66.

[3] C. FOSSEY, Etudes assyriennes (*Journal asiatique*, 1917, I, p. 473).

[4] Cf. Notes on Turquois, p. 55; *T'oung Pao*, 1916, p. 465. MUSS-ARNOLT (*Transactions Am. Phil. Assoc.*, Vol. XXIII, 1892, p. 139) states erroneously that both the Greek and the Semitic words are independently derived from Sanskrit. In the attempt to trace the history of loan-words it is first of all necessary to ascertain the history of the objects.

[5] As intimated by me in *American Anthropologist*, 1916, p. 589. *Tien-tse* as the product of Pan-ta-li are mentioned in the *Tao i ši lio*, written in 1349 by Wan Ta-yüan (ROCKHILL, *T'oung Pao*, 1915, p. 464).

[6] Essays on the Chinese Language, p. 352.

[7] In the *Pen ts'ao kan mu* (Ch. 8, p. 17 b) is mentioned a stone *p'iao pi lü* 縹 碧 綠, explained as a precious stone (*pao ši*) of *pi* 碧 color. This is possibly the foundation of Watters' statement.

[8] *Yüan ši*, Ch. 16, p. 10 b. See, further, Notes on Turquois, pp. 58–59.

Yün-nan Province.[1] In this text, the term *pi t'ien-tse* 碧墳子 is employed. T'an Ts'ui[2] says that turquoises (*pi t'ien*) are produced in the Moṅ-yaṅ t'u-se 孟養土司 of Yün-nan. In the *Hiṅ-ṅan fu či* 興安府志,[3] the gazetteer of the prefecture of Hiṅ-ṅan in southern Šen-si, it is said that *pi t'ien* (written 墳) were formerly a product of this locality, and mined under the T'ang and Sung, the mines being closed in the beginning of the Ming. This notice is suspicious, as we hear of *pi-tien* or *tien-tse* neither under the T'ang nor the Sung; the term comes into existence under the Yüan.[4]

88. 金精 *kin tsiṅ* ("essence of gold") appears to have been the term for lapis lazuli during the T'ang period. The stone came from the famous mines of Badaxšān.[5]

At the time of the Yüan or Mongol dynasty a new word for lapis lazuli springs up in the form *lan-č'i* 蘭赤. The Chinese traveller Č'aṅ Te, who was despatched in 1259 as envoy by the Mongol Emperor Mangu to his brother Hulagu, King of Persia, and whose diary, the *Si ši ki*, was edited by Liu Yu in 1263, reports that a stone of that name is found on the rocks of the mountains in the south-western countries of Persia. The word *lan-č'i* is written with two characters meaning "orchid" and "red," which yields no sense; and BRETSCHNEIDER[6] is therefore right in concluding that the two elements represent the transcription of a foreign name. He is inclined to think that "it is the same as *landshiwer*, the Arabic name for lapis lazuli." In New Persian it is *lāžvard* or *lājvard* (Arabic *lāzvard*). Another Arabic word is *linej*, by which the *cyanos* of Dioscorides is translated.[7] An Arabic form *lanjiver* is not known to me.

"There is also in the same country [Badashan] another mountain, in which azure is found; 'tis the finest in the world, and is got in a vein like silver. There are also other mountains which contain a great amount of silver ore, so that the country is a very rich one." Thus runs

---

[1] *Ta Miṅ i t'uṅ či*, Ch. 86, p. 8.

[2] *Tien hai yü heṅ či*, 1799, Ch. 1, p. 6 b (ed. of *Wen yiṅ lou yü ti ts'uṅ šu*). See above, p. 228. *T'u-se* are districts under a native chieftain, who himself is subject to Chinese authority.

[3] Ch. 11, p. 11 b (ed. of 1788).

[4] The turquois has not been recognized in a text of the *Wei si wen kien ki* of 1769 by G. SOULIÉ (*Bull. de l'Ecole française*, Vol. VIII, p. 372), where the question is of coral and turquois used by the Ku-tsuṅ (a Tibetan tribe) women as ornaments; instead of *yüan-song*, as there transcribed, read *lü suṅ ši* 綠松石.

[5] CHAVANNES, Documents sur les Tou-kiue, p. 159; and *T'oung Pao*, 1904, p. 66.

[6] *Chinese Recorder*, Vol. VI, p. 16; or Mediæval Researches, Vol. I, p. 151.

[7] LECLERC, Traité des simples, Vol. III, p. 254.

Marco Polo's account.[1] YULE comments as follows: "The mines of Lájwurd (whence l'Azur and Lazuli) have been, like the ruby mines, celebrated for ages. They lie in the upper valley of the Kokcha, called Korán, within the tract called Yamgán, of which the popular etymology is Hamah-Kán, or 'All-Mines,' and were visited by Wood in 1838.[2] The produce now is said to be of very inferior quality, and in quantity from thirty to sixty pud (thirty-six lbs. each) annually. The best quality sells at Bokhara at thirty to sixty tillas, or 12 *l.* to 24 *l.* the pud (Manphúl)."[3] In the Dictionary of Four Languages,[4] lapis lazuli is styled *ts'in kin ši* 青金石; in Tibetan *mu-men,* Mongol and Manchu *nomin.*

The diamond is likewise attributed by the Chinese to Sasanian Persia, and I have formerly shown that several Iranian tribes were acquainted with this precious stone in the beginning of our era.[5] Diamond-points were imported from Persia into China under the T'ang dynasty.[6]

89. The first mention of amber in Chinese records is the reference to amber in Ki-pin (Kashmir).[7] Then we receive notice of the occurrence of amber in Ta Ts'in (the Hellenistic Orient)[8] and in Sasanian Persia.[9] The correctness of the latter account is confirmed by the Bündahišn, in which the Pahlavi term for amber, *kahrupāi,* is transmitted.[10] This word corresponds to New Persian *kāhrubā,* a compound formed with *kāh* ("straw") and *rubā* ("to lift, to attract").[11] The Arabs derived their *kahrubā* (first in Ibn el-Abbās) from the Persians; and between the

---

[1] YULE's edition, Vol. I, p. 157.

[2] This refers to WOOD, Journey to the Oxus, p. 263.

[3] See, further, M. BAUER, Precious Stones, p. 442.

[4] Ch. 22, p. 65.

[5] The Diamond, p. 53.

[6] *Ta T'an leu tien,* Ch. 22, p. 8.

[7] *Ts'ien Han šu,* Ch. 96 A, p. 5.

[8] In the *Wei lio* and *Hou Han šu* (cf. CHAVANNES, *T'oung Pao,* 1907, p. 182).

[9] *Nan ši,* Ch. 79, p. 8; *Wei šu,* Ch. 102, p. 5 a; *Sui šu,* Ch. 83, p. 7 b. The *Sui šu* has altered the name *hu-p'o* into *šou-p'o* 獸魄. in order to observe the tabu of the name Hu in Li Hu 李虎, the father of the founder of the T'ang dynasty. Amber (also coral and silver) is attributed to Mount Ni 尼山 in the country Fu-lu-ni 伏盧尼 to the north of Persia, also to the country Hu-se-mi 呼似密 (*Wei šu,* Ch. 102, p. 6 b).

[10] WEST, Pahlavi Texts, Vol. I, p. 273.

[11] Analogies occur in all languages: Chinese *ši-kiai* 拾芥 ("attracting mustard-seeds"); Sanskrit *tṛiṇagrāhin* ("attracting straw"); Tibetan *sbur len* or *sbur loṅ,* of the same meaning: French (obsolete) *tire-paille.* Another Persian word for amber is *šahbarī.*

ninth and the tenth century, the word penetrated from the Arabic into Syriac.[1] In Armenian it is *kahribā* and *kahribar*. The same word migrated westward: Spanish *carabe*, Portuguese *carabe* or *charabe*, Italian *carabe*, French *carabé*; Byzantine κεραβέ; Cumanian *charabar*. Under the Ming, amber is listed as a product of Herat, Khotan, and Samarkand.[2] A peculiar variety styled "gold amber" (*kin p'o* 金珀) is assigned to Arabia (T'ien-fan).[3]

The question arises, From what sources did the Persians derive their amber? G. JACOB,[4] from a study of Arabic sources, has reached the conclusion that the Arabs obtained amber from the Baltic. The great importance of Baltic amber in the history of trade is well known, but, in my estimation, has been somewhat exaggerated by the specialists, whereas the fact is easily overlooked that amber is found in many parts of the world. I do not deny that a great deal of amber secured by the Arabs may be credited to the Baltic sources of supply, but I fail to see that this theory (for it is no more) follows directly from the data of Arabic writers. These refer merely to the countries of the Rūs and Bulgār as the places of provenience, but who will guarantee that the amber of the Russians hailed exclusively from the Baltic? We know surely enough that amber occurs in southern Russia and in Rumania. Again, Ibn al-Baiṭār knows nothing about Rūs and Bulgār in this connection, but, with reference to al-Jafiki, speaks of two kinds of amber, one coming from Greece and the Orient, the other being found on the littoral and underground in the western portion of Spain.[5] Pliny informs us that, according to Philemon, amber is a fossil substance, and that it is found in Scythia in two localities, one white and of waxen color, styled *electrum;* while in the other place it is red, and is called *sualiternicum*.[6] This Scythian or South-Russian amber may have been traded by the Iranian Scythians to Iran. In order to settle definitely the question of the provenience of ancient Persian and Arabic amber, it would be necessary, first of all, to obtain a certain number of authentic, ancient Persian and Arabic ambers, and to subject them to a chemical analysis. We know also that several ancient amber supplies were

---

[1] Cf. E. SEIDEL, Mechithar, p. 146; and G. JACOB, ZDMG, Vol. XLIII, 1889, p. 359.

[2] *Ta Min i t'un či*, Ch. 89, pp. 23, 24 b, 25 (ed. of 1461).

[3] *Ibid.*, Ch. 91, p. 20.

[4] *L. c.*, and Arabische Handelsartikel, p. 63.

[5] LECLERC, Traité des simples, Vol. III, p. 209.

[6] Philemon fossile esse et in Scythia erui duobus locis, candidum atque cerei coloris quod vocaretur electrum, in alio fulvum quod appellaretur sualiternicum (XXXVII, 11, § 33).

exhausted long ago. Thus Pliny and the ancient Chinese agree on the fact that amber was a product of India, while no amber-mines are known there at present.[1] Amber was formerly found in the district of Yuṅ-č'aṅ in Yün-nan, and even on the sacred Hwa-šan in Šen-si.[2]

G. JACOB[3] has called attention to the fact that the supposition of a derivation of the Chinese word from Pahlavi *kahrupāi* is confronted with unsurmountable difficulties of a chronological character. The phonetic difficulties are still more aggravating; for Chinese *hu-p'o* 琥珀 was anciently *gu-bak, and any alleged resemblance between the two words vanishes. Still less can Greek *harpax*[4] come into question as the foundation of the Chinese word, which, in my opinion, comes from an ancient Šan or T'ai language of Yün-nan, whence the Chinese received a kind of amber as early at least as the first century A.D. Of the same origin, I am inclined to think, is the word *tun-mou* 頓牟 for amber, first and exclusively used by the philosopher Waṅ Č'uṅ.[5]

Uigur *kubik* is not the original of the Chinese word, as assumed by Klaproth; but the Uigur, on the contrary (like Korean *xobag*), is a transcription of the Chinese word. Mongol *xuba* and Manchu *xôba* are likewise so, except that these forms were borrowed at a later period, when the final consonant of Chinese *bak* or *bek* was silent.[6]

90. Coral is a substance of animal origin; but, as it has always been conceived in the Orient as a precious stone,[7] a brief notice of it, as far as Sino-Persian relations are concerned, may be added here. The

---

[1] Cf. *Ts'ien Han šu*, Ch. 96 A, p. 5 (amber of Kashmir); *Nan ši*, Ch. 78, p. 7.

[2] Cf. *Hwa yo ři* 華嶽志, Ch. 3, p. 1 (ed. of 1831).

[3] *L. c.*, p. 355.

[4] Proposed by HIRTH, China and the Roman Orient, p. 245. This was merely a local Syriac name, derived from Greek ἁρπάζω (In Syria quoque feminas verticillos inde facere et vocare harpaga, quia folia paleasque et vestium fimbrias rapiat.— Pliny, XXXVII, 11, § 37).

[5] Cf. A. FORKE, Lun-heng, pt. II, p. 350. This is not the place for a discussion of this problem, which I have taken up in a study entitled "Ancient Remains from the Languages of the Nan Man."

[6] For further information on amber, the reader may be referred to my Historical Jottings on Amber in Asia (*Memoirs Am. Anthr. Assoc.*, Vol. I, pt. 3). I hope to come back to this subject in greater detail in the course of my Sino-Hellenistic studies, where it will be shown that the Chinese tradition regarding the origin and properties of amber is largely influenced by the theories of the ancients.

[7] The proof of the animal character of coral is a recent achievement of our science. Peyssonel was the first to demonstrate in 1727 that the alleged coral-flowers are real animals; Pallas then described the coral as *Isis nobilis;* and Lamarck formed a special genus under the name *Corallium rubrum* (cf. LACAZE-DUTHIERS, Histoire naturelle du corail, Paris, 1864; GUIBOURT, Histoire naturelle des drogues, Vol. IV, p. 378). The common notion in Asia was that coral is a marine tree.

Chinese learned of the genuine coral through their intercourse with the Hellenistic Orient: as we are informed by the *Wei lio* and the Han Annals,[1] Ta Ts'in produced coral; and the substance was so common, that the inhabitants used it for making the king-posts of their habitations. The T'ang Annals[2] then describe how the marine product is fished in the coral islands by men seated in large craft and using nets of iron wire. When the corals begin to grow on the rocks, they are white like mushrooms; after a year they turn yellow, and when three years have elapsed, they change into red. Their branches then begin to intertwine, and grow to a height of three or four feet.[3] Hirth may be right in supposing that this fishing took place in the Red Sea, and that the "Coral Sea" of the Nestorian inscription and the "sea producing corals and genuine pearls" of the *Wei lio* are apparently identical with the latter.[4] But it may have been the Persian Gulf as well, or even the Mediterranean. Pliny[5] is not very enthusiastic about the Red-Sea coral; and the Periplus speaks of the importation of coral into India, which W. H. SCHOFF[6] seems to me to identify correctly with the Mediterranean coral. Moreover, the Chinese themselves correlate the above account of coral-fishing with Persia, for the *Yi wu či* 異物志 is cited in the *Čen lei pen ts'ao*[7] as saying that coral is produced in Persia, being considered by the people there as their most precious jewel; and the *Pen ts'ao yen i* speaks of a coral-island in the sea of Persia,[8] going on to tell the same story regarding coral-fishing as the T'ang Annals with reference to Fu-lin (Syria). Su Kun of the T'ang states that coral grows in the Southern Sea, but likewise comes from Persia and Ceylon, the latter statement being repeated by the *T'u kin pen ts'ao* of the Sung. It is interesting that the *Pen ts'ao* of the T'ang insists on the holes in coral, a characteristic which in the Orient is still regarded (and justly so) as a mark of authenticity. Under the T'ang, coral was first introduced into the materia medica. In the Annals, coral is ascribed to

---

[1] HIRTH, China and the Roman Orient, pp. 41, 73.

[2] *Ibid.*, p. 44.

[3] *Ibid.*, p. 59.

[4] *Ibid.*, p. 246.

[5] XXXII, 11.

[6] The Periplus of the Erythræan Sea, p. 128.

[7] Ch. 4, p. 37.

[8] Ch. 5, p. 7 (ed. of Lu Sin-yüan). The coral island where the coral-tree grows is also mentioned by an Arabic author, who wrote about A.D. 1000 (G. FERRAND, Textes relatifs à l'Extrême-Orient, Vol. I, p. 147). See, further, E. WIEDEMANN, Zur Mineralogie im Islam, p. 244.

Sasanian Persia;[1] and it is stated in the T'ang Annals that Persia produces coral not higher than three feet.[2] There is no doubt that Persian corals have found their way all over Asia; and many of them may still be preserved by Tibetans, who prize above all coral, amber, and turquois. The coral encountered by the Chinese in Ki-pin (Kashmir)[3] may also have been of Persian origin. Unfortunately we have no information on the subject from ancient Iranian sources, nor do we know an ancient Iranian name for coral. Solinus informs us that Zoroaster attributed to coral a certain power and salubrious effects;[4] and what Pliny says about coral endowed with sacred properties and being a preservative against all dangers, sounds very much like an idea emanating from Persia. Persian infants still wear a piece of coral on the abdomen as a talisman to ward off harm;[5] and, according to Pliny, this was the practice at his time, only that the branches of coral were hung at the infant's neck.

The Chinese word for coral, 珊瑚 šan-hu, *san-gu (Japanese san-go), possibly is of foreign origin, but possibly it is not.[6] For the present there is no word in any West-Asiatic or Iranian language with which it could be correlated. In Hebrew it is ra'moṭ, which the Seventy transcribes ραμοθ or translates μετέωρα. The common word in New Persian is marjān (hence Russian maržan); other designations are birbāl, xuruhak or xurohak, bussad or bissad (Arabic bessed or bussad). In Armenian it is bust.[7]

91. The identification of Chinese 婆娑 p'o-so (*bwa-sa) with Persian pāzahr or pādzahr[8] ("bezoar," literally, "antidote"), first proposed by HIRTH,[9] in my opinion, is not tenable, although it has been indorsed

---

[1] Čou Šu, Ch. 50, p. 6; Sui Šu, Ch. 83, p. 7 b; regarding coral in Fu-lu-ni, see above, p. 521, note 9.

[2] T'an Šu, Ch. 221 B, p. 6 b. The Lian Šu (Ch. 54, p. 14 b) attributes to Persia coral-trees one or two feet high.

[3] Ts'ien Han Šu, Ch. 96 A, p. 5. This passage (not Hou Han Šu, Ch. 118, as stated by HIRTH, Chau Ju-kua, p. 226, after Bretschneider) contains the earliest mention of the word šan-hu.

[4] Habet enim, ut Zoroastres ait, materia haec quandam potestatem, ac propterea quidquid inde sit, ducitur inter salutaria (II, 39, § 42).

[5] SCHLIMMER, Terminologie, p. 166.

[6] According to BRETSCHNEIDER (Chinese Recorder, Vol. VI, p. 16), "it seems not to be a Chinese name."

[7] Cf. PATKANOV, The Precious Stones according to the Notions of the Armenians (in Russian), p. 52.

[8] Pāzand pādazahar (see HÜBSCHMANN, Persische Studien, p. 193). STEINGASS gives also pānzahr. The derivation from bād "wind" (H. FÜHNER, Janus, Vol. VI, 1901, p. 317) is not correct.

[9] Länder des Islam, p. 45.

by Pelliot.[1] Pelliot, however, noticed well that what the Chinese describe as *p'o-so* or *mo-so* 摩娑 is not bezoar, and that the transcription is anomalous.[2] This being the case, it is preferable to reject the identification, and there are other weighty reasons prompting us to do so. There is no Chinese account that tells us that Persia had bezoars or traded bezoars to China. The Chinese were (and are) well acquainted with the bezoar[3] (I gathered several in China myself), and bezoars are easy to determine. Now, if *p'o-so* or *mo-so* were to represent Persian *pāzahr* and a Persian bezoar, the Chinese would not for a moment fail to inform us that *p'o-so* is the *Po-se niu-hwaṅ* or Persian bezoar; but they say nothing to this effect. On the contrary, the texts cited under this heading in the *Pen ts'ao kaṅ mu*[4] do not make any mention of Persia, but agree in pointing to the Malay Archipelago as the provenience of the *p'o-so* stone. Ma Či of the Sung assigns it to the Southern Sea (Nan Hai). Li Ši-čen points to the *Keṅ sin yü ts'e* 庚辛玉册, written about 1430, as saying that the stone comes from San-fu-ts'i (Palembang on Sumatra).[5] F. de Mély designates it only as a "pierre d'épreuve," and refers to an identification with aventurine, proposed by Rémusat.[6] Bezoar is a calculus concretion found in the stomachs of a number of mammals, and Oriental literatures abound in stories regarding such stones extracted from animals. Not only do the Chinese not say that the *p'o-so* stone is of animal origin, but, on the contrary, they state explicitly that it is of mineral origin. The *Keṅ sin yü ts'e* relates how mariners passing by a certain mountain on Sumatra break this stone with axes out of the rock, and that the stone when burnt emits a sulphurous odor. Ma Či describes this stone as being green in color and without speckles; those with gold stars, and when rubbed yielding a milky juice, are the best. All this does not fit the bezoar. Also the description in the *Pen ts'ao yen i*[7] refers only to a stone of mineral origin.

---

[1] *T'oung Pao*, 1912, p. 438.

[2] The initial of the Persian word would require a labial surd in Chinese. Whether the *p'o-sa* 婆薩 of the *Pei hu lu* belongs here is doubtful to me; it is not explained what this stone is. As admitted in the *Pen ts'ao yen i* (Ch. 4, p. 4 b), the form *mo-so* is secondary.

[3] It is first mentioned in the ancient work *Pie lu*, then in the *Wu ši pen ts'ao* of the third century, and by T'ao Huṅ-kiṅ.

[4] Ch. 10, p. 10 b.

[5] This text is cited in the same manner in the *Tuṅ si yaṅ k'ao* of 1618 (Ch. 3, p. 10). Cf. F. de Mély, Lapidaire chinois, p. 120.

[6] *Ibid.*, pp. LXIV, 260.

[7] Ch. 4, p. 4 b (ed. of Lu Sin-yüan).

Even as early as the T'ang period, the term *p'o-so* merely denotes a stone. It is mentioned in a colophon to the *P'in ts'üan šan kü ts'ao mu ki* 平泉山居草木記 by Li Te-yü 李德裕 (A.D. 787–849) as a curious stone preserved in the P'o-so Pavilion south of the Č'aṅ-tien 長殿 in Ho-nan.

Yada or *jada*, as justly said by Pelliot, is a bezoar; but what attracted the Chinese to this Turkish-Mongol word was not its character as a bezoar, but its rôle in magic as a rain-producing stone. Li Ši-čen[1] has devoted a separate article to it under the name 鮓荅 *ča-ta*, and has recognized it as a kind of bezoar; in fact, it follows immediately his article on the Chinese bezoar (*niu-hwaṅ*).[2]

The Persian word was brought to China as late as the seventeenth century by the Jesuits. Pantoja and Aleni, in their geography of the world, entitled *Či fan wai ki*,[3] and published in 1623, mention an animal of Borneo resembling a sheep and a deer, called *pa-tsa'r* 把雜爾,[4] in the abdomen of which grows a stone capable of curing all diseases, and highly prized by the Westerners. The Chinese recognized that this was a bezoar.[5] Bezoars are obtained on Borneo, but chiefly from a monkey (*Simia longumanis*, Dayak *buhi*) and hedgehog. The Malayan name for bezoar is *gulīga*; and, as far as I know, the Persian word is not used by the Malayans.[6] The Chinese Gazetteer of Macao mentions "an animal like a sheep or goat, in whose belly is produced a stone capable

---

[1] *Pen ts'ao kaṅ mu*, Ch. 50 B, p. 15 b.

[2] There is an extensive literature on the subject of the rain-stone. The earliest Chinese source known to me, and not mentioned by Pelliot, is the *K'ai yüan t'ien pao i ši* 開元天寶遺事 by Waṅ Žen-yü 王仁裕 of the T'ang (p. 20 b). Cf. also the *Sü K'ien šu* 續黔書, written by Čaṅ Ču 張澍 in 1805 (Ch. 6, p. 8, ed. of *Yüe ya t'aṅ ts'uṅ šu*). The Yakut know this stone as *sata* (BOEHTLINGK, Jakut. Wörterbuch, p. 153); Pallas gives a Kalmuk form *sädan*. See, further, W. W. ROCK-HILL, Rubruck, p. 195; F. v. ERDMANN, Temudschin, p. 94; G. OPPERT, Presbyter Johannes, p. 102; J. RUSKA, Steinbuch des Qazwīnī, p. 19, and *Der Islam*, Vol. IV, 1913, pp. 26–30 (it is of especial interest that, according to the Persian mineralogical treatise of Mohammed Ben Mansur, the rain-stone comes from mines on the frontier of China, or is taken from the nest of a large water-bird, called *surxab*, on the frontier of China; thus, after all, the Turks may have obtained their bezoars from China); VÁMBÉRY, Primitive Cultur, p. 249; POTANIN, Tangutsko-Tibetskaya Okraina Kitaya, Vol. II, p. 352, where further literature is cited.

[3] Ch. 1, p. 11 (see above, p. 433).

[4] This form comes very near to the *pajar* of Barbosa in 1516.

[5] Cf. the *Lu čaṅ kuṅ ši k'i* (above, p. 346), p. 48.

[6] Regarding the Malayan beliefs in bezoars, see, for instance, L. BOUCHAL in *Mitt. Anthr. Ges. Wien*, 1900, pp. 179–180; BECCARI, Wanderings in the Great Forests of Borneo, p. 327; KREEMER in *Bijdr. taal- land- en volkenkunde*, 1914, p. 38; etc.

of curing any disease, and called *pa-tsa'r*" (written as above);[1] cf. Portuguese *bazar, bazoár, bezoar.*

On the other hand, bezoars became universal in the early middle ages, and the Arabs also list bezoars from China and India.[2] From the Persian word *fādaj*, explained as "a stone from China, bezoar," it appears also that Chinese bezoars were traded to Persia. In Persia, as is well known, bezoars are highly prized as remedies and talismans.[3]

---

[1] *Ao-men ĕi lio,* Ch. B, p. 37.

[2] J. RUSKA, Steinbuch des Aristoteles, p. 148.

[3] C. ACOSTA (Tractado de las drogas, pp. 153–160, Burgos, 1578), E. KAEMPFER (Amoenitates exoticae, pp. 402–403), GUIBOURT (Histoire naturelle des drogues simples, Vol. IV, pp. 106 *et seq.*), and G. F. KUNZ (Magic of Jewels and Charms, pp. 203–220) give a great deal of interesting information on the subject. See also YULE, Hobson-Jobson, p. 90; E. WIEDEMANN, Zur Mineralogie im Islam, p. 228; D. HOOPER, *Journal As. Soc. Bengal,* Vol. VI, 1910, p. 519.

92. 薩寳 sa-pao, *saδ(sar)-pav. Title of the official in charge of the affairs of the Persian religion in Si-ṅan, an office dating back to the time when temples of the celestial god of fire were erected there, about A.D. 621. In an excellent article PELLIOT has assembled all texts relative to this function.[1] I do not believe, however, that we are justified in accepting Devéria's theory that the Chinese transcription should render Syriac sābā ("old man"). This plainly conflicts with the laws of transcription so rigorously expounded and upheld by Pelliot himself: it is necessary to account for the final dental or liquid in the character sa, which regularly appears in the T'ang transcriptions. It would be strange also if the Persians should have applied a Syriac word to a sacred institution of their own. It is evident that the Chinese transcription corresponds to a Middle-Persian form traceable to Old Persian xšaθra-pāvan (xšçpava, xšaçapāvā), which resulted in Assyrian axšadar-apān or axšadrapān, Hebrew axašdarfnim,[2] Greek σατράπης (Armenian šahapand, Sanskrit kṣatrapa). The Middle-Persian form from which the Chinese transcription was very exactly made must have been *šaθ-pāv or *xšaθ-pāv. The character sa renders also Middle and New Persian sar ("head, chief").[3]

93. 庫薩和 K'u-sa-ho, *Ku-saδ(r)-γwa, was the title 字 of the kings of Pārsa (Persia).[4] This transcription appears to be based on an Iranian xšaθva or xšarva, corresponding to Old Iranian *xšáyavan-, *xšaivan, Sogdian xšēvan ("king").[5] It is notable that the initial spirant x is plainly and aptly expressed in Chinese by the element k'u,[6] while in the preceding transcription it is suppressed. The differentiation in time may possibly account for this phenomenon: the transcription sa-pao comes down from about A.D. 621; while K'u-sa-ho, being con-

---

[1] Le Sa-pao, Bull. de l'Ecole française, Vol. III, pp. 665–671.

[2] H. POGNON, Journal asiatique, 1917, I, p. 395.

[3] R. GAUTHIOT, Journal asiatique, 1911, II, p. 60.

[4] Sui šu, Ch. 83, p. 7 b.

[5] R. GAUTHIOT, Essai sur le vocalisme du sogdien, p. 97. See also the note of ANDREAS in A. Christensen, L'Empire des Sassanides, p. 113. I am unable to see how the Chinese transcription could correspond to the name Khosrou, as proposed by several scholars (CHAVANNES, Documents sur les Tou-kiue occidentaux, p. 171; and HIRTH, Journal Am. Or. Soc., Vol. XXXIII, 1913, p. 197).

[6] In the Manichæan transcriptions it is expressed by 呼 *xu (hu); see CHAVANNES and PELLIOT, Traité manichéen, p. 25.

529

tained in the Sui Annals, belongs to the latter part of the sixth century. According to SALEMANN,[1] Iranian initial *xš*- develops into Middle-Persian *š*-; solely the most ancient Armenian loan-words show *ašx*- for *xš*-, otherwise *š* appears regularly save that *šx* takes the place of intervocalic *xš*.[2] In view of our Sino-Iranian form, this rule should perhaps be reconsidered, but this must remain for the discussion of Iranian scholars.

94. 殺 野 *ša-ye*, **šat (šaδ)-ya*. Title of the sons of the king of Persia (*Wei šu*, Ch. 102, p. 6; *T'ai p'iñ hwan yü ki*, Ch. 185, p. 17). It corresponds to Avestan *xšaθrya* ("lord, ruler").[3] The princes of the Sasanian empire were styled *saθraδārān*.[4] According to Sasanian custom, the sons of kings ruled provinces as "kings."[5] Regarding 殺 in transcriptions of Iranian names, cf. the name of the river Yaxartes 藥 殺 (*Sui šu*, Ch. 83, p. 4b) Yao-ša, that is **Yak-šaδ(šar)*. As the Middle-Persian name is Xšārt or Ašārt (Pāzend Ašārd),[6] we are bound to assume that the prototype of the Chinese transcription was **Axšārt* or **Yaxšārt*.

95. 腎 嚕 *i-tsan*, but, as the *fan-ts'ie* of the last character is indicated by 才 割, the proper reading is *i-ts'at*, **i-džaδ, i-dzaδ*, designation of the king of Pārsa (國 人 號 or 謂 王 曰 腎 嚕: *Wei šu*, Ch. 102, p. 6; *T'ai p'iñ hwan yü ki*, Ch. 185, p. 17). The Chinese name apparently represents a transcription of Ixšeδ, the Ixšīdh of al-Bērūnī, title of the kings of Sogd and Fergana, a dialectic form of Old Persian *xšāyaθiya*.[7] Ixšeδ is the Avestan *xšaeta* ("brilliant"), a later form being *šēdah*. It must be borne in mind that Sogdian was the *lingua franca* and international language of Central Asia, and even the vehicle of civiliza-

---

[1] Grundriss der iran. Phil., Vol. I, pt. 1, p. 262.

[2] Cf. also GAUTHIOT, *op. cit.*, p. 54, § 61.

[3] K. Hori's identification with New Persian *šāh* (Spiegel Memorial Volume, p. 248) must be rejected. The time of the *Wei šu* plainly refers to Sasanian Persia; that is, to the Middle-Persian language.

[4] A. CHRISTENSEN, *op. cit.*, p. 20. Cf. Old Persian *xšçm, xšaçam* ("royalty, kingdom"), Avestan *xšaθrem*, Sanskrit *kṣatram* (A. MEILLET, Grammaire du vieux perse, p. 143); *xšaθrya* corresponds to Sanskrit *kṣatriya*.

[5] NÖLDEKE, Tabari, p. 49; Grundriss, Vol. II, p. 171. I think that H. POGNON (*Journal asiatique*, 1917, I, p. 397) is right in assuming that "satrap" was a purely honorific title granted by the king not only to the governors of the provinces, but also to many high functionaries.

[6] WEST, Pahlavi Texts, Vol. I, p. 80.

[7] See SACHAU, Chronology of Ancient Nations, p. 109; F. JUSTI, Iranisches Namenbuch, p. 141; A. MEILLET, Grammaire du vieux perse, pp. 77, 167 (*xšāyaθiya pārsaiy*, "king in Persia"); F. W. K. MÜLLER, Ein Doppelblatt aus einem manichäischen Hymnenbuch, p. 31.

tion.[1] The suggestion offered by K. HORI,[2] that the Chinese transcription should represent the Persian word *izad* ("god"), is not acceptable: first, New Persian cannot come into question, but only Middle Persian; second, it is not proved that *izad* was ever a title of the kings of Persia. On the contrary, as stated by NÖLDEKE,[3] the Sasanians applied to themselves the word *bag* ("god"), but not *yazdān*, which was the proper word for "god" even at that time.

96. 防步率 *faṅ-pu-šwai*, *pwaṅ-bu-zwiδ, designation of the queen of Pārsa (*Wei šu*, Ch. 102, p. 6; *T'ai p'iṅ hwan yü ki*, Ch. 185, p. 17). The foundation of this transcription is presented by Middle Persian *bānbušn, bānbišn* (Armenian *bambišn*), "consort of the king of Persia."[4] The Iranian prototype of the Chinese transcription seems to have been *bānbuzwiδ. The latter element may bear some relation to Sogdian *wáδu* or *wyδyšth* ("consort").[5]

97. 摸胡壇 *mo-hu-t'an*, *mak-ku(mag-gu)-dan. Officials of Persia in charge of the judicial department 掌國內獄訟 (*Wei šu*, Ch. 102, p. 6). K. HORI[6] has overlooked the fact that the element *t'an* forms part of the transcription, and has simply equalized *mo-hu* with Avestan *moγu*. The transcription *mak-ku (mag-gu) is obviously founded on Middle Persian *magu*, and therefore is perfectly exact. The later transcription 穆護 *muk-gu (*mu-hu*) is based on New Persian *muγ*, *mōγ*.[7] The ending *dan* reminds one of such formations as *herbeδān* ("judge") and *mobeδān mōbeδ* ("chief of the Magi"), the latter being Old Persian *magupati*, Armenian *mogpet*, Pahlavi *maupat*, New Persian *mūbid* (which, according to the Persian Dictionary of Steingass, means also "one who administers justice, judge"). Above all, compare the Armenian loan-word *movpetan* (also *movpet, mogpet, mog*).[8] Hence it

---

[1] R. GAUTHIOT, Essai sur le vocalisme du sogdien, p. x; P. PELLIOT, Les influences iraniennes en Asie centrale et en Extrême-Orient, p. 11.

[2] Spiegel Memorial Volume, p. 248.

[3] Tabari, p. 452.

[4] HÜBSCHMANN, Armen. Gram., p. 116. In his opinion, the form *bānbušn*, judging from the Armenian, is wrong; but its authenticity is fully confirmed by the Chinese transcription.

[5] R. GAUTHIOT, Essai sur le vocalisme du sogdien, pp. 59, 112. The three aforementioned titles had already been indicated by ABEL-RÉMUSAT (Nouvelles mélanges asiatiques, Vol. I, p. 249) after Ma Twan-lin, but partially in wrong transcription: "Le roi a le titre de Yi-thso; la reine, celui de Tchi-sou, et les fils du roi, celui de Cha-ye."

[6] Spiegel Memorial Volume, p. 248.

[7] CHAVANNES and PELLIOT, Traité manichéen, p. 170. Accordingly this example cannot be invoked as proving that *muk* might transcribe also *mak*, as formerly assumed by PELLIOT (Bull. de l'Ecole française, Vol. IV, p. 312).

[8] HORN, Neupersische Etymologie, No. 984; and HÜBSCHMANN, Persische Studien, p. 123.

may justly be inferred that there was a Middle-Persian form *ma-gutan or *magudan, from which the Chinese transcription was exactly made.

98. 泥忽汗 *ni-hu-han*, *ni-hwut-γan. Officials of Persia who have charge of the Treasury (*Wei šu*, Ch. 102, p. 6). The word, in fact, is a family-name or title written by the Greek authors Ναχοραγάν, Ναχοεργάν, Σαρναχοργάνης (prefixed by the word *sar*, "head, upper"). Firdausī mentions repeatedly under the reign of Khosrau II a Naxwāra, and the treasurer of this king is styled "son of Naxwāra."[1] The treasury is named for him al-Naxīrajān. The Chinese transcription is made after the Pahlavi model *Nixurγan or Nexurγan; and, indeed, the form Nixorakan is also found.[2]

99. 地卑勃 *ti-pei-p'o*, *di-pi-bwiδ(bir, wir). Officials of Persia who have charge of official documents and all affairs (*Čou šu*, Ch. 50, p. 5b). In the parallel passage of the *Wei šu* (Ch. 102, p. 6), the second character is misprinted 早 *tsao*,[3] *tsaw; *di-tsaw would not correspond to any Iranian word. From the definition of the term it becomes obvious that the above transcription *di-pi answers to *dipi* ("writing, inscription"),[4] Middle Persian *dipīr* or *dapīr*, New Persian *dibīr* or *dabīr* (Armenian *dpir*); and that *di-pi-bwiδ corresponds to Middle Persian *dipīvar*, from *dipi-bara, the suffix *-var* (anciently *bara*) meaning "carrying, bearing."[5] The forms *dipīr* and *dibīr* are contractions from *dipīvar*. This word, as follows from the definition, appears to have comprised also what was understood by *dēvān*, the administrative chanceries of the Sasanian empire.

100. 遏羅訶地 *ño-lo-ho-ti*, *at(ar)-la-ha-di. Officials of Persia who superintended the inner affairs of the king (or the affairs of the royal household — *Wei šu*, Ch. 102, p. 6). Theophylactus Simocatta[6] gives the following information on the hereditary functions among the seven high families in the Sasanian empire: "The family called Artabides possesses the royal dignity, and has also the office of placing

---

[1] NÖLDEKE, Tabari, pp. 152–153, 439.

[2] JUSTI, Iran. Namenbuch, p. 219. In Naxuraqan or Naxīrajān *q* and *j* represent Pahlavi *g*. The reconstructions attempted by MODI (Spiegel Memorial Volume, p. LIX) of this and other Sino-Iranian words on the basis of the modern Chinese pronunciation do not call for any discussion.

[3] This misprint is not peculiar to the modern editions, but occurs in an edition of this work printed in 1596, so that in all probability it was extant in the original issue. It is easy to see how the two characters were confounded.

[4] In the Old-Persian inscriptions, where it occurs in the accusative form *dipim* and in the locative *dipiyā* (A. MEILLET, Grammaire du vieux perse, pp. 147, 183).

[5] C. SALEMAN, Grundriss iran. Phil., Vol. I, pt. 1, pp. 272, 282.

[6] III, 8.

the crown on the king's head. Another family presides over military affairs, another superintends civil affairs, another settles the litigations of those who have a dispute and desire an arbiter. The fifth family commands the cavalry, the sixth collects the taxes and supervises the royal treasures, and the seventh takes care of armament and military equipment." Artabides ('Ἀρταβίδης), as observed by Nöldeke,[1] should be read Argabides ('Ἀργαβίδης), the equivalent of Argabeδ. There is also a form ἀργαπέτης in correspondence with Pahlavi arkpat. This title originally designated the commandant of a castle (arg, "citadel"), and subsequently a very high military rank.[2] In later Hebrew we find this title in the forms alkafta, arkafta, or arkabta.[3] The above transcription is apparently based on the form *Argade ('Ἀργαδη) = Argabeδ.

101. 薛波勃 sie-po-p'o, *sit-pwa-bwiδ. Officials of Persia in charge of the army (infantry and cavalry, pāiγan and aswārān), of the four quarters, the four pātkōs (pāt, "province"; kōs, "guarding") 掌四方兵馬: Wei šu, Ch. 102, p. 6. The Čou šu (Ch. 50, p. 5ᵇ) has 薩 *sat, sar, in the place of the first character. The word corresponds to Middle Persian spāhbeδ ("general"); Pahlavi pat, New Persian -bad, -bud ("master"). Ērānspāhbeδ was the title of the generalissimo of the army of the Sasanian empire up to the time of Khusrau I. The Pahlavi form is given as spāhpat;[4] the Chinese transcription, however, corresponds better to New Persian sipahbaδ, so that also a Middle-Persian form *spāhbaδ (-beδ or -buδ) may be inferred.

102. 五思達 ńu-se-ta, *u-se-daδ, used in the Chinese inscription dated 1489 of the Jews of K'ai-foň fu in Ho-nan, in connection with the preceding name 列微 Lie-wei (Levi).[5] As justly recognized by G. Devéria, this transcription represents Persian ustād,[6] which means "teacher, master."[6] The Persian Jews availed themselves of this term for the rendering of the Hebrew title Rāb (Rabbi), although in Persian the name follows the title. The Chinese Jews simply adopted the Chinese mode of expression, in which the family-name precedes the title, Ustad Lie-wei meaning as much as "Rabbi Levi." The transcription itself appears to be of much older date than the Ming, and was doubtless recorded at a time when the final consonant of ta was still articulated. In a former article I have shown from the data of the Jewish inscriptions that the Chinese Jews emigrated from Persia and appeared in China not earlier than in the era of the Sung. This historical proof is signally confirmed by a piece of linguistic evidence. In the Annals of the Yüan Dynasty (Yüan ši, Ch. 33, p. 7 b; 43, p. 11 b) the Jews are styled Šu-hu (Ju-hud)

---

[1] Tabari, p. 5.

[2] Christensen, op. cit., p. 27; Nöldeke, op. cit., p. 437; Hübschmann, Persische Studien, pp. 239, 240.

[3] M. Jastrow, Dictionary of the Targumim, p. 73.

[4] Hübschmann, Armen. Gram., p. 240.

[5] J. Tobar, Inscriptions juives de K'ai-fong-fou, p. 44.

[6] Regarding this word, see chiefly H. Hübschmann, Persische Studien, p. 14.

尤 忿 or Ču-wu 圭 兀. This form can have been transcribed only on the basis of New Persian Juhūδ or Jahūδ with initial palatal sonant. As is well known, the change of initial *y* into *j* is peculiar to New Persian.[1] In Pahlavi we have Yahūt, as in Hebrew Yehūdī and in Arabic Yahūd. A Middle-Persian Yahūt would have been very easy for the Chinese to transcribe. The very form of their transcription shows, however, that it was modelled on the New-Persian type, and that it cannot be much older than the tenth century or the age of the Sung.

---

[1] Cf. HORN, Grundr. iran. Phil., Vol. I, pt. 2, p. 73.

IRANO–SINICA

After dealing with the cultural elements derived by the Chinese from the Iranians, it will be only just to look also at the reverse of the medal and consider what the Iranians owe to the Chinese.

1. Some products of China had reached Iranian peoples long before any Chinese set their foot on Iranian soil. When Čaṅ K'ien in 128 B.C. reached Ta-hia (Bactria), he was amazed to see there staves or walking-sticks made from bamboo of Kiuṅ 邛竹杖[1] and cloth of Šu (Se-č'wan) 蜀布. What this textile exactly was is not known.[2] Both these articles hailed from what is now Se-č'wan, Kiuṅ being situated in Žuṅ čou 榮州 in the prefecture of Kia-tiṅ, in the southern part of the province. When the Chinese envoy inquired from the people of Ta-hia how they had obtained these objects of his own country, they replied that they purchased them in India. Hence Čaṅ K'ien concluded that India could not be so far distant from Se-č'wan. It is well known how this new geographical notion subsequently led the Chinese to the discovery of Yün-nan. There was accordingly an ancient trade-route running from Se-č'wan through Yün-nan into north-eastern India; and, as India on her north-west frontier was in connection with Iranian territory, Chinese merchandise could thus reach Iran. The bamboo of Kiuṅ, also called 筇, has been identified by the Chinese with the so-called square bamboo (*Bambusa* or *Phyllostachys quadrangularis*).[3] The cylindrical form is so universal a feature in bamboo, that the report of the existence in China and Japan of a bamboo with four-angled stems was first considered in Europe a myth, or a pathological abnormity. It is now well assured that it represents a regular and normal species, which grows wild in the north-eastern portion of Yün-nan, and is cultivated chiefly as an ornament in gardens and in temple-courts, the longer stems being used

---

[1] He certainly did not see "a stick of bamboo," as understood by Hirth (*Journal Am. Or. Soc.*, Vol. XXXVII, 1917, p. 98), but it was a finished product imported in a larger quantity.

[2] Assuredly it was not silk, as arbitrarily inferred by F. v. Richthofen (China, Vol. I, p. 465). The word *pu* never refers to silk materials.

[3] For an interesting article on this subject, see D. J. Macgowan, *Chinese Recorder*, Vol. XVI, 1885, pp. 141–142; further, the same journal, 1886, pp. 140–141. E. Satow, Cultivation of Bamboos in Japan, p. 92 (Tokyo, 1899). The square bamboo (Japanese *šikaku-dake*) is said to have been introduced into Japan from Liukiu. Forbes and Hemsley, *Journal Linnean Soc.*, Vol. XXXVI, p. 443.

for staves, the smaller ones for tobacco-pipes. The shoots of this species are prized above all other bamboo-shoots as an esculent.

The *Pei hu lu*[1] has the following notice on staves of the square bamboo: "Č'eṅ čou 澄州 (in Kwaṅ-si) produces the square bamboo. Its trunk is as sharp as a knife, and is very strong. It can be made into staves which will never break. These are the staves from the bamboo of K'iuṅ 筇, mentioned by Čaṅ K'ien. Such are produced also in Yuṅ čou 融州,[2] the largest of these reaching several tens of feet in height. According to the *Čeṅ šeṅ tsi* 正聲集, there are in the southern territory square bamboo staves on which the white cicadas chirp, and which Č'en Čeṅ-tsie 陳貞節 has extolled. Moreover, Hai-yen 海晏[3] produces rushes (*lu* 蘆, *Phragmites communis*) capable of being made into staves for support. P'an čou 潘州[4] produces thousand-years ferns 千歲蕨 and walking-sticks which are small and resemble the palmyra palm 貝多 (*Borassus flabelliformis*). There is, further, the *su-tsie* bamboo 疎節竹, from which staves are abundantly made for the Buddhist and Taoist clergy,— all singular objects. According to the *Hui tsui* 會最, the *t'uṅ* 通 bamboo from the Čen River 湊川 is straight, without knots in its upper parts, and hollow."

The *Ko ku yao lun*[5] states that the square bamboo is produced in western Se-č'wan, and also grows on the mountain Fei-lai-fuṅ 飛來峯 on the West Lake in Če-kiaṅ; the knots of this bamboo are prickly, hence it is styled in Se-č'wan *tse ču* 刺竹 ("prickly bamboo").

According to the *Min siao ki* 閩小記,[6] written by Čou Liaṅ-kuṅ 周亮工 in the latter part of the seventeenth century, square bamboo and staves made from it are produced in the district of Yuṅ-tiṅ 永定 in the prefecture of T'iṅ-čou and in the district of T'ai-niṅ 泰寧 in the prefecture of Šao-wu, both in Fu-kien Province.[7]

---

[1] Ch. 3, p. 10 b (ed. of Lu Sin-yüan); see above, p. 268.

[2] In the prefecture of Liu-čou, Kwaṅ-si.

[3] Explained in the commentary as the name of a locality, but its situation is not indicated and is unknown to me.

[4] The present Mou-miṅ hien, forming the prefectural city of Kao-čou fu, Kwaṅ-tuṅ.

[5] Ch. 8, p. 9 (ed. of Si yin hüan ts'uṅ šu).

[6] Ed. of Šwo liṅ, p. 17.

[7] The *Šan hai kiṅ* mentions the "narrow bamboo (*hia ču* 狹竹) growing in abundance on the Tortoise Mountain"; and Kwo P'o (A.D. 276-324), in his commentary to this work, identifies with it the bamboo of Kiuṅ. According to the *Kwaṅ či*, the Kiuṅ bamboo occurred in the districts of Nan-kwaṅ 南廣 (at present Nan-k'i 南溪) and Kiuṅ-tu in Se-č'wan. The Memoirs of Mount Lo-fou (*Lo-fou šan ki*) in Kwaṅ-tuṅ state that the Kiuṅ bamboo was originally produced on Mount Kiuṅ, being identical with that noticed by Čaṅ K'ien in Ta-hia, and that village-elders use it as a staff. A treatise on bamboo therefore calls it the "bamboo supporting the old" 扶老竹. These texts are cited in the *T'ai p'iṅ yü lan* (Ch. 963, p. 3).

It is said to occur also in the prefecture of Teṅ-čou 登 州, Šan-tuṅ Province, where it is likewise made into walking-sticks.[1] The latter being much in demand by Buddhist monks, the bamboo has received the epithet "Lo-han bamboo" (bamboo of the Arhat).[2]

It is perfectly manifest that what was exported from Se-č'wan by way of Yūn-nan into India, and thence forwarded to Bactria, was the square bamboo in the form of walking-canes. India is immensely rich in bamboos; and only a peculiar variety, which did not exist in India, could have compensated for the trouble and cost which this long and wearisome trade-route must have caused in those days. ·For years, I must confess, it has been a source of wonder to me why Se-č'wan bamboo should have been carried as far as Bactria, until I encountered the text of the *Pei hu lu*, which gives a satisfactory solution of the problem.[3]

2. The most important article by which the Chinese became famously known in ancient times, of course, was silk. This subject is so extensive, and has so frequently been treated in special monographs, that it does not require recapitulation in this place. I shall only recall the fact that the Chinese silk materials, after traversing Central Asia, reached the Iranian Parthians, who acted as mediators in this trade with the anterior Orient.[4] It is assumed that the introduction of seri-culture into Persia, especially into Gilan, where it still flourishes, falls in the latter part of the Sasanian epoch. It is very probable that the acquaintance of the Khotanese with the rearing of silkworms, introduced by a Chinese princess in A.D. 419, gave the impetus to a further growth of this new industry in a western direction, gradually spreading to Yarkand, Fergana, and Persia.[5] Chinese brocade (*dībā-i čīn*) is fre-quently mentioned by Firdausī as playing a prominent part in Persian decorations.[6] He also speaks of a very fine and decorated Chinese silk under the name *parniyān*, corresponding to Middle Persian *parnīkān*.[7] Iranian has a peculiar word for "silk," not yet satisfactorily explained: Pahlavi *aprēšum, *aparēšum; New Persian *abrēšum, abrēšam* (Arme-

---

[1] *Šan tuṅ t'uṅ či*, Ch. 9, p. 6.

[2] See *K'ien šu* 黔 書, Ch. 4, p. 7 b (in *Yüe ya t'aṅ ts'uṅ šu, t'ao* 24) and *Sü K'ien šu*, Ch. 7, p. 2 b (*ibid.*). Cf. also *Čú p'u siaṅ lu* 竹 譜 詳 錄, written by Li K'an 李 衎 in 1299 (Ch. 4, p. 1 b; ed. of *Či pu tsu čai ts'uṅ šu*).

[3] The speculations of J. MARQUART (Eranšahr, pp. 319–320) in regard to this bamboo necessarily fall to the ground. There is no misunderstanding on the part of Čaṅ K'ien, and the account of the *Si ki* is perfectly correct and clear.

[4] HIRTH, Chinesische Studien, p. 10.

[5] SPIEGEL, Eranische Altertumskunde, Vol. I, p. 256.

[6] J. J. MODI, Asiatic Papers, p. 254 (Bombay, 1905).

[7] HÜBSCHMANN, Persische Studien, p. 242.

nian, loan-word from Persian, *aprišum*); hence Arabic *ibarīsam* or *ibrīsam;* Pamir dialects *waršum, waršüm,* Šugni *wrežŏm,* etc.; Afghan *wrēšam.*[1] Certain it is that we have here a type not related to any Chinese word for "silk." In this connection I wish to register my utter disbelief in the traditional opinion, inaugurated by KLAPROTH, that Greek *ser* ("silk-worm"; hence Seres, Serica) should be connected with Mongol *širgek* and Manchu *sirge* ("silk"), the latter with Chinese *se* 絲.[2] My reasons for rejecting this theory may be stated as briefly as possible. I do not see how a Greek word can be explained from Mongol or Manchu,—languages which we merely know in their most recent forms, Mongol from the thirteenth and Manchu from the sixteenth century. Neither the Greek nor the Mongol-Manchu word can be correlated with Chinese *se.* The latter was never provided with a final consonant. Klaproth resorted to the hypothesis that in ancient dialects of China along the borders of the empire a final *r* might (*peut-être*) have existed. This, however, was assuredly not the case. We know that the termination *'r* 兒, so frequently associated with nouns in Pekingese, is of comparatively recent origin, and not older than the Yüan period (thirteenth century); the beginnings of this usage may go back to the end of the twelfth or even to the ninth century.[3] At any rate, it did not exist in ancient times when the Greek *ser* came into being. Moreover, this suffix *'r* is not used arbitrarily: it joins certain words, while others take the suffix *tse* 子, and others again do not allow any suffix. The word *se,* however, has never been amalgamated with *'r.* In all probability, its ancient phonetic value was *si, sa. It is thus phonetically impossible to derive from it the Mongol-Manchu word or Korean *sir,* added by Abel-Rémusat. I do not deny that this series may have its root in a Chinese word, but its parentage cannot be traced to *se.* I do

---

[1] HÜBSCHMANN, Arm. Gram., p. 107; HORN, Neupers. Etymologie, No. 65. The derivation from Sanskrit *kṣauma* is surely wrong. Bulgar *ibrišim,* Rumanian *ibrišin,* are likewise connected with the Iranian series.

[2] Cf. KLAPROTH, Conjecture sur l'origine du nom de la soie chez les anciens (*Journal asiatique,* Vol. I, 1822, pp. 243–245, with additions by ABEL-RÉMUSAT, 245–247); Asia polyglotta, p. 341; and Mémoires relatifs à l'Asie, Vol. III, p. 264. Klaproth's opinion has been generally, but thoughtlessly, accepted (HIRTH, *op. cit.,* p. 217; F. v. RICHTHOFEN, China, Vol. I, p. 443; SCHRADER, Reallexikon, p. 757). PELLIOT (*T'oung Pao,* 1912, p. 741), I believe, was the first to point out that Chinese *se* was never possessed of a final consonant.

[3] See my note in *T'oung Pao,* 1916, p. 77; and H. MASPERO, Sur quelques textes anciens de chinois parlé, p. 12. Maspero encountered the word *mao'r* ("cat") in a text of the ninth century. It hardly makes any great difference whether we conceive *'r* as a diminutive or as a suffix. Originally it may have had the force of a diminutive, and have gradually developed into a pure suffix. Cf. also P. SCHMIDT, K istorii kitaiskago razgovornago yazyka, in Sbornik stat'ei professorov, p. 19 (Vladivostok, 1917).

not believe, either, that Russian *šolk* ("silk"), as is usually stated (even by Dal'), is derived from Mongol *širgek:* first of all, the alleged phonetic coincidence is conspicuous by its absence; and, secondly, an ancient Russian word cannot be directly associated with Mongol; it would be necessary to trace the same or a similar word in Turkish, but there it does not exist; "silk" in Turkish is *ipäk, torgu, torka,* etc. It is more probable that the Russian word (Old Slavic *šelk,* Lithuanian *szilkaĩ*), in the same manner as our *silk,* is traceable to *sericum.* There is no reason to assume that the Greek words *ser, Sera, Seres,* etc., have their origin in Chinese. This series was first propagated by Iranians, and, in my opinion, is of Iranian origin (cf. New Persian *sarah,* "silk"; hence Arabic *sarak*).

· Persian *kimxāw* or *kamxāb, kamxā, kimxā* (Arabic *kīmxāw,* Hindustānī *kamxāb*), designating a "gold brocade," as I formerly explained,[1] may be derived from Chinese 錦 花 *kin-hwa,* *kim-xwa.

3-4. Of fruits, the West is chiefly indebted to China for the peach (*Amygdalus persica*) and the apricot (*Prunus armeniaca*). It is not impossible that these two gifts were transmitted by the silk-dealers, first to Iran (in the second or first century B.C.), and thence to Armenia, Greece, and Rome (in the first century A.D.). In Rome the two trees appear as late as the first century of the Imperium, being mentioned as *Persica* and *Armeniaca arbor* by Pliny[2] and Columella. Neither tree is mentioned by Theophrastus, which is to say that they were not noted in Asia by the staff of Alexander's expedition.[3] De Candolle has ably pleaded for China as the home of the peach and apricot, and Engler[4] holds the same opinion. The zone of the wild apricot may well extend from Russian Turkistan to Sungaria, south-eastern Mongolia, and the Himalaya; but the historical fact remains that the Chinese have been the first to cultivate this fruit from ancient times. Previous authors have justly connected the westward migration of peach and apricot with the lively intercourse of China and western Asia following Čan K'ien's mission.[5] Persian has only descriptive names for these fruits, the peach being termed *šaft-ālu* ("large plum"), the apricot *zard-ālu*

---

[1] *T'oung Pao,* 1916, p. 477; Yule, Hobson-Jobson, p. 484.

[2] xv, 11, 13.

[3] De Candolle (Origin of Cultivated Plants, p. 222) is mistaken in crediting Theophrastus with the knowledge of the peach. Joret (Plantes dans l'antiquité, p. 79) has already pointed out this error, and it is here restated for the benefit of those botanists who still depend on de Candolle's book.

[4] In Hehn, Kulturpflanzen, p. 433.

[5] Joret, *op. cit.,* p. 81; Schrader in Hehn, p. 434.

("yellow plum").[1] Both fruits are referred to in Pahlavi literature (above, pp. 192, 193).

As to the transplantation of the Chinese peach into India, we have an interesting bit of information in the memoirs of the Chinese pilgrim Hüan Tsaṅ.[2] At the time of the great Indo-Scythian king Kaniṣka, whose fame spread all over the neighboring countries, the tribes west of the Yellow River (Ho-si in Kan-su) dreaded his power, and sent hostages to him. Kaniṣka treated them with marked attention, and assigned to them special mansions and guards of honor. The country where the hostages resided in the winter received the name Cīnabhukti ("China allotment," in the eastern Panjāb). In this kingdom and throughout India there existed neither pear nor peach. These were planted by the hostages. The peach therefore was called cīnanī ("Chinese fruit"); and the pear, cīnarājaputra ("crown-prince of China"). These names are still prevalent.[3] Although Hüan Tsaṅ recorded in A.D. 630 an oral tradition overheard by him in India, and relative to a time lying back over half a millennium, his well-tested trustworthiness cannot be doubted in this case: the story thus existed in India, and may indeed be traceable to an event that took place under the reign of Kaniṣka, the exact date of which is still controversial.[4] There are mainly two reasons which prompt me to accept Hüan Tsaṅ's account. From a botanical point of view, the peach is not a native of India. It occurs there only

---

[1] In the Pamir languages we meet a common name for the apricot, Minjan *čerī*, Waxī *čiwān* or *čoān* (but Sariqolī *nōš*, Signi *naž*). The same type occurs in the Dardu languages (*jui* or *ji* for the tree, *jarote* or *jorote* for the fruit, and *juru* for the ripe fruit) and in Kāçmīrī (*tser, tser-kul*); further, in West-Tibetan *ču-li* or *čo-li*, Balti *su-ri*, Kanaurī *čul* (other Tibetan words for "apricot" are *k'am-bu, a-šu,* and *ša-rag*, the last-named being dried apricots with little pulp and almost as hard as a stone). KLAPROTH (*Journal asiatique*, Vol. II, 1823, p. 159) has recorded in Bukhāra a word for the apricot in the form *tserduli*. It is not easy to determine how this type has migrated. TOMASCHEK (Pamir-Dialekte, p. 791) is inclined to think that originally it might have been Tibetan, as Baltistan furnishes the best apricots. For my part, I have derived the Tibetan from the Pamir languages (*T'oung Pao*, 1916, p. 82). The word is decidedly not Tibetan; and as to its origin, I should hesitate only between the Pamir and Dardu languages.

[2] *Ta T'aṅ Si yü ki*, Ch. 4, p. 5.

[3] There are a few other Indian names of products formed with "China": *cīnapiṣṭa* ("minium"), *cīnaka* ("Panicum miliaceum, fennel, a kind of camphor"), *cīnakarpūra* ("a kind of camphor"), *cīnavaṅga* ("lead").

[4] Cf. V. A. SMITH, Early History of India, 3d ed., p. 263 (I do not believe with Smith that "the territory of the ruler to whose family the hostages belonged seems to have been not very distant from Kashgar"; the Chinese term Ho-si, at the time of the Han, comprised the present province of Kan-su from Lan-čou to An-si); T. WATTERS, On Yüan Chwang's Travels, Vol. I, pp. 292–293 (his comments on the story of the peach miss the mark, and his notes on the name Cīna are erroneous; see also PELLIOT, *Bull. de l'Ecole française*, Vol. V, p. 457).

in a cultivated state, and does not even succeed well, the fruit being mediocre and acid.[1] There is no ancient Sanskrit name for the tree; nor does it play any rôle in the folk-lore of India, as it does in China. Further, as regards the time of the introduction, whether the reign of Kaniṣka be placed in the first century before or after our era, it is singularly synchronous with the transplantation of the tree into western Asia.

5. As indicated by the Persian name *dār-čīnī* or *dār-čīn* ("Chinese wood" or "bark"; Arabic *dār ṣīnī*), cinnamon was obtained by the Persians and Arabs from China.[2] Ibn Khordādzbeh, who wrote between A.D. 844 and 848, is the first Arabic author who enumerates cinnamon among the products exported from China.[3] The Chinese export cannot have assumed large dimensions: it is not alluded to in Chinese records, Čao Žu-kwa is reticent about it.[4] Ceylon was always the main seat of cinnamon production, and the tree (*Cinnamomum zeylanicum*) is a native of the Ceylon forests.[5] The bark of this tree is also called *dar-čīnī*. It is well known that cassia and cinnamon are mentioned by classical authors, and have given rise to many sensational speculations as to the origin of the cinnamon of the ancients. Herodotus[6] places cinnamon in Arabia, and tells a wondrous story as to how it is gathered. Theophrastus[7] seeks the home of cassia and cinnamomum, together with frankincense and myrrh, in the Arabian peninsula about Saba, Hadramyt, Kitibaina, and Mamali. Strabo[8] locates it in the land of the Sabæans, in Arabia, also in Ethiopia and southern India; finally he has a "cinnamon-bearing country" at the end of the habitable countries of the south, on the shore of the Indian ocean.[9] Pliny[10] has cinnamomum or cinnamum grow in the country of the Ethiopians, and it is carried over sea on rafts by the Troglodytae.

---

[1] C. JORET, Plantes dans l'antiquité, Vol. II, p. 281.

[2] LECLERC, Traité des simples, Vol. II, pp. 68, 272. The loan-word *daričenik* in Armenian proves that the word was known in Middle Persian (*dār-i čēnik); cf. HÜBSCHMANN, Armen. Gram., p. 137.

[3] G. FERRAND, Textes relatifs à l'Extrême-Orient, p. 31.

[4] SCHOFF (Periplus, p. 83) asserts that between the third and sixth centuries there was an active sea-trade in this article in Chinese ships from China to Persia. No reference is given. I wonder from what source this is derived.

[5] DE CANDOLLE, Origin of Cultivated Plants, p. 146; WATT, Commercial Products of India, p. 313.

[6] III, 107, 111.

[7] Hist. plant., IX. IV, 2.

[8] XV. IV, 19; XVI. IV, 25; XV. I, 22.

[9] I. IV, 2.

[10] XII, 42.

The descriptions given of cinnamon and cassia by Theophrastus[1] show that the ancients did not exactly agree on the identity of these plants, and Theophrastus himself speaks from hearsay ("In regard to cinnamon and cassia they say the following: both are shrubs, it is said, and not of large size. . . . Such is the account given by some. Others say that cinnamon is shrubby or rather like an under-brush, and that there are two kinds, one black, the other white"). The difference between cinnamon and cassia seems to have been that the latter possessed stouter branches, was very fibrous, and difficult to strip off the bark. This bark was used; it was bitter, and had a pungent odor.[2]

Certain it is that the two words are of Semitic origin.[3] The fact that there is no cinnamon in Arabia and Ethiopia was already known to GARCIA DA ORTA.[4] An unfortunate attempt has been made to trace the cinnamon of the ancients to the Chinese.[5] This theory has thus been formulated by MUSS-ARNOLT:[6] "This spice was imported by Phœnician merchants from Egypt, where it is called *khisi-t*. The Egyptians, again, brought it from the land of Punt, to which it was imported from Japan, where we have it under the form *kei-chi* ('branch of the cinnamon-tree'), or better *kei-shin* ('heart of the cinnamon') [read *sin*, *sim]. The Japanese itself is again borrowed from the Chinese *kei-ši* [?]. The *-t* in the Egyptian represents the feminine suffix." As may be seen from O. SCHRADER,[7] this strange hypothesis was first put forward in 1883 by C. SCHUMANN. Schrader himself feels somewhat sceptic about it, and regards the appearance of Chinese merchandise on the markets of Egypt at such an early date as hardly probable. From a sinological viewpoint, this speculation must be wholly rejected, both in its linguistic and its historical bearings. Japan was not in existence in 1500 B.C., when cinnamon-wood of the country Punt is spoken of in the Egyptian inscriptions; and China was then a small agrarian inland community restricted to the northern part of the present empire, and

---

[1] Hist. plant., IX. v, 1–3.

[2] Theophrastus, IX. v, 3.

[3] Greek κασία is derived from Hebrew *qeṣṭ'ā*, perhaps related to Assyrian *kasu*, *kasiya* (POGNON, *Journal asiatique*, 1917, I, p. 400). Greek *kinnamomon* is traced to Hebrew *qinnamōn* (Exodus, XXX, 23).

[4] MARKHAM, Colloquies, pp. 119–120.

[5] Thus also FLÜCKIGER and HANBURY (Pharmacographia, p. 520), whose argumentation is not sound, as it lacks all sense of chronology. The Persian term *dar-čīnī*, for instance, is strictly of mediæval origin, and cannot be invoked as evidence for the supposition that cinnamon was exported from China many centuries before Christ.

[6] *Transactions Am. Phil. Assoc.*, Vol. XXIII, 1892, p. 115.

[7] Reallexikon, p. 989.

not acquainted with any *Cassia* trees of the south. Certainly there was no Chinese navigation and sea-trade at that time. The Chinese word *kwei* 桂 (*kwai, kwi) occurs at an early date, but it is a generic term for *Lauraceae;* and there are about thirteen species of *Cassia*, and about sixteen species of *Cinnamomum*, in China. The essential point is that the ancient texts maintain silence as to cinnamon; that is, the product from the bark of the tree. *Cinnamomum cassia* is a native of Kwaṅ-si, Kwaṅ-tuṅ, and Indo-China; and the Chinese made its first acquaintance under the Han, when they began to colonize and to absorb southern China. The first description of this species is contained in the *Nan fan ts'ao mu čwaṅ* of the third century.[1] This work speaks of large forests of this tree covering the mountains of Kwaṅ-tuṅ, and of its cultivation in gardens of Kiao-či (Tonking). It was not the Chinese, but non-Chinese peoples of Indo-China, who first brought the tree into cultivation, which, like all other southern cultivations, was simply adopted by the conquering Chinese. The medicinal employment of the bark (*kwei p'i* 桂皮) is first mentioned by T'ao Huṅ-kiṅ (A.D. 451–536), and probably was not known much earlier. It must be positively denied, however, that the Chinese or any nation of Indo-China had any share in the trade which brought cinnamon to the Semites, Egyptians, or Greeks at the time of Herodotus or earlier. The earliest date we may assume for any navigation from the coasts of Indo-China into the Indian Ocean is the second century B.C.[2] The solution of the cinnamon problem of the ancients seems simpler to me than to my predecessors. First, there is no valid reason to assume that what our modern botany understands by *Cassia* and *Cinnamomum* must be strictly identical with the products so named by the ancients. Several different species are evidently involved. It is perfectly conceivable that in ancient times there was a fragrant bark supplied by a certain tree of Ethiopia or Arabia or both, which is either extinct or unknown to us, or, as Fée inclines to think, a species of *Amyris*. It is further legitimate to conclude, without forcing the evidence, that the greater part of the cinnamon supply came from Ceylon and India,[3] India being expressly included by Strabo. This, at least, is infinitely more reasonable than acquiescing in the wild fantasies of a Schumann or Muss-Arnolt, who lack the most elementary knowledge of East-Asiatic history.

6. The word "China" in the names of Persian and Arabic products,

---

[1] The more important texts relative to the subject are accessible in BRETSCHNEIDER, Bot. Sin., pt. III, No. 303.

[2] Cf. PELLIOT, *T'oung Pao*, 1912, pp. 457–461.

[3] The Malabar cinnamon is mentioned by Marco Polo (YULE's ed., Vol. II, p. 389) and others.

or the attribution of certain products to China, is not always to be understood literally. Sometimes it merely refers to a far-eastern product, sometimes even to an Indian product,[1] and sometimes to products handled and traded by the Chinese, regardless of their provenience. Such cases, however, are exceptions. As a rule, these Persian-Arabic terms apply to actual products of China.

SCHLIMMER[2] mentions under the name *Killingea monocephala* the zedoary of China: according to Piddington's Index Plantarum, it should be the plant furnishing the famous root known in Persia as *jadwāre xitāi* ("Chinese jadvār"); genuine specimens are regarded as a divine panacea, and often paid at the fourfold price of fine gold. The identification, however, is hardly correct, for *K. monocephala* is *kin niu ts'ao* 金牛草 in Chinese,[3] which hardly holds an important place in the Chinese pharmacopœia. The plant which Schlimmer had in mind doubtless is *Curcuma zedoaria*, a native of Bengal and perhaps of China and various other parts of Asia.[4] It is called in Sanskrit *nirviṣā* ("poisonless") or *ṣida*, in Kuča or Tokharian B *viralom* or *wiralom*,[5] Persian *jadvār*, Arabic *zadvār* (hence our *zedoary*, French *zedoaire*). Abu Mansur describes it as *zarvār*, calling it an Indian remedy similar to Costus and a good antidote.[6] In the middle ages it was a much-desired article of trade bought by European merchants in the Levant, where it was sold as a product of the farthest east.[7] Persian *zarumbād*, Arabic *zeronbād*, designating an aromatic root similar to zedoary, resulted in our *zerumbet*.[8] While it is not certain that *Curcuma zedoaria* occurs in China (a Chinese name is not known to me), it is noteworthy that the Persians, as indicated above, ascribe to the root a Chinese origin: thus also *kažūr* (from Sanskrit *karcūra*) is explained in the Persian Dictionary of

---

[1] Such an example I have given in *T'oung Pao*, 1915, p. 319: *bīš*, an edible aconite, does not occur in China, as stated by Damīrī, but in India. In regard to cubebs, however, GARCIA DA ORTA (C. MARKHAM, Colloquies, p. 169) was mistaken in denying that they were grown in China, and in asserting that they are called *kabāb-čīnī* only because they are brought by the Chinese. As I have shown (*ibid.*, pp. 282–288), cubebs were cultivated in China from the Sung period onward.

[2] Terminologie, p. 335.

[3] Also this identification is doubtful (STUART, Chinese Materia Medica, p. 228).

[4] W. ROXBURGH, Flora Indica, p. 8; WATT, Commercial Products of India, p. 444, and Dictionary, Vol. II, p. 669.

[5] S. LÉVI, *Journal asiatique*, 1911, II, pp. 123, 138.

[6] ACHUNDOW, Abu Mansur, p. 79. See also LECLERC, Traité des simples, Vol. I, p. 347.

[7] W. HEYD, Histoire du commerce du levant, Vol. II, p. 676.

[8] YULE, Hobson-Jobson, p. 979.

Steingass as "zedoary, a Chinese root." Further, we read under *māh-parwār* or *parwīn*, "zedoary, a Chinese root like ginger, but perfumed."

7. Abu Mansur distinguishes under the Arabic name *zanjabīl* three kinds of ginger (product of *Amomum zingiber*, or *Zingiber officinale*),— Chinese, Zanzibar, and Melinawi or Zurunbāj, the best being the Chinese.[1] According to STEINGASS,[2] Persian *anqala* denotes "a kind of China ginger."[3] The Persian word (likewise in Arabic) demonstrates that the product was received from India: compare Prākrit *singabēra*, Sanskrit *çṛṅgavera* (of recent origin),[4] Old Arabic *zangabīl*, Pahlavi *šangavīr*, New Persian *šankalīl*, Arabic-Persian *zanjabīl*, Armenian *snrvēl* or *snkrvil* (from *singivēl), Greek ζιγγίβερις, Latin *zingiberi;* Madagasy *šakavīru* (Indian loan-word).[5]

The word *galangal*, denoting the aromatic rhizome of *Alpinia galanga*, is not of Chinese origin, as first supposed by D. HANBURY,[6] and after him by HIRTH[7] and GILES.[8] The error was mainly provoked by the fact that the Arabic word from which the European name is derived was wrongly written by Hanbury *khalanjān*, while in fact it is *khūlanjān* (*xūlandžān*), Persian *xāwalinjān*. The fact that Ibn Khordādzbeh, who wrote about A.D. 844–848, mentions *khūlanjān* as one of the products of China,[9] does not prove that the Arabs received this word from China; for this rhizome is not a product peculiar to China, but is intensively grown in India, and there the Arabs made the first acquaintance of it. Ibn al-Baiṭār[10] states expressly that *khūlanjān* comes from India; and, as was recognized long ago, the Arabic word is derived from Sanskrit *kulañja*,[11] which denotes *Alpinia galanga*. The European forms with *ng* (*galangan, galgan*, etc.) were suggested by the older Arabic pronunciation *khūlangān*.[12] In Middle Greek we have

[1] ACHUNDOW, Abu Mansur, p. 76.

[2] Persian Dictionary, p. 113.

[3] Concerning ginger among the Arabs, cf. LECLERC, Traité des simples, Vol. II, p. 217; and regarding its preparation, see G. FERRAND, Textes relatifs à l'Extrême-Orient, p. 609.

[4] Cf. the discussion of E. HULTZSCH and F. W. THOMAS in *Journal Roy. As. Soc.,* 1912, pp. 475, 1093. See also YULE, Hobson-Jobson, p. 374.

[5] The curious word for "ginger" in Kuča or Tokharian B, *lvāṅkaro* (S. LÉVI, *Journal asiatique,* 1911, II, pp. 124, 137), is not yet explained.

[6] Science Papers, p. 373.

[7] Chinesische Studien, p. 219.

[8] Glossary of Reference, p. 102.

[9] G. FERRAND, Textes relatifs à l'Extrême-Orient, p. 31.

[10] *Ibid.,* p. 259. Cf. also ACHUNDOW, Abu Mansur, p. 60.

[11] ROEDIGER and POTT, Z. K. d. Morgenl., Vol. VII, 1850, p. 128.

[12] E. WIEDEMANN (*Sitzber. Phys.-Med. Soz. Erl.,* Vol. XLV, 1913, p. 44) gives as Arabic forms also *xaulangād* and *xalangān*.

κολούτζία, χαυλιζέν, and γαλαγγά; in Russian, *kalgán*. The whole group has nothing to do with Chinese *kao-lian-kian*.[1] Moreover, the latter refers to a different species, *Alpinia officinarum;* while *Alpinia galanga* does not occur in China, but is a native of Bengal, Assam, Burma, Ceylon, and the Konkan. GARCIA DA ORTA was already well posted on the differences between the two.[2]

8. Abu Mansur mentions the medical properties of *māmīrān*.[3] According to ACHUNDOW,[4] a rhizome originating from China, and called in Turkistan *momiran*, is described by Dragendorff, and is regarded by him as identical with the so-called mishmee (from *Coptis teeta* Wall.), which is said to be styled *mamiračin* in the Caucasus. He further correlates the same drug with *Ranunculus ficaria* (χελιδόνιον τὸ μικρόν), subsequently described by the Arabs under the name *mamirun*. Al-Jafiki is quoted by Ibn al-Baiṭār as saying that the *māmīrān* comes from China, and that its properties come near to those of *Curcuma;*[5] these roots, however, are also a product of Spain, the Berber country, and Greece.[6] The Sheikh Daūd says that the best which comes from India is blackish, while that of China is yellowish. Ibn Baṭūṭa[7] mentions the importation of *māmīrān* from China, saying that it has the same properties as *kurkum*. Hajji Mahomed, in his account of Cathay (ca. 1550), speaks of a little root growing in the mountains of Succuir (Su-čou in Kan-su), where the rhubarb grows, and which they call Mambroni Cini (*māmīrān-i Čīnī*, "mamiran of China"). "This is extremely dear, and is used in most of their ailments, but especially where the eyes are affected. They grind it on a stone with rose-water, and anoint the eyes with it. The result is wonderfully beneficial."[8] In 1583 LEONHART RAUWOLF[9] mentions

---

[1] Needless to say that the vivisections of Hirth, who did not know the Sanskrit term, lack philological method.

[2] MARKHAM, Colloquies, p. 208. Garcia gives *lavandou* as the name used in China; this is apparently a corrupted Malayan form (cf. Javanese *laos*). In Java, he says, there is another larger kind, called *lancuaz;* in India both are styled *lancuaz*. This is Malayan *leṅkūwas*, Makasar *laṅkuwasa*, Čam *lakuah* or *lakuak*, Tagalog *laṅkuas*. The Arabic names are written by Garcia *calvegiam, chamligiam,* and *galungem;* the author's Portuguese spelling, of course, must be taken into consideration.

[3] ACHUNDOW, Abu Mansur, p. 138.

[4] *Ibid.*, p. 268.

[5] LECLERC, Traité des simples, Vol. II, p. 441. Dioscorides remarks that the sap of this plant has the color of saffron.

[6] In Byzantine Greek it is μαμηρέ or μεμηρέν, derived from the Persian-Arabic word.

[7] Ed. of DEFRÉMERY and SANGUINETTI, Vol. II, p. 186.

[8] YULE, Cathay, new ed., Vol. I, p. 292.

[9] Beschreibung der Raiss inn die Morgenländer, p. 126.

the drug *mamirani tchini* for eye-diseases, being yellowish like *Curcuma*.

Bernier mentions mamiran as one of the products brought by the caravans from Tibet. Also according to a modern Mohammedan source, mamiran and rhubarb are exported from Tibet.[1]

Mamīra is a reputed drug for eye-diseases, applied to bitter roots of kindred properties but of different origin. By some it is regarded as the rhizome of *Coptis teeta* (*tīta* being the name of the drug in the Mishmi country); by others, from *Thalictrum foliosum*, a tall plant common throughout the temperate Himalaya and in the Kasia Hills.[2] In another passage, however, YULE[3] suggests that this root might be the ginseng of the Chinese, which is highly improbable.

It is most likely that by *mamira* is understood in general the root of *Coptis teeta*. This is a ranunculaceous plant, and the root has sometimes the appearance of a bird's claw. It is shipped in large quantities from China (Chinese *hwaṅ-lien* 黃連) *via* Singapore to India. The Chinese regard it as a panacea for a great many ills; among others, for clearing inflamed eyes.

9. Abu Mansur discriminates between two kinds of rhubarb,— the Chinese (*rīwand-i sīnī*) and that of Khorasan, adding that the former is most employed.[4] Accordingly a species of rhubarb (probably *Rheum ribes*) must have been indigenous to Persia. Yāqūt says that the finest kind grew in the soil of Nīšāpūr.[5] According to E. BOISSIER,[6] *Rheum ribes* occurs near Van and in Agerowdagh in Armenia, on Mount Pir Omar Gudrun in Kurdistan, in the Daëna Mountain of eastern Persia, near Persepolis, in the province Aderbeijan in northern Persia, and in the mountains of Baluchistan. There is a general Iranian name for "rhubarb": Middle Persian *rēwās*, New Persian *rēwās, rēwand, rīwand* (hence Armenian *erevant*), Kurd *rīwās, rībās;* Balučī *ravaš;* Afghan *rawāš.*[7] The Persian name has penetrated in the same form into Arabic

---

[1] CH. SCHEFER, Histoire de l'Asie centrale par Mir Abdoul Kerim Boukhary, p. 239. Cf. also R. DOZY, Supplément aux dictionnaires arabes, Vol. II, p. 565.

[2] YULE, Hobson-Jobson, p. 548.

[3] Cathay, Vol. I, p. 292.

[4] ACHUNDOW, Abu Mansur, p. 74. Chinese rhubarb is also called simply *čīnī* ("Chinese") in Persian, *ṣīnī* in Arabic.

[5] BARBIER DE MEYNARD, Dict. géogr. de la Perse, p. 579.

[6] Flora Orientalis, Vol. IV, p. 1004. *Rheum ribes* does not occur in China or Central Asia.

[7] The Afghan word in particular refers to *Rheum spiciforme*, which grows wild and abundantly in many parts of Afghanistan. When green, the leaf-stalks are called *rawāš;* and when blanched by heaping up stones and gravel around them, *čukri;* when fresh, they are eaten either raw or cooked (WATT, Dictionary, Vol.VI, p. 487). The species under notice occurs also in Kan-su, China: FORBES and

and Turkish, likewise into Russian as *reven'* and into Serbian as *reved*. It is assumed also that Greek ῥῆον (from *rewon) and ῥᾶ are derived from Iranian, and it is more than likely that Iran furnished the rhubarb known to the ancients. The two Greek names first appear in Dioscorides,[1] who states that the plant grows in the regions beyond the Bosporus, for which reason it was subsequently styled *rha ponticum* or *rha barbarum* (hence our *rhubarb*, Spanish *ruibarbo*, Italian *rabarbaro*, French *rhubarbe*),— an interesting case analogous to that of the Hu plants of the Chinese. In the fourth century, Ammianus Marcellinus[2] states that the plant receives its name from the River Rha ('Pᾱ, Finnish Rau, Rawa), on the banks of which it grows. This is the Volga, but the plant does not occur there. It is clear that Ammianus' opinion is erroneous, being merely elicited by the homophony of the names of the plant and the river. Pliny[3] describes a root termed *rhacoma*, which when pounded yields a color like that of wine but inclining to saffron, and which was brought from beyond the Pontus. Certain it is that this drug represents some species of *Rheum*, in my opinion identical with that of Iran.[4] There is no reason to speculate, as has been done by some authors, that the rhubarb of the ancients came from China; for the Chinese did not know rhubarb, as formerly assumed, from time immemorial. This is shown at the outset by the composite name *ta hwan* 大黃 ("the great yellow one") or *hwan lian* 黃良 ("the yellow good one"), merely descriptive attributes, while for all genuinely ancient plants there is a root-word of a single syllable. The alleged mention of rhubarb in the *Pen kiṅ* or *Pen ts'ao*, attributed to the mythical Emperor Šen-nuṅ, proves nothing; that work is entirely spurious, and the text in which we have it at present is a reconstruction based on quotations in the preserved *Pen-ts'ao* literature, and teems with interpolations and anachronisms.[5] All that is certain is that rhubarb was known to the

HEMSLEY, *Journal Linnean Soc.*, Vol. XXVI, p. 355. There is accordingly no reason to seek for an outside origin of the Iranian word (cf. SCHRADER, Reallexikon, p. 685). The Iranian word originally designated an indigenous Iranian species, and was applied to *Rheum officinale* and *palmatum* from the tenth century onward, when the roots of these species were imported from China.

[1] III, 2. Theophrastus is not acquainted with this genus.

[2] XXII. VIII, 28.

[3] XXVII, 105.

[4] FLÜCKIGER and HANBURY (Pharmacographia, p. 493) state, "Whether produced in the regions of the Euxine (Pontus), or merely received thence from remoter countries, is a question that cannot be solved." The authors are not acquainted with the Iranian species, and their scepticism is not justified.

[5] It is suspicious that, according to Wu P'u of the third century, Šen Nuṅ and Lei Kuṅ ascribed poisonous properties to *ta hwan*, while this in fact is not true. The *Pen kiṅ* (according to others, the *Pie lu*) states that it is non-poisonous.

Chinese in the age of the Han, for the name *ta hwan* occurs on one of the wooden tablets of that period discovered in Turkistan by Sir A. Stein and deciphered by CHAVANNES.[1]

Abu Mansur, as cited above, is the first Persian author who speaks of Chinese rhubarb. He is followed by a number of Arabic writers. It is therefore reasonable to infer that only in the course of the tenth century did rhubarb develop into an article of trade from China to western Asia. In 1154 Edrīsī mentions rhubarb as a product of China growing in the mountains of Buthink (perhaps north-eastern Tibet).[2] Ibn Sa'īd, who wrote in the thirteenth century, speaks of the abundance of rhubarb in China.[3] Ibn al-Baiṭār treats at great length of *rawend*, by which he understands Persian and Chinese rhubarb,[4] and of *rībās*, "very common in Syria and the northern countries," identified by LECLERC with *Rheum ribes*.[5]

MARCO POLO relates that rhubarb is found in great abundance over all mountains of the province of Sukchur (Su-čou in Kan-su), and that merchants go there to buy it, and carry it thence all over the world.[6] In another passage he attributes rhubarb also to the mountains around the city of Su-čou in Kian-su,[7] which, Yule says, is believed by the most competent authorities to be quite erroneous. True it is that rhubarb has never been found in that province or anywhere in middle China; neither is there an allusion to this in Chinese accounts, which restrict the area of the plant to Šen-si, Kan-su, Se-č'wan, and Tibet. Nevertheless it would not be impossible that at Polo's time a sporadic attempt was made to cultivate rhubarb in the environs of Su-čou. Friar Odoric mentions rhubarb for the province Kansan (Kan-su), growing in such abundance that you may load an ass with it for less than six groats.[8]

Chinese records tell us very little about the export-trade in this article. Čao Žu-kwa alone mentions rhubarb among the imports of

---

[1] Documents chinois découverts dans les sables du Turkestan oriental, p. 115, No. 527.

[2] W. HEYD, Histoire du commerce du levant, Vol. II, p. 665. See also FLÜCKIGER and HANBURY, Pharmacographia, pp. 493–494.

[3] G. FERRAND, Textes relatifs à l'Extrême-Orient, p. 350.

[4] LECLERC, Traité des simples, Vol. II, pp. 155–164.

[5] *Ibid.*, p. 190. This passage was unknown to me when I identified above the Persian term *rīwand* with this species, arriving at this conclusion simply by consulting Boissier's Flora.

[6] YULE, Marco Polo, Vol. I, p. 217.

[7] *Ibid.*, Vol. II, p. 181.

[8] YULE, Cathay, Vol. II, p. 247.

San-fu-ts'i (Palembang) and Malabar.[1] In vain also should we look in Chinese books for anything on the subject that would correspond to the importance attached to it in the West.

GARCIA DA ORTA (1562) held it for certain that "all the rhubarb that comes from Ormuz to India first comes from China to Ormuz by the province of Uzbeg which is part of Tartary. The fame is that it comes from China by land, but some say that it grows in the same province, at a city called Çamarcander (Samarkand).[2] But this is very bad and of little weight. Horses are purged with it in Persia, and I have also seen it so used in Balagate. It seems to me that this is the rhubarb which in Europe we called *ravam turquino*, not because it is of Turkey but from there." He emphasizes the point that there is no other rhubarb than that from China, and that the rhubarb coming to Persia or Uzbeg goes thence to Venice and to Spain; some goes to Venice by way of Alexandria, a good deal by Aleppo and Syrian Tripoli, all these routes being partly by sea, but chiefly by land;[3] the rhubarb is not so much powdered, for it is more rubbed in a month at sea than in a year going by land.[4] As early as the thirteenth century at least, as we see from Ibn al-Baitār, what was known to the Arabs as "rhubarb of the Turks or the Persians," in fact hailed from China. In the same manner, it was at a later time that in Europe "Russian, Turkey, and China rhubarb" were distinguished, these names being merely indicative of the various routes by which the drug was conveyed to Europe from China.[5] Also CHRISTOVAL ACOSTA notes the corruption of rhubarb at sea and its overland transportation to Persia, Arabia, and Alexandria.[6]

---

[1] HIRTH, Chau Ju-kua, pp. 61, 88.

[2] Probably *Rheum ribes*, mentioned above.

[3] LEONHART RAUWOLF (Beschreibung der Raiss inn die Morgenländer, 1583, p. 461) reports that large quantities of rhubarb are shipped from India to Aleppo both by sea and by land.

[4] Cf. MARKHAM, Colloquies, pp. 390–392.

[5] In regard to the Russian trade in rhubarb see G. CAHEN, Le livre de comptes de la caravane russe à Pékin, p. 108 (Paris, 1911).

[6] Reobarbaro (medicina singular, y digna de ser de todo el linage humano venerada) se halla solamente dentro de la China, de donde lo traen a vender a Cätaon (que es el puerto de mas comercio de la China, donde estan los Portugueses) y de alli vienẽ por mar a la India: y deste que viene por mar no se haze mucho caso, por venir, por la mayor parte corrõpido (por quanto el Reobarbaro se corrõpe cõ mucha facilidad enla mar) y dela misma tierra d^etro de la China, lo lleuan a la Tartaria, y por la prouincia de Vzbeque lo lleuã a Ormuz, y a toda la Persia, Arabia, y Alexãdria: de dõde se distribuye por toda la Europa (Tractado de las drogas, y medicinas de las Indias Orientales, p. 287, Burgos, 1576). Cf. also LINSCHOTEN (Vol. II, p. 101, ed. of Hakluyt Society), who, as in most of his notices of Indian products, exploits Garcia.

JOHN GERARDE[1] illustrates the rhubarb-plant and annotates, "It is brought out of the countrie of Sina (commonly called China) which is towarde the east in the upper part of India, and that India which is without the river Ganges: and not at all Ex Scenitarum provincia, (as many do unadvisedly thinke) which is in Arabia the happie, and far from China," etc. "The best rubarbe is that which is brought from China fresh and newe," etc.

WATT[2] gives a Persian term *rēvande-hindi* ("Indian rhubarb") for *Rheum emodi*. Curiously, in Hindustānī this is called *Hindi-rēvand čīnī* ("Chinese rhubarb of India"), and in Bengālī *Bangla-rēvan čīnī* ("Chinese rhubarb of Bengal"), indicating that the Chinese product was preëminently in the minds of the people, and that the Himalayan rhubarbs were only secondary substitutes.

10. Abu Mansur[3] mentions under the Arabic name *ratta* a fruit called "Indian hazel-nut" (*bunduq-i hindī*), also Chinese *Salsola kali*. It is the size of a small plum, contains a small blackish stone, and is brought from China. It is useful in chronic diseases and in cases of poisoning, and is hot and dry in the second degree. This is *Sapindus mukorossi*, in Chinese *wu* (or *mu*)*-hwan-tse* 無 (or 木) 患 子 (with a number of synonymes), the seeds being roasted and eaten.

11. Arabic *suk*, a drug composed of several ingredients, according to Ibn Sina, was originally a secret Chinese remedy formed with *amlaj* (Sanskrit *āmalaka*, *Phyllanthus emblica*, the emblic myrobalan).[4] It is the 菴 摩 勒 *an-mo-lo*, **an-mwa-lak*, of the Chinese.[5] In Persian it is *amala* or *amula*.

12. Persian *guli xairā* (*xairū*) is explained as Chinese and Persian hollyhock (*Althæa rosea*).[6] This is the *šu k'wei* 蜀 葵 ("mallow of Se-č'wan") of the Chinese, also called *žuṅ k'wei* ("mallow of the Žuṅ"). It is the common hollyhock, which STUART[7] thinks may have been originally introduced into China from some western country.

13. Ibn al-Baiṭār[8] speaks of a "rose of China" (*ward sīnī*), usually called *nisrīn*. According to Leclerc, this is a malvaceous plant. In Persian we find *gul-čīnī* ("rose of China"), the identification of which,

---

[1] The Herball or Generall Historie of Plantes, p. 317 (London, 1597).

[2] Dictionary, Vol. VI, p. 486.

[3] ACHUNDOW, Abu Mansur, p. 74.

[4] E. SEIDEL, Mechithar, p. 215.

[5] *Pen ts'ao kaṅ mu*, Ch. 30, p. 5 b; *Fan yi miṅ yi tsi*, Ch. 8, p. 1. STUART (Chinese Materia Medica, p. 421) wrongly identifies the name with *Spondias amara*.

[6] STEINGASS, Persian Dictionary, p. 1092.

[7] Chinese Materia Medica, p. 33.

[8] LECLERC, Traité des simples, Vol. III, pp. 369, 409.

judging from what Steingass says, is not exactly known. The Arabic author, further, has a *šah-ṣīnī* ("Chinese king"), described as a drug in the shape of small, thin, and black tabloids prepared from the sap of a plant. It is useful as a refrigerant for feverish headache and inflamed tumors. It is reduced to a powder and applied to the diseased spot.[1] Leclerc annotates that, according to the Persian treatises, this plant originating from China, as indicated by its name, is serviceable for headache in general. Dimaškī, who wrote about 1325, ascribes *šah-čīnī* to the island of Čankhay in the Malayan Archipelago, saying that its leaves are known under the name "betel."[2] Steingass, in his Persian Dictionary, explains the term as "the expressed juice of a plant brought from China, good for headaches." I do not know what plant is understood here.

14. According to Ibn al-Baiṭār, the mango (Arabic *anbā*) is found only in India and China.[3] This is *Mangifera indica* (family *Anacardiaceae*), a native of India, and the queen of the Indian fruits, counting several hundreds of varieties. Its Sanskrit name is *āmra*, known to the Chinese in the transcription 菴羅 *an-lo*, *am-la(ra)*. Persian *amba* and Arabic *anbā* are derived from the same word. During the T'ang period the fruit was grown in Fergana.[4] Malayan *maṅga* (like our *mango*) is based on Tamil *maṅgas*, and is the foundation of the Chinese transcription *muṅ* 樣. The *an-lo* tree is first mentioned for Čen-la (Camboja) in the Sui Annals,[5] where its leaves are compared with those of the jujube (*Zizyphus vulgaris*), and its fruits with those of a plum (*Prunus triflora*).

15. Išak Ibn Amrān says, "Sandal is a wood that comes to us from China."[6] *Santalum album* is grown in Kwaṅ-tuṅ to some extent, but it is more probable that the sandal-wood used in western Asia came from India (cf. Persian *čandān*, *čandal*, Armenian *čandan*, Arabic *ṣandal*, from Sanskrit *candana*).

16. Anṭākī notes the *xaleṅ* tree ("birch") in India and China; and Ibn al-Kebīr remarks that it is particularly large in China, in the country of the Rūs (Russians) and Bulgār, where are made from it vessels and plates which are exported to distant places; the arrows made of this wood are unsurpassed. According to Qazwīnī and Ibn

---

[1] *Ibid.*, p. 314.

[2] G. Ferrand, Textes relatifs à l'Extrême-Orient, p. 381.

[3] Leclerc, Traité des simples, Vol. II, p. 471. Cf. Ibn Baṭūṭa, ed. of Defrémery and Sanguinetti, Vol. III, p. 127; Yule, Hobson-Jobson, p. 553.

[4] *T'ai p'iṅ hwan yü ki*, Ch. 181, p. 13 b.

[5] *Sui šu*, Ch. 82, p. 3 b.

[6] Leclerc, *op. cit.*, p. 383.

Faḍlān, the tree occurred in Tabaristān, whence its wood reached the comb-makers of Rei.[1] The Arabic xaleṅ, Persian xadaṅ or xadanj, is of Altaic origin: Uigur qadaṅ, Koibal, Soyot and Karagas kadeṅ, Čuwaš xoran, Yakut xatyṅ, Mordwinian kileṅ, all referring to the birch (Betula alba). It is a common tree in the mountains of northern China (hwa 樺), first described by C'en Ts'aṅ-k'i of the eighth century.[2] The bark was used by the Chinese for making torches and candles filled with wax, as a padding or lining of underclothes and boots, for knife-hilts and the decoration of bows, the latter being styled "birch-bark bows."[3] The universal use of birch-bark among all tribes of Siberia for pails, baskets, and dishes, and as a roof-covering, is well known.

17. It would be very desirable to have more exact data as to when and how the consumption of Chinese tea (Camellia theifera) spread among Mohammedan peoples. The Arabic merchant Soleiman, who wrote about A.D. 851, appears to be the first outsider who gives an accurate notice of the use of tea-leaves as a beverage on the part of the Chinese, availing himself of the curious name sāx.[4] It is strange that the following Arabic authors who wrote on Chinese affairs have nothing to say on the subject. In the splendid collection of Arabic texts relative to the East, so ably gathered and interpreted by G. FERRAND, tea is not even mentioned. It is likewise absent in the Persian pharmacology of Abu Mansur and in the vast compilation of Ibn al-Baiṭār. On the other hand, Chinese mediæval authors like Čou K'ü-fei and Čao Žu-kwa do not note tea as an article of export from China. As far as we can judge at present, it seems that the habit of tea-drinking spread to western Asia not earlier than the thirteenth century, and that it was perhaps the Mongols who assumed the rôle of propagators. In Mongol, Turkish, Persian, Indian, Portuguese, Neo-Greek, and Russian, we equally find the word čai, based on North-Chinese č'a.[5] Ramu-

---

[1] G. JACOB, Handelsartikel der Araber, p. 60.

[2] Pen ts'ao kaṅ mu, Ch. 35 B, p. 13.

[3] Ko ku yao lun, Ch. 8, p. 8 b. Cf. also O. FRANKE, Beschreibung des Jehol-Gebietes, p. 77.

[4] REINAUD, Relation des voyages, Vol. I, p. 40 (cf. YULE, Cathay, new ed., Vol. I, p. 131). Modern Chinese č'a was articulated *ja (dža) in the T'ang period; but, judging from the Korean and Japanese form sa, a variant sa may be supposed also for some Chinese dialects. As the word, however, was never possessed of a final consonant in Chinese, the final spirant in Soleiman's sāx is a peculiar Arabic affair (provided the reading of the manuscript be correct).

[5] The Tibetans claim a peculiar position in the history of tea. They still have the Chinese word in the ancient form ja (dža), and, as shown by me in T'oung Pao (1916, p. 505), have imported and consumed tea from the days of the T'ang. In fact, tea was the dominant economic factor and the key-note in the political relations of China and Tibet.

sio, in the posthumous introduction to his edition of Marco Polo published in 1545, mentions having learned of the tea beverage from a Persian merchant, Hajji Muhammed.[1] A. DE MANDELSLO,[2] in 1662, still reports that the Persians, instead of *Thè*, drink their *Kahwa* (coffee). In the fifteenth century, A-lo-tiṅ, an envoy from T'ien-fan (Arabia), in presenting his tribute to an emperor of the Ming, solicited tea-leaves.[3]

The Kew Bulletin for 1896 (p. 157) contains the following interesting information on "White Tea of Persia:" —

"In the Consular Report on the trade of Ispahan and Yezd (Foreign Office, Annual Series, 1896, No. 1662) the following particulars are given of the tea trade in Persia: 'Black or Calcutta tea for Persian consumption continues to arrive in steady quantities, 2,000,000 pounds representing last year's supply. White tea from China, or more particularly from Tongking, is consumed only in Yezd, and, therefore, the supply is limited.' Through the courtesy of Mr. John R. Preece, Her Majesty's Consul at Ispahan, Kew received a small quantity of the 'White tea' above mentioned for the Museum of Economic Botany. The tea proved to be very similar to that described in the *Kew Bulletin* under the name of P'u-erh tea (*Kew Bulletin*, 1889, pp. 118 and 139). The finest of this tea is said to be reserved for the Court of Peking. The sample from Yezd was composed of the undeveloped leaf buds so thickly coated with fine hairs as to give them a silvery appearance. Owing to the shaking in transit some of the hairs had been rubbed off and had formed small yellow pellets about ⅜ inch diameter. Although the hairs are much more abundant than usual there is little doubt that the leaves have been derived from the Assam tea plant (*Camellia theifera*, Griff.) found wild in some parts of Assam and Burma but now largely cultivated in Burma, Tongking, etc. The same species has been shown to yield Lao tea (*Kew Bulletin*, 1892, p. 219), and Leppett tea (*Kew Bulletin*, 1896, p. 10). The liquor from the Persian white tea was of a pale straw colour with the delicate flavour of good China tea. It is not unknown but now little appreciated in the English market."

18. The Arabic stone-book sailing under the false flag of Aristotle distinguishes several kinds of onyx (*jiza'*), which come from two places, China and the country of the west, the latter being the finest. Qazwīnī gives Yemen and China as localities, telling an anecdote that the Chinese disdain to quarry the stone and leave this to specially privileged slaves, who have no other means of livelihood and sell the stone only outside of China.[4] As formerly stated,[5] this may be the *pi yü* 碧玉 of the Chinese.

19. Qazwīnī also mentions a stone under the name *husyat iblīs* ("devil's testicles") which should occur in China. Whoever carries it is

---

[1] YULE, Cathay, new ed., Vol. I, p. 292; or Hobson-Jobson, p. 906.

[2] Travels, p. 15.

[3] BRETSCHNEIDER, Mediæval Researches, Vol. II, p. 300.

[4] J. RUSKA, Steinbuch des Aristoteles, p. 145; and Steinbuch des Qazwīnī, p. 12; LECLERC, Traité des simples, Vol. I, p. 354.

[5] Notes on Turquois, p. 52.

not held up by bandits; also his baggage in which the stone is hidden is safe from attack, and its wearer rises in the esteem of his fellow-mates.[1] I do not know what Chinese stone is understood here.

20. It is well known that the Chinese have a peculiar alloy of copper consisting of copper 40.4, zinc 25.4, nickel 31.6, iron 2.6, and occasionally some silver and arsenic. It looks white or silver-like in the finish, and is hence called *pai-t'un* ("white copper"). In Anglo-Indian it is *tootnague* (Tamil *tutunāgum*, Portuguese *tutanaga*).[2] It is also known to foreigners in the East under the Cantonese name *paktung*. It is mentioned as early as A.D. 265 in the dictionary *Kwan ya* 廣 雅,[3] where the definition occurs that *pai-t'un* is called *wu* 鋈.

This alloy was adopted by the Persians under the name *xār-čīnī* (Arabic *xār-ṣīnī*).[4] The Persians say that the Chinese make this alloy into mirrors and arrowheads, a wound from which is mortal.[5] Vullers cites a passage from the poet Abu al Maʻānī, "One who rejects and spurns his friend pierces his heart with *xār-ṣīnī*." Qazwīnī speaks of very efficient lance-heads and harpoons of this metal. The Persians have further the term *isfīdruj*, which means "white copper," and which accordingly represents a literal rendering of Chinese *pai-t'un*. Moreover, there is Persian *sepīdrūi* (Arabic *isbiadāri*, *isbādārīh*); that is, "whitish in appearance." English *spelter* (German *spiauter*, *speauter*, *spialter*, Russian *špiauter*), a designation of zinc, is derived from this word.[6] Dimašqī, who wrote about 1325, explains *xār-ṣīnī* as a metal from China, the yellow color of copper being mixed with black and white; the mirrors imported from China, called "mirrors of distortion," are made from this alloy. It is an artificial product, hard, and fragile; it is injured by fire, after being wrought. Qazwīnī adds that no other metal yields a ring equalling that of this alloy, and that none is so suitable for the manufacture of large and small bells.[7]

21. In the thirteenth century the Arabs became acquainted with saltpetre, which they received from China; for they designate it as

---

[1] Ruska, *ibid.*, p. 21.

[2] Cf. Yule, Hobson-Jobson, p. 932. This, of course, is a misnomer, as the Indian word, connected with Persian *tūtiya* (above, p. 512), in fact refers to zinc.

[3] Ch. 8 A, p. 16 (ed. of *Ki fu ts'un šu*).

[4] Literally, "stone of China." Spanish *kazini* is derived from the Arabic word.

[5] Steingass, Persian Dictionary, p. 438.

[6] It seems also that the Persian word is the source of the curious Japanese term *sabari* or *sahari*, which denotes the white copper of the Chinese. The foreign character of this product is also indicated by the writing 胡 銅 器.

[7] Cf. E. Wiedemann, *Sitzber. Phys.-Med. Soz. Erl.*, Vols. XXXVII, 1905, pp. 403–404; and XLV, 1913, p. 46; R. Dozy, Supplément, Vol. I, p. 857.

thelg as-sīn ("Chinese snow"), and the rocket as *sahm xatāī* ("Chinese arrow").[1]

22. Ibn al-Faqīh extols the art-industries of the Chinese, particularly pottery, lamps, and other such durable implements, which are admirable as to their art and permanent in their execution.[2] Kaolin is known to the Persians as *xāk-i čīnī* ("Chinese earth"). In excellent quality it is found in Kermanshah, but the art of making porcelain there is now lost.[3] The Persian term for porcelain is *fagfūrī* or *fagfūr-i čīnī*.[4] Fagfūr (Sogdian vaγvūr, "Son of Heaven"), as far as I know, is the only sinicism to be found in Iranian, being a literal rendering of Chinese *t'ien-tse* 天子.

23. Persian *čūbi čīnī* ("China root"), Neo-Sanskrit *cobacīnī* or *copacīnī* (*kub-čīnī* in the bazars of India), is the root of *Smilax pseudo-china*, so-called Chinese sarsaparilla (*t'u-fu-liṅ* 土茯苓), a famous remedy for the treatment of Morbus americanus, first introduced into Europe by the returning sailors of Columbus, and into India by the sailors of Vasco da Gama (Sanskrit *phiraṅgaroga*, "disease of the Franks"). It is first mentioned, together with the Chinese remedy, in Indian writings of the sixteenth century, notably the Bhāvaprakāça.[5] Good information on this subject is given by GARCIA DA ORTA, who says, "As all these lands and China and Japan have this *morbo napolitano*, it pleased a merciful God to provide this root as a remedy with which good doctors can cure it, although the majority fall into error. As it is cured with this medicine, the root was traced to the Chinese, when there was a cure with it in the year 1535."[6] Garcia gives a detailed description of the shrub which he says is called *lampatam* by the Chinese.[7] This transcription corresponds to Chinese *leṅ-fan-t'wan* 冷飯團 (literally, "cold rice ball"), a synonyme of *t'u-fu-liṅ;* pronounced at

---

[1] G. JACOB, Oriental Elements of Culture in the Occident (*Smithsonian Report for* 1902, p. 520). See also LECLERC, Traité des simples, Vol. I, pp. 71, 333; and QUATREMÈRE, *Journal asiatique*, 1850, I, p. 222.

[2] E. WIEDEMANN, Zur Technik bei den Arabern, *Sitzber. Phys.-Med. Soz. Erl.*, Vol. XXXVIII, 1906, p. 355.

[3] SCHLIMMER, Terminologie, p. 334.

[4] See Beginnings of Porcelain, p. 126.

[5] J. JOLLY, Indische Medicin, p. 106.

[6] C. MARKHAM, Colloquies, p. 379. Cf. also FLÜCKIGER and HANBURY, Pharmacographia, p. 712. F. PYRARD (Vol. I, p. 182; ed. of Hakluyt Society), who travelled in India from 1601 to 1610, observes, "Venereal disease is not so common, albeit it is found, and is cured with China-wood, without sweating or anything else. This disease they call *farangui baescour* (Arabic *bāsūr*, 'piles'), from its coming to them from Europe." A long description of the remedy is given by LINSCHOTEN (Vol. II, pp. 107–112, ed. of Hakluyt Society).

[7] C. ACOSTA (Tractado de las drogas, p. 80) writes this word *lampatan.*

Canton *laṅ-fan-t'ŭn*, at Amoy *liṅ-hoan-toan*. It must be borne in mind that final Portuguese *m* is not intended for the labial nasal, but indicates the nasalization of the preceding vowel, *am* and *ã* being alternately used. The frequent final guttural nasal *ṅ* of Chinese has always been reproduced by the Portuguese by a nasalized vowel or diphthong; for instance, *tufão* ("typhoon"), given by Fernão Pinto as a Chinese term, where *fão* corresponds to Chinese *fuṅ* ("wind"); *tutão*, reproducing Chinese *tu-t'uṅ* 都統 ("Lieutenant-General"). Thus the transcription *lampatam* moves along the same line. The Portuguese designation of the root is *raiz da China* ("root of China").

There is an overland trade in this root from China by way of Turkistan to Ladākh, and probably also to Persia.[1] The plant has been known to the Chinese from ancient times, being described by T'ao Huṅ-kiṅ.[2] The employment of the root in the treatment of Morbus americanus (*yaṅ mei tu čwaṅ* 楊梅毒瘡) is described at length by Li Ši-čen, who quotes this text from Waṅ Ki 汪機, a celebrated physician, who lived during the Kia-tsiṅ period (1522–66), and author of the *Pen ts'ao hui pien* 本草會編. This is an excellent confirmation of the synchronous account of Garcia.[3] Li Ši-čen states expressly, "The *yaṅ-mei* ulcers are not mentioned in the ancient recipes, neither were there any people afflicted with this disease. Only recently did it arise in Kwaṅ-tuṅ, whence it spread to all parts of China."

24. Of Chinese loan-words in Persian, Horn[4] enumerates only *čāi* ("tea"), *čādān* ("teapot"), *čāu* ("paper money"), and perhaps also *kāgaδ* or *kāgiδ* ("paper"). As will be seen, there are many more Chinese loans in Persian; but the word for "paper" is not one of them, although the Persians received the knowledge of paper from the Chinese. This theory was first set forth by Hirth,[5] who asserts, "The Arabic word *kāghid* for paper, derived from the Persian,[6] can without great difficulty be traced to a term *ku-chih* 穀紙 (ancient pronunciation *kok-dz'*), which means 'paper from the bark of the mulberry-tree,' and was already used in times of antiquity." This view has been accepted by

---

[1] *T'oung Pao*, 1916, p. 477.

[2] *Pen ts'ao kaṅ mu*, Ch. 8 B, p. 2; also Ch. 4 B, p. 6 b; Bretschneider, Bot. Sin., pt. III, p. 320.

[3] I have sufficient material to enable me to publish at some later date a detailed history of the disease from Chinese sources.

[4] Grundriss der iran. Phil., Vol. I, pt. 2, p. 7.

[5] *T'oung Pao*, Vol. I, 1890, p. 12; or Chines. Studien, p. 269.

[6] In my opinion, the word is of Uigur origin (*kagat, kagas*), and was subsequently adopted by the Persians, and from the Persians by the Arabs. In Persian we have the forms *kāγad, kāγid, kāγaz*, and *kāgiz* (Baluči *kāgad*). Aside from this vacillating mode of spelling, the word is decidedly non-Persian. See, further, below, p. 558.

KARABACEK and HOERNLE.[1] Let us assume for a moment that the premises on which this speculation is based are correct: how could the Uigur, Persians, and Arabs make *kāgaδ* out of a Chinese *kok-či* (or *dzi*)? How may we account for the vocalization *ā*, which persists wherever the word has taken root (Hindi *kāgad*, Urdu *kāgaz*, Tamil *kāgidam*, Malayalam *kāyitam*, Kannada *kāgada*)?[2] The Uigur and Persians, according to their phonetic system, were indeed capable of reproducing the Chinese word correctly if they so intended; in fact, Chinese loan-words in the two languages are self-evident without torturing the evidence. For myself, I am unable to see any coincidence between *kok-či* and *kāgad*. But this alleged *kok-či*, in fact, does not exist. The word *ku*, as written by Hirth, is known to every one as meaning "grain, cereals;" and none of our dictionaries assigns to it the significance "mulberry." It is simply a character substituted for *kou* 構 (anciently *ku, without a final consonant), which refers exclusively to the paper-mulberry (*Broussonetia papyrifera*), expressed also (and this is the most common word) by *č'u* 楮. The *Pen ts'ao kan mu*[3] gives the character *ku* 穀 on the same footing with *č'u*, quoting the former from the ancient dictionary *Si min*,[4] and adding expressly that it has the phonetic value of 媾, and is written also 構. The character *ku*, accordingly, to be read *kou*, is merely a graphic variant, and has nothing to do with the word *ku* (*kuk), meaning "cereals."

According to Li Ši-čen, this word *kou* (*ku) originates from the language of Č'u 楚, in which it had the significance "milk" (*žu* 乳); and, as the bark of this tree contained a milk-like sap, this word was transferred to the tree. It is noteworthy in this connection that Ts'ai Lun, the inventor of paper in A.D. 105, was a native of Č'u. The dialectic origin of the word *kou* shows well how we have two root-words for exactly the same species of tree. This is advisedly stated by Li Ši-čen, who rejects as an error the opinion that the two words should refer to two different trees; he also repudiates expressly the view that the word *kou* bears any relation to the word *ku* in the sense of cereals or rice. According to T'ao Hun-kin, the term *kou či* was used by the people of the south, who, however, said also *č'u či;* the latter word,

---

[1] *Journal Roy. As. Soc.*, 1903, p. 671.

[2] According to BÜHLER (Indische Palãographie, p. 91), paper was introduced into India by the Mohammedans after the twelfth century. The alleged Sanskrit word for "paper," *kāyagata*, ferreted out by HOERNLE (*Journal Roy. As. Soc.*, 1911, p. 476), rests on a misunderstanding of a Sanskrit text, as has been shown by Lieut.-Col. WADDELL on the basis of the Tibetan translation of this text ((*ibid.*, 1914, pp. 136–137).

[3] Ch. 36, p. 4.

[4] See above, p. 201.

indeed, has always been more common. Hirth's supposition of a former pronunciation *kok* cannot be accepted; but, even did this alleged *kok* exist, I should continue to disbelieve in the proposed etymology of the Persian-Arabic word. There is no reason to assume that, because paper was adopted by the Arabs and Persians from the Chinese, their designation of it should hail from the same quarter. I do not know of a foreign language that was willing to adopt from the Chinese any designation for paper. Our word comes from the Greek-Latin *papyrus;* Russian *bumaga* originally means "cotton," being ultimately traceable to Middle Persian *pambak*.[1] The Tibetans learned the technique of paper-making from the Chinese, but have a word of their own to designate paper (*šog-bu*). So have the Japanese (*kami*) and the Koreans (*muntsi*). The Mongols call paper *tsagasun* (Buryat *tsāraso*, *sārahan*), a purely Mongol word, meaning "the white one." Among the Golde on the Amur I recorded the word *xausal*. The Lolo have *t'o-i*, the Annamese *bia*, the Čam *baa, baar*, or *biar*, the Khmer *credas*, which, like Malayan *kertas*, is borrowed from Arabic *kirtas* (Greek χάρτης).[2] As stated, the Persian-Arabic word is borrowed from a Turkish language: Uigur *kagat* or *kagas;* Tuba, Lebed, Kumandu, Comanian *kagat;* Kirgiz, Karakirgiz, Taranči, and Kazan *kagaz*. The origin of this word can be explained from Turkish; for in Lebed, Kumandu, and Šor, we have *kagaš* with the significance "tree-bark."

I need not repeat here the oft-told story of how the manufacture of paper was introduced into Samarkand by Chinese captives in A.D. 751. Prior to this date, as has been established by Karabacek, Chinese paper was imported to Samarkand as early as 650–1, again in 707.[3] Under the Sasanians, Chinese paper was known in Persia; but it was a very rare article, and reserved for royal state documents.[4]

25. Another form in which paper reached the Persians was paper money. It is well known that the Chinese were the originators of

---

[1] See above, p. 490.

[2] S. FRAENKEL, Die aramäischen Fremdwörter im Arabischen, p. 245.

[3] Cf. HOERNLE, *Journal Roy. As. Soc.*, 1903, p. 670. I regret being unable to accept his general result that the Arabs or Samarkandis should be credited with the invention of pure rag-paper (p. 674). This had already been accomplished in China, and indeed was the work of Ts'ai Lun. I expect to come back to this problem on another occasion. With all respect for the researches of Karabacek, Wiesner, and Hoernle, I am not convinced that the far-reaching conclusions of these scholars are all justified. We are in need of more investigations (and less theorizing), especially of ancient papers made in China. There are numerous accounts of many sorts of paper, hitherto unnoticed, in Chinese records, which should be closely studied.

[4] According to Masudi (B. DE MEYNARD, Les Prairies d'or, Vol. II, p. 202); see also E. DROUIN, Mémoire sur les Huns Ephthalites, p. 53 (reprint from *Le Muséon*, 1895).

paper bank-notes.[1] The Mongol rulers introduced them into Persia, first in 1294. The notes were direct copies of Kubilai's, even the Chinese characters being imitated as part of the device upon them, and the Chinese word *č'ao* 鈔 being employed. This word was then adopted by the Persians as *čāu* or *čāv*.[2] The most interesting point about this affair is that in that year (1294) the Chinese process of block-printing was for the first time practised in Tabriz in connection with the printing of these bank-notes.

In his graphic account describing the utilization of paper money by the Great Khan, MARCO POLO[3] makes the following statement: "He makes them take of the bark of a certain tree, in fact of the mulberry tree, the leaves of which are the food of the silkworms,— these trees being so numerous that whole districts are full of them. What they take is a certain fine white bast or skin which lies between the wood of the tree and the thick outer bark, and this they make into something resembling sheets of paper, but black. When these sheets have been prepared they are cut up into pieces of different sizes." In the third edition of Yule's memorable work, the editor, HENRI CORDIER,[4] has added the following annotation: "Dr. Bretschneider (History of Botanical Discoveries, Vol. I, p. 4) makes the remark: 'Polo states that the Great Khan causeth the bark of great mulberry trees, made into something like paper, to pass for money.' He seems to be mistaken. Paper in China is not made from mulberry-trees, but from the *Broussonetia papyrifera*, which latter tree belongs to the same order of *Moraceae*. The same fibres are used also in some parts of China for making cloth, and Marco Polo alludes probably to the same tree when stating that 'in the province of Cuiju (Kuei-chou) they manufacture stuff of the bark of certain trees, which form very fine summer clothing.'"

This is a singular error of Bretschneider. Marco Polo is perfectly correct: not only did the Chinese actually manufacture paper from the bark of the mulberry-tree (*Morus alba*), but also it was this paper which was preferred for the making of paper money. Bretschneider is certainly right in saying that paper is made from the *Broussonetia*, but

---

[1] KLAPROTH, Sur l'origine du papier-monnaie (in his Mémoires relatifs à l'Asie, Vol. I, pp. 375–388); YULE, Marco Polo, Vol. I, pp. 426–430; ANONYMUS, Paper Money among the Chinese (*Chin. Repository*, Vol. XX, 1851, pp. 289–296); S. SABURO, The Origin of the Paper Currency (*Journal Peking Or. Soc.*, Vol. II, 1889, pp. 265–307); S. W. BUSHELL, Specimens of Ancient Chinese Paper Money (*ibid.*, pp. 308–316); H. B. MORSE, Currency in China (*Journal China Branch Roy. As. Soc.*, Vol. XXXVIII, 1907, pp. 17–31); etc.

[2] For details consult YULE, *l. c.*

[3] H. YULE, The Book of Ser Marco Polo, Vol. I, p. 423.

[4] *Ibid.*, p. 430.

he is assuredly wrong in the assertion that paper is not made in China from mulberry-trees. This fact he could have easily ascertained from S. JULIEN,[1] who alludes to mulberry-tree paper twice, first, as "papier de racines et d'écorce de mûrier;" and, second, in speaking of the bark paper from *Broussonetia*,—"On emploie aussi pour le même usage l'écorce d'*Hibiscus Rosa sinensis* et de mûrier; ce dernier papier sert encore à recueillir les graines de vers à soie." What is understood by the latter process may be seen from plate I in Julien's earlier work on sericulture,[2] where the paper from the bark of the mulberry-tree is likewise mentioned.

The *Či p'u* 紙譜, a treatise on paper, written by Su Yi-kien 蘇易簡 toward the close of the tenth century, enumerates, among the various sorts of paper manufactured during his lifetime, paper from the bark of the mulberry-tree (*san p'i* 桑皮) made by the people of the north.[3]

Chinese paper money of mulberry-bark was known in the Islamic world in the beginning of the fourteenth century; that is, during the Mongol period. Accordingly it must have been manufactured in China during the Yūan dynasty. Ahmed Šibab Eddin, who died in Cairo in 1338 at the age of ninety-three, and left an important geographical work in thirty volumes, containing interesting information on China gathered from the lips of eye-witnesses, makes the following comment on paper money, in the translation of CH. SCHEFER:[4] "On emploie dans le Khita, en guise de monnaie, des morceaux d'un papier de forme allongée fabriqué avec des filaments de mûriers sur lequel est imprimé le nom de l'empereur. Lorsqu'un de ces papiers est usé, on le porte aux officiers du prince et, moyennant une perte minime, on reçoit un autre billet en échange, ainsi que cela a lieu dans nos hôtels des monnaies, pour les matières d'or et d'argent que l'on y porte pour être converties en pièces monnayées."

And in another passage: "La monnaie des Chinois est faite de billets fabriqués avec l'écorce du mûrier. Il y en a de grands et de

---

[1] Industries anciennes et modernes de l'empire chinois, pp. 145, 149 (Paris 1869).

[2] Résumé des principaux traités chinois sur la culture des mûriers et l'éducation des vers à soie, p. 98 (Paris, 1837). According to the notions of the Chinese, JULIEN remarks, everything made from hemp, like cord and weavings, is banished from the establishments where silkworms are reared, and our European paper would be very harmful to the latter. There seems to be a sympathetic relation between the silkworm feeding on the leaves of the mulberry and the mulberry paper on which the cocoons of the females are placed.

[3] *Ko ti kin yüan*, Ch. 37, p. 6.

[4] Relations des Musulmans avec les Chinois (Centenaire de l'Ecole des langues orientales vivantes, Paris, 1895, p. 17).

petits. . . . On les fabrique avec des filaments tendres du mûrier et, après y avoir apposé un sceau au nom de l'empereur, on les met en circulation."[1]

The bank-notes of the Ming dynasty were likewise made of mulberry-pulp, in rectangular sheets one foot long and six inches wide, the material being of a greenish color, as stated in the Annals of the Dynasty.[2] It is clear that the Ming emperors, like many other institutions, adopted this practice from their predecessors, the Mongols. KLAPROTH[3] is wrong in saying that the assignats of the Sung, Kin, and Mongols were all made from the bark of the tree *ču* (*Broussonetia*), and those of the Ming from all sorts of plants.[4]

In the *Hui kian či* 回 彊 誌, an interesting description of Turkistan by two Manchu officials Surde and Fusambô, published in 1772,[5] the following note, headed "Mohammedan Paper" 回 子 紙, occurs: "There are two sorts of Turkistan paper, black and white, made from mulberry-bark, cotton 棉 布, and silk-refuse equally mixed, resulting in a coarse, thick, strong, and tough material. It is cut into small rolls fully a foot long, which are burnished by means of stones, and are then fit for writing."

Sir AUREL STEIN[6] reports that paper is still manufactured from mulberry-trees in Khotan. Also J. WIESNER,[7] the meritorious investigator

---

[1] *Ibid.*, p. 20.

[2] *Min ši*, Ch. 81, p. 1 (以桑穰爲料其制方高一尺廣六寸質青色). The same text is found on a bill issued in 1375, reproduced and translated by W. VISSERING (On Chinese Currency, see plate at end of volume), the minister of finance being expressly ordered to use the fibres of the mulberry-tree in the composition of these bills.

[3] Mémoires relatifs à l'Asie, Vol. I, p. 387.

[4] This is repeated by ROCKHILL (Rubruck, p. 201). I do not deny, of course, that paper money was made from *Broussonetia*. The Chinese numismatists, in their description of the ancient paper notes, as far as I know, make no reference to the material (cf., for instance, *Ts'üan pu t'un či* 泉 布 統 志, Ch. 5, p. 42; 6 A, p. 2; 6 B, p. 44). The *Yüan ši* (Ch. 97, p. 3) does not state, either, the character of the paper employed in the Mongol notes. My point is, that the Mongols, while they enlisted *Broussonetia* paper for this purpose, used mulberry-bark paper as well, and that the latter was exclusively utilized by the Ming.

[5] A. WYLIE, Notes on Chinese Literature, p. 64. The John Crerar Library of Chicago owns an old manuscript of this work, clearly written, in 4 vols. and chapters, illustrated by nine ink-sketches of types of Mohammedans and a map. The volumes are not paged.

[6] Ancient Khotan, Vol. I, p. 134.

[7] Mikroskopische Untersuchung alter ostturkestanischer Papiere, p. 9 (Vienna, 1902). I cannot pass over in silence a curious error of this scholar when he says (p. 8) that it is not proved that *Cannabis sativa* (called by him "genuine hemp") is cultivated in China, and that the so-called Chinese hemp paper should be intended for China grass. Every tyro in things Chinese knows that hemp (*Cannabis sativa*)

of ancient papers, has included the fibre of *Morus alba* and *M. nigra* among the materials to which his researches extended.

Mulberry-bark paper is ascribed to Bengal in the *Si yan č'ao kun tien lu* 西洋朝貢典錄 by Hwan Sin-ts'en 黃省曾, published in 1520.[1] Such paper is still made in Corea also, and is thicker and more solid than that of China.[2] The bark of a species of mulberry is utilized by the Shan for the same purpose.[3]

As the mulberry-tree is eagerly cultivated in Persia in connection with the silk-industry, it is possible also that the Persian paper in the bank-notes of the Mongols was a product of the mulberry.[4] At any rate, good Marco Polo is cleared, and his veracity and exactness have been established again.

Before the introduction of rag-paper the Persians availed themselves of parchment as writing-material. It is supposed by Herzfeld that Darius Hystaspes introduced the use of leather into the royal archives, but this interpretation has been contested.[5] A fragment of Ctesias preserved by Diodorus[6] mentions the employment of parchment (διφθέρα) in the royal archives of Persia. The practice seems to be of Semitic, probably Syrian, origin. In the business life of the Romans, parchment (*membrana*) superseded wooden tablets in the first century A.D.[7] The Avesta and Zend written on prepared cow-skins with gold ink is mentioned in the Artāi-vīrāf-nāmak (1, 7). The Iranian word *pōst* ("skin") resulted in Sanskrit *pusta* or *pustaka* ("volume, book"),[8] from which Tibetan *po-ti* is derived.[9] On the other hand, the Persians have borrowed from the Greek διφθέρα ("skin, parchment") their word *daftar* or *defter* ("book," Arabic *daftar, diftar*), which likewise

---

belongs to the oldest cultivated plants of the Chinese (see above, p. 293), and that hemp paper is already listed among the papers invented by Ts'ai Lun in A.D. 105 (cf. CHAVANNES, Les Livres chinois avant l'invention du papier, *Journal asiatique*, 1905, p. 6 of the reprint).

[1] Ch. B., p. 10 b (ed. of *Pie hia čai ts'un šu*).

[2] C. DALLET, Histoire de l'église de Corée, Vol. I, p. CLXXXIII.

[3] J. G. SCOTT and J. P. HARDIMAN, Gazetteer of Upper Burma and the Shan States, pt. I, Vol. II, p. 411.

[4] The Persian word for the mulberry, *tūδ*, is supposed to be a loan-word from Aramaic (HORN, Grundriss iran. Phil., Vol. I, pt. 2, p. 6); but this is erroneous (see below, p. 582).

[5] Cf. V. GARDTHAUSEN, Buchwesen im Altertum, p. 91.

[6] II, 32.

[7] K. DZIATZKO, Ausgewählte Kapitel des antiken Buchwesens, p. 131.

[8] R. GAUTHIOT in *Mémoires Soc. de Linguistique*, Vol. XIX, 1915, p. 130.

[9] *T'oung Pao*, 1916, p. 452.

spread to Central Asia (Tibetan deb-t'er, Mongol debter, Manchu debtelin).[1]

The use of parchment on the part of the people of Parthia (An-si) has already been noted by the mission of Čaṅ K'ien, who placed it on record that "they make signs on leather, from side to side, by way of literary records." It is accordingly certain that parchment was utilized in Iran as early as the second century B.C. There are also later references to this practice; for instance, in the Nan ši,[2] where it is said that the Hu (Iranians) use sheep-skin 羊 皮 as paper. The Chinese have hardly ever made use of parchment for writing-purposes, but they prepare parchment (from the skins of sheep, donkeys, or oxen) for the making of shadow-play figures. The only parchment manuscripts ever found in China were the Scriptures of the Jews of K'ai-foṅ, which are also mentioned in their inscriptions.[3]

26. Most of the Chinese loan-words in Persian were imported by the Mongol rulers in the thirteenth century (the so-called Il-Khans, 1265–1335), being chiefly terms relative to official and administrative institutions. The best known of these is pāizah, being a reproduction of Chinese p'ai-tse 牌子, an official warrant or badge containing imperial commands, letters of safe-conduct, permits of requisition, according to the rank of the bearer, made of silver, brass, iron, etc. They were taken over by the Mongols from the Liao and Kin,[4] and are mentioned by Rubruck, Marco Polo,[5] and Rašid-eddin.

27. Titles like waṅ 王 ("king, prince"), t'ai waṅ 太王 ("great prince"), kao waṅ 高王 ("great general"), t'ai hu 大后 ("empress"), fu žen (Persian fučin) 夫人 (title for women of rank), and kuṅ ču 公主 ("princess") were likewise adopted in Mongol Persia.[6] Persian jinksānak, title of a Mongol prefect or governor, transcribes Chinese č'eṅ siaṅ 承相 ("minister of state").[7]

28. From Turkish tribes the Persians have adopted the word toγ

---

[1] T'oung Pao, 1916, p. 481.

[2] Ch. 79, p. 7.

[3] Cf. J. TOBAR, Inscriptions juives de K'ai-fong-fou, pp. 78, 86, 96 (note 2).

[4] CHAVANNES, Journal asiatique, 1898, I, p. 396.

[5] YULE's edition, Vol. I, p. 351, which consult for a history of the p'ai-tse; see, further, LAUFER, Keleti Szemle, 1907, pp. 195–196; ŽAMTSARANO, Paiza among the Mongols at the Present Time (Zapiski Oriental Section Russian Archæol. Soc., Vol. XXII, 1914, pp. 155–159).

[6] E. BLOCHET, Introduction à l'histoire des Mongols de Rashid Ed-din, p. 183; and Djami el-Tévarikh, p. 473. Regarding the title waṅ, see also J. J. MODI, Asiatic Papers, p. 251.

[7] Cf. my notes in T'oung Pao, 1916, p. 528.

(togh) or *tuγ*,[1] which designates the tassels of horse-hair attached to the points of a standard or to the helmet of a Pasha (in the latter case a sign of rank). Among the Turks of Central Asia, the standard of a high military officer is formed by a yak's tail fastened at the top of a pole. This is said also to mark the graves of saintly personages.[2] In the language of the Uigur, the word is *tuk*.[3] As correctly recognized by ABEL-RÉMUSAT,[4] who had recourse only to Osmanli, the Turkish word is derived from Chinese 纛 *tu*, anciently *duk, that occurs at an early date in the *Čou li* and *Ts'ien Han šu*. Originally it denoted a banner carried in funeral processions; under the Han, it was the standard of the commander-in-chief of the army, which, according to Ts'ai Yun 蔡邕 (A.D. 133–192), was made of yak-tails.[5] Yak-tails (Sanskrit *cāmara*, Anglo-Indian *chowry*) were anciently used in India and Central Asia as insignia of royalty or rank.[6]

29. The *Čou šu*[7] states that in respect to the five cereals and the fauna Persia agrees with China, save that rice and millet are lacking in Persia. The term "millet" is expressed by the compound *šu šu* 黍秫; that is, the glutinous variety of *Panicum miliaceum* and the glutinous variety of the spiked millet (*Setaria italica glutinosa*). Now, we find in Persian a word *šušu* in the sense of "millet." It remains to study the history of this word, in order to ascertain whether it might be a Chinese loan-word.

SCHLIMMER[8] notes *erzen* as Persian word for *Panicum miliaceum*.

30. Persian (also Osmanli) *čänk* ("a harp or guitar, particularly played by women") is probably derived from Chinese *čen* 箏 ("a harpsichord with twelve brass strings").

31. One of the most interesting Chinese loan-words in Persian is *xutu* (khutu), from Chinese *ku-tu* (written in various ways), principally denoting the ivory tooth of the walrus. This subject has been dis-

---

[1] In Šugnan, a Pamir language, it occurs as *tux* (SALEMANN, in Vostočnye Zam'átki, p. 286).

[2] SHAW, Turki Language, Vol. II, p. 76.

[3] RADLOFF, Wört. der Türk-Dial., Vol. III, col. 1425.

[4] Recherches sur les langues tatares, p. 303.

[5] See K'an-hi *sub* 糸.

[6] YULE, Hobson-Jobson, p. 214. Under the Emirs of the Khanat Bukhara there was the title *toksaba:* he who received this title had the privilege of having a *tug* carried before him; hence the origin of the word *toksaba* (VÉLIAMINOF–ZERNOF, *Mélanges asiatiques*, Vol. VIII, p. 576). Cf. also a brief note by PARKER (*China Review*, Vol. XVII, p. 300).

[7] Ch. 50, p. 6.

[8] Terminologie, p. 420.

cussed by me in two articles.[1] VULLERS[2] gives no less than seven definitions of the Persian word: (1) cornu bovis cuiusdam Sinensis; (2) secundum alios cornu rhinocerotis; (3) secundum alios cornu avis cuiusdam permagnae in regno vastato, quod inter Chinam et Aethiopiam situm est, degentis, e quo conficiunt anulos osseos et manubria cultri et quo res venenatae dignosci possunt; (4) secundum alios cornu serpentis, quod mille annos natus profert; (5) secundum alios cornu viperae; (6) secundum alios cornu piscis annosi; (7) secundum alios dentes animalis cuiusdam. Of these explanations, No. 3 is that of al-Akfānī, and the bird in question is the buceros. No. 4 is a reproduction of the definition of *ku-tu-si* in the Liao Annals ("the horn of a thousand-years-old snake"). How the Persians and Arabs arrived at the other definitions will be easily understood from my former discussion of the subject. In the Ethiopic version of the Alexander Romance are mentioned, among the gifts sent to Alexander by the king of China, twenty (in the Syriac version, ten) snakes' horns, each a cubit long.[3]

Meanwhile I have succeeded in tracing a new Chinese definition of *ku-tu*. Čou Mi 周密 (1230-1320), in his *Či ya t'an tsa č'ao*,[4] states, "According to Po-ki 伯幾,[5] what is now styled *ku-tu si* 骨觸犀 is a horn of the earth (*ti kio* 地角, 'a horn found underground'?)." He refers again to its property of neutralizing poison and to knife-hilts made of the substance.

In the edition of the *Ko ku yao lun*,[6] the text regarding *ku-tu-si* is somewhat different from that quoted by me in *T'oung Pao* (1913, p. 325). *Ku-tu-si* is not identified there with *pi-si*, as appears from the text of the *P'ei wen yün fu* and *Pen ts'ao kan mu*, but *pi-si* is a variety of *ku-tu-si* of particularly high value.

---

[1] Arabic and Chinese Trade in Walrus and Narwhal Ivory (*T'oung Pao*, 1913, pp. 315-364, with Addenda by P. PELLIOT, pp. 365-370); and Supplementary Notes on Walrus and Narwhal Ivory (*ibid.*, 1916, pp. 348-389). Regarding objects of walrus ivory in Persia, see pp. 365-366.

[2] Lexicon Persico-Latinum, Vol. I, p. 659.

[3] E. A. W. BUDGE, Life and Exploits of Alexander the Great, p. 180; likewise his translation of the Syriac version, p. 112 (Syriac edition, p. 200). In the Syriac occurs another gift from China, "a thousand talents of *mai-kâsī*" (literally, "waters of cups"). Budge leaves this problem unsolved. Apparently we face the transcription of a Chinese word, which I presume is *mak, mag 墨 (at present *mo*), "China ink." In Mongol and Manchu we find this word as *bexe*, in Kalmuk as *beke*.

[4] Ch. A, p. 29 b (ed. of *Yüe ya t'an ts'un šu*).

[5] Surname of Sien-yü Č'u 鮮于樞, calligraphist and poet at the end of the thirteenth century (see PELLIOT, T'oung Pao, 1913, p. 368).

[6] Ch. 6, p. 9 b (ed. of *Si yin hüan ts'un šu*).

The Chinese Gazetteer of Macao[1] contains the following notice of the walrus (*hai ma*): "Its tooth is hard, of a pure bright white with veins as fine as silk threads or hair. It can be utilized for the carving of ivory beads and other objects."

Finally I have found another document in which the fish-teeth of the Russians are identified with the tusks of the walrus (morse). This is contained in the work of G. FLETCHER, "The Russe Common Wealth," published in London, 1591,[2] and runs as follows: "Besides these (which are all good and substantiall commodities) they have divers other of smaller account, that are natural and proper to that country: as the fishe tooth (which they cal *ribazuba*), which is used both among themselves and the Persians and Bougharians, that fetcht it from thence for beads, knives, and sword hafts of noblemen and gentlemen, and for divers other uses. Some use the powder of it against poyson, as the unicornes horne. The fish that weareth it is called a morse, and is caught about Pechora. These fishe teeth, some of them are almost two foot of length, and weigh eleven or twelve pound apiece."[3]

---

[1] *Ao-men ĕi lio*, Ch. B, p. 37.

[2] Ed. of E. A. BOND, p. 13 (Hakluyt Society, 1856).

[3] The following case is interesting as showing how narwhal ivory could reach India straight from the Arctic. PIETRO DELLA VALLE (Vol. I, p. 4, Hakluyt Soc. ed.), travelling on a ship from the Persian Gulf to India in 1623, tells this story: "On Monday, the Sea being calm, the Captain, and I, were standing upon the deck of our Ship, discoursing of sundry matters, and he took occasion to show me a piece of Horn, which he told me himself had found in the yar 1611 in a Northern Country, whither he then sail'd, which they call Greenland, lying in the latitude of seventy-six degrees. He related how he found this horn in the earth, being probably the horn of some Animal dead there, and that, when it was intire, it was between five and six feet long, and seven inches in circumference at the root, where it was thickest. The piece which I saw (for the horn was broken, and sold by pieces in several places) was something more than half a span long, and little less than five inches thick; the color of it was white, inclining to yellow, like that of Ivory when it is old; it was hollow and smooth within, but wreath'd on the outside. The Captain saw not the Animal, nor knew whether it were of the land or the sea, for, according to the place where he found it, it might be as well one as the other; but he believed for certain, that it was of a Unicorn, both because the experience of its being good against poyson argu'd so much, and for that the signes attributed by Authors to the Unicorn's horn agreed also to this, as he conceiv'd. But herein I dissent from him, inasmuch as, if I remember aright, the horn of the Unicorn, whom the Greeks call'd Monoceros, is, by Pliny, describ'd black, and not white. The Captain added that it was a report, that Unicorns are found in certain Northern parts of America, not far from that Country of Greenland; and so not unlikely but that there might be some also in Greenland, a neighbouring Country, and not yet known whether it be Continent or Island; and that they might sometimes come thither from the contiguous lands of America, in case it be no Island. . . . The Company of the Greenland Merchants of England had the horn, which he found, because Captains of ships are their stipendiaries, and, besides their salary, must make no other profit of their Voyages; but whatever they gain or find, in case it be known, and they conceal it not, all accrues

The term *pi-si* has been the subject of brief discussions on the part of PELLIOT[1] and myself.[2] The *Ko ku yao lun,* as far as is known at present, appears to be the earliest work in which the expression occurs. Hitherto it had only been known as a modern colloquialism, and Pelliot urged tracing it in the texts. I am now in a position to comply with this demand. T'an Ts'ui 檀萃, in his *Tien hai yü heṅ či,*[3] published in 1799, gives an excellent account of Yün-nan Province, its mineral resources, fauna, flora, and aboriginal population, and states that *pi-hia-si* 碧霞璽 or *pi-hia-pi* 碧霞玭 or *pi-si* 碧洗 are all of the class of precious stones which are produced in the Moṅ-mi t'u-se 猛密土司 of Yün-nan.[4] It is obvious that these words are merely transcriptions of a non-Chinese term; and, if we were positive that it took its starting-point from Yün-nan, it would not be unreasonable to infer that it hails from one of the native T'ai or Shan languages. T'an Ts'ui adds that the best *pi-si* are deep red in color; that those in which purple, yellow, and green are combined, and the white ones, take the second place; while those half white and half black are of the third grade. We are accordingly confronted with a certain class of precious stones which remain to be determined mineralogically.

32. The Persian name for China is Čīn, Čīnistān, or Čīnastān. In Middle Persian we meet Sāini in the Farvardin Yašt and Sini in the Būndahišn,[5] besides Čēn and Čēnastān.[6] The form with initial palatal is confirmed, on the one hand, by Armenian Čen-k', Čenastan, Čenbakur ("emperor of China"), *čenazneay* ("originating from China"), *čenik* ("Chinese"), and, on the other hand, by Sogdian Čynstn (Čīna-

to the Company that employes them. When the Horn was intire it was sent to Constantinople to be sold, where two thousand pounds sterling was offer'd for it: But the English Company, hoping to get a greater rate, sold it not at Constantinople, but sent it into Muscovy, where much about the same price was bidden for it, which, being refus'd, it was carry'd back into Turkey, and fell of its value, a much less sum being now proffer'd than before. Hereupon the Company conceiv'd that it would sell more easily in pieces then intire, because few could be found who would purchase it at so great a rate. Accordingly they broke it, and it was sold by pieces in sundry places; yet, for all this, the whole proceed amounted onely to about twelve hundred pounds sterling. And of these pieces they gave one to the Captain who found it, and this was it which he shew'd me."

[1] *T'oung Pao,* 1913, p. 365.

[2] *Ibid.,* 1916, p. 375.

[3] Ch. 1, p. 6 (ed. of *Wen yiṅ lou yü ti ts'uṅ šu*). Title and treatment of the subject are in imitation of the *Kwei hai yü heṅ či* of Fan Č'eṅ-ta of the twelfth century.

[4] *T'u-se* are districts under the jurisdiction of a native chieftain, who himself is more or less subject to the authority of the Chinese.

[5] Cf. J. J. MODI, References to China in the Ancient Books of the Parsees, reprinted in his Asiatic Papers, pp. 241 *et seq.*

[6] HÜBSCHMANN, Armen. Gram., p. 49.

stān).[1] The parallelism of initial *č* and *s* corresponds exactly to the Greek doublet Σῖναι and Θῖναι (=Čīnai), and the Iranian forms with *č* meet their counterpart in Sanskrit Cīna (Čīna). This state of affairs renders probable the supposition that the Indian, Iranian, and Greek designations for China have issued from a common source, and that this prototype may be sought for in China itself. I am now inclined to think that there is some degree of probability in the old theory that the name "China" should be traceable to that of the dynasty Ts'in. I formerly rejected this theory, simply for the reason that no one had as yet presented a convincing demonstration of the case;[2] nor did I become converted by the demonstration in favor of Ts'in then attempted by PELLIOT.[3] Pelliot has cited several examples from which it appears that even under the Han the Chinese were still designated as "men of the Ts'in" in Central Asia. This fact in itself is interesting, but does not go to prove that the foreign names Čīna, Čĕn, etc., are based on the name Ts'in. It must be shown phonetically that such a derivation is possible, and this is what Pelliot failed to demonstrate: he does not even dwell for a moment on the question of the ancient pronunciation of the character *ts'in* 秦. If in ancient times it should have had the same articulation as at present, the alleged phonetic coincidence with the foreign designations would amount to nothing. The ancient phonetic value of 秦 was *din, *dzin, *džin (jin), *dž'in, with initial dental or palatal sonant;[4] and it is possible, and in harmony with phonetic

---

[1] R. GAUTHIOT, *T'oung Pao*, 1913, p. 428.

[2] *T'oung Pao*, 1912, pp. 719–726.

[3] *Ibid.*, pp. 727–742. The mention of the name Cīna in the Arthaçāstra of Cāṇakya or Kauṭilya, and Jacobi's opinion on the question, did not at all prompt me to my view, as represented by Pelliot. I had held this view for at least ten years previously, and Jacobi's article simply offered the occasion which led me to express my view. Pelliot's commotion over the date of the Sanskrit work was superfluous. I shall point only to the judgment of V. A. SMITH (Early History of India, 3d ed., 1914, p. 153), who says that "the Arthaçāstra is a genuine ancient work of Maurya age, and presumably attributed rightly to Cāṇakya or Kauṭilya; this verdict, of course, does not exclude the possibility, or probability, that the existing text may contain minor interpolations of later date, but the bulk of the book certainly dates from the Maurya period," and to the statement of A. B. KEITH (*Journal Roy. As. Soc.*, 1916, p. 137), "It is perfectly possible that the Arthaçāstra is an early work, and that it may be assigned to the first century B.C., while its matter very probably is older by a good deal than that." The doubts as to the Ts'in etymology of the name "China" came from many quarters. Thus J. J. MODI (Asiatic Papers, p. 247), on the supposition that the Farvardin Yašt may have been written prior to the fourth or fifth century B.C., argued, "If so, the fact that the name of China as Saini occurs in this old document, throws a doubt on the belief that it was the Ts'in dynasty of the third century B.C. that gave its name to China. It appears, therefore, that the name was older than the third century B.C."

[4] In the dialect of Shanghai it is still pronounced *dziñ*.

laws, that a Chinese initial *dž* was reproduced in Iranian by the palatal surd *č*. It is this phonetic agreement on the one hand, and the coincidence of the Sanskrit, Iranian, and Greek names for China on the other, which induce me to admit the Ts'in etymology as a possible theory; that the derivation has really been thus, no one can assert positively. The presence of the designation Ts'in for Chinese during the Han is an historical accessory, but it does not form a fundamental link in the evidence.

33. The preceding notes should be considered only as an outline of a series of studies which should be further developed by the co-operation of Persian scholars and Arabists familiar with the Arabic sources on the history and geography of Iran. A comprehensive study of all Persian sources relating to China would also be very welcome. Another interesting task to be pursued in this connection would be an attempt to trace the development of the idealized portrait which the Persian and Arabic poets have sketched of the Chinese. It is known that in the Oriental versions of the Alexander Romance the Chinese make their appearance as one of the numerous nations visited by Alexander the Great (Iskandar). In Firdausī's (935–1025) version he travels to China as his own ambassador, and is honorably received by the Fagfūr (Son of Heaven), to whom he delivers a letter confirming his possessions and dignities, provided he will acknowledge Iskandar as his lord and pay tribute of all fruits of his country; to this the Fagfūr consents. In Nizāmī's (1141–1203) *Iskandarnāme* ("Book of Alexander"), Iskandar betakes himself from India by way of Tibet to China, where a contest between the Greek and Chinese painters takes place, the former ultimately carrying the day.[1] In the Ethiopic version of the Alexander story, "the king of China commanded that they should spread out costly stuffs upon a couch, and the couch was made of gold ornamented with jewels and inlaid with a design in gold; and he sat in his hall, and his princes and nobles were round about him, and when he spake they made answer unto him and spake submissively. Then he commanded the captain to bring in Alexander the ambassador. Now when I Alexander had come in with the captain, he made me to stand before the King, and the men stood up dressed in raiment of gold and silver; and I stood there a long time and none spake unto me."[2] The Kowtow (*k'o-t'ou*) question was evidently not raised. It is still more amusing to read farther on that the king of China made the ambassador sit by his side upon the couch,— an impossible situation. The Fagfūr sent to Alexander garments of finely woven stuff, one hundred pounds

---

[1] Cf. F. SPIEGEL, Die Alexandersage bei den Orientalen, pp. 31, 46.

[2] E. A. W. BUDGE, Life and Exploits of Alexander the Great, p. 173.

in weight, two hundred tents, men-servants and maid-servants, two hundred shields of elephant-hide, as many Indian swords mounted in gold and ornamented with gold and precious stones of great value, as many horses suitable for kings, and one thousand loads of the finest gold and silver, for in this country are situated the mountains wherefrom they dig gold. The wall of that city is built of gold ore, and likewise the habitations of the people; and from this place Solomon, the son of David, brought the gold with which he built the sanctuary, and he made the vessels and the shields of the gold of the land of China.[1] In the history of Alexander the Great contained in the "Universal History" of al-Makīn, who died at Damascus in 1273–74, a distinction is made between the kings of Nearer China and Farther China.[2]

The most naïve version of Alexander's adventures in China is contained in the legendary "History of the Kings of Persia," written in Arabic by al-Ta'ālibī (961–1038).[3] Here, the king of China is taken aback, and loses his sleep when Alexander with his army enters China. Under cover of night he visits Alexander, offering his submission in order to prevent bloodshed. Alexander first demands the revenue of his kingdom for five years, but gradually condescends to accept one third for one year. The following day a huge force of Chinese troops surrounds the army of Alexander, who believes his end has come, when the king of China appears, descending from his horse and kissing the soil (!). Alexander charges him with perfidy, which the king of China denies. "What, then, does this army mean?"—"I wanted to show thee," the king of China replied, "that I did not submit from weakness or owing to the small number of my forces. I had observed that the superior world favored thee and allowed thee to triumph over more powerful kings than thou. Whoever combats the superior world will be vanquished. For this reason I wanted to submit to the superior world by submitting to thee, and humbly to obey it by obeying thee and complying with thy orders." Alexander rejoined, "No demand should be made of a man like thee. I never met any one more qualified as a sage. Now I abandon all my claims upon thee and depart." The king of China responded, "Thou wilt lose nothing by this arrangement." He then despatched rich presents to him, like a thousand pieces of silk, painted silk, brocade, silver, sable-skins, etc., and pledged himself to pay an annual tribute. Although the whole story, of course, is pure invention, Chinese methods of overcoming an enemy by superior diplomacy are not badly characterized.

---

[1] *Ibid.*, p. 179.

[2] *Ibid.*, pp. 369, 394.

[3] H. ZOTENBERG, Histoire des rois des Perses, pp. 436–440.

IRANIAN ELEMENTS IN MONGOL

On the preceding pages, as well as in my "Loan-Words in Tibetan," I had occasion to point out a number of Mongol words traceable to Iranian; and, as this subject has evoked some interest since the discoveries made in Turkistan, I deem it useful to treat it here in a coherent notice and to sum up our present knowledge of the matter.

1. Certain relations of the Mongol language to Iranian were known about a century ago to I. J. SCHMIDT,[1] the real founder of Mongol philology. It was Schmidt who, as far back as 1824, first recognized in the Mongol name Xormusda (Khormusda) the Iranian Ormuzd or Ahuramazdāh of the Avesta. Even Schmidt's adversary, J. KLAPROTH, was obliged to admit that this theory was justified.[2] Rémusat's objections were refuted by SCHMIDT himself.[3] At present we know that the name in question was propagated over Central Asia by the Sogdians in the forms Xūrmazṯā (Wurmazṯ) and Ōharmīzd.[4] What we are still ignorant of is how the transformation of the supreme Iranian god into the supreme Indian god was effected; for in the Buddhist literature of the Mongols the name Xormusda strictly refers to the god Indra. Also in the polyglot Buddhist dictionaries the corresponding terms of Chinese, Tibetan, etc., relate to Indra.

2. Esroa, Esrua, or Esrun, is in the Buddhist literature of the Mongols the designation of the Indian god Brahma. The Iranian origin of this word has been advocated by A. SCHIEFNER.[5] Although taken for a corruption of Sanskrit *içvara* ("lord"), it seems, according to Schiefner, to be in closer relation to Avestan *çraosha* (*sraoša*) or *çravañh*. Certain it is that the Mongol word is derived from the Uigur

---

[1] Forschungen im Gebiete der Bildungsgeschichte der Völker Mittel-Asiens, p. 148.

[2] "Cette hypothèse mérite d'être soigneusement examinée et nous invitons M. Schmidt à recueillir d'autres faits propres à lui donner plus de certitude" (*Nouveau Journal asiatique*, Vol. VII, 1831, p. 180).

[3] Geschichte der Ost-Mongolen, p. 353.

[4] F. W. K. MÜLLER, Die "persischen" Kalenderausdrücke, pp. 6, 7; Handschriftenreste, II, pp. 20, 94.

[5] In his introduction to W. RADLOFF's Proben der Volkslitteratur der türkischen Stämme, Vol. II, p. XI. Schiefner derives also Kurbustu of the Soyon from Ormuzd.

57²

Äzrua, which in the Manichean texts of the Uigur appears as the name of an Iranian deity. C. SALEMANN[1] has promised a discussion of this word, but I have not yet seen this article. Meanwhile GAUTHIOT[2] has solved this problem on the basis of the Sogdian form 'zrw' (= azrwa), which appears as the equivalent of Brahma in the Sogdian Buddhist texts. The Sogdian word, according to him, is the equivalent of Avestan zrvan.

3. Mongol suburgan, tope, Stūpa, is derived from Uigur supurgan. The latter may be of Iranian origin, and, as suggested by GAUTHIOT,[3] go back to spur-xān ("house of perfection").

4. Mongol titim, diadem, crown (corresponding in meaning to and rendering Sanskrit mukuṭa). This word is traceable to Sogdian δiδim.[4] The prototype is Greek διάδημα (whence our "diadem"), which has been preserved in Iran since Macedonian times.[5] In New Persian it is dāhīm or dēhīm, developed from an older *dēδēm. Mongol titim, accordingly, cannot be derived from New Persian, but represents an older form of Iranian speech, which is justly correlated with the Sogdian form.

5. Mongol šimnus, a class of demons (in Buddhist texts, translation of Sanskrit Māra, "the Evil One"), is doubtless derived from Uigur šmnu, the latter from Sogdian šmnu.[6] Cf. also Altaic and Teleutic šulumys ("evil spirit").

6. In view of the Sogdian loan-words in Mongol, it is not impossible that, as suggested by F. W. K. MÜLLER,[7] the termination -ntsa (-nča) in šibagantsa, čibagantsa, or šimnantsa ("bhikṣuṇī, nun;" Manchu čibahanči) should be traceable to the Sogdian feminine suffix -nč (presumably from inč, "woman"). The same ending occurs in Uigur upasanč (Sanskrit upāsikā, "Buddhist lay-woman") and Mongol ubasantsa. R. GAUTHIOT[8] is certainly right in observing that it is im-

---

[1] Bull. de l'Acad. de St.-Pét., 1909, p. 1218.

[2] In CHAVANNES and PELLIOT, Traité manichéen, p. 47.

[3] Ibid., p. 132.

[4] MÜLLER, Uigurica, p. 47.

[5] NÖLDEKE, Persische Studien, II, p. 35; cf. also HÜBSCHMANN, Persische Studien, p. 199.

[6] F. W. K. MÜLLER, Uigurica, p. 58; Soghdische Texte, I, pp. 11, 27. In Sogdian Christian literature, the word serves for the rendering of "Satan." According to MÜLLER (SPAW, 1909, p. 847), also Mongol nišan ("seal") and badman (not explained) should be Middle Persian, and have found their way into Mongol through the medium of the Uigur.

[7] Uigurica, p. 47.

[8] Essai sur le vocalisme du sogdien, p. 112.

possible to prove this interdependence; yet it is probable to a high degree and seems altogether plausible.

7. Textiles made from cotton are designated in Mongol *büs* (Kalmuk *bös*), in Jurči (Jučen or Niüči) *busu*, in Manchu *boso*. This series, first of all, is traceable to Uigur *böz*.[1] The entire group is manifestly connected, as already recognized by SCHOTT,[2] with Greek βύσσος (*byssos*), which itself goes back to Semitic (Hebrew *būṣ*, Assyrian *būṣu*). But how the Semitic word advanced to Central Asia is still obscure; its presence in Uigur might point to Iranian mediation, but it has not yet been traced in any Iranian language. Perhaps it was transmitted to the Uigur directly by Nestorian missionaries. The case would then be analogous to Mongol *nom* (Manchu *nomun*), from Uigur *nom, num* ("a sacred book, law"), which ABEL-RÉMUSAT[3] traced through Semitic to Greek νόμος.

Cotton itself is styled in Mongol *kübeň* or *kübüň*, in Manchu *kubun*. SCHOTT (*l.c.*) was inclined to derive this word from Chinese *ku-pei*, but this is impossible in view of the labial surd. Nevertheless it may be that the Mongol term is connected with a vernacular form based on Sanskrit *karpāsa*, to which also Chinese *ku-pei* is indirectly traceable (above, p. 491). This form must be sought for in Iranian; true it is, in Persian we have *kirpās* (correspondingly in Armenian *kerpas*) and in Arabic *kirbās*. In Vaxī, a Pamir dialect, however, we find *kubas*,[4] which, save the final *s*, agrees with the Mongol form. The final nasals in the Mongol and Manchu words remain to be explained.

8. Mongol *anar*, pomegranate, is doubtless derived from Persian *anār* (above, p. 285). In the Chinese-Uigur Dictionary we meet the form *nara*.[5] In this case, accordingly, Uigur cannot be held responsible as the mediator between Persian and Mongol. In all probability, the fruit was directly transmitted by Iranians to the Mongols, who thus adopted also the name for it.

9. Mongol *turma*, radish, is derived from Persian *turma* (also *turub, turb, turf*).[6]

---

[1] F. W. K. MÜLLER, Uigurica, II, p. 70.

[2] Altaisches Sprachengeschlecht, p. 5; and *Abh. Berl. Akad.*, 1867, p. 138.

[3] Recherches sur les langues tartares, p. 137.

[4] HJULER, The Pamir Languages, p. 38.

[5] KLAPROTH, Sprache und Schrift der Uiguren, p. 14; and RADLOFF, Türk. Wört., Vol. III, col. 648.

[6] Cf. *T'oung Pao*, 1916, p. 84. The derivation from Persian escaped MUNKACSI and GOMBOCZ (*Mém. Soc. finno-ougrienne*, Vol. XXX, p. 131), who erroneously seek the foundation of the word in Turkish.

10. Mongol *xasini*, asafœtida, from Persian *kasnī* ("product of Ghazni"). Cf. above, p. 361.

11. Mongol *bodso*, an alcoholic beverage made from barley-meal or milk, is connected by KOVALEVSKI in his Mongol Dictionary with Persian *boza*, a beverage made from rice, millet, or barley.

12. Mongol *bolot*, steel, is derived from New Persian *pūlād*, whether directly or through the medium of Turkish languages is not certain. The Persian word is widely diffused, and occurs in Tibetan, Armenian, Ossetic, Grusinian, Turkish, and Russian.[1]

13. Mongol *bägdär*, coat-of-mail, armor, goes back to Persian *bagtar* (Jagatai *bäktär*, Tibetan *beg-tse*).

14. Mongol *sagari* and *sarisu*, shagreen.[2] From Persian *sagrī*. In Tibetan it is *sag-ri*;[3] in Manchu *sarin* (while Manchu *šempi* is a transcription of Chinese *sie-p'i* 斜皮).[4]

15. Mongol *kukur, kugur*, sulphur. From Persian *gugurd*, Afghan *kokurt* (Arabic *kibrīt*, Hebrew *gafrit*, Modern Syriac *kugurd*).

16. Other Persian loan-words in Mongol have come from Tibetan, thus: Mongol *nal*, spinel, balas ruby. From Tibetan *nal;* Persian *lāl* (Notes on Turqois, p. 48). Mongol *zira*, cummin. From Tibetan *zi-ra;* Persian *zīra, žīra* (above, p. 383).

17. In some cases the relation of Mongol to Persian is not entirely clear. In these instances we have corresponding words in Turkish, and it cannot be decided with certainty whether the Mongol word is traceable to Turkish or Persian.

Thus Mongol *böriyä*, trumpet (cf. Manchu *buren* and *buleri*), Turkish *boru*, Uigur *börgü*,[5] Persian *būrī*.

18. Mongol *dsärän* (*dsägärän*), a species of antelope (*Procapra subgutturosa*); Altaic *järän*, wild goat of the steppe; Jagatai *jiren*, gazelle; Persian *jīrān*, gazelle.

19. Mongol *tōs* (written *tagus, togos*, to indicate the length of the vowel), peacock. From Persian *ṭāwus* (Turkī *ta'us*).

20. Mongol *toti*, parrot. From Persian *totī* (Uigur and Turkī *totī*).

21. Mongol *bag*, garden. This word occurs in a Mongol-Chinese inscription of the year 1314, where the corresponding Chinese term signifies "garden," and, as recognized by H. C. V. D. GABELENTZ,[6] doubtless represents Persian *bāγ* ("garden").

---

[1] Cf. *T'oung Pao*, 1916, pp. 82, 479.
[2] K'ien-luń's Polyglot Dictionary, Ch. 24, pp. 38, 39.
[3] *T'oung Pao*, 1916, p. 478.
[4] This term is not noted in the Dictionary of Giles.
[5] PELLIOT, *T'oung Pao*, 1915, p. 22.
[6] *Z. K. d. Morg.*, Vol. II, 1839, p. 12.

22. Mongol *šikär*, *šikir*, sugar. From Persian *šakar*.

23. Mongol *šitara*, Kalmuk *šatar*, chess. From Persian *šatranj*.

E. Blochet's derivation of Mongol *bogda* from Persian *bokhta* is a pseudo-Iranicum. The Mongol term is not a loan-word, but indigenous.[1] BOEHTLINGK, in his Yakut Dictionary, has justly compared it with Yakut *bogdo*.

---

[1] Cf. *T'oung Pao*, 1916, p. 495.

CHINESE ELEMENTS IN TURKI

On the preceding pages I had occasion to make reference in more than one instance to words of the Turkī language spoken in Chinese Turkistan. A. v. Le Coq[1] has appended an excellent Turkī vocabulary to a collection of texts recorded by him in the territory of Turfan. This list contains a certain percentage of Chinese loan-words which I wish briefly to discuss here.

In general, these have been correctly recognized and indicated by Le Coq, though not identified with their Chinese equivalents. But several pointed out as such are not Chinese; while there are others which are Chinese, but are not so designated; and a certain number of words put down as Chinese are left in doubt by the addition of an interrogation-mark. To the first class belongs *γań-za* ("tobacco-pipe"), alleged to be Chinese; on the contrary, this is a thoroughly Altaic word, no trace of which is to be discovered in Chinese.[2] It is *khamsa* or *xamsa* in Yakut, already indicated by Boehtlingk.[3] It is *gangsa* or *gantsa* in Mongol;[4] *gansa* in the Buryat dialect of Selengin.[5] The word has further invaded the Ugrian territory: Wogul *qansa*, Ostyak *xońsa*, and Samoyed *xansa*.[6] It is noteworthy that the term has also found its way into Tibetan, where its status as a loan-word has not yet been recognized. It is written in the form *gań-zag* (pronounced *gań-za;* Kovalevski writes it *gań-sa*, and Ramsay gives it as *kanzak* for West-Tibetan); this spelling is due to popular assimilation of the word with Tibetan *gań-zag* ("man, person").

In *jū-xai gül* ("narcissus") I am unable, as suggested by the author, to recognize a Chinese-Turkish formation. The narcissus is styled in

---

[1] Sprichwörter und Lieder aus der Gegend von Turfan, *Baessler-Archiv*, Beiheft I, 1910.

[2] The Chinese word for a tobacco-pipe, (*yen-*) *tai*, is found as *dai* in Golde and other Tungusian languages, because the Tungusian tribes receive their pipes from China.

[3] Jakutisches Wörterbuch, p. 79.

[4] Kovalevski, Dictionnaire mongol, pp. 980, 982.

[5] Castrén, Burjatische Sprachlehre, p. 130.

[6] A. Ahlquist (*Journal de la Société finno-ougrienne*, Vol. VIII, 1890, p. 9), who regards the Ugrian words as loans from Turkish.

577

Chinese *šwi-hien* 水仙 ("water-fairy").[1] *Gül*, of course, is Persian *gul* ("flower"). *Jüsäi* ("garlic") is not Chinese either. *Mäjäzä* ("chair")· is hardly Chinese, as suggested.

To the second class belong *toṅ* ("cold, frozen"), which is apparently identical with Chinese *tuṅ* 凍 of the same meaning, and *tung* ("wooden bucket"), which is the equivalent of Chinese *t'uṅ* 桶 ("tub, barrel"). There are, further, *pän* ("board"), from Chinese *pan* 板; *yangza* ("sort, kind"), from *yaṅ-tse* 樣子; *qäwá* ("gourd"), from *kwa* 瓜.

The word *toṅ-kai* ("donkey's knuckle-bones employed in a game") is tentatively marked Chinese. This term is mentioned, with a brief description of the game, in the Manchu Polyglot Dictionary[2] as Chinese (colloquial) *tan čen'r kun'r* 彈針兒轄兒 and Tibetan *t'e-k'ei-gan;* the latter is not Tibetan, and without any doubt represents a transcription. The Chinese term, however, may be so likewise. In Manchu, the word *toxai* denotes the smooth side of the knuckle-bone, and is apparently related to Turkī *toṅkai*.

The Chinese origin of *lä-zä* ("red pepper, pimento") is not to be questioned. It is Chinese *la-tse* 辣子.[3] Still less can the Chinese character of *ïr-žïn* ("two men," that is, descendant of a Chinese and a Turkish woman) be called into doubt; this, of course, is *er žen* 二人.

The following Chinese words indicated by Le Coq may be identified, only those of special interest being selected:

dän, inn, bungalow, from *tien* 店. This word has been carried by the Chinese all over Central Asia. It has also been traced in Sogdian in the form *ẓïm*.[4]

gō-sï, official placards posted in a public place, from *kao-ši* 告示.

sai-puṅ, tailor, from *ts'ai-fuṅ* 裁縫.

maupaṅ, miller, mill, from *mo-faṅ (ču)* 磨坊主.

yaṅ-xō, match, from *yaṅ hwo* 洋火.

tuṅči bäk, interpreter; the first element from *t'uṅ-ši* 通事 (see Loan-Words in Tibetan, No. 310; and *Journal Am. Or. Soc.*, 1917, p. 200).

čän, money, from *č'ien* 錢.

tï-za, banknotes issued by the Governor of Urumči, from *t'i-tse* 提子.

jōzä, table (Le Coq erroneously "chair"), from *čo-tse* 棹子.

čaṅ, bed, from *čwaṅ* 牀.

dä-dïr, kind of horse-bean, perhaps from *ta-tou* 大豆.

daṅ-za, notebook, from *čaṅ-tse* 賬子.

šum-pō, title of the Chinese governor, from *sün fu* 巡撫(?).

lä-täi, candlestick, from *la t'ai* 蠟臺.

miṅ-läṅ-zä, door-curtain, from *men-liṅ-tse* 門綾子.

yaṅ-yō, potato, from *yaṅ yao* 洋藥.

---

[1] See, further, above, p. 427.

[2] Cf. K. HIMLY, *T'oung Pao*, Vol. VI, 1895, p. 280.

[3] Cf. Loan-Words in Tibetan, No. 237.

[4] F. W. K. MÜLLER, Soghdische Texte, I, p. 104.

In the Turkī collectanea of G. RAQUETTE[1] I note the following Chinese words:

čiṅ-säy, celery, from *č'in ts'ai* 芹 菜.
manto, meat-dumpling, from *man-t'ou* 饅 頭.
čizä, a Chinese foot (measure), from *č'i-tse* 尺 子.
lobo, a long turnip, from *lo-po* 蘿 蔔.
jiṅ, a Chinese pound, from *čin* 斤.

A few other remarks on Turkī words recorded by Le Coq may follow here:

nähäl ("ruby") is apparently Persian *läl* (above, p. 575).
zummurät ("emerald") is not Arabic-Turkish, but Persian (above, p. 519).
There is no reason to question the Persian origin of *palas* ("cloth, sail"); it is identical with Persian *bäläs* (above, p. 495).
döwä ("hill") is identical with Turkish *deve, teve* ("camel"); cf. *T'oung Pao*, 1915, p. 21.
yilpis ("snow-leopard") is identical with Mongol *irbis* ("panther").

---

[1] Eastern Turki Grammar, *Mitt. Sem. Or. Spr.*, 1914, II, pp. 170–232.

THE INDIAN ELEMENTS IN THE PERSIAN PHARMA-COLOGY OF ABU MANSUR MUWAFFAQ

On the preceding pages reference has repeatedly been made to the work of Abu Mansur as proving that the Persians were acquainted with certain plants and products, or as demonstrating the inter-relations of Persia and India, or of Persia and China. Abu Mansur's "Principles of Pharmacology" is a book of fundamental importance, in that it is the first to reveal what Persian-Arabic medicine and pharmacology owe to India, and how Indian drugs were further conveyed to Europe. The author himself informs us that he had been travelling in India, where he became acquainted with her medical literature. It therefore seems to me a useful task to collect here what is found of Indian elements in his work, and thus present a complete summary of the influence exerted by India on the Persia of the tenth century. It is not my object to trace merely Indian loan-words in Persian, although several not hitherto recognized (as, for instance, *balādur, turunj, dand, pūpal,* etc.) have been identified by me; but I wish to draw up a list of all Indian drugs or products occurring in Abu Mansur, regardless of their designations, and to identify them with their Indian equivalents. Abu Mansur gives the names in Arabic; the Persian names are supplied from Achundow's commentary or other sources. The numbers in parentheses refer to those in Achundow's translation.

J. Jolly has added to the publication of Achundow a few observations on Indian words occurring in the work of Abu Mansur; but the real Indian plants and drugs are not noticed by him at all, while his alleged identifications are mere guesswork. Thus he proposes for *armāk* or *armal* Skr. *amlaka, amlikā,* and *āmra,* three entirely different plants, none of which corresponds to the description of *armak,* which is a bark very similar to *kurfa* (*Winterania canella*), the best being brought from Yemen; it is accordingly an Arabic, not an Indian plant. *Harbuwand* (No. 576) is described as a grain smaller than pepper, somewhat yellow-ish, and smelling like *Aloëxylon agallochum;* according to Jolly, this should be derived from Skr. *kharva-vindhyā* ("small cardamom"), but the question is not of cardamoms, and there is no phonetic coincidence of the words. The text says that *kader* (No. 500) is a wholesome remedy to soften the pustules of small-pox. Jolly proposes no less

580

than four Sanskrit plant-names,— *kadara, kadala, kandara,* and *kandata,* while the Tohfat states that *kader* is called *kawi* in India, being a tree similar to the date-palm, the flower being known as *kaburah* (p. 197); *kader,* accordingly, is an Arabic word, while *kawi* is the supposed Indian equivalent and may correspond to Sanskrit *kapi* (*Emblica officinalis, Pongamia glabra,* or *Olibanum*). These examples suffice: the twenty-one identifications proposed by Jolly are not convincing. Many of these have also been rejected by Achundow.

The Indian loan-words in Persian should occasionally be made the subject of an exhaustive study. A few of these are enumerated by P. HORN.[1] *Kurkum* ("saffron"), however, is not of Indian origin, as stated by him (cf. above, p. 321). Skr. *surā,* mentioned above, occurs in Persian as *sur* ("rice-wine"). Middle Persian *kapīk,* Persian *kabī* ("monkey"), is derived from Skr. *kapi.*[2]

1(1). *aruz,* P. *birinj,* rice (*Oryza sativa*). Cf. above, p. 373.

2(5). *utruj,* P. *turunj,* citron (*Citrus medica*). From Skr. *mātulunga* (above, p. 301), also *mātulanga, -lānga,* and *-linga.*

3(11). *ihlilaj,* P. *halīla,* myrobalan (*Terminalia chebula*). Skr. *harītakī* (above, p. 378).

4(76). *balīlaj,* P. *balīla, Terminalia belerica,* Skr. *vibhītaka* (cf. *T'oung Pao,* 1915, p. 275).

5(12). *amlaj,* P. *amīla* (*amela, amula*), *Emblica officinalis* or *Phyllanthus emblica.* Skr. *amala* (also *dhātrī*), provided the botanical identification is correct; phonetically, P. *āmila* would rather point to Skr. *āmla* or *amlikā* (*Tamarindus indica*), Chinese transcription 菴弭羅 *an-mi-lo,* *am-mi-la. Abu Mansur states that "there is a variety *sīr-amlaj;* some physicians erroneously read this name *šīr-amlaj,* believing that it was administered in milk (*šīr*); but this is a gross error, for it is *sīr,* and this is an Indian word, and *amlaj* signifies 'without stone.' I was there where *amlaj* grows, and have seen it with my own eyes." The etymology given is fantastic, but may have been communicated to the author in India.

6(33). *atmat, Nelumbium speciosum* or *Nelumbo nucifera* (p. 205). "It is a kernel like an Indian hazel-nut. Its effect is like that of *Orchis morio.* It is the seed of *Nymphæa alba indica,* and is as round as the Indian hazel-nut." Both the botanical identification and the translation appear to me somewhat questionable. Cf. No. 47.

7(36). *āzādraxt, āzādiraxt, Melia azadiracta.* Abu Mansur adds *šīšiān* as the Arabic name of the plant. Ibn al-Baiṭār (LECLERC, Vol. I,

---

[1] Grundr. iran. Philol., Vol. I, pt. 2, p. 7.

[2] HÜBSCHMANN, Pers. Studien, p. 87.

p. 54) explains the Persian word as "free tree," and Leclerc accordingly derives it from *azād-diraxt.* Skr. *nimba, nimbaka, mahānimba.*

8(40). *ušnān, Herba alkali,* chiefly species of *Salsola.* "There are four kinds of alkali herb, a white, yellow, green, and an Indian kind which occurs as Indian hazel-nut (*funduq-i hindī*), also called *xurs-i sīnī* ('Chinese *xurs'*) and *rutta.*" Cf. *T'oung Pao,* 1916, p. 93; above, p. 551.

9(54). *bitīx ul-hindī,* P. *hindewāne,* water-melon (above, p. 443).

10(73). *belādur, balādur,* the marking-nut tree (*Semecarpus anacardium*). Cf. above, p. 482.

11(77). *birinj-i kābilī,* "rice of Kabul" (*Embelia ribes*). Skr. *viḍaṅga* (cf. *T'oung Pao,* 1915, pp. 282–288; 1916, p. 69).

12(78). *bang,* henbane (*Hyoscyamus*), a narcotic prepared from hemp-seeds. The seed was used as a substitute for opium (Abu Mansur, No. 59). Skr. *bhaṅgā,* hemp (*Cannabis sativa*). The Persian word is also traced to Avestan *baṅha,* "a narcotic," but it seems to me preferable to assume direct derivation from Skr. in historical times. Arabic *banj,* Portuguese *bango,* French *bangue.* P. *šabībī,* "a narcotic root; also the inebriating hemp-seed."

13(85). *bīš, halahil,* aconite (*Aconitum*). Hindī *bīš,* Skr. *viṣā* (*Aconitum ferox*), from *viṣa,* "poison;" Skr. *hālāhala,* a species of aconite and a strong poison prepared from it. Cf. *T'oung Pao,* 1915, pp. 319–320, note.

14(87). *tūt,* mulberry (*Morus alba*), a native of China. The opinion of NÖLDEKE (Pers. Studien, II, p. 43), that the Persian word is traceable to Semitic, is entirely erroneous, as this species spread from the far east and India to Iran and Europe, and began to be cultivated in the Mediterranean area only from the twelfth century. Skr. *tūda* and *tūla,* Bengalī and Hindustānī *tūl, tūt, Morus alba* or *indica* (ROXBURGH, Flora Indica, p. 658); cf. SCHRADER in Hehn, Kulturpflanzen, p. 393. *Morus nigra,* the black mulberry, is a native of Persia.

15(90). *tamr ul-hindī,* P. *tamar-i hindī,* tamarind (*Tamarindus indica*), cultivated throughout India and Burma. Skr. *tintiḍa, tintiḍīka, tintilikā,* etc., *jhābuka, amlīkā.*

16(94). *tanbūl,* P. *pān, barge-tanbōl,* betel (*Piper betle*). Skr. *tāmbūla, nāgavallikā.*

17(111). *jūz-i buwwā,* P. *jūz-i būya,* nutmeg (*Myristica moschata, officinalis,* or *fragrans*). Skr. *jāti, jātikoça, jātisāra, jātiphala.*

18(112). *jūz-i mātil,* P. *tātūra, dātūra, Datura metel.* Skr. *mātula, dhatūra.* Cf. *T'oung Pao,* 1917, p. 23.

19(142). *habb ul-qilqil* (*qulqul*), seeds of *Cassia tora* (the fœtid cassia). Skr. *prapunāḍa, prapunāṭa, prapumnāla, tubarīçimba;* Singhalese *peti-*

tora (also cultivated in Indo-China, China, and Japan: PERROT and HURRIER, p. 146; STUART, p. 96; Japanese *ebisu-gusa*).

20(248). *duhn ul-amlaj*, oil of myrobalan (*oleum emblicae*). Cf. No. 5.

21(251). *duhn ul-sunbul*, Indian nard-oil (*oleum Valerianae jatamansi*). Cf. No. 32.

22(253). *dār-ṣīnī*, P. *dār-čīnī*, cinnamon (*Laurus cinnamomum, Cinnamomum tamala*). Arabic also *sadāj*. Skr. *tvaca*.

23(254). *dār-filfil*, P. *pipal, pilpil*, long pepper (*Piper longum*). Skr. *pippalī*.

24(260). *dand, dend, dund*, Croton tiglium. From Skr. *dantī, Croton polyandrus* (also called *Baliospermum montanum*). Abu Mansur adds that this plant is called in Indian *čeipal*. This is Skr. *jayapāla, Croton jamalgota* (the latter from Hindustānī *jamālgōta*), styled also *sāraka*. Arabic also *dend ṣīnī* (Löw, Aram. Pflanzennamen, p. 170). Cf. above, p. 448. In Tibetan we have *dan-da* and *dan-rog*.

25(261). P. *dīvdār, dēvdār*, Pinus or Cedrus *devdara, deodara*, or *deodora*. Skr. *devadāru* ("tree of the gods"). In Persian also *sanōbar-i hindī, naštar*; Arabic *šajratud-dēvdār, sanōbarul-hind*.

26(272). *zarīra*, sweet flag (*Acorus calamus*). Achundow (p. 192) identifies Arabic *zarīra* with an alleged Indian word *dhsarirah*, indicated by Berendes; I cannot trace such an Indian word. *Zarīra* appears to be identical with Arabic *dirira* (GARCIA) or *darira* ("aroma"); cf. also Löw, *l.c.*, p. 342. Skr. *vacā*, conveyed to Persian and Arabic as *vāj* (GARCIA: Guzerat *vaz*, Deccan *bache*, Malabar *vazabu*, Concan *vaicam*, employed by Abu Mansur in No. 564, where Achundow identifies it with *Iris pseudacorus*, and on p. 272 also with *Acorus calamus*), *ugragandha*, and *ṣaḍgranthā*.

27(281). *ratta*, P. *bunduq-i hindī* ("Indian hazel-nut"), *Sapindus mukorossi* and *trifoliatus* (not in Watt); Achundow's identification is apparently erroneous. The question evidently is of *Guilandina bonduc* (cf. LECLERC, Vol. I, p. 276), also called *Cæsalpinia bonducella*, the fever-nut or physic-nut, Skr. *kuberākṣī* ("eye of Kubera"), *latākarañja;* P. *xāyahe-i iblīs*; Arabic *akitmakit, kitmakit*.

28(288). *šangalūl* (Middle Persian *šangavīr*), Arabic-Persian *zanjabīl*, ginger (*Zingiber officinale*). Three kinds—Chinese, Zanzibar, and Melinawi or *zurunbāj*—are distinguished. The word is based on an Indian vernacular form *s(š)angavīra, corresponding to Pāli *singivera*, Skr. *çṛṅgavera; ārdraka* (the fresh root).

29(292). *zurunbād*, P. *zarambād, Curcuma zedoaria*. Cf. YULE, Hobson-Jobson, p. 979.

30(304). *zarwār, Curcuma aromatica* or *zedoaria*. "This is an Indian remedy." Achundow (p. 193) suspects a clerical error for *zadwār* (also *jadwār*). Skr. *nirviṣa, vanaharidrā*. Cf. above, p. 544.

31(311). *sukkar*, P. *šakar, šakkar*, sugar-cane, sugar (*Saccharum officinarum*). Prakrit and Pāli *sakkharā*, Skr. *çarkarā*.

32(315). *sunbul*, P. *sunbul-i hindī, Valeriana jatamansi*. Skr. *jatāmāmsī*.

33(316). *salīxa, Laurus cassia*. Skr. *tvaca* Cf. No. 22.

34(324). *saqmūniyā, Convolvulus scammonia*. "There are three kinds, an Indian, that from Čarmgān, and that from Antiochia; the latter being the best, the Indian ranking next. The Indian kind is the gum of *Convolvulus* (or *Ipomœa*) *turpethum*." The latter is Skr. *triputa*, or *trivṛt;* hence Hindustani *tarbud*, P. *turbid*, Arabic *turbund*. *C. scammonia* is a native of Syria, Asia Minor, and Greece, and is cultivated in some parts of India.

35(333). *sātil*. "It is an Indian remedy which resembles a *Tuber terrae* (fungus), and purges the corrupted humours." It is also called *šatil* and in Persian *rōšanak*.

36(361). *šal* (*šul*), "Indian quince (*Cydonia indica*)." In the commentary (p. 245), Achundow cites also a Persian *bih-i hindī* ("Indian quince"), and adds that Schlimmer mentions merely a *Cydonia vulgaris*. What this *Cydonia indica* is supposed to be is a mystery: neither Roxburgh nor Watt knows such an Indian species. A. de Candolle already knew that there is no Sanskrit name for the quince. The Persian quince is mentioned by Abu Mansur (No. 309) as *safarjal* (P. *bih* or *beh*, and *ābī*).

37(368). *sandal* (Arabic), *čandan, čandal* (Persian), sandal-wood (*Lignum santalinuṃ*). Red (from *Pterocarpus santalinus*) and white (from *Santalum album*) are distinguished. Skr. *candana*.

38(386). *tālīsfar*, alleged to be *Myristica moschata;* on p. 247, however, Achundow withdraws this interpretation. According to Daud, it is the bark of the mulberry coming from the Dekkan. The word, at all events, appears to be Indian: cf. Skr. *tālīçapattra*, "leaf of *Flacourtia cataphracta*."

39(422). *fulful*, also *filfil*, black pepper (*Piper nigrum*). Skr. *pippalī, marica*.

40(434). *fūfal*, P. *pūpal*, areca-nut palm (*Areca catechu*). Skr. *pūgaphala;* Singhalese *puvak*.

41(450). *qust*, P. *kust*, *Costus amarus* or *speciosus* (cf. also p. 254). Skr. *kuṣṭha, idem* and *Saussurea lappa*.

42(456). *qāqula*, P. *hīl-i buzurg*, grains of paradise seeds, greater seeds of cardamom (*Amomum granum paradisi*, or *melegueta*).

43(457). *qaranful*, P. *mexak*, cloves (*Caryophyllus aromaticus*). Skr. *lavaṅga*.

44(459). *qūlāni*, a kind of barley brought from India. JOLLY (p. 196), without giving an Indian name, regards this as *Glycine labialis* (ROXBURGH, Flora Indica, p. 565); Watt does not give this species for India. Cf. No. 572, where it is described under the name *hāl*.

45(480). *kundur*, incense (*Boswellia thurifera*). Skr. *kunduru*, *kundura*, *kundu*, *kunduruka*. Achundow does not mention a Persian form *kundurū*, as asserted by HÜBSCHMANN (Armen. Gram., p. 172). Pahlavi **kundurūk* and Armenian *kndruk* are directly traceable to Skr. *kunduruka*.

46(483). *kāfūr* (Arabic and Persian), camphor (*Laurus camphora*). The same word appears already in Middle Persian. Skr. *karpūra*.

47(512). *lāk*, *ränglāk*, lac (*Gummi laccae*). Cf. above, p. 476.

48(517). *māš*, mungo bean (*Phaseolus mungo*). Skr. *māṣa* (*Phaseolus radiatus*). This Indian word is widely diffused over Asia: Tibetan *ma-ša*, Mongol *maša*, Turkī *māš* ("a small kind of bean"), Taranči *maš* ("bean"), Sart *maš* ("lentil"), Osmanli *maš*.

49(525). *mušktirāmušīr*, *mušktirāmšī*, *Origanum dictamnus*. "The best is that of India." The name is said to come from the Syriac (p. 267). AINSLEE (Materia Indica, Vol. I, p. 112) calls it dittany of Crete, and says that he has never seen it in India. Indeed it does not occur there, hence the Indian variety of Abu Mansur must be *O. marjorana*, the sweet marjoran, Skr. *phaṇijjhaka*, Arabic *mardakuš* or *mizunjuš*.

50(550). *nargīl* (Arabic *nārjīl*), coco-nut (*Cocos nucifera*). Avicenna: *juz hindī* ("Indian nut"). Skr. *nārikela*, *nārikera*, etc.

51(552). *nīlūfar*, P. *nīlūpar*, *Nymphæa alba*, *N. lotus*, etc. Skr. *nīlōtpala* (*Nymphæa lotus*); also *kumuda*, *kamala*, etc. Cf. LOEW, *l.c.*, p. 313.

52(557). *nīl*, *līla*, indigo (*Indigofera tinctoria*). Skr. *nīla* (above, p. 370).

53(572). *hāl*, P. *hīl-i xurde*, lesser cardamom (*Cardamomum minus* or *malabaricum*, or *Elettaria cardamomum*). Skr. *elā*.

54(583). *yabrūh*, mandrake (*Atropa mandragora*). "Two kinds are distinguished, an Indian, called *yabrūh ul-sanam*, and a Nabathæan." As the genus *Atropa* does not occur in India, with the exception of *A. belladonna*, which, however, is restricted to the territory stretching from Simla to Kashmir, it is obvious that a species of *Datura* is to be understood by the Indian mandrake of Abu Mansur. This case is interesting, in that it shows again the identical employment of the mandrake and the datura (cf. LAUFER, La Mandragore, *T'oung Pao*, 1917, pp. 1–30).

Appendix IV

THE BASIL

I propose to treat here briefly of the history of a genus of plants which has not yet been discussed by historians,— *Ocimum*, an extensive genus of the order *Labiatae*. I do not share the common opinion of most commentators of Theophrastus and Pliny, that their ὤκιμον or *ocimum* is identical with the *Ocimum basilicum* of Linné. Theophrastus touches on *okimon* in several passages; but what he describes is a shrub, not an herb, nor does he emphasize any of the characteristic properties of *Ocimum basilicum*. Fée justly comments on Pliny (xx, 48) that this species is not understood by him, it being originally from India (or rather, as will be seen, from Iran), and never found in a wild state. From what Varro says, he infers that Pliny's *ocimum* must be sought among the leguminous plants, the genus *Hedysarum*, *Lathyrus*, or *Medicago*.[1] Positive evidence of this conclusion comes from Ibn al-Baiṭār, whose vast compilation is principally based on the work of Dioscorides, with the addition of annotations of Arabic authors. Ibn al-Baiṭār, in his discussion of the plant which we call *Ocimum*, does not fall back on the *okimon* of Dioscorides (ii, 171), and, in fact, does not cite him at all.[2] He merely reproduces the data of Arabic writers: this is decisive, and leads us to reject any connection between the *ocimum* of the ancients and the species coming from the Orient and known to our science of botany as *Ocimum*.[3]

There is good reason to assume that at least one species, if not several, is a native of Persia, and was diffused from there to India and China, probably also to the West. This is *Ocimum basilicum*, the sweet or common basil. The name βασιλικόν ("royal") as the designation of an *Ocimum* first occurs in Byzantine literature, in Aetius (sixth century) and Symeon Seth; and, since the king of Persia was known to the Greeks simply as "the king" (βασιλεύς), it is more than probable that the Greek term is reproduced after the model of Persian *šāh-siparam* (*spram*) or *šāh-i sfaram*, which means as much as "fragrant .

---

[1] Cf. BOSTOCK and RILEY, Natural History of Pliny, Vol. IV, p. 249.

[2] Cf. LECLERC, Traité des simples, Vol. II, p. 186; Vol. III, p. 191.

[3] Leclerc upholds the opposite opinion, although Sprengel, Fée, and Littré argue in the same manner as here proposed.

586

leaf of the king," and denotes the basil.[1] The plant is esteemed for its leaves, which serve for culinary purposes to season soups or other dishes, and which have a flavor somewhat like cloves. The juice of the leaves is employed medicinally.

Indeed, as shown by our word "basil," it was under this Middle-Greek name, which did not exist in the period of classical antiquity, that the plant became known to the herbalists of Europe. Thus the celebrated JOHN GERARDE[2] says, "The latter Grecians have called it *basilikon:* in shops likewise *Basilicum,* and *Regium:* in Spanish *Albahaca:*[3] in French Basilic: in English Basill, Garden Basill, the greater Basill royall, the lesser Basill gentle, and Bush Basill." D. REMBERT DODOENS[4] speaks of the basill royall or great basill, and says, "In this countrey the Herboristes do plante it in their gardens." There is much in favor of Sickenberger's supposition that the introduction of the basil into Europe may be due to the returning crusaders,[5] while the Arabic name adopted in Spain and Portugal suggests a Moorish transplantation into western Europe.

Two varieties are common throughout Persia and Russian Turkistan, — one with green and another with dark-red leaves.[6] According to Avicenna, it grows in the mountains of Ispahan.[7] Abu Mansur sets forth its medicinal properties.[8] It is further cultivated throughout India, Malaya, and China.[9]

W. ROXBURGH[10] states that *Ocimum basilicum* is a native of Persia, and was thence sent to the Botanic Garden at Calcutta under the Persian names *deban-šah* and *deban-macwassi.* According to W.

---

[1] POTT, *Z. f. K. Morg.,* Vol. VII, 1850, p. 145. Osmanli *fesligen* or *fesliyen* is likewise based on the Greek word. According to the Century Dictionary, the word *basil* is of unknown origin. The Oxford Dictionary cites from Prior, "perhaps because the herb was used in some royal unguent, bath, or medicine,"— a baseless speculation, as in fact it was never used in this way.

[2] The Herball or Generall Historie of Plantes, p. 547 (London, 1597).

[3] Also *alfabega, alhabega, alabega,* Portuguese *alfabaca* (French *fabrègue*), from Arabic *al-habak* (*rīxāni*); the latter occurs in LECLERC, Traité des simples, Vol. I, p. 404.

[4] Nievve Herball, translation of HENRY LYTE, p. 239 (London, 1578).

[5] Cited in ACHUNDOW, Abu Mansur, p. 211.

[6] KORŽINSKI, Očerki rastitelnosti Turkestana, p. 51. SCHLIMMER mentions the two species *Ocimum album* and *basilicum* as occurring in Persia.

[7] LECLERC, Traité des simples, Vol. III, p. 191.

[8] ACHUNDOW, Abu Mansur, pp. 66, 90, 103.

[9] FORBES and HEMSLEY, *Journ. Linn. Soc.,* Vol. XXVI, p. 266; KING and GAMBLE, Materials for a Flora of the Malayan Peninsula, p. 702 (Perak, Penang, Malacca, perhaps only cultivated).

[10] Flora Indica, p. 464.

AINSLIE,[1] the plant was brought to India from Persia, where it is common, by Sir John Malcolm. This is quite possible; but the fact cannot be doubted that the basil was known in India at a much earlier date, for we have a variety of Sanskrit names for it. Also G. WATT[2] holds that the herb is indigenous in Persia and Sind. It is now cultivated throughout tropical India from the Panjab to Burma.

The Chinese name of *Ocimum basilicum* is *lo-lo* 羅勒 (*la-lak). It is first described in the *Ts'i min yao šu* of the sixth century, where it is said that Ši Lo (273–333) tabooed the name (on account of the identity of the second character with that in his own name, cf. above, p. 298) and changed it into *lan hian* 蘭香; but T'ao Huṅ-kiṅ (451–536) mentions it again as *lo-lo*, and gives as popular designation *Si-waṅ-mu ts'ai* 西王母菜 ("vegetable of the goddess Si-waṅ-mu"). The *Ts'i min yao šu* cites an older work *Wei huṅ fu sü* 草弘賦叙 ("Preface to the Poems of Wei Huṅ") to the effect that the plant *lo-lo* grows on the hills of the K'un-lun and comes from the primitive culture of the Western Barbarians (出西蠻之俗). This appears to be an allusion to foreign origin; nevertheless an introduction from abroad is not hinted at in any of the subsequent herbals. Of these, the *Pen ts'ao* of the Kia-yu period (1056–64) is the first which speaks of the basil as introduced into the materia medica. The name *lo-lo* has no meaning in Chinese, and at first sight conveys the impression of a foreign word. Each of the two elements is most frequent in transcriptions from the Sanskrit. In fact, one of the Sanskrit names of the basil is *karālaka* (or *karāla*), and Chinese *la-lak (*ra-lak) corresponds exactly; the first syllable *ka-* is sometimes dropped in the Indian vernaculars.[3] If this coincidence is fortuitous, the accident is extraordinary; but it is hardly possible to believe in an accident of this kind.

There is, further, a plant 浮爛羅勒 *fou-lan-lo-lo*, *fu (bu)-lan-la-lak, solely mentioned by Č'en Ts'aṅ-k'i of the eighth century as growing in Sogdiana (K'aṅ) and resembling the *hou-p'o* 厚朴 (*Magnolia hypoleuca*), Japanese *hō-no-ki*.[4] The *Pen ts'ao kaṅ mu* has therefore placed this notice as an appendix to *hou-p'o*. This Sogdian plant and its name remain unidentified. At the outset it is most improbable that a *Magnolia* is involved; this is a typical genus of the far east, which to my knowledge has not yet been traced in any Iranian region. BOISSIER's

---

[1] Materia Indica, Vol. II, p. 424.

[2] Dictionary, Vol. V, p. 441.

[3] Cf. for instance *kakinduka* ("*Diospyros tomentosa*")— Uriya *kendhu*, Bengal, *kend*.

[4] *Čeṅ lei pen ts'ao*, Ch. 12, p. 56 b; *Pen ts'ao kaṅ mu*, Ch. 35 A, p. 4; STUART Chinese Materia Medica, p. 255.

"Flora Orientalis" does not contain any *Magnolia*. The foreign name is apparently a compound, the second element of which, *lo-lo*, is identical with the Indian-Chinese name of the basil, so that it is justifiable to suppose that the entire name denotes an Iranian variety of the basil or another member of the genus *Ocimum*.

The basil is styled in Middle Persian *palangamušk*, in New Persian *palaṅmišk*, Arabic-Persian *falanjmušk*, *faranjmušk*, Abu Mansur: *faranjamušk* (Armenian *p'alangamušk*),[1] the second element *mušk* or *mišk* meaning "musk," and the first component denoting anything of a motley color, like a panther or giraffe. The significance of the word, accordingly, is "spotted and musky." This definition is quite plausible, for the leaves of some basils are spotted. JOHN PARKINSON,[2] discussing the various names of the basil, remarks, "The first is usually called *Ocimum vulgare*, or *vulgatius*, and *Ocimum Citratum*. In English, Common or Garden Basill. The other is called *Ocimum minimum*, or *Gariophyllatum*, Clove Basill, or Bush Basill. The last eyther of his place, or forme of his leaves, being spotted and curled, or all, is called *Ocimum Indicum maculatum*, *latifolium* and *crispum*. In English according to the Latine, Indian Basill, broade leafed Basill, spotted or curled Basill, which you please."[3] The Arabic forms are phonetically developed from Persian *palaṅ*; and it is somewhat surprising that R. DOZY[4] explains Arabic *faranjmušk* as "musk of the Franks," although he refers to the variants *baranj* and *falanj*.

While there is a certain resemblance between the Middle-Persian name and our Chinese transcription, I do not believe that the two can be identified. The Chinese calls for an initial sonant and a *u*-vowel; whereas the Iranian form, as positively corroborated by the Armenian loan-word, is possessed of an initial surd with following *a*. I am rather inclined to regard *bu-lan as a Sogdian word, and to derive it from Sogdian *bōδa*, *bōδan* ("perfume").[5] The name *bu-lan ra-lak would accordingly signify "aromatic basil" (corresponding to our "sweet basil"), the peculiar aroma being the prominent characteristic of the

---

[1] HÜBSCHMANN, Armen. Gram., p. 254. According to others, this word would refer to *Ocimum gratissimum*, the shrubby basil, but practically this makes no difference, as the properties and employment of the herbs are the same.

[2] Paradisi in sole paradisus terrestris, p. 450 (London, 1629). The technical term of the botanists in describing the leaves is *subtus punctata* (G. BENTHAM, Labiatarum genera, p. 5; DE CANDOLLE, Prodromus, pars XII, p. 32).

[3] LINNÉ (Species plantarum, Vol. I, p. 597, Holmiae, 1753) has *Ocymum latifolium maculatum sive crispum*.

[4] Supplément aux dictionnaires arabes, Vol. II, p. 262.

[5] R. GAUTHIOT, Essai sur le vocalisme du sogdien, pp. 45, 101, 102; F. W. K. MÜLLER, Handschriften-Reste in Estrangelo-Schrift, II, p. 35.

herb. As it is localized in Sogdiana, it is perfectly justifiable to regard
the term as Sogdian; it may be, however, that the second component did
not form part of the Sogdian word, and is an addition of Č'en Ts'aṅ-k'i;
it is also possible that the term applies to another species of *Ocimum* or
to a peculiar variety of *Ocimum basilicum*, differentiated by cultivation. It is well known that the New-Persian word *bōi*, *bō* ("scent, perfume") enters into composition with a number of aromatics;[1] and
Persian *nāz-bō* is indeed a designation of the basil, and means "having
an agreeable odor." In the same manner we have Sanskrit *gandhapatra*
("fragrant leaf, basil").

From India one or more species of *Ocimum* (*basilicum*, *sanctum*,
and *gratissimum*) spread into the Malayan Archipelago. The Sanskrit
term *surasī* or *surasā* has been adopted by Malayan *sulasi*, Javanese
selasih or *sulasih*, Sunda *salasih*. Javanese has likewise received *tulasih*
or *telasih* from Sanskrit *tulasī*.[2] The two *surasā*, the white and black
varieties of the Tulsī-plant, appear in the Bower Manuscript.[3] In the
folk-lore of India the plant plays an extensive rôle.[4] ODORIC OF POR-
DENONE relates, "In this country every man hath before his house a
plant of twigs as thick as a pillar would be here, and this never withers
as long as it gets water." YULE[5] justly comments that this plant is the
sacred tulasi (*Ocimum sanctum*). It is widely employed in the pharmacopœia of the Persians and Arabs.[6] Arabic terms are: *badrūj*, *xauk*,
rixān, *kebīr*, *aqīn*, *xamāxim*.

---

[1] HÜBSCHMANN, Armen. Gram., p. 123. Cf. also above, p. 462; and HORN,
Neupers. Etymol., No. 240.

[2] Cf. H. KERN, *Bijdragen tot de taal-, land- en volkenkunde*, 1880, p. 564.

[3] HOERNLE's edition, p. 22. There are also the forms *suravallī*, *surasāgraṇī*,
and *surasāgraja*, the two last-named relating to the white variety.

[4] YULE, Hobson-Jobson, p. 931.

[5] Cathay, new ed. by Cordier, Vol. II, p. 116.

[6] LECLERC, Traité des simples, Vol. I, pp. 92, 367, 403, 404, 456, 474; Vol. II,
pp. 100, 104, 191, 375, 390.

ADDITIONAL NOTES ON LOAN–WORDS IN TIBETAN

In my "Loan-Words in Tibetan" (*T'oung Pao*, 1916, pp. 403–552) I was obliged to deal succinctly with some of the problems which are discussed at greater length in this volume. The brief notes given there on saffron, cummin, almond, alfalfa, coriander, etc., are now superseded by the contributions here inserted. A detailed history of Guinea pepper (No. 237) is now ready in manuscript, and will appear as a chapter in my "History of the Cultivated Plants of America." The numbers of the following additions refer to those of the former article.

Note the termination -*e* in the loan-words derived from the Indian vernaculars: *braṁ-ze, neu-le, ma-he, seṅ-ge, ban-de, bhaṅ-ge*. This -*e* appears to be identical with the nominative -*e* of Māgadhī.

49. *ga-bur*, camphor. Sir GEORGE A. GRIERSON (see below) observes, "The softening of initial *k* to *g* is, I think, certainly not Indian." The Tibetan form has always been a mystery to me: it is not only the initial *g*, but also the labial sonant *b*, which are striking as compared with the surds in Skr. *karpūra*. As is well known, this word has migrated westward, the initial *k* being retained everywhere: Persian-Arabic *kāfūr* (GARCIA: *capur* and *cafur*), Spanish *alcanfor* (ACOSTA: *canfora*). These forms share the loss of the medial *r* with Tibetan. This phenomenon pre-existed in Indian; for in Hindustānī we have *kapūr*, in Singhalese *kapuru*, in Javanese and Malayan *kāpur*. The Mongols have adopted from the Tibetans the same word as *gabur;* but, according to KOVALEVSKI (p. 2431), there is also a Tibeto-Mongol spelling *gad-pu-ra:* this can only be a transcription of the Chinese type 羯布羅 *kie-pu-lo*, anciently *g'iaδ-bu-la, based on an Indian original *garpūra, or *garbūra. Tibetan *ga-bur*, of course, cannot be based on the Chinese form; but the latter doubtless demonstrates that, within the sphere of Indian speech, there must have been a dialectic variant of the word with initial sonant.

54. The *Pol. D.* (27, p. 31) gives *naliśam* (printed *aliśam*) as a Mongol word; assuredly it is not Tibetan. The corresponding Manchu word is *xalxôri*.

58. Regarding *śiṅ-kun*, see above, p. 362.

60. With respect to the Chinese transcription *su-ki-mi-lo-si*, PELLIOT (*T'oung Pao*, 1912, p. 455) had pointed out that the last element *si*

591

does not form part of the transcription. This is most likely, but the
Sino-Indian word is thus recorded in the *Pen ts'ao kaṅ mu*.

64. Add: Skr. also *bilāla, birāla*.

65. Sikkim *noile*, Dhimal *nyūl*, Bodo *nyūlai* ("ichneumon").

74. *ban-de*, as suggested by my friend W. E. Clark of the University of Chicago, is connected with Pāli and Jaina Prakrit *bhante*, Skr.
bhadanta ("reverend").

79. I have traced Tibetan *sendha-pa* to Sanskrit *sindhuja*. This, as
a matter of fact, is correct, but from a philological viewpoint the Tibetan
form is based on Sanskrit *saindhava* with the same meaning ("relating
to the sea, relating to or coming from the Indus, a horse from the Indus
country, rock-salt from the Indus region"). The same word we find in
Chinese garb as 先 陀 婆 *sien-t'o-p'o*, *siän-da-bwa, explained as "rock-
salt" (*Fan yi miṅ yi tsi*, section 25). Tokharian has adopted it in the
form *sindhāp* or *sintāp* (S. LÉVI, *Journal asiatique*, 1911, II, pp. 124, 139).

158. The recent discussion opened in the *Journal of the Royal
Asiatic Society* (1917, p. 834) by Mr. H. BEVERIDGE in regard to the
title *tarxan* (*tarkhan*, originally *tarkan*), then taken up by Dr. F. W.
THOMAS (*ibid.*, 1918, p. 122), and resumed by BEVERIDGE (1918, p. 314),
induces me to enlarge my previous notes on this subject, and to trace
the early history of this curious term as accurately as in the present state
of science is possible.

The word *tarkan* is of Old-Turkish, not of Mongol, origin. It is first
recorded during the T'ang dynasty (A.D. 618–906) as the designation of
a dignity, usually preceded by a proper name, both in the Old-Turkish
inscriptions of the Orkhon (for instance, Apa Tarkan) and in the Chinese
Annals of the T'ang (cf. THOMSEN, Inscriptions de l'Orkhon, pp. 59,
131, 185; RADLOFF, Alttürk. Inschriften, p. 369, and Wörterb. Türk-
Dialecte, Vol. III, col. 851; MARQUART, Chronologie d. alttürk. Inschriften, p. 43; HIRTH, Nachworte zur Inschrift des Tonjukuk,
pp. 55–56). An old Chinese gloss relative to the significance of the
title does not seem to exist, or has not yet been traced. According to
Hirth, the title was connected with the high command over the troops.
The modern Chinese interpretation is "ennobled:" the title is bestowed only on those who have gained merit in war (WATTERS, Essays,
p. 372). The Tibetan gloss indicated by me, "endowed with great
power, or empowered with authority," inspires confidence. The subsequent explanation, "exempt from taxes," seems to be a mere makeshift and to take too narrow a view of the matter. A lengthy dissertation on the meaning of the title is inserted in the Ain-i Akbari of 1597
(translation of BLOCHMANN, p. 364); but it must not be forgotten that
what holds good for the Mongol and Mogul periods is not necessarily

valid for the Turkish epoch under the T'ang. According to the T'ang Annals (*T'aṅ šu*, Ch. 217 B, p. 8), the officials of the Kirgiz were divided into six classes, the sixth being called *tarkan*. The other offices are designated by purely Chinese names, and refer to civil and military grades. Among the Kirgiz, therefore, *tarkan* denoted a high military rank and function.

The title has been traced by E. CHAVANNES and SYLVAIN LÉVI in the Itinerary of Wu K'uṅ (751–790). The Chinese author relates that the kingdom of Ki-pin (Gandhāra and territory adjoining in the west) sent in 750, as envoy to the court of China, the great director Sa-po ta-kan 薩 波 達 幹 (or 干), anciently *Sat or Sar-pa dar-kan (cf. *Journal asiatique*, 1895, II, p. 345). Chavannes and Lévi have recognized a Turkish dynasty in the then reigning house of Ki-pin, and have regarded the title *ta-kan* also as Turkish, without, however, identifying it (*ibid.*, p. 379). In 1903 Chavannes noted the identity of the Chinese transcription with Turkish *tarkan* (Documents sur les Tou-kiue occidentaux, p. 239). The Chinese transcription *dar-kan does not allow us to presuppose a Turkish model *darkan;* but the Old-Turkish form was indeed *tarkan*, as is also confirmed by New Persian *tarxān* and Armenian *t'arxan* (HÜBSCHMANN, Armen. Gram., p. 266). Tarsā, the Persian designation of the Christians, is transcribed in Chinese by the same character, 達 婆 *ta-so*, anciently *dar-sa. The complex phonetic phenomenon which is here involved will be discussed by me in another place. Wherever the Chinese mention the title, it regularly refers to Turkish personages: thus the pilgrim Hüan Tsaṅ is accompanied by an officer Mo-tu tarkan, assigned to him by the Turkish Kagan (WATTERS, On Yuan Chwang's Travels, Vol. I, pp. 75, 77); for examples in the Chinese Annals, see HIRTH, *l.c.*

In the Vita S. Clementis (XVI), a Bori-tarkános appears as commander of Belgrad; this may be Turkish *büri* ("wolf"). Among the Bulgars, Bulias tarkános (Old Turkish *boila tarkan*) was one of the titles of the oldest two princes (cf. MARQUART, *l.c.*, pp. 41, 42). As a Hunnic title, *tarxan* occurs in the Armenian History of Albania by Moses Kalankatvaci (HÜBSCHMANN, *l.c.*, p. 516). The word has survived in the name of the Russian city Astrakhan, originally Haj or Hajji Tarkhan, as it was still called by Ibn Baṭūṭa (ed. DEFRÉMERY, Vol. II, pp. 410, 458), who adds that *tarkhan* among the Turks designates a place exempt from any taxation. PEGOLETTI calls the city Gintarchan (YULE, Cathay, Vol. III, p. 146). Our word does not occur in Marco Polo, as supposed by H. Beveridge, nor do the Mongols know it in the form *tarkan*, but they have only *darkan* or *darxan* (KOVALEVSKI, p. 1676), which has two different meanings,—"workman, artist," and "exempt from taxes." GOLSTUNSKI, in his Mongol-Russian Dictionary (Vol. III, p. 63), defines it as "smith, master; exempt from taxes and obligations." There is no association between these two meanings, as wrongly deduced by E. BLOCHET (Djami el-Tévarikh, Vol. II, p. 58). In Karakirgiz we have *darkan* in the sense of "smith, artist," while the same word in Kirgiz means "favorite of the Khan" and "liberty." Perhaps *darkan* was an independent Mongol-Turkish word, which was subsequently amalgamated with Old Turkish *tarkan*.

The Tibetan forms *dar-k'a-č'e* and *dar-rgan* lead to Uigur *darkači* (-*či* being a suffix) and *dargan* or *darkan*. Tibetan tradition itself assigns these words to the Uigur language; thus it is legitimate to conclude that Mongol, on its part, derived the words from the Uigur, and that the initial dental sonant is peculiar or due to the latter. The Tibetan transcriptions, further, are decisive in reconstructing the Uigur forms; for an Uigur (or Mongol) *tarkan* would have been transcribed by the Tibetans only *t'ar-k'an*. Among the Mongols, the title never had an extensive application; it does not occur in the chronicle of Sanaṅ Setsen. Also the fact that the Manchu and other Tungusian languages did not adopt it from the Mongols is apt to show that it is of comparatively recent date among the Mongols. Neither was it the Mongols who conveyed the word to Persia, as is evidenced by the Persian form *tarxān*. The form *dargan* paves the way to *daruga*, which, although a different word, that has assumed a development of its own, in its foundation is doubtless related to *darkan, tarkan*. Both words start with the common significance "official, governor, commander, high authority," and gradually depreciate in value, *daruga* simply becoming a chief, mayor, superintendent, manager, and *tarkan* a favorite of the Khan.

There is no evidence of the existence of the title on Asiatic soil prior to the seventh or eighth century A.D. The Chinese do not ascribe it to the Hiuṅ-nu or any of the numerous early Turkish tribes with which they came in contact, while they have preserved many titles and offices in their languages. We have no right to assume an unlimited antiquity for any historical or linguistic phenomenon; nor can it be argued with Mr. Beveridge that "the antiquity of the name is evidenced by the fact that its etymology is unknown, and that Oriental writers are obliged to make absurd guesses on the subject." There are a great many ancient words the etymology of which is perfectly known, and there are many words of recent origin the etymology of which is shrouded in mystery or dubious. I have no judgment on the point raised by Mr. Beveridge, that the names Tarchon, Tarquin, and Tarkhan may be identical; but for chronological and ethnographical reasons this theory does not seem very probable. At any rate, both detailed phonetic and historical investigations are necessary in order to establish such an identity; a merely apparent coincidence of words proves little or nothing.

170. The Turkish origin of *tupak* is also maintained by W. GEIGER (Lautlehre des Balučī, p. 66): Balūčī *tūpak, tupaṅ, tūfaṅ, tōpak;* Yidgā *tufuk.*

171. The word *čākū* occurs also in Kurd *čaku, čaxo,* etc. (J. DE MORGAN, Mission en Perse, Vol. V, p. 140).

183. The word *se-mo-do* occurs in the Tibetan translation of the Amarakoṣa (p. 166).

198. *pir-t'i* ("quick-match") is also connected with Turkī *piltä* (LE COQ, p. 86 b).

207. Another Sanskrit term for *Panicum miliaceum* is *cīnaka* ("Chinese") and *cinna.*

279. *k'ra-rtse,* pronounced *t'ar-tse,* is perhaps merely a bad spelling of Persian *tarāzū* (No. 128).

299. *t'ai rje* is possibly connected with Mongol *taiji* (cf. O. FRANKE, Jehol, p. 30).

On p. 421 it is stated that the animal *kun-ta* is not yet traced to its Sanskrit original. Boehtlingk's Dictionary, however, has Sanskrit *kunta* with the meaning "a small animal, a worm"; but this entry may be simply based on the Tibetan *mDzaṅs-blun.* The Chinese transcription calls for a prototype *kunda.

To the Persian loan-words add *šo-ra* (above, p. 503).

To the Arabic loan-words add *šeg* ("chieftain, elder"), from Arabic *šaix.*

To the Turkī loan-words add *gaṅ-zag* (above, p. 577).

Sir GEORGE A. GRIERSON, editor of the "Linguistic Survey of India," has done me the honor to look over my Loan-Words in Tibetan, and to favor me with the following observations, which are herewith published with his kind permission:

The Kāshmīrī for "egg" (p. 405) is *t'ül.*

15. I cannot think that *andañil is a possible Apabhraṃça (using the word in its technical sense) word. The presence of *ñ* seems to point to Kāshmīrī, in which *ni* has a tendency to change to *ñi.* The Ksh. equivalent of Skr. *nīla-* is *nīlu,* pronounced *nyūl,* and it is a common-place that *ny* and *ñ* in that language have the same sound. In fact, original medial *ny* is written *ñ* (e.g. *dāña,* from Skr. *dhānya-,* "paddy"), in this following Paiçācī Prakrit.

17. 'Ārya-pa-lo. This is typical Piçāca, which changes *ry* to *r(i)y* and *v(b)* to *p.* In all Indian Prakrits, *ārya* would become *ajja-,* with short initial *a.*

18. *pōt'ī* is the common word for "book" all over North India. The Ksh. form is *pūt'i*.

21. *sĕndūra-* is the regular Prakrit form of Skr. *sindūra-*.

28. I do not see how *ba-dan* can represent *patāka*. The change of initial *p* to *b* is, I think, impossible in any Prakrit or modern Indian language. Of course, the change might have occurred in Tibetan.[1]

29. *sāccha*, with a long *ā*, is impossible in Prakrit. Compare Hindostānī *sācā* ("a mould").

30. In true Apabhraṃça, medial *k* often becomes *g* (Hemacandra, iv, 396). This accounts for the *g* in *mu-tig*. But the Ap. form would be *mu(ō)ttiga-*, not *mukt-* or *mut-*.

45. Is not Tibetan *k'a-ra* = Hindostānī *khàr*, "coarse sugar?" I should be inclined to derive the Tibetan word *śa-ka-ra* from the Persian word *śakar*, not from Skr. *śarkarā*. If the Tibetan word came from India, it would be *sa-ka-ra*. In regular Prakrit, and in all the modern Indo-Aryan vernaculars except Bengali, Sanskrit *ś(ç)* becomes *s*. The Persian word is in regular use in Kāshmīrī *śakar*, and could thus have got into Tibet.

68. The regular Prakrit form is *vidduma-*, which is quite common. See, e.g., the index to the *Sĕtubandha*. I have never met any form such as *viruma-*, or the like.

113. Although *dār-cīnī* is the dictionary word, *dāl-cīnī* is universal all over North India.

118. I have not come across *cob-cīnī* in Kāshmīrī, but in that language other compounds with *cōb* are common, to indicate the roots of various plants. This leads me to think that the word probably got into Tibetan through Kashmir.

122. The word *tsādar*, a shawl, is pure Kāshmīrī. It came into that language from India.

143. *Araq* is, of course, common all over North India. It is even used by Hindus, and appears in Hindī. In Kāshmīrī, *arak* means "sweat." It is the same word.

143–156. I think it is certain that all these Arabic words came *via* India. They are all in common use in North India and Kashmir. The only exception is No. 148. I do not remember coming across this corruption of *masjid* anywhere in India proper. But, curiously enough,

---

[1] It should be borne in mind that the derivation of *ba-dan* from *patāka* is proposed by the Tibetan grammarians; whether this is objectively correct, is another question. At any rate, *ba-dan* is not a Tibetan word, and the object which it denotes came from India with Buddhism.—[B.L.]

masīt occurs in the Ormuṛī language spoken in Afghanistan. Of course, the form *bagšis* with *g* (No. 145) does not occur in India.[1]

173. *Argon* occurs in Kāshmīrī in the same sense.

---

[1] The final *g* (pronounced *k*) is a purely graphic, not a phonetic phenomenon; Tibetan writing has no final *k*.—[B.L.]

GENERAL INDEX

The Index contains also additional information.

A-lo-yi-lo, 378 note 2, 511.
Abel-Rémusat, see Rémusat.
Abu Dulaf, 351.
Abu Mansur, 194, 209, 298, 301, 306,
307, 315, 320, 332, 350, 354, 364, 366,
369, 370, 373, 380, 383, 396, 399, 405,
425, 443, 446, 449, 453, 455, 459, 481,
483, 507, 509, 544–547, 549, 551, 553,
587, 589; Indian elements in pharmacology of, 580–585.
Abulfeda, 351.
Achundow, A. C., 194, 209, 253, 298,
301, 304, 306, 307, 315, 320, 327, 332,
350, 354, 364, 366, 367, 370, 373, 380,
383, 396, 399, 402, 405, 425, 443, 446,
449, 453–455, 459, 478, 483, 507, 509,
544–547, 551, 580, 583–585, 587.
Aconite, 582.
Acorn, in Persia, 246.
Acosta, C., 356, 528, 550, 556, 591.
Aden, almonds of, 405.
Aeschylus, 320.
Aetius, 586.
Africa, aloes of, 480; date-palm intro-
duced into eastern, 389 note 1; ebony
from, 485, 486; home of Ricinus, 404;
home of sesame cultivation, 290;
home of water-melon, 438; myrrh
from East, 461.
Ahlquist, A., 577.
Ahmed Sibab Eddin, 561.
Ai-lao, 489.
Ain-i Akbari, 222, 282, 319, 502, 592.
Ainslie, W., 241, 254, 266, 364, 367, 453,
484, 514, 585, 588.
Aitchison, 343.
Akbar, promoter of viticulture, 240.
al-Akfānī, 566.
Albertus Magnus, 395 note 6, 411.
Alcohol, Chinese allusion to, 237.
Aleni, Giulio, S. J., 433, 527.
Alexander Romance, Chinese in, 570–
571; Ethiopic version of, 566.
Alexandria, 550.
Alfalfa, cultivation of, in Fergana, 210;
history of, 208–219; wild species of, in
China, 217–218.— Alfalfa is culti-
vated in Arabia, being styled *gadhūb*
on the South-Arabian coast. The
Arabs also received the plant from
Persia. In Egypt it became only
known during the nineteenth century
under the name "Arabian clover"

(*bersīm hegiāsi*); cf. G. Schweinfurth,
Z. Ethn., 1891, p. 658.
Almeria, 492, 497.
Almond, 193, 405–409.
Altabas, altobas, term for brocades,
derivation of, 492.
Alum, 336, 474–475.
Amber, 521–523; of Samarkand, 251.
Ammianus Marcellinus, 355, 548.
Amomum, 481–482.
An-si, Chinese name of the dynasty of
the Arsacides or Parthia, 187, 221,
457; cotton stuffs of, 488.
Anabasis, 223, 224.
Andamans, Memecylon on, 315.
Anderson, J., 266, 286.
Andreas, 529.
Anglo-Saxons, cultivation of carrot by,
451, 452; cultivation of coriander by,
299.
Annam, pepper of, 375; Psoralea of, 484;
styled Yavana, 212; Styrax benjoin
of, 465.
Antimony, 509.
Ao-men či lio, 434, 501.
Apricot, in India, 240, 408; transmitted
from China to the west, 539.
Arabia, alleged home of fig-culture, 411;
amber from, 522; costus of, 463;
manna of, 346 note 3; myrrh from,
461; saffron from, 310; turmeric exported from India to, 314.
Arabs, activity in sugar-industry of,
377; date of, 390; gold-dust of, 510;
grapes of, 223; grape-wine of, 239;
importers of asbestos into China, 500;
nux-vomica of, 449; rape-turnip of,
381; symbolism of pomegranate
among, 287; trading brocades with
Kirgiz, 488–489; viticulture of, 241;
yüe no textiles of, 494.
Areca palm, 584.
Argentine, alfalfa in, 219.
Aristobulus, 239, 372.
Aristophanes, 208.
Aristotle, 411, 512.
Armenia, alfalfa in, 218; grape-wine in,
220; peach and apricot in, 539; rhu-
barb of, 547.
Armenian apple, Greek term for apricot,
203, 209.
Aromatics, 455–467.
Arrian, 455.

599

Moṅ-ku či, 295.
Moṅ liaṅ lu, 229, 282.
Moṅ Šen, 233, 238, 265, 292, 297, 303, 376.
Moṅ-tse, 216.
Monardes, N. de, 342.
Mongol dynasty, cultivation of alfalfa, encouraged by, 217.
Mongol, Iranian Elements in, 572–576.
Mongolia, Brassica rapa in, 381; flax in, 295.
Morange, M., 449, 450.
Morbus americanus, 556.
Morga, A. de, 283.
Morgan, J. de, 343, 369, 435, 444, 595.
Morse, H. B., 560.
Moses of Khorene, Armenian historian, 310 note 1, 369, 377.
Mosul, manna of, 344.
Mu-kū-lan, Mekrān, 355.
Mu-lu, Chinese name of a city on the eastern frontier of Parthia, 187.
Mukerji, N. G., 261, 397, 452.
Mulberry, 339, 582.
Müller, F. W. K., 267, 290, 417, 461, 490 530, 572–574, 578, 589.
Muṅ ts'üan tsa yen, 227, 229.
Mungo bean, 585.
Munkacsi, B., 345, 574.
Muqaddasī, 255, 377, 425.
Musil, A., 287.
Musk, of China, 310 note 1; traded in Yūn-nan, 469.
Musk flower, 193.
Muss-Arnolt, 226, 285, 459, 519, 542, 543.
Myrobalan, 378, 583.
Myrrh, 460–462.
Myrtle, 461.

Nagasaki, figs introduced into, 414.
Nan-čao, 469; cotton in, 491; peculiar variety of pomegranate in, 286; se-se in, 517; wild walnut in, 270.
Nan čao ye ši, 413, 471.
Nan čou i wu či, 317, 417, 464.
Nan čou ki, 247, 248, 250, 460–462, 480, 482, 483.
Nan Fan, Southern Barbarians, 358, 375, 491.
Nan faṅ ts'ao mu čwaṅ, 263, 329, 330–332, 334, 375, 376, 388, 417, 464, 486, 543.
Nan hai yao p'u, 327.
Nan i či, 469.
Nan Man, se-se among women of the, 517.
Nan ši, 490, 491, 493, 521, 523, 564.
Nan-tou, vine in, 222.
Nan Ts'i šu, 282, 376.
Nan Yūe či, 491, 510.
Nan yūe hiṅ ki, 330.
Nanjio, Bunyiu, 254, 303, 457.

Narcissus, 427–428; mentioned in Pah-lavi literature, 192.
Needham, J. F., 492.
Needles, of gold, silver, and brass, 512.
Nepal, spinach introduced into China from, 393.
Nicolaus of Damaskus, 247.
Nizāmī, 570.
Nöldeke, T., 209, 390, 391, 427, 461, 493, 495, 530–533, 573, 582.
Nonsuch, 218.
Numerals of Malayan-Pose (Pasa) language, 472–473.
Nuṅ čen ts'üan šu, 336.
Nuṅ šu, 307.
Nux-vomica, 448–450.

Oak-galls, 367–369.
Oak manna, 349.
Oakley, 218.
Odoric of Pordenone, 346, 352, 549, 590.
Oil, from walnuts, 266.
Okada, K., 501.
Olearius, A., 277, 337.
Olive, 415–419; absent in Bactria, 223; in India, 239; in Pahlavi literature, 193.— No other text regarding the olive is known than that of the *Yu yaṅ tsa tsu*. Li Ši-čen (*Pen ts'ao kaṅ mu*, Ch. 31, p. 10b) cites this single text only, and is at a loss as to what to make of this plant. He has added this note as an appendix to the article on *mo-č'u* (*mwa-džu), saying that the *ts'i-tun* fruit is of the same kind. G. Ferrand (*Journal asiatique*, 1916, II, p 523) has identified the term *mo-č'ū* with Javanese *maja*, the fruit of the *Aegle marmelos*.
Ono Ranzan, 204, 250, 260, 273, 293.
Onyx, 554.
Oppert, G., 527.
Oranges, method of storing, 231.
Ormuz, 346.
Osbeck, P., 238.
Ouseley, W., 372, 374, 479, 485, 507.

Pa-lai, locality in southern India, 240.
Pai piṅ faṅ, 381.
Pālaka, Pālakka, name of country, 397, 398.
Palembang, 470; p'o-so stone of, 526; storax-oil of, 459.
Palestine, coriander in, 299.
Palladius, 315, 436, 509, 511.
Pallas, P. S., 523, 527.
Pallegoix, 299, 323, 332, 443, 476.
Pandanus, 192.
Pantoja, S. J., 433, 527.
Pao či lun, 197.
Pao p'u tse, 279.
Pao ts'aṅ lun, 509, 515.

BOTANICAL INDEX

Abrus precatorius 215
Acacia catechu 481
Aconitum ferox 582
Aconitum fischeri 379
Acorus calamus 583
Actea spicata 400
Agallochum 463
Aleurites triloba 263, 408
Alhagi camelorum 346, 347
Alhagi maurorum 347
Allium ascalonicum 304
Allium fistulosum 303
Allium odorum 483, 484
Allium porrum 304
Allium sativum 302, 427
Allium scorodoprasum 205, 259, 302
Aloe abyssinica 480
Aloe perryi 480
Aloe vulgaris 480
Aloëxylon agallochum 580
Alpinia galanga 545, 546
Alpinia globosum 242
Alpinia officinarum 546
Althaea rosea 551
Altingia excelsa 459
Amarantus 195
Amomum 482
Amomum granum paradisi 584
Amomum melegueta 584
Amomum villosum 481
Amomum xanthioides 481
Amomum zingiber 545
Amygdalus cochinchinensis 407
Amygdalus communis 405, 406
Amygdalus coparia 405
Amygdalus persica 539
Amyris 543
Amyris gileadensis 429, 430
Andropogon nardoides 455
Angelica anomala 358
Angelica decursiva 196.
Antiaris 450
Apium graveolens 401, 402
Apium petroselinum 102
Aplotaxis auriculata 464
Apocynum syriacum 349
Areca catechu 584
Aristolochia kaempferi 464
Artocarpus integrifolia 479
Astragalus adscendens 348
Astragalus florulentus 348
Atraphaxis spinosa 347
Atriplex L. 397
Atropa belladonna 585
Atropa mandragora 585
Aucklandia costus 462

Averrhoa carambola 415

Baliospermum montanum 583
Balsamodendron giliadense 429
Balsamodendron mukul 462, 467
Balsamodendron pubescens 462, 467
Bambusa arundinacea 350
Bambusa quadrangularis 535
Barkhausia 200
Barkhausia repens 199
Basella rubra 324–328, 336
Benincasa cerifera 439, 443, 445
Beta bengalensis 397
Beta maritima 397
Beta vulgaris 399, 400
Betula alba 553
Bombax malabaricum 491
Borassus flabelliformis 536
Boswellia 470
Boswellia serrata 467
Boswellia thurifera 585
Brassica capitata 381
Brassica caulozapa 381
Brassica cypria 380
Brassica marina 380
Brassica napus 381
Brassica rapa 199, 381
Brassica rapa-depressa 381, 429
Brassica silvestris 380
Broussonetia papyrifera 558, 560
Brunella vulgaris 200
Bupleurum falcatum 196
Butea frondosa 328

Caesalpinia bonducella 583
Camellia oleifera 251
Camellia theifera 553, 554
Canarium album 417
Canarium commune 269, 479
Canarium pimela 417
Canavallia ensiformis 426
Cannabis sativa 289, 291, 403, 562, 582
Capsella bursa-pastoris 427
Cardamomum malabaricum 585
Cardamomum minus 585
Carthamus tinctorius 309, 310, 312, 318, 324, 325, 327, 393
Carum bulbocastanum 383
Carum carui 383, 384
Carya cathayensis 271
Caryophyllus aromaticus 222, 584
Cassia fistula 421–426, 472
Cassia tora 582
Castanea vulgaris 369
Catalpa bungei 271
Cathartocarpus 425

617

INDEX OF WORDS

Iranian, Indian, Mongol and other words reconstructed on the basis of Chinese transcriptions are provided with an asterisk.

Alphabetical Index of Languages

Chinese

a-lo-p'o 420, 421
a-sa-na hian 455, 456
a-t'i-mu-to-k'ie 290
a-wei 358, 361
a-yü-tsie 359, 361
a-yüe 247, 248
a-yüe-hun 247, 248
a-ži 410
an-lo 552

ča-ta 527
čen-t'ou-kia 215
či ma 293
čo-pi 493
ču-čô 376
ču-mu-la 518

č'a-kû-mo 318
č'a mu 250
č'ui-hu-ken 196

fan mu-pie 448
fań-pu-šwai 531
fei-žań 260
fou-lan-lo-lo 588
fu lo-po 451 note 3
fu-t'u ts'ai 402

hai liu 284 note 2
hai-na 336
han-hûe 210
hei-ńan 473
hian ts'ai 298
hiń-kû 361
hiuń-k'iuń 200
ho-li-lo 378
ho t'ao, hu t'ao 256
hu fen 201

hu hien 195
hu hwań lien 199
hu kan kiań 201
hu kiai 380
hu k'iań ši če 199
hu k'in 196, 400
hu kiu-tse 483
hu kwa 300
hu-lo 503
hu lo-po 451
hu-lu-pa 202, 446
hu ma 288, 290–292
hu-man 196
hu-mań 385
hu mien mań 195
hu-na 496
hu pa-ho 198
hu-ša 305
hu šeń 195
hu-swi 202, 297, 298
hu tou 197, 305, 307
hu ts'ai 199, 202, 381
hu ts'uń 303
hu t'ui-tse 197
hu t'uń lei 202, 339
hu wań ši če 199
hu-ye-yen-mo 420, 423
hu yen 201
hu yen-či 327, 328
hui-hu tou 305
hui-hui tou 197, 307
hui-hui ts'uń 303
hun 248
hun-t'i 303, 304
hun-t'o ts'ai 304
huń hwa 310
hün-kû 358
hwań kwa 300
hwań-lien 547

hwań-p'o-nai 197
hwo mao siu 499
hwo-ši-k'o pa-tu 448
hwo-ši-la 448

i-lan 404
i-muk-i 486
i-ts'at 530

kan hian 455
kan-lan 417, 460
kan-suń hian 215, 428
ken ta ts'ai 399
kiań-hwań 313
kiao ma 300 note 4
kin-hwa 539
kin tsiń 520
ko 471
ku-čuń 491
ku-pei 491
ku-pu-p'o-lû 479 note 1
ku-sui-pu 195
ku-tu 565
kû-liń-kia 216
kû-šeń 290–292
kûn-t'a 399
kwo tou 306

k'iań hwo 199
k'iań t'ao 259
k'iań ts'iń 199
k'ie-p'o-lo 343
k'u-lu-ma 385
k'u-mań 385
k'u-mi-č'e 215
K'u-sa-ho 529
k'u ši pa tou 448

lan-č'i 520

621

jarak 404
kalgán 546
kamiñan 465
kanari 269
kápas 491
kápor-bárus 479 note 1
kertas 559
korma 386
leña 290 note 9
sulasi 590
tingkal 503

Arabic

abruh 447
afs 367
akitmakit 583
amlaj 581
anbā 552
aqīn 590
araq 237, 596
aruz 581
atmat 581
azādiraxt 581

badrūj 590
baladur 582
balīlaj 581
bang 582
banj 582
beladur 482
birinj-i kābilī 582
bīš 582
bitīx ul-hindī 582
bussad 525

dar-čīnī 583
dar-filfil 583
dar ṣīnī 541, 583
dauku 453
dībadž 489, 492
duhn az-zanbaq 332
duhn ul-amlaj 583
duhn ul-sunbul 583

falanjmušk 589
filfil, fulful 374 note 3
fiṣfiṣa 209
fistaq, fustaq 252
fūfal 584
fulful, filfil 584

habb ul-qilqil 582
hāl 585
halahil 582
halīlaj 378
hindubā 402
hinna 336
hulba 446
husyat iblīs 554

ibarīsam 538

ihlilaj 581
isbiadari 555
isfenāh 395
isfist 209

jauz ul-qei 449
jiza' 555
jōz 256
julbar 306
jūz-i buwwā 582
jūz-i mātil 582

kafūr 585, 591
kahrubā 521
kammūn 383
karnab 380
kebīr 590
kibrīt 575
kirbās 574
kundur 585
kurkum 321

lak 478 note 5
lak 585
lauz, lewze 405
lazvard 520
līnej 520
luban jawī 465 note 1

mamirun 546
mann 343
mardakuš 585
mastaki 252
māš 585
mī'a 459
murdāsanj 509
murr 461
mušktirāmušīr 585

na-ho tou 307
nakhl 386
nārjīl 585
nehsel 453
nīl, līla 585
nīlej 370
nīlūfar 585
nisrīn 551

pāzahr 525

qanbīt 381
qāqula 584
qaranful 584
quinna 364
qitta 301
qūlāni 585
qurtum 327
qust 584
qutun 491

ranej 240 note 7
ratba 209
ratta 551, 583

rixān 590
rummān 285
rutta 582

sabāhīa 453
sadaj 583
safarjal 584
saidalāni 425
sakbīnaj 366
salīxa 584
sandal 552, 584
saqmūniyā 584
sarak 539
sātil 584
sāx 553
sefanariya 453
suk 551
sukkar 584
sunbul 584

šābuni 425
šah-ṣīnī 552
šal 584
šaljam 381
šīšīan 581

ṣīnī 547 note 4

tabašir 351
talīsfar 584
tamr 386
tamr ul-hindī 582
tanbūl 582
terenjobīn 345
tīn, tima 411
turbund 584
tūt 582
tūtiya 513

ušnān 582
utruj 581

vaj 583

wars 315, 316

xaleń 552
xamāxim 590
xār-ṣīnī ("stone of China"), Arabic term for Chinese tootnague, 555. The designation "stone" corresponds to the t'ou-ši ("tou stone" of the Chinese, which denotes the zinc and brass of the Persians.
xarnub, xarrūb 424
xarnub hindi 422
xarva 404
xauk 590

THE GEOGRAPHICAL REVIEW

PUBLISHED BY

THE AMERICAN GEOGRAPHICAL SOCIETY
OF NEW YORK

JANUARY

1922

The fourth volume contains 175 excellent plates representing finds of every description. A considerable number are in color, the workmanship here being of the same high standard as all else about the book.

The final "volume" is in reality a portfolio of maps, most adequately arranged for convenience of reference. In addition to one general map and an index map (the latter a most useful feature), both on a scale of 1 : 3,000,000, there are ninety-four others on a scale of 1 : 253,440, or 1 inch = 4 miles; many of these are of no little importance geographically.

A most praiseworthy feature is the list of titles on the region under consideration, together with the abbreviations under which they are cited (pp. xxv–xxviii). No less to be commended are the detailed descriptions of finds, inserted in the text as additional sections appended to the chapters with which they have to do; in this way the text is relieved of much detail that would render it distinctly less readable, while at the same time this descriptive material is placed just where it is most convenient for reference. Finally, there is a most excellent index.

Among the many things for which "Serindia" deserves particular approbation is the care taken in the transcriptions of Chinese names. Sir Thomas Wade's system, as exemplified in Gile's Dictionary, is that employed; and, while most students will probably agree that it is far from being an ideal one, it is at least the one best known and most generally used and therefore to that extent the standard, so far as writers in English are concerned. The way in which the transcriptions are accompanied by the corresponding Chinese characters is also deserving of notice and, in so far as possible, of emulation.

It would be difficult to point out any particular in which "Serindia" falls short of being all that a report of this kind should be. Sir Aurel presents the results of his labors in straightforward, succinct fashion, without weaving any theories or engaging in polemics of any sort. He has, furthermore, the happy gift, whether he is writing a personal narrative or compiling a formal scientific report, of knowing how to be vivid and interesting as well as painstaking and accurate.

                                                        C. W. BISHOP

ANCIENT CULTURAL RELATIONS BETWEEN CHINA AND IRAN

BERTHOLD LAUFER. Sino-Iranica: Chinese Contributions to the History of Civilization in Ancient Iran, with Special Reference to the History of Cultivated Plants and Products. iv and pp. 185–630; indexes. *Field Museum of Nat. Hist. Publ. 201 (Anthropol. Ser., Vol. 15, No. 3)*. Chicago, 1919.

As we all know, the success achieved by Dr. Laufer in the past two decades and more in throwing light upon Far Eastern culture development has placed him in the foremost rank among students in that field. His recent work, "Sino-Iranica," fully maintains the high standard of scholarship which he has set for himself, and for others as well, along those lines.

The object of the book, the author tells us (p. 207), is "to present . . . a synthetic and comprehensive picture of a great and unique plant-migration in the sense of a cultural movement, and simultaneously . . . to determine the Iranian stratum in the structure of Chinese civilization." In the pursuance of this aim no less than one hundred and thirty-five subjects are treated of. These are concerned mainly, as the subtitle informs us, with the diffusion of various cultivated plants. Such things, however, as Persian textiles and Iranian minerals, metals, and precious stones are also discussed; while a section is devoted to the titles employed by the Sassanian Government, and another to the cultural debt owed by the Persians to China.

In connection with his plant studies Dr. Laufer lays special emphasis upon the historical fact, so often ignored by later writers both native and foreign, that the great Chinese traveler, Chang K'ien, who first established a direct and conscious contact between his own land and western Asia, brought back with him two new plants and only two, viz., alfalfa and the vine (p. 190 and *passim*). The point is further made (pp. 220 ff.) that there can have been but one center for the origin of grape growing and that viticulture and wine making and the use of alfalfa as well were found by the Chinese among Aryan-speaking peoples and not at all among the Turks.

Another interesting and significant fact pointed out (p. 293) is that one of the fundamental differences between the Chinese and the Mediterranean civilizations was the use of hemp by the former and of flax by the latter as the material for clothing. As Dr. Laufer says (p. 294), their failure to acquire flax in ancient times "is a clear index of the fact that the Chinese never were in direct contact with the Mediterranean culture-area, and that even such cultivated plants of this area as reached them were not transmitted from there directly, but solely through the medium of Iranians."

It is a trifle puzzling at first glance to see why the section on rice (pp. 372–373) should have been included, since it is unlikely that either of the two regions in question, China or Iran, received that food plant from the other. Nevertheless rice appears to have penetrated to Central Asian regions extremely early. The Chinese authorities cited on this point in "Sino-Iranica" carry us back to the second century B. C., while if the statement in the "Muh T'ien-tsü Chuen," or "Journeys of the Emperor Muh," can be believed, that ruler found a fine quality of rice growing in the west, apparently in the Tarim or Turfan regions, as far back as the tenth century B. C. Hence it seems not impossible that at least the easternmost of the Iranian-speaking peoples got their rice culture, some time during the prehistoric period, from the valley of the Hwang Ho instead of from that of the Indus or of the Euphrates.

It is interesting to note that Dr. Laufer now admits (p. 569) "some degree of probability in the old theory that the name 'China' should be traceable to that of the dynasty Ts'in." Those who have hitherto combated this hypothesis have done so mainly because they failed fully to grasp the importance of the rôle played by the principality of Ts'in (or Ch'in according to the more general pronunciation) long before her conquest of the rest of China late in the third century B.C. To say that "China" cannot be derived from "Ch'in" merely because the former name seems to have appeared in western lands in one form or another before the prince of Ch'in became emperor of China is wholly to overlook two cardinal facts: firstly, that from the eighth century B.C. onward Ch'in was in full control of the Central Asian gateway from China to the West; and secondly, that in the fourth century B.C. she also secured command of the Yünnan-India route by her conquest of eastern Szechwan and the upper Han valley. Thus it was wholly through Ch'in in those days, before the commencement of overseas intercourse between China and the West, that any knowledge of what we now call China proper could possibly reach the latter. Furthermore, in the fourth and third centuries B.C. Ch'in was one of the largest, most powerful, and best organized states not merely of eastern Asia but of the whole world. Hence it is not remarkable that the peoples of India, Persia, and lands farther to the west should have extended her name to the regions lying beyond her, whose products they received only through her and which she did actually in the latter half of the third century B.C. incorporate within her dominions.

Dr. Laufer rightly emphasizes (p. 186) the importance of the application of the laws of ancient Chinese phonology to the elucidation of Iranian tribal and place names and points out (p. 187) that the ancient Chinese scholars had developed "a rational method and a fixed system in reproducing words of foreign languages."

Undoubtedly well founded is the criticism of the prevalent system of transliterating Chinese words. Dr. Laufer might have said even more than he does in this regard; not only is it "clumsy and antiquated," but it is arbitrary and in many respects hopelessly incomprehensible to the lay reader without special instruction. Unfortunately it is too generally and firmly entrenched and represents too many vested interests to render likely its speedy replacement even by a system so consistent, so transparently simple, and so accurately phonetic, as that suggested on page 188. What is needed and what will be required before any real advance in this direction can be hoped for is a general and explicit agreement upon an improved system among students the world over. No man, even one of the standing in this field enjoyed by Dr. Laufer, can hope to do much singlehanded toward this end.

The primary usefulness of "Sino-Iranica" lies undoubtedly in the vast fund of information with which it provides us regarding a great variety of definite and specific Oriental culture objects and products. From this point of view it can scarcely be praised too highly. Yet it also performs another service, no less important because it is of a more general and less immediately striking kind. Most of us were taught at school that the Persian Empire was a vast, tyrannical, barbarian, "Asiatic" power whose function in the development of civilization was, like that of the Mongols or the Turks, destructive merely. Dr. Laufer, in this as in others of his writings, does much toward enabling us to correct such a view and to realize that in point of fact Persia in the days both of the

Achaemenides and of the Sassanians was a center of the world's very highest civilization, playing in the cultural development of all the lands about her a part whose importance we are only just beginning to understand.

C. W. BISHOP

POLITICAL AND SOCIAL IMPRESSIONS OF THE FAR EAST

J. O. P. BLAND. **China, Japan, and Korea.** x and 327 pp.; ills., index. Charles Scribner's Sons, New York, 1921. $5.00. 9 x 6 inches.

Mr. Bland's former residence and practical experience in the Far East and his wide official acquaintance entitle him to serious hearing. His book is divided into two parts, the first being a political survey and the second a series of chapters entitled "Studies and Impressions." The political and social life of the people of the Far East is in many respects closely interwoven with the physical circumstances of the region, a point of view which was developed in a scholarly and convincing manner in the January number of the *Geographical Review* ("The Geographical Factor in the Development of Chinese Civilization," by Carl Whiting Bishop, *Geogr. Rev.*, Vol. 12, 1922, pp. 19–41). We find ample evidence of this throughout Mr. Bland's book, particularly in the portions that deal with the density of population and the fine shades of adjustment which may be observed as the population, particularly of China and Japan, steadily increases, making the food supply a more intense question and bringing into play modern political and social forces of deep importance.

The second half of Mr. Bland's book is most delightfully written. He is never a dull writer, but in the later chapters the style rises to a very high level of literary perfection without at any time losing the quality of restraint that bespeaks judgment and integrity. In the field of political geography there are few contributions but rather the reaffirmation of points of view and of facts made familiar through western writings on oriental policies and practices during the past ten years.

One marked effect of the overcrowding of China, in the view of Mr. Bland, is the widespread desire to better living conditions; and this desire has steadily molded the character of the Chinese people through many centuries. Thus there is perpetual striving to raise one's self above the common level. Men of ambition seek power in order that they may gather to themselves forces and revenues that free them from the bondage of the soil or the shop and give them a standing comparable to that of the foreigner in the treaty ports. In Young China the author finds no saving grace, believing that the teachings of liberalism among the pupils of missionaries and colleges founded by foreign associations are quite superficial and that the blundering of the Tuchuns and the divisions of the past decade represent a constant force in Chinese life. To him there seems no way to escape but through a powerful government—if necessary a restoration of the Manchu Dynasty— and the holding of the great mass of the people and the local governors in a strong grasp. He believes that to let democracy come in is to open the door to disintegrating influences and bribery in a more extreme form than it has existed in China during the whole period of foreign domination.

The case of the crowded population in the homeland of Japan is argued skillfully and carefully, and an important and necessary distinction is drawn between two forms of expansion by Japan, one form being an actual movement of population overseas as a result of the crowding at home, and the other a kind of economic penetration which is equally efficient in relieving pressure because the flow of capital into a new region may control the labor of the region and its output of raw materials, thus fostering industry at home and providing additional support or better support for the crowding millions.

One is tempted to say of Mr. Bland's conclusions that he is too ruthless a logician. It can be almost demonstrated that every new cause is a failure—until it has achieved a state of pronounced success! A thousand visible and invisible forces fight a new idea. Moreover, the followers of the new (as well as the old) always include a large number of foolish persons and "optimists *de metier*," as Mr. Bland happily phrases it. We must recognize, however, that this is a changing world and that the force of new ideas cannot be measured in terms of old customs and characteristics. Finally, it may be said with assurance that it is not the masses of the people, to whom Mr. Bland refers political and social doctrines at every turn, that have ever controlled the fate of revolutionary changes. The mass of the people is inert, neutral, and lacking in initiative. In all lands and times

ISIS

International Review devoted to the History
::   of Science and Civilization   ::

EDITED BY

GEORGE SARTON and CHARLES SINGER

N° 8. — VOL. III (2), Autumn 1920

CONTAINING :

All papers and books relating to the biological and medical
sciences should be sent to CHARLES SINGER, Westbury Lodge,
Oxford, England. Those relating to the other sciences to GEORGE
SARTON, Harvard University, Cambridge, Mass, U. S. A.

Soc. An. M. WEISSENBRUCH	PAUL HAUPT
Imprimeur du Roi	Librairie Académique.
49, rue du Poinçon	(ci-devant MAX DRECHSEL)
Bruxelles (Belgique).	Berne (Suisse).

Prix : t. I, 50 fr. belges, t. II et suivants 30 fr. (payables par anticipation)
Edition sur Hollande, t. II et suivants : 50 fr.

multitude of degenerate and deteriorate forms in the groups with which he has to deal. The great lesson of history is the continuity of the human mind and the work that does not seek to deal with events as continuous is not history even though it treat of antiquity.

The truth is that nearly all works of science are accounted of little worth by the following generation. Thirty short years is a long life for a scientific treatise and fortunate indeed is the author of that work the leaves of which his grandson will care to turn. It is rightt hat this should be so and it is the scientist's part to maintain the growing edge of active advance on the dead mass of irrelevant and imperfect observation and outworn and untenable hypothesis. But yet the living is in an ecological relation with the dead that it would be death to disturb. Well might the historian of science place at the head of his work the ancient epitaph :

> Reader, thou that passest by,
> As thou art so once was I,
> As I am so shalt thou be,
> Wherefore, reader, pray for me!

Prof. HARVEY-GIBSON has given us a valuable, sincere, and reliable piece of work for which every biological teacher should be grateful, but we hope that in future editions he will remove these strictures on his predecessors and delete the suggestion, specially dangerous in a book intended for students, that any period of history can be summed up without study.   CHARLES SINGER.

Berthold Laufer. — Sino-Iranica. Chinese contributions to the History of Civilization in Ancient Iran. With special reference to the history of cultivated plants and products. Field Museum of Natural History, Anthropological Series, vol. XV, p. 185-630. Chicago, 1919.

There can be no greater pleasure for a reviewer than to welcome a book like this, which he can admire and praise almost without restriction. LAUFER's new book is a fundamental contribution not simply to the history of botany but as well to the history of Eastern civilization. To appreciate its full value it will suffice to remember that for the history of cultivated plants, we were entirely dependent upon the works of A. DE CANDOLLE, HEHN and CHARLES JORET (Paris, 1897-1904), admirable works to be sure, but unable to give us much information on the Asiatic migrations of plants For example, DE CAN-DOLLE's information on Chinese plants was entirely derived from BRETSCHNEIDER's Study and value of Chinese botanical works (Chinese Recorder 1871-1881) which, says LAUFER, « teems with misunderstandings and errors ». Now LAUFER comes in offering us large and important material based upon a first hand study of Chinese and other oriental sources.

The special problem which he set himself to solve was to determine on the basis of Chinese tradition, which plants were introduced by them from Iran and cultivated in their own country, also which other plants of Iran, cultivated or wild, were noticed and described by Chinese authors, finally which drugs and aromatics of vegetable origin were imported from Iran to China. Such a study is specially important because plant migrations are inconceivable without other cultural exchanges. LAUFER's studies — the present one and others which are forthcoming — will thus enable us to measure the Iranian stratum in the structure of Chinese civilization. This is the more important that the Iranians were « the great mediators between the West and the East, conveying the heritage of Hellenistic ideas to central and eastern Asia and transmitting valuable plants and goods of China to the Mediterranean area ». — Iranian literature gives one but little botanical information. LAUFER reproduces (p. 192-194) an interesting disquisition on plants extracted from the Būndahishn (chap. 27) which proves that the ancient Persians were interested in botany, — but no other botanical text of theirs has come down to us. He has also used the *Materia Medica* written about A. D. 970 by ABŪ MANSŪR MUWAFFAQ BEN 'ALî of Herat. (By the way this is not only the earliest Persian work on the subject, but the oldest extant New-Persian prose. The MS. dated Jan. 1056 from which this text was published is also the oldest extant Persian MS.). However LAUFER's more important sources by far are Chinese sources. Happily Chinese scholars soon developed a rational and fixed method in there production of foreignwords, so that « in almost every instance it is possible to restore with a high degree of certainty the original Iranian forms from which the Chinese transcriptions evere accurately made ». I cannot insist on this, but I have said enough to suggest that LAUFER's studies are of supreme interest also for the philologist and more especially the Oriental phonologist.

We are more interested in the significance of LAUFER's work from the point of view of the history of civilization. Let us listen again to the author (p. 189) : « Stress is laid on the point that the Chinese furnish us with immensely useful material for elaborating a history of cultivated plants... The Chinese merit our admiration for their far-sighted economic policy in making so many useful foreign plants tributary to themselves and amalgamating them with their sound system of agriculture. The Chinese were thinking, sensible, and broad-minded people, and never declined to accept gratefully whatever good things foreigners had to offer. In plant-economy they are the foremost masters in the world, and China presents a unique spectacle in that all useful plants of the universe are cultivated there ». — The two first foreign plants introduced in China were two plants of Iranian origin, the alfalfa and the grape-vine. This occurred in the latter part of the second century B. C. A great number of other plants followed, originating most of them from Iran and Central Asia, but also from other parts of the world. LAUFER lays stress on the point that the Iranian plant migrations extend over a period of a millennium and a half. In other words, he takes pains throughout his book to destroy the legend according to which, most of these Iranian plants were acclimatized in China during the Han period, thanks to the efforts of the famous general CHANG K'IEN. It is only for the two plants named above that positive evidence permits to give credit for their importation to CHANG K'IEN. For a few other plants, one has no better evidence than that supplied by the *Ts'i min yao shu*, a VIth century book, which we know only in a very corrupted state.

LAUFER has used all the Chinese and Japanese *Pen ts'ao* (herbals) available, also Japanese botanical works. He also studied some early Chinese medical books, chiefly those of CHANG CHUNG-KING, or CHANG KI, a second century physician. Unfortunately these books have suffered from various interpolations which it is now impossible to unravel. Apropos of this, LAUFER remarks (p. 205) : « A critical bibliography of early Chinese medical literature is an earnest desideratum. » To the information gathered from Chinese and Iranian sources, which form the substance of his volume, LAUFER has added all that was pertinent in the classical, Semitic and Indian traditions.

His book is built in the following way. After a short introduction (24 p.) follow a series of chapters, each of which is a monograph devoted to one or more plants, to wit : alfalfa ; grape-vine ; pistacchio ; walnut ; pomegranate ; sesame and flax ; coriander ; cucumber ; chive, onion and shallot ; garden pea and broad pea ; saffron and turmeric ; safflower ; jasmine ; henna ; balsam-poplar ; manna ; asafoetida ; galbanum ; oak-galls ; indigo ; rice ; pepper ; sugar ; myrobalan ; « gold peach » ; fu-tse ; brassica ; cummin ; date-palm ; spinach ; sugar beet and lettuce ; ricinus ; almond ; fig ; olive ; cassia pods and carob ; narcissus ; balm of gilead ; water melon ; fenugreek ; nux-vomica ; carrot ; aromatics. To this encyclopaedia of Chinese botany are added a few other notes of philological or historical interest, to wit : The Malayan Po-se and its products (Po-se in Chinese denotes two countries, a certain Malayan country, also Persia, for ex. the date is called *Po-se tsao, i.e.* Persian jujube); Persian textiles;

Iranian minerals, metals and precious stones ; Titles of the Sasanian government ; Irano-sinica, (miscell.) ; Iranian elements in Mongol ; Chinese elements in Turki ; Indian elements in the Persian pharmacology of ABU MANSUR MUWAFFAQ (that is the important tenth century work quoted above ; it is partly based on Indian knowledge : the author says himself that he had been travelling in India) ; the Basil ; additional notes on loan words in Tibetan.

This encyclopaedia is fittingly completed by very elaborate indexes : a general index which contains also additional information ; a botanical index and *thirty* indexes (one for each language !) of the words quoted, including those reconstructed on the basis of Chinese transcriptions. The only criticism which I make bold to suggest concerns the transliteration of Chinese names. BERTHOLD LAUFER introduces here a new system. Whatever the merit of his system be, I believe that individual initiatives of this kind can but increase the present confusion. Indeed, a satisfactory and universal system of transliteration is one of the most urgent needs of the Republic of Letters; but such need can only be fulfilled by an international understanding.

GEORGE SARTON.

Edmund O. von Lippmann. — Entstehung und Ausbreitung der Alchemie. Mit einem Anhange : Zur älteren Geschichte der Metalle. Ein Beitrag zur Kulturgeschichte, gr. in-8° XVI + 742 p. Berlin, JULIUS SPRINGER, 1919.

Some twenty years ago, Dr. VON LIPPMANN to whom we owe already many important contributions to the history of science and chiefly to the history of chemistry, began to accumulate materials for the solution of the following problems : « when did the belief in the artificial production of Gold and Silver originate? which circumstances favoured the growth of this superstition and how did it develop? » The more he advanced in his study, examining in each case as far as was in his power the original documents, the more did the whole subject become complex. Yet he managed to master and to clarify and classify the immense amount of data which he had gathered and the results of his long studies are now offered to the public. Truly a monumental work, the greatest single addition to our knowledge of ancient chemistry — (and also to our knowledge of human superstition) — since the days of KOPP. This work fills in an important gap in our historical literature, for since the publication of KOPP's book — I am thinking now chiefly of his « Die Alchemie in älterer und neuerer Zeit », Heidelberg 1886 — many new fundamental data have been brought to light. BERTHELOT's work is also partly out of date. By the way, VON LIPPMANN, while admiring greatly the activity of his great French predecessor,

THE

JOURNAL

OF THE

ROYAL ASIATIC SOCIETY

OF

GREAT BRITAIN AND IRELAND

FOR

1920

QUOT RAMI TOT
ARBORES

A.D. MDCCCXXIII. INST.

PUBLISHED BY THE SOCIETY

74 GROSVENOR STREET, LONDON, W.I.

M DCCCC X X

Mahābhārata. They are all well reproduced, and are preceded by a brief preface, some introductory remarks, and a table giving explanatory details. The only criticism I can make is to express a regret that the English text of this fine work has not been revised by an Englishman, who could have corrected its style and language in a good many places with advantage. That, however, is a point of quite secondary importance, entirely overshadowed by the excellence of the plates themselves, which constitute the real essence of the book.

Indian scholars will be interested to see how the personages of one of their great epics have been treated in this outlying part of the Indian sphere of influence. It cannot be too often repeated that in the history of Asiatic civilizations the part played by India will never be fully understood or adequately valued if we confine our attention to India proper and neglect Indo-China and the Indian Archipelago. I therefore welcome this work as an instalment which will, I hope, interest British students in these matters, and I anticipate the thanks due to the Dutch editors for having issued an edition of the work in English for the benefit of a wide circle of readers to whom Dutch is an unfamiliar tongue.

C. O. BLAGDEN.

SINO-IRANICA : CHINESE CONTRIBUTIONS TO THE HISTORY OF CIVILIZATION IN ANCIENT IRAN. With special reference to the History of Cultivated Plants and Products. By BERTHOLD LAUFER, Curator of Anthropology. Chicago ,1919.

Whenever a new work from Dr. Laufer's various and voluminous pen appears, there come into my mind two absurd lines from an ancient music-hall song and dance, " You should see me dance the Polka, You should see me cover the ground ! " For Dr. Laufer does cover a great deal of ground, in fact, the greater part of the continent of Asia, and who knows where he will stop? In truth, he is not so much, or perhaps so little, a mere Sinologist as an Asianist, for his researches range from the extreme North-East, China, which suffices for most men, Tibet, on which he is now a leading authority, Siberia, and now going west (*unberufen*) within the cultural and linguistic limits of Iran. Wherever he settles for the time, his wide reading, well-grounded and solid learning, coupled with his critical intelligence, ensure the dissipation of much error, and the establishment of new and valuable information.

In this area of Iran Dr. Laufer has aimed at showing the influences of Iranian products—plants, minerals, textiles—upon the culture of Eastern Asia, and towards the end of the volume he devotes a section, for the book is not divided into chapters, to " the reverse of the medal ", and to the consideration of " what the Iranians owe to the Chinese ". Nor is this all. There are also five Appendices, to Iranian Elements in Mongol, Chinese Elements in Turki, The Indian Elements in the Persian Pharmacology of Abu Mansur Muwaffaq, The Basil, and to Additional Notes on Loan-words in Tibetan. The volume ends with a General Index, a Botanical Index, and an Index of Words occurring in the text, arranged under no less than thirty languages in order. There Indexes will greatly facilitate the use by workers of what is bound to be a book of reference of a most valuable kind, rather than a work to be perused from end to end.

I regret that the space available renders an adequate review of this sound and stimulating volume impossible, but a quotation from the Introduction (p. 188) will at least show the author's aim in his own words :—

" The linguistic phenomena . . . form merely a side-issue of this investigation. My main task is to trace the history of all objects of material culture, pre-eminently cultivated plants, drugs, minerals, metals, precious stones, and textiles, in their migration from Persia to China (Sino-Iranica), and others transmitted from China to Persia (Irano-Sinica)."

Animals, games, and musical instruments are left for future treatment.

Perhaps I can best use the space remaining to me by a précis of one of the author's sections, chosen at random, that on the spinach. Dr. Laufer begins by quoting the Russian author Bretschneider that spinach is said by the Chinese to come from Persia, and that the various European names are derived probably from the Persian term *esfinadsch*. "The problem," observes Laufer, "is not quite so simple." He proceeds to quote a passage from a Chinese work, which he describes as not only the earliest datable mention of spinach in Chinese records, but in general the earliest reference to it that we thus far possess. Here it is stated that in A.D. 647 the vegetable *po-ling* (spinach) was sent to the Chinese Court from Nepal, and Laufer points out that the present Chinese colloquial name *po-ts'ai*, or " *po* vegetable " is a contraction of *po-ling*, and does *not* mean " Persian vegetable ". Laufer next approves De Candolle's belief that spinach was first raised as a vegetable in Persia, but cites Leclerc's *Traité des simples* to prove that the Arabs carried the plant to Spain —contrary to De Candolle's statement—where it was cultivated towards the end of the eleventh century. From Spain it spread to the rest of Europe, and was well known and generally eaten in England in the sixteenth century.

In Persian literature the earliest mention of spinach occurs, again according to Leclerc, in the pharmacopœia of Abu Mansur, and is apparently of the thirteenth century.

Dr. Laufer concludes that " we are compelled to admit that the spinach was introduced into Nepal from some Iranian region, and thence transmitted to China in A.D. 647 ". And further that " the Persian cultivation can be but of comparatively recent origin, and is not older than the sixth century or so ".

L. C. HOPKINS.